D0821701

CLIMATE
AND
MAN'S ENVIRONMENT

Climate and Man's Environment
An Introduction to Applied Climatology

JOHN E. OLIVER
Department of Geography
Columbia University

JOHN WILEY & SONS, INC.
New York • London • Sydney • Toronto

Library of Congress Cataloging in Publication Data

Oliver, John E.
    Climate and man's environment.

    Bibliography: p.
1. Climatology.   2. Man—Influence of environment.
3. Man—Influence of climate.   I. Title.

QC981.039   551.6   73-5707
ISBN 0-471-65338-1

Printed in the United States of America

10-9 8 7 6 5 4 3 2 1

Preface

This text is the outcome of my varied experiences in teaching courses in climatology and applied climatology. While there are a number of excellent introductory texts covering the dynamics of climate and regional climatology, they frequently do not contain material concerning the varied applications of the concepts and principles expressed. Because of the multidisciplinary nature of applied climatology, it was necessary to scour the literature for appropriate readings so that a meaningful course could be taught. Unfortunately, such readings almost always assumed a basic knowledge of the subject to which climate is being related. In many instances, students do not have such a background and much valuable class time is spent in supplying the basic concepts needed. By bringing together much of this varied material, presenting it in a rational conceptual context, this void will hopefully be filled. Furthermore, the text should supply a suitable starting point for the discussion of the many fascinating aspects of applied climatology.

The impact of climate upon man and his environment crosses many disciplines; no one person can assume expertise in all these areas. Accordingly, I have drawn heavily upon the thoughts and ideas of numerous authors, have corresponded with many, and have belabored my colleagues with endless questions. To all of these persons I express my gratitude. I am also grateful to the many publishers who permitted me to use or adapt their diagrams; these are acknowledged in the text. Similarly, I would like to thank the following for granting permission to use their work: W. S. Broecker, Jen-hu Chang, J. F. Cronin, J. R. Goldsmith, L. Greenburg, P. P. Micklin, R. E. Munn, R. E. Murphy, W. R. D. Sewell, H. Suzuki, W. Van Royen, and especially A. N. Strahler who allowed me to use many of his excellent diagrams.

No author with a young family could possibly complete a manuscript without an enormous amount of patience and fortitude on the part of his wife; it is certainly to the credit of my wife that this book was ultimately completed.

John E. Oliver
Columbia University

November 1972.

Contents

CLIMATE
AND
MAN'S ENVIRONMENT

Introduction

In the 1951 *Compendium of Meteorology*, Landsberg and Jacobs state: "If we consider climate as the statistical collective of individual conditions of weather, we can define *applied climatology* as the scientific analysis of this collective in the light of a useful application for an operational purpose. . . . The term operational is . . . broadly interpreted as any useful endeavor, such as industrial, manufacturing, agricultural, or technical pursuits." Such a definition still holds true and is, in part, followed in this text. However, in view of the current and growing interest in study of the environment as a functioning entity of many interacting systems, I have seen fit to extend the concept of "applied climatology." Few would deny that the study of the environment and environmental deterioration crosses the boundaries of established disciplines and that the modern environmental scientist needs a broad background in quite diverse disciplines. Climate-environment relationships certainly form part of this required background so that it is necessary to identify a wider scope of applied climatology than that defined previously. It might instead be considered as application of the principles and concepts of climatology to spheres of endeavor that concern man and his past and present environments. The aim of such study is to facilitate comprehension of the relationships that exist and to promote a rational interpretation of climatic concepts as they relate to both natural and man-modified environments.

In keeping with this aim, Part I treats relationships between climate and environment.[1] It essentially concerns the application of climatic principles and climatic interpretation to areas of study directly related to the natural environment. The approach used is systematic but it is hoped that the interconnectivity between each "Climate

and . . ." chapter becomes clear. Part II covers applied climatology within the context of the Landsberg-Jacobs definition. It assesses the role of climate in the study of man and his pursuits, and the operational purposes including bioclimatology, industry, and agriculture.

Part III serves a dual function. First, it provides a summary of the changes in climate over time; second, it is an exercise in the utilization of applied climatology as it is defined here. To reconstruct climates of the past many interpretive tools are needed and, for the most part, the nature of past climates must be derived from their impact upon other environmental systems. Thus a reciprocal relationship exists; to interpret past climates one must understand other environmental systems—but to understand those systems it is necessary to comprehend the role of climate in shaping them.

With applied climatology so widely defined, its scope is obviously enormous. This poses problems concerning the selection of material presented and the level at which it is to be treated. When it is considered that entire journals may be devoted to a subject that is treated in a few pages here, the problem is clearly evident. To meet this, I have chosen to provide a survey of ideas and interpretations rather than select isolated examples and study them in depth. The value of this approach is that once basic principles are covered, follow-up studies can be completed more easily; to facilitate this approach an extensive bibliography is given.

I hope that the text will be useful in a number of ways. For students in climatology courses it provides a source that allows them to comprehend the significance of climatology beyond the basic coverage of the workings of climate, and why and where climatic regimes are so located. To students in other disciplines, ranging from geomorphology to human ecology, it provides a climatic viewpoint of aspects of their studies that may have been overlooked in a single-discipline approach. Finally, to the growing

---

[1] Since climate is, of course, environment, the relationship suggested here is one of emphasis; in effect, the climatic environment is viewed as one conditioner of the total environment and the nature of environmental interaction.

1

number of students studying environmental science, it provides a perspective on one of the most important inputs in their field of study.

In keeping with the survey nature of the text, statistics and mathematics have been kept to a minimum. Equations presented are, for the most part, symbolic representations of worldwide exchanges: Data presented in the text are given in metric units except where, in cited references, the original units are retained. For those interested in analytic use of climatic data, the Appendices include data sources and statistical methods.

PART I

Climate and Environment

CHAPTER 1

Climate and the Ecosphere

In recent years the necessity of treating environmental problems holistically has meant that problem solving has become both interdisciplinary and multidisciplinary. To facilitate such treatment, many ecological terms and concepts have become widely used in disparate disciplines. The concepts of the biosphere (that part of the earth-atmosphere system in which life exists) and the ecosystem (a self-sustaining community of organisms taken together with their physical environment) provide two widely used examples. To express a combination of these two concepts simultaneously, LaMont Cole (1958) introduced —in somewhat apologetic terms—the concept of the ecosphere. It has proved a useful addition to the literature, because, in essence, it provides a single concept that allows treatment of the worldwide system in ecologic terms.

The interaction between living (biotic) and nonliving (abiotic) components of the ecosphere can be represented in a variety of ways. Figure 1-1, for example, shows the basic components that combine to produce a terrestrial ecosystem. The system depends upon the nature of the climate, biota, and surface materials, and the manner in which they interact. To analyse such a system, Odum (1962) suggests that it is necessary to deal with both its structure and its function. The structure of the system involves:

1. The composition of the biological community.
2. The quantity and distribution of abiotic materials.
3. The range, or gradient, of conditions of existence, for example, that of temperature or light.

Used in conjunction with Figure 1-1, such a framework would allow a relatively static aspect of the system to be assessed. To explain how the system works, it is necessary to analyse the factors that allow it to function, where function concerns:

1. The rate of energy flow through the system.
2. The rate of material and nutrient cycling.
3. The environmental (e.g., varying daylength) and biologic (e.g., nitrogen fixation) regulations that occur.

Thus, to place both structure and function in perspective, it becomes necessary to redesign the schematic representation shown in Figure 1-1, for the active processes of, and inputs into, the system need to be explored.

Figure 1-2 represents, in schematic form, how this might be conceived; superimposed upon the original ecosystem are the realms that contribute toward its function. Comprising processes associated with the atmosphere, biosphere, and terresphere, the full complexity of the system becomes apparent. As indicated by the intersecting circles, each of the designated realms is both directly and indirectly related to the others.

It is clear that atmospheric processes, especially on a long-term climatological basis, form a significant part of the system, playing a role in both structural and functional aspects. Furthermore, it is evident that the system, be it a local

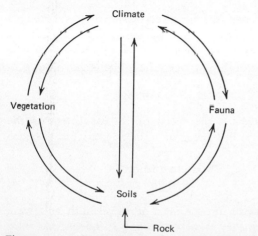

Figure 1-1. Components of a terrestrial ecosystem.

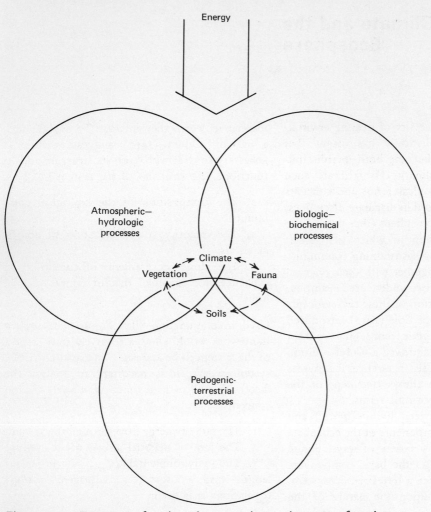

Figure 1–2. Ecosystem function shown as interacting sets of environmental processes.

ecosystem or the entire ecosphere, requires energy to function. Since this energy is almost totally derived from the sun, the study of solar radiation—its nature and its distribution over time and space—forms the very core of understanding the energetics of the ecosphere.

ENERGY FOR THE SYSTEM

The ultimate source of almost all energy available on earth is the sun. A giant nuclear fission device, converting hydrogen to helium, the sun radiates energy to surrounding space. As electromagnetic radiation, solar energy is described in terms of wavelike disturbances that, like any wave, are characterized by wavelength (the

distance separating successive crests—or troughs —of the wave) and wave frequency (the number of waves passing a fixed point each second). The relationship between these two is given by

$$c = \lambda v$$

where c = the speed of the wave [equals the speed of light 299,800 km (186,000 miles)/sec]

v = wave frequency

λ = wavelength

Because the product of wavelength and wave frequency equal a constant, it follows that short wavelengths are characterized by high frequency and long wavelengths by low frequency (Figure

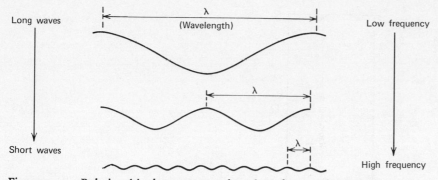

Long waves

λ
(Wavelength)

Low frequency

λ

Short waves

λ

High frequency

Figure 1-3. Relationship between wavelength and wave frequency.

1-3). Wavelengths can be measured in centimeters (cm), microns (μ), with 1μ equal to 10^{-4} cm or Angstrom units (Å), where 1 Å equals 10^{-8} cm.

Since waves may have a wide range of lengths, an entire spectrum of electromagnetic radiation can be identified. Figure 1-4 shows the spectrum and indicates how different bands are characterized by different types of radiation. The diagram shows that solar radiation is confined to a limited portion of the spectrum, with a peak occurring in the part comprising visible light. The reason for this localization is explained through examination of the laws governing radiation.

All bodies, not at absolute zero ($-273°C$, ($0°K$), the temperature at which all molecular motion ceases), radiate energy to surrounding space. The efficiency at which they radiate is highly variable, with the perfect radiator being a blackbody. This has the property of absorbing all radiation that falls upon it and then emitting radiation in a manner that depends solely upon its temperature. Blackbody radiation is the highest value that a body can theoretically emit at a given temperature.

The sun may be considered a blackbody and, through analysis of the solar beam, it can be established that the wavelength of maximum emission (λ_{max}) is about 0.5 μ. Using Wien's displacement law, given as

$$\lambda_{max} = K/T$$

(T = absolute temperature, K = constant = 2897), the surface temperature of the sun is calculated to be about 6000°C. The flux of energy of a blackbody at this temperature is

derived using Stefan's formula, which relates the flux to the fourth power of the temperature of the radiating body:

$$F = \sigma T^4$$

where F = flux of radiation [ly (langley)/min]
T = absolute temperature
σ = Stefan-Boltzmann constant, 0.813×10^{-10} cal/cm²/min (°K)⁴

The flux is evaluated at 100,000 ly/min (langleys are radiation units where 1 ly = 1 cal/cm²).

The earth, at a mean distance of 93,000,000 miles (149.5×10^6 km) from the sun intercepts but a minute fraction of this output, about 0.0025%. In fact, the amount of radiation falling on a surface perpendicular to the solar beam, at the outer edge of the atmosphere, is estimated at 2.0 ly/min. This value, although it does vary, is called the solar constant. Note that this value refers to solar energy incident on a plane; the earth is a sphere with only half its surface exposed to solar radiation at any one time. The value must be decreased accordingly.[1]

As a sphere, rotating on an inclined axis at a variable distance from the sun, the earth's surface (excluding the effects of the atmosphere for the moment) receives energy that is highly variable over both time and space. Effects of these

[1] The incident solar radiation averaged over the globe will be given by

$$\text{solar constant} \times \frac{\text{area of circular plane}}{\text{area of sphere}}$$

$$= \text{solar constant} \times \frac{\pi R^2}{4\pi R^2}$$

that is, $\frac{1}{4}$ of the value.

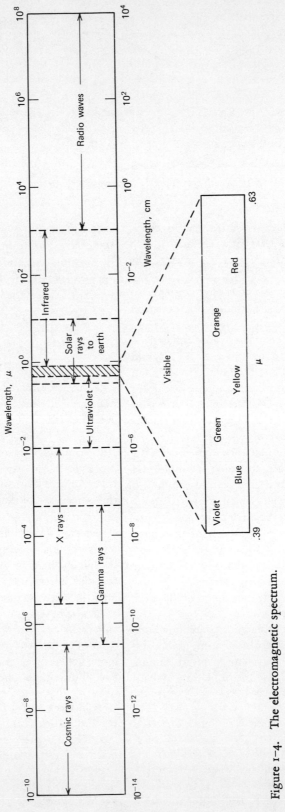

Figure 1-4. The electromagnetic spectrum.

astronomical features are summarized in Figure 1–5. In revolving around the sun, the earth is in its closest position on January 3 (perihelion), while it is most distant on July 4 (aphelion). This variable earth-sun distance means that the amount of solar radiation intercepted by the earth in January (perihelion) is about 7% more than in July.

A spherical earth also means that only one point on earth can have the sun directly overhead at any one time. That is, assuming no atmosphere, the intensity of radiation on a surface (I_h) is equal to that incident radiation (I_o) when the sun is at an elevation of 90°. As the angle between the surface and the overhead sun decreases, so I_h will decrease. The relationship is given by

$$I_h = I_o \sin a$$

where a is the sun's altitude, that is, the angle between the surface (horizon) and the overhead sun. The significance of this relationship is best seen by example.

If the sun is directly overhead, then $a = 90°$, and

$$I_h = I_o \sin 90°$$

The sine of 90° is 1; thus

$$I_h = I_o$$

Consider now the intensity of insolation at 60°N at noon on the day of the equinox. The angle of the sun in the sky is 30°.

$$I_h = I_o \sin 30,$$

the sine of 30° is 0.5, so

$$I_h = 0.5 I_o$$

The value of radiation intensity is decreased by half. The altitude of the sun in the sky at different times of the year, a function of latitude, plays a very important role in determining the amount of radiation received.

In the same way, the fact that the earth's axis maintains a fixed tilt causes the lengths of day and night to vary over the globe. Obviously, long hours of darkness decrease the amount of radiation, with the result that the seasonal variation in length of day and night over the

earth's surface plays a significant part in determining radiation received.

The result of the interplay of these factors is shown in Figure 1–6, which gives the latitudinal distribution of solar radiation over a year. Note the highest values occur at perihelion and that annual variation is least in equatorial latitudes. At the poles, the values are most extreme. In the winter season, no radiation is received; in the summer, despite the low angle of the sun in the sky, the values are quite large, reflecting the fact that the sun is in the sky continuously.

The Role of the Atmosphere

The envelope of gases that encircle the earth modify the amount of solar radiation that is received at the surface. At about 88 km (55 miles) above the earth's surface, the solar beam differs little from its original form. At this level, irradiation of oxygen causes photodissociation of the oxygen molecule:

$$O_2 \rightarrow O + O$$

At this altitude, low atmospheric density means that little recombination occurs. This takes place largely between 30 and 60 km above the surface where ozone is formed:

$$O_2 + O \rightarrow O_3$$

Below this level, wavelengths less than 0.29 μ have been absorbed and the ozone reverts again to oxygen. The formation of the ozone layer plays an exceptionally important role in the ecosphere, for it acts to screen out much of the ultraviolet light, a form of radiation that is harmful to life on earth.

In the denser parts of the atmosphere, many other modifications occur. A small amount of solar radiation in the longer wavelength band is absorbed by water vapor and carbon dioxide. Gas molecules, dust, and clouds in this part both scatter and reflect solar radiation. Scattering occurs from very small particles and is most effective for short wavelengths. Were it not for this scattering effect, the sun would appear as a fiery ball in a black sky, for selective scattering of shorter wavelengths is responsible for the blue color of the sky. Larger atmospheric particles reflect rather than scatter the solar beam, with less discrimination as to wavelength.

Figure 1–5. Earth-sun relationships. (*a*) Earth's orbit around the sun and illumination characteristics at the equinoxes and solstices. (*b*) The angle of the sun in the sky at different latitudes at the equinox (upper) and the northern hemisphere winter solstice (lower). (From Strahler, A.N., *Physical Geography*, Copyright © 1969 by John Wiley and Sons. Reproduced by permission.)

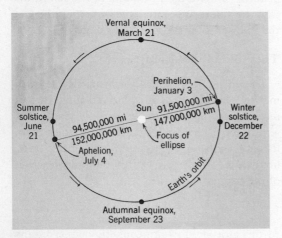

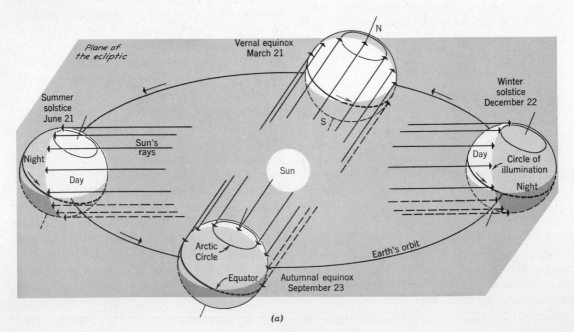

(*a*)

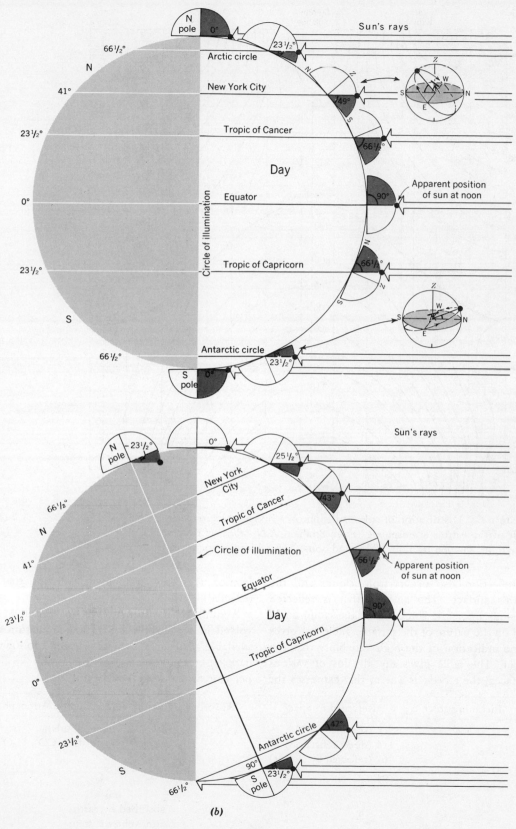

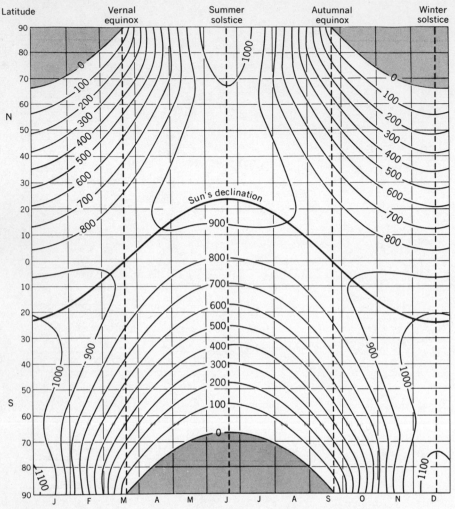

Figure 1-6. Distribution of solar radiation on a horizontal surface at the limit of the earth's atmosphere. (From Strahler, A.N., *Physical Geography*, Copyright © 1969 by John Wiley and Sons. Reproduced by permission.)

Reflection also occurs from clouds and the earth's surface. The amount that is reflected depends upon the thickness of the cloud cover and on the nature of the surface. Table 1-1 gives some indication of the high variability that can occur. The table gives the albedos of various surfaces; the albedo is a term that expresses the fraction (percent) of incident radiation that is reflected.

A number of estimates of the amounts involved in the disposition of solar radiation are available. Sellers (1965) provides the following data [figures in kilolangleys (kilolangley = kly), per year, where 1 kly = 1000 cal/cm^2]:

Incoming solar	Returned to Space	Absorbed by Earth	Absorbed by Atmosphere
263	15 (scattered)	124	7 (clouds)
	16 (direct reflection)		38 (dust, CO_2, H_2O)
	63 (reflected by clouds)		
	94	124	45
			169 absorbed by earth-atmosphere system

Table 1-1

ALBEDOS FOR SHORT-WAVE PORTION ($<4.0\mu$) OF ELECTROMAGNETIC SPECTRUM[a]

Surface	Albedo (%)	Surface	Albedo (%)
Fresh snow	75–95	Desert	25–30
Sea ice	30–40		
		Savanna	
Dune sand (dry)	35–45	Dry	25–30
Dark soil	5–15	Wet	15–20
Road top (black)	5–10	Green meadows	10–20
Concrete	17–27	Deciduous forest	10–20
		Coniferous forest	5–15
Clouds			
Cumuliform	70–90	Tundra	15–20
Stratus	59–84	Crops	15–25
Cirrostratus	44–50		
Water (see below)			

Reflectivity over water as related to sun's zenith angle:

Zenith distance (degrees)	0	40	60	80	90
Percentage reflectivity	2.0	2.5	6.0	34.8	100.0

[a]After Sellers (1965) and Smithsonian tables.

In the modification of the solar beam it can be seen that the atmosphere and earth lose a significant portion of solar radiation, which is returned back to space. The high proportion reflected by clouds provides the major amount lost; in the scattering figure it must be remembered that solar radiation is returned to space and down to earth. This downscatter adds to the total received by the earth, occurring as diffuse radiation. The amount returned to space comprises the earth-atmosphere albedo and it can be seen that it is approximately 35% (94/100 × 263%).

The short-wave (solar) radiation that arrives at the surface of the earth forms the base for the metabolism of plant and animal life. Through photosynthesis, radiant energy is converted to chemical energy by autotrophic plants. These provide almost all of the energy entering the food chains of the ecosphere. The passage of energy through the system can be traced through a number of trophic levels (Figure 1–7). At the first level, autotrophs are estimated to use 1×10^{21} cal/yr of solar energy, a ridiculously small percentage (less than 1%) of that which is available. It is enough, however, to sustain the prolific life on earth. The significance of solar radiation in relation to plant life is considered more fully in Chapter 5.

Terrestrial Radiation

The energy that is required to heat the earth and atmosphere and the energy to drive the "atmospheric engine" are derived from solar

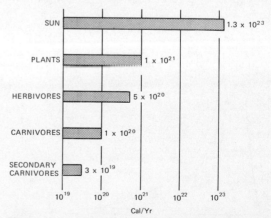

Figure 1–7. Utilization of solar radiation decreases at each trophic level. Autotrophs use only 0.08% of energy from the sun while herbivores and carnivores use only a fraction of this (note logarithmic scale). (After Cole, 1958.)

energy that is first absorbed and then reradiated by the earth. Recall that the maximum wavelength of emission for a blackbody is inversely related to the temperature of the radiating body ($\lambda_{max} = K/T$). The sun, a hot body, produces short-wave radiation, with a peak—per unit wavelength—at about 0.5μ. The earth, at an appreciably lower temperature, radiates at a longer wavelength, with maximum emission falling in the infrared band of the electromagnetic spectrum (Figure 1–8).

While the atmosphere is relatively transparent to short-wave, solar radiation, it is largely opaque to long-wave terrestrial radiation. Carbon dioxide, water vapor, and ozone absorb long-wave radiation at various bands, leaving "windows"

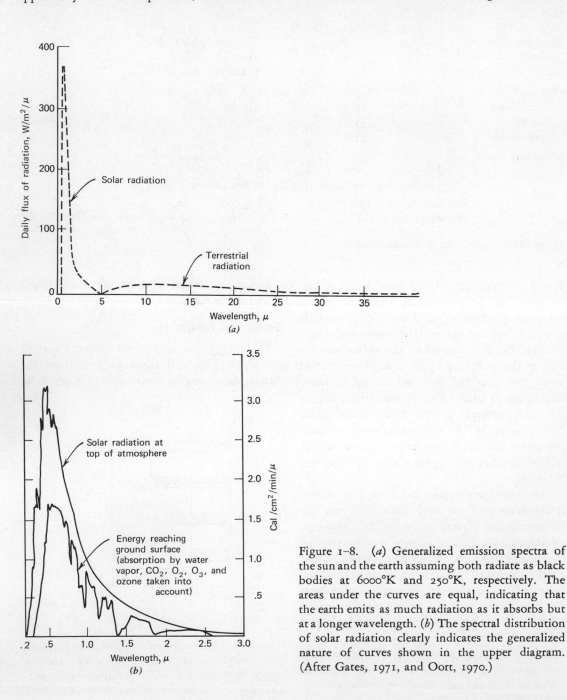

Figure 1–8. (*a*) Generalized emission spectra of the sun and the earth assuming both radiate as black bodies at 6000°K and 250°K, respectively. The areas under the curves are equal, indicating that the earth emits as much radiation as it absorbs but at a longer wavelength. (*b*) The spectral distribution of solar radiation clearly indicates the generalized nature of curves shown in the upper diagram. (After Gates, 1971, and Oort, 1970.)

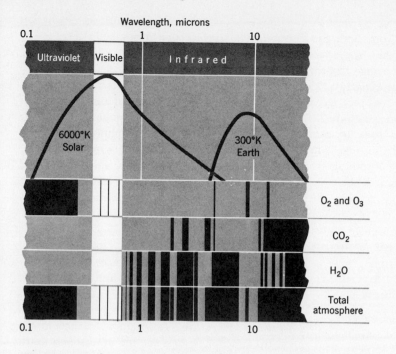

Figure 1–9. Absorption of radiation by the atmosphere as a function of wavelength. Note that "windows" occur through which some infrared radiation can pass directly to space. Absorption of terrestrial radiation in the bands shown give rise to the "Greenhouse Effect." (From Weyl, P.K., *Oceanography*, Copyright © 1970 by John Wiley and Sons. Reproduced by permission.)

through which some can pass directly back to space (Figure 1–9). This absorbed energy is then reemitted, some passing to space and some back to earth. The water vapor and carbon dioxide of the lower atmosphere thus tend to act as a blanket that restricts the direct outward flow of radiation. This then acts to warm the lower portions of the atmosphere through a process that is likened to the heating of a greenhouse and which is thus termed the "greenhouse effect." This effect raises the surface temperature

to a higher level than would occur if radiation escaped directly to space. Were it not for this effect, the mean global temperature would average between −22 and −26°C (−7 to −15°F). Instead, it is in the order of 15°C (59°F).

Just as values for the disposition of incoming solar radiation have been estimated, so have those for outgoing terrestrial radiation. The following again uses data given by Sellers (values in kilolangleys per year):

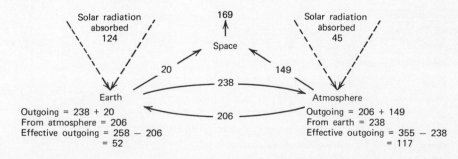

At first glance, what appear as discrepancies are shown. Radiation values from both earth (258) and atmosphere (355) are appreciably higher than that shown as being absorbed annually (124 and 45 kly respectively). The difference, of course, relates to the factors already outlined. Earth radiation is absorbed by the atmosphere and reradiated back to earth. An equilibrium is established, so that while effective outgoing radiation and incoming solar radiation absorbed are in balance, there is a considerable exchange of energy between earth and atmosphere. It is evident that to maintain this exchange other earth-atmosphere transfer mechanisms must exist. These are appropriately considered in the following section.

The Radiation and Heat Balances

If the net exchange between incoming and outgoing radiation is calculated simultaneously at a station, a balance is found to exist. This is represented by the equation

$$R = (Q + q)(1 - a) + I\downarrow - I\uparrow$$

where R = net all wave radiation

$(Q + q)$ = the sum of both direct and diffuse solar radiation

$(1 - a)$ = the fraction of short wave radiation reflected (a = albedo)

$I\uparrow$ = the outgoing long-wave radiation

$I\downarrow$ = the counterradiation

Figure 1-10 shows the mean annual values of net radiation (R) over the globe. The relative influence of each of the components of the equation can be seen in the resulting distribution. As would be expected, highest values are found in low latitudes while in polar realms (although not shown on the map), negative values occur. The differences in radiation characteristics between land and sea are clearly indicated. The sharp breaks in the isolines as they pass from land to sea indicate the differences.

It was shown that, on an annual basis, the earth absorbs 124 kly and effectively radiates 52 kly to the atmosphere. There is an annual surplus of 72 kly. To retain a balance in the earth-atmosphere system, the atmosphere must lose as much as the earth gains. This is the case (45 kly absorbed, 117 kly radiated = net loss of 72 kly) as it must be. Were it not, then the earth would get progressively warmer.

It is evident, however, that to maintain this balance there must be a continuous transfer of energy from the earth to the atmosphere. The

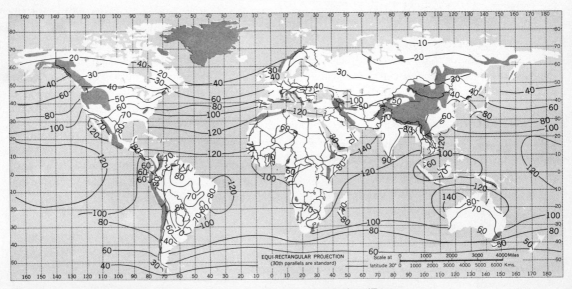

Figure 1-10.   The radiation balance. Values given kg cal/sq cm/yr. (From Trewartha, G.T., *Introduction to Climate*, 4th ed., Used with permission of McGraw-Hill Book Company.)

way in which this occurs might be conceived by considering a column of the earth's surface extending down to where vertical heat exchange no longer occurs. The net rate at which heat in this column changes, be it land or water, depends upon many variables. Obviously the net radiation, already defined as the relationship between insolation, albedo, and effective outgoing radiation, is of prime importance. But consideration must be also given to the transfer of sensible heat within the column (H), the heat exchange that occurs through evaporation (LE, where E is evaporation and L represents the latent heat of vaporization), and the horizontal transfer of heat into or out of the column (ΔF). The heat change that occurs (G) can thus be represented:

$$G = (Q + q)\,(1 - a)$$
$$+ \,I{\downarrow} \,-\, I{\uparrow} \,-\, H \,-\, LE \,-\, \Delta F$$

Included in this equation are components of the net radiation, R, because

$$R = (Q + q)\,(1 - a) + I{\downarrow}\, - \,I{\uparrow}$$

so that the equation can be written

$$G = R \,-\, H \,-\, LE \,-\, \Delta F$$

or, in more customary form

$$R = H + LE + \Delta F + G$$

This equation represents the heat or energy budget. The values of the terms in the equation vary depending upon the nature of the surface column in question and the time period over which it is considered. For example, over land the subsurface horizontal flux of heat (F), is negligible and can be omitted. Over water surfaces it assumes importance. In treating the equation on an annual basis the change in heat storage in the system (G), can be dropped, for an equilibrium is attained over the annual period. The equation becomes

$$R = H + LE$$

The mean latitudinal values of the components of the equation are shown in Table 1-2. A number of facts are evident from the table.

1. The exchange of heat through evaporation (LE) is greatest at the equator and decreases poleward.

2. Values for LE are twice as high over oceans as over land, and it is seen that more than 90% of the energy over the oceans is used in the evaporation process.

3. The sensible heat flux, H, is highest over land where it decreases with latitude. Over the oceans, values increase with latitude. Generally, the ratio H/LE (the Bowen ratio), increases with latitude for the global system as a whole.

Of the 72 kly that are exchanged over the year, 59 kly are accounted for by evaporation and transfer by latent heat while only 13 kly are transferred through the sensible heat flux by turbulent transfer and conduction. The role of the hydrologic cycle in the transfer of heat is an extremely important one.

The importance of the radiation and heat balances in the study of the ecosphere is demonstrated by the fact that it is possible to use it as a base to study the entire system. Such an approach has been outlined by Miller (1968), who, in an introduction to his study writes: "The student who makes a budget analysis (mass and energy) should become more aware of the earth's surface, so close and often so ignored, as a place where conflicts and divisions continuously occur and where physical laws are worked out and chemical reactions take place." In a similar way, Lowry (1969) uses the energy budget approach in his treatment of biometeorology. In dealing with this aspect he states ". . . biometeorology deals primarily with the atmospheric part of the ecosphere, and though the biologist's interest is centered on organismic responses to atmospheric variables, knowledge of energy transfer serves in the study of all parts of our environment."

Modification of the Energy Budget

Concern over the deterioration of the environment has led to the publication of a great deal of literature, ranging from scholarly works to "Sunday supplements," in recent years. Intermingled among the multitude of problems dealt with is the often-voiced concern that man may be altering the atmosphere and that this, depending upon the interpretation given, might cause a new ice age or the melting of the ice caps with subsequent inundation of low-lying land.

Table 1-2

MEAN LATITUDINAL VALUES OF THE COMPONENTS OF THE ENERGY BUDGET EQUATION
FOR THE EARTH'S SURFACE (IN KILOLANGLEY YEAR^{-1})[a]

Latitude Zone	Oceans				Land			Earth			
	R	LE	H	F	R	LE	H	R	LE	H	F
80–90 N								−9	3	−10	−2
70–80								1	9	−1	−7
60–70	23	33	16	−26	20	14	6	21	20	10	−9
50–60	29	39	16	−26	30	14	11	30	28	14	−12
40–50	51	53	14	−16	45	24	21	48	38	17	−7
30–40	83	86	13	−16	60	23	37	73	59	24	−10
20–30	113	105	9	−1	69	20	49	96	73	24	−1
10–20	119	99	6	14	71	29	42	106	81	16	9
0–10	115	80	4	31	72	48	24	105	72	11	22
0–90 N								72	55	16	1
0–10 S	115	84	4	27	72	50	22	105	76	10	19
10–20	113	104	5	4	73	41	32	104	90	11	3
20–30	101	100	7	−6	70	28	42	94	83	16	−5
30–40	82	80	8	−6	62	28	34	80	74	11	−5
40–50	57	55	9	−7	41	21	20	56	53	10	−7
50–60	28	31	10	−13	31	20	11	28	31	11	14
60–70								13	10	11	−8
70–80								−2	3	−4	−1
80–90								−11	0	−11	0
0–90 S								72	62	11	−1
Globe	82	74	8	0	49	25	24	72	59	13	0

[a] From Sellers (1965).

As is discussed in detail in Part III, such concern is not without basis. This can be demonstrated by taking inventory of the way in which the radiation budget can be modified.

Consider the net radiation equation defined earlier:

$$R = (Q + q)(1 - a) + I\downarrow - I\uparrow$$

The amount of solar radiation that is received at a surface, $(Q + q)(1 - a)$, can be altered by changing any one of the terms. We have found that incident solar radiation is reflected by particles in the atmosphere. By increasing the particles more reflection could occur, with the result that less radiation is received. It was shown, too, that clouds are important reflectors. The addition of small particles, the results of incomplete combustion, can promote cloud formation by providing condensation nuclei for water vapor. The recent controversy in the United States over the building of the supersonic transport planes (SSTs) made this a well-publicized point. Condensation trails of these aircraft, flying at altitudes in excess of 50,000 ft, could ultimately form a high veil of cirrus clouds that might interfere with the passage of solar radiation through the atmosphere.

The expression could also be modified by altering the surface albedo. Recent experiments have shown that when a snow or ice surface is covered with a thin layer of black material (ashes or soot), the energy balance is so modified that the snow below will melt.

$I\downarrow - I\uparrow$ can also be modified in a variety of ways. Increased cloud cover will increase the value of counterradiation, but the most widely publicized effect is that of increasing the carbon dioxide in the atmosphere. This gas, an important absorber of long-wave radiation, is increasing in amount as a result of the burning of fossil fuels by man.

Obviously there is a problem in attempting to project what the ultimate result of all these changes might be. Note that some of the changes are contributing toward a cooling of the earth (e.g., increased earth-atmosphere albedo) while others (e.g., carbon dioxide) could result in a warming trend. As shown later, sorting out what actually might occur is no easy problem.

Beyond the modification of the preceding factors, man's activities are also modifying the heat balance directly. Prodigious amounts of fossil fuel are being consumed at the present time and their consumption is adding stored energy to the system. In some cities, more energy is produced at times than is received from solar radiation. In New York City, Bornstein (1968) has estimated that in winter the amount produced by the heat of combustion alone is two and one-half times that received from solar radiation.

In an estimation of the effect of the addition of heat to the system, Holdren (1971) suggests that if the amount produced on a worldwide basis reached 1% of that received from solar radiation, then world temperature would rise 1.3°F (0.7°C). On a global scale, such an increase would have widespread effects. At present, solar energy received by the earth each year is about 21,200 times that produced by man; in order for man's input to be 1% of this, the amount produced must be 212 times that of today. As Holdren points out, this is actually eight doublings of present-day output. Since the amount of energy production by man increases between 4 and 5% annually, then the amount doubles every 14 to 17 yr. If this doubling rate were to continue, then the 1% value would be reached in from 107 to 134 yr. Whether such exponential growth can be assumed is a point of disagreement.

ATMOSPHERE AND CYCLES OF THE ECOSPHERE

In a study of the ecosphere, much understanding of the interaction between processes that occur can be derived through analysis of biogeochemical cycles. While climate plays an important part in many of the cycles, it is clear that its role will be most evident in those cycles in which there is a gaseous phase.

The stable, or nonvarying, and variable constituents of the atmosphere are given in Table 1-3. Nitrogen and oxygen together comprise about 99% of the total volume, with the ever-important carbon dioxide comprising only about 0.033%. These constituents and their relative abundance are testimony to the intricacy of interaction between atmosphere, biosphere, and terresphere. The present-day equilibrium is a function of the common evolution of these three great realms and points to the necessity of their simultaneous treatment.

Table 1-3

COMPOSITION OF THE ATMOSPHERE (DRY) UP TO ABOUT 90 km[a]

Constituent	Percent by Volume	Percent by Weight
Nitrogen (N_2)	78.088	75.527
Oxygen (O_2)	20.949	23.143
Argon (A)	0.93	1.282
Carbon dioxide (CO_2)	0.03	4.56×10^{-2}
Neon (Ne)	1.8×10^{-3}	1.25×10^{-3}
Helium (He)	5.24×10^{-4}	7.24×10^{-5}
Methane (CH_4)	1.4×10^{-4}	7.25×10^{-5}
Krypton (Kr)	1.14×10^{-4}	3.30×10^{-4}
Nitrous oxide (N_2O)	5×10^{-5}	7.6×10^{-5}
Xenon (Xe)	8.6×10^{-6}	3.9×10^{-5}
Hydrogen (H_2)	5×10^{-5}	3.48×10^{-6}

[a] From Air Force Geophysics Research Directorate (1960).

It is suggested that the early atmosphere of the earth consisted essentially of carbon dioxide (CO_2), water vapor (H_2O), and nitrogen (N_2), a result of the outgassing of the earth subsequent to its formation. No free oxygen was available. Such a composition might be compared with the atmosphere found on Venus today, for as the Venera 4 space probe indicated, CO_2 forms perhaps 90% of the atmosphere of that planet.

Clearly, over the course of time, atmospheric composition has undergone radical changes. From an abundance of CO_2, the amount is only a fraction of the total volume; oxygen, which did not occur, now forms some 21% of the present-day volume. The relative increase and decrease of these gases can only be explained by considering their role in the ecosphere, both in the past and as they function today.

To explain the amount of oxygen present today, a number of hypotheses have been put forward. It has been suggested that thermal dissociation of water vapor ($2H_2O \rightarrow 2H_2 + O_2$) could have occurred in primordial times, but if this had happened then at the high temperatures that prevailed, the oxygen would most certainly have combined with hot surface rocks. Alternatively, photochemical dissociation of water vapor might have produced oxygen, but the amount produced could not account for all that exists today. Perhaps the most plausible explanation of the increase in oxygen content is its formation as a product of photosynthesis. In this process, plants combine energy and CO_2 to produce organic matter and oxygen. Obviously, a problem does exist in that it is necessary to explain how plants could come into being in an atmosphere without oxygen in the first place. This criticism is countered by the observation that some living cells exist without oxygen; sulfur bacteria for example, produce organic matter in a CO_2 atmosphere.

In outlining a possible evolution of atmospheric oxygen, Gates (1970) evokes a number of the preceding ideas. Drawing upon the fact that the ozone layer of the atmosphere absorbs ultraviolet light—a form of radiation harmful to life— he relates oxygen content to the evolution of life. It is assumed that the primordial atmosphere was transparent to ultraviolet radiation and that the oxygen content was less than 0.001, the present atmospheric level (PAL). Photodissociation of water vapor slowly, over a vast period of time, brought the level to 0.01 PAL, at which time ozone formation would screen out some of the damaging short wave radiation. This might have allowed life in the sea to develop, perhaps explosively, with the result that extensive photosynthesis would occur. The increased rate of oxygen production would then bring the level to 0.1 PAL. At such a level much more ultraviolet light would be screened out and conditions were such that life could pass from the sea to the land. Land life, both flora and fauna, then underwent the long process of evolution until eventually an equilibrium was attained. Gates does note, however, that the oxygen content has probably varied over geologic time and during the Carboniferous, when prolific vegetation occurred, the level may have been as much as 10 PAL.

Other authors have outlined a similar evolution. The example shown in Figure 1-11 relates oxygen levels of the atmosphere to the evolution of the lithosphere, biosphere, and hydrosphere. Each of these forms a reservoir of oxygen and exchanges between them take place continuously.

Recently, some concern has been voiced over the possibility of depletion of atmospheric oxygen through man's activities. This results from the multitude of ways in which man has interrupted or modified the oxygen cycle (Figure 1-12). Vegetation has been modified over vast areas, largely by the replacement of diverse natural systems by simplified agricultural ones. Cities and urban sprawl have replaced forests and grasslands; modified surfaces interrupt the rate of mineral oxidation at the surface. Fossil fuels, drawing upon oxygen for their combustion, have been extensively burnt; photosynthesis by marine organisms may have been reduced by pesticides and herbicides added to the oceans. In all, man has done much to modify the oxygen cycle.

To obtain a clear view of what might be happening to man's oxygen supply, Machta and Hughes (1970) measured the concentration of oxygen at large numbers of clean-air locations. The mean concentration was found to be 20.946% by volume, with little geographic variation. This value was compared with measurements taken since 1900 and it was found that no change in concentration was evident. Further, it was estimated that if all known reserves of fossil fuels were used by man, the atmospheric oxygen content would be reduced to 20.800%, a negligible reduction. As the SCEP (1970) study suggests, atmospheric oxygen presents an unusual case in modern environmental science in that it represents a "nonproblem." Unfortunately, such a statement cannot be made about the changing content of carbon dioxide in the atmosphere.

The decrease of carbon dioxide since the early history of the earth is attributed to its entrance into fixed form in organic matter and in materials of the hydrosphere and terresphere. An inventory

Years before present	Lithosphere	Biosphere	Hydrosphere	Atmosphere
20 Million	Glaciation	Mammals diversify Grasses appear		Oxygen approaches PAL.
50 Million	Coal formation Volcanism			
100 Million		Social insects, flowering plants Mammals		Atmospheric oxygen increases at fluctuating rate
200 Million	Great volcanism Coal formation	Insects appear Land plants appear	Oceans continue to increase in volume	
500 Million	Glaciation sedimentary calcium sulfate	Metazoa appear rapid increase in phytoplankton	Surface waters opened to phytoplankton	Oxygen at 3–10 percent of present atmospheric level Oxygen at 1 percent of present atmospheric level, ozone screen effective
1 Billion	Volcanism			Oxygen increasing, carbon dioxide decreasing
2 Billion	Red beds	Advanced oxygen–medianting enzymes First oxygen–generating photosynthetic cells	Oxygen diffuses into atmosphere Start of oxygen generation with ferrous iron as oxygen sink	Oxygen in atmosphere
	Glaciation Banded iron formations Oldest sediments Oldest earth rocks	Abiogenic evolution		No free oxygen
5 Billion	(Origin of solar system)			

Figure 1–11. Chronology relating the evolution of the lithosphere, biosphere and hydrosphere to atmospheric content of oxygen. (After Cloud and Gibor, 1970.)

of carbon dioxide in the ecosphere shows enormous amounts in these reservoirs (Figure 1–13). There is, though, a constant exchange between each of these: increases in the atmospheric content occurring through volcanic activity, combustion of fossil fuels and through atmosphere-hydrosphere exchanges when there is a decrease in ocean volume. In the lithosphere, the carbon dioxide becomes tied up in the form of carbonate minerals through precipitation from solution or direct synthesis by marine life. The oceans, which contain about 60 times more carbon dioxide than the atmosphere, alter in content as a result of changes in oceanic temperatures. High carbon dioxide concentration in the atmosphere can cause the gas to be directly

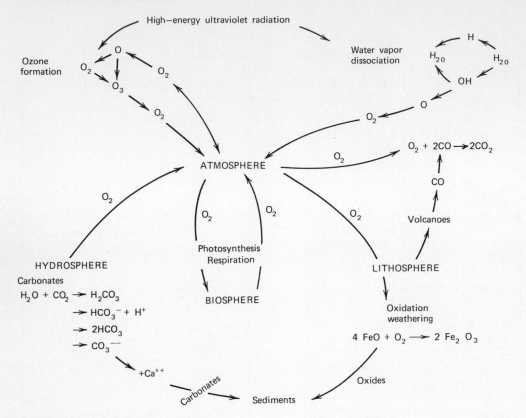

Figure 1-12.   The oxygen cycle. The cycle is highly complex and, because oxygen appears in so many chemical forms, is intimately interrelated to other cycles. (After Cloud and Gibor, 1970.)

absorbed by the oceans, while low atmospheric concentration results in the reverse process. A constant interplay, often extending over long periods of time, takes place.

While atmospheric carbon dioxide forms only a fraction of the total atmospheric volume, it plays an extremely important role. As noted earlier, it absorbs long-wave terrestrial radiation and contributes significantly toward the greenhouse effect. It is because of this role that variations in carbon dioxide in the atmosphere have been the center of intensive study. Indeed,

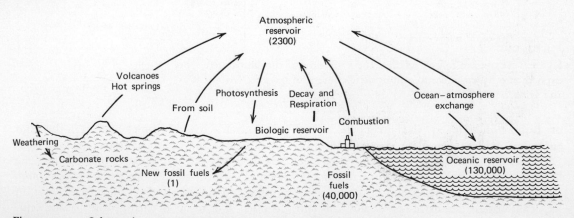

Figure 1-13.   Schematic representation showing carbon dioxide reservoirs and exchanges. Data in millions of tons. (After Plass, 1959.)

one theory of climatic change in geologic times is based upon natural changes that have occurred. More recently, the burning of fossil fuels by man, which produces large quantities of carbon dioxide, has been cited as the cause for a rise in world temperatures. This assumption was based upon the fact that the rise in carbon dioxide in the atmosphere between 1860 and 1940 [from 285 ppm to 312 ppm according to data given by Callendar (1961)], correlated strongly with a worldwide warming trend over the same period. Since the mid-1940s, however, while carbon dioxide continues to increase at about 0.2%/yr, the world temperatures have decreased slightly. This complex and somewhat conflicting problem is covered in more detail in Part III, which

places it in perspective of the entire problem of changing climates.

While the volume of carbon dioxide in the atmosphere is given as 0.033%, its cycle within the atmosphere and biosphere causes the mean concentration to vary over both time and space. The absorption of carbon dioxide, its fixation by photosynthesis, and eventual oxidation of organic compounds produced, cause its concentration to vary relative to photosynthetic activity. In vegetated areas, during the day carbon dioxide content may fall well below the mean value; at night it will be above. In a forest, carbon dioxide content may be six times that of the average value and, as illustrated in Figure 1-14, the diurnal changes can be quite drastic.

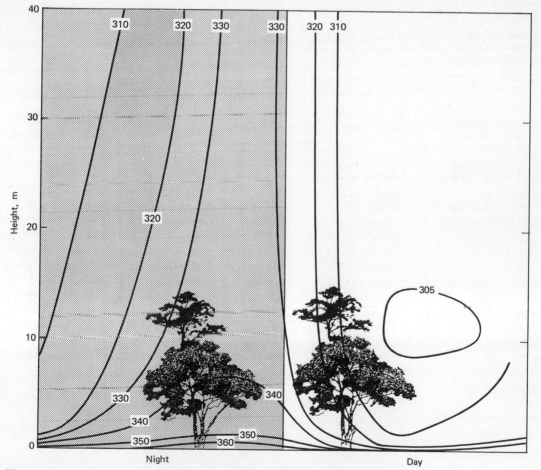

Figure 1-14. The vertical distribution of carbon dioxide in a forested area. The amount varies over time; at night, without ongoing photosynthesis, the levels are higher than during the day. The relatively high levels of concentration near the ground at night are related to soil respiration. (After Bolin, 1970.)

Carbon dioxide forms, of course, only one part of the total carbon cycle. Not only is it but one part of the total complex, but it is not the only carbon gas that has caused problems as a result of man's activities. The highly noxious carbon monoxide has also become the focus of much attention. The ambient concentration of this gas is in the order of 0.04 to 0.90 ppm; at places remote from a pollution source it is usually in the order of 0.1 ppm. This low concentration has been found despite the fact that man—the main producer of carbon monoxide—is adding about 200 million tons each year to the atmosphere. Since no marked increase in the atmospheric content has been found, it must be assumed that the gas has a short residence time (variously given as 2.7 to 4.0 yr) and that there must be a sink that absorbs large quantities of the gas.

While the soil reservoir had been suggested as a possible sink, it remained for Inman et al.

(1971) to show just how important it was. These investigators found that by placing a nonsterile potting soil in a test atmosphere, the rate at which carbon monoxide was depleted could be obtained. They found that carbon monoxide concentration decreased from 120 ppm to practically zero in about 3 hr. Tests were carried out with various soils, both sterile and nonsterile, and some of the results are summarized in Table 1-4. Generally, it was found that soils high in organic matter were the most active sinks.

Using the derived data as a base, it was found that the average rate of carbon monoxide absorption rate was 8.44 mg/hr/m^2, a value equivalent to 191.1 metric tons/mile^2/yr. This implies that the soil sink of the United States alone has the ability to deplete 569 million metric tons of carbon monoxide per year. This is about 6½ times the amount that is produced annually by man.

The role of the soil as a carbon monoxide

Table 1-4

RATE OF REMOVAL OF CO FROM TEST ATMOSPHERES AT 25°C BY VARIOUS SOILS[a]

Location of Soil[b]	Vegetation	pH	Sand:Silt:Clay (%)	Organic Matter (%)	CO uptake[c] (mg/hr per Square Meter of Soil)
Eureka-Arcata	Coast redwoods	5.7	53:34:13	25.1	16.99
H. Cowell State Park	Oak	5.3	73:12:15	11.2	15.92
H. Cowell State Park	Coast redwoods	5.7	57:26:17	13.6	14.39
Lake Arrowhead[d]	Ponderosa pine	6.2	65:24:11	17.4	13.89
Redding	Grass-legume pasture	5.1	53:32:15	21.0	11.94
Riverside[d]	Grapefruit[e]	6.6	75:14:11	4.3	11.48
Yosemite Valley	Grass meadow	5.05	49:42:9	20.6	10.52
Kauai, Hawaii	Forest	4.74	58:18:24	22.8	9.90
San Bernardino Freeway[d]	None	7.2	55:30:15	2.2	6.89
Mojave Desert	Chaparral	7.9	79:6:15	2.4	6.46
Woodland	Oak stubble[e]	6.6	33:32:35	2.1	6.23
Riverside (desert)[d]	Chaparral	7.35	85: 4:11	1.0	4.31
Yosemite wall	White fir	5.1	65:18:17	5.7	3.48
Corcoran	Cotton (fallow)[e]	7.1	57:22:21	2.8	3.48
Hanford	Almonds	6.95	53:26:21	3.5	2.82
Boynton Beach, Florida	Weeds (fallow)[e]	6.0	80: 0:14	1.4	2.65
Oahu, Hawaii		4.93	40:26:34	15.3	2.16

[a] From Inman et al. (1971).
[b] All soils were collected in California unless otherwise noted.
[c] Average rate at the end of the test period: two to three determinations.
[d] Locations where high levels of air pollution occur as a result of the combustion of fossil fuels and photochemical smog.
[e] Land under cultivation or with recent history of cultivation.

sink answered many problems relating to this gas. Earlier, in looking for an answer, other interesting facts were found. For example, Swinnerton et al. (1970), following the general notion that the oceans formed sinks for carbon monoxide, found that precisely the reverse was true. Surface waters were found to be super-saturated in carbon monoxide with respect to partial pressure of carbon monoxide in the atmosphere. Indeed, they showed that the ocean was a source, rather than a sink, of carbon monoxide.

Just as man is the major producer of carbon monoxide, it seems too that soon he will be the main producer of nitrogen. The amount of nitrogen fixed by fertilizers today is almost equal to that of biological action, and it is thought that

by the year 2000, it will exceed it. This rather startling fact is because nitrogen, while the most abundant gas in the atmosphere, cannot be used directly by the great majority of living organisms. In some ways, gaseous nitrogen might almost be considered an inert gas!

An inventory of nitrogen in the ecosphere shows that the atmospheric reservoir contains approximately 3,800,000 billion metric tons while the crustal and surface sedimentary deposits together contain about 18,000,000 billion metric tons (Figure 1–15). Despite these large quantities, the incorporation of nitrogen into a chemical form that can be used directly by plants and animals is minute. The most important contribution to fixation comes through a few legumes, some primitive plants and free-living algae and

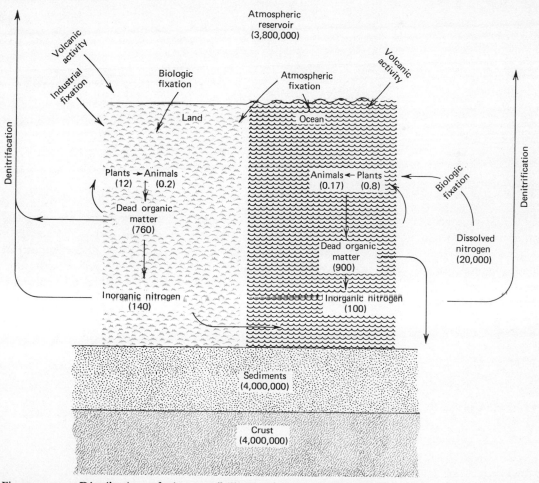

Figure 1–15. Distribution of nitrogen (billions of metric tons) and the exchanges that occur between the various parts of the biosphere. (After Delwiche, 1970.)

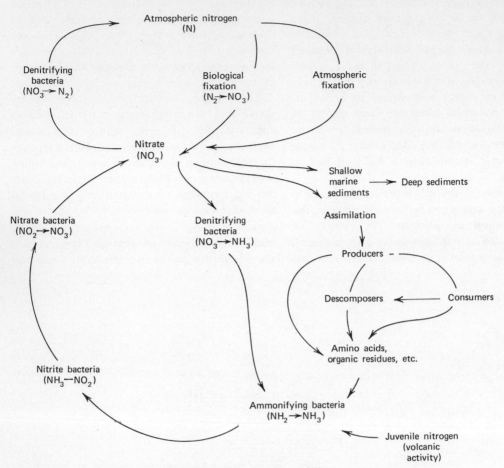

Figure 1–16.   The nitrogen cycle. (After Kormondy, 1969.)

bacteria. Some nitrogen compounds are also produced when cosmic radiation, meteor trails, or lightning provide concentrated energy required for nitrogen to react with hydrogen (from H_2O) and oxygen in the atmosphere. A summary of the nitrogen cycle is shown in Figure 1–16.

While atmospheric processes are directly concerned with the oxygen, carbon dioxide, and nitrogen cycles, it also plays a very important role in other biogeochemical cycles that do not have a gaseous phase. This will become apparent in the chapters that follow.

CHAPTER 2

Climatic Aspects of the Hydrologic Cycle

The hydrologic cycle is a conceptual representation through which the exchange of water over the earth's surface can be integrated into a comprehensible framework. Figure 2-1 shows the cycle in diagrammatic form. The amounts of water involved in the exchanges shown have been estimated by a number of authors; the estimates for the United States and for the world shown in Figure 2-2 illustrate the variable manner of representation. Clearly, while the study of the cycle and its component parts crosses many disciplines, climatological analysis plays an important role, for much of the cycle is intimately concerned with the state of the atmosphere. So significant is the role that it has developed into the specialized field of hydroclimatology.[1]

Hydroclimatology is defined as the ". . . study of the influence of climate upon the waters of the land." (Langbein, 1968). While the hydrologic cycle is the best expression of the global exchange of water, a good idea of the disposition of water over a land surface, where it is of most significance to man, can be expressed by the following hydrologic equation:

$$P = E + T + I + SW + RO + GW + \Delta ST$$

where P is precipitation, E is evaporation, T is transpiration, I is interception, RO is runoff, SW and GW are rates of change in soil water and ground water, respectively, and ΔST is a residual change in storage in the system.

Each of the parameters in the equation falls into the realm of climatic studies. Precipitation represents the main input into the system and a study of the spatial and temporal distribution of precipitation is basic to its entire analysis. The return of moisture to the air is achieved through the combined processes of evaporation and transpiration, the rates at which these occur

being largely dependent upon the prevailing heat and moisture conditions. Moisture intercepted by vegetation cover can, in heavily forested regions, reach large proportions. Much of this is returned to the atmosphere by the evaporation process. While the study of precipitation and the combined evaporation-transpiration processes dominate hydroclimatic studies, due consideration must also be given to runoff, soil, and ground water. This is evident from the balanced hydrologic equation, for any factor influencing one of the variables, directly or indirectly, influences the others.

To facilitate analysis of the components of the hydrologic equation, aspects of each are considered separately in the following discussion. Despite such systematic treatment, remember that all parts of the equation are intimately interrelated and occur simultaneously.

PRECIPITATION

Precipitation, a term that encompasses all forms of moisture deposited at the earth's surface, is the source of almost all fresh water available to man. Consequently, the distribution of precipitation over time and space is a prerequisite to the study of the hydrologic characteristics of any area. The whole notion of the use of water by man depends upon its distribution. Resource planning, in terms of overcoming water shortages or excesses, ultimately depends upon the nature of the precipitation that occurs in any area.

Water vapor is always present in the lower layers of the atmosphere (Table 2-1). The amount that occurs, however, is highly variable. For precipitation to occur, the water vapor present must condense or sublimate to form water droplets or ice crystals, the constituents of clouds. The water droplets or ice crystals must then attain a large enough size to enable them to fall to the ground as precipitation. The precipitation process thus consists of the cooling

[1] This term is less widely used than hydrometeorology, which also encompasses long-term aspects of the hydrologic cycle.

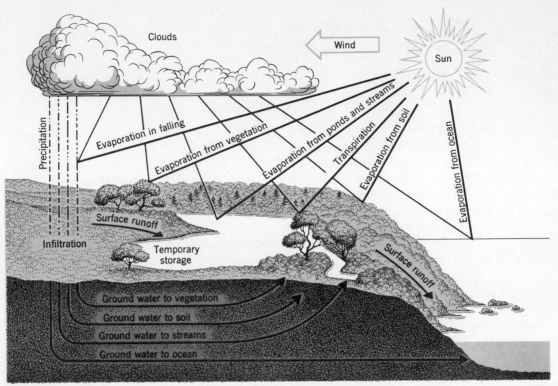

Figure 2-1. The hydrologic cycle. (From Dasmann, R.F., *Environmental Conservation*, Copyright © 1968 by John Wiley and Sons. Reproduced by permission.)

Table 2-1

GEOGRAPHICAL DISTRIBUTION OF RELATIVE HUMIDITY
(Percent of Saturation Vapor Content)[a]

Latitude[b]	N. Winter S. Summer	N. Spring S. Autumn	N. Summer S. Winter	N. Autumn S. Spring	Year
70°–60°N	86	81	77	84	82
60°–50°N	83	74	76	80	78
50°–40°N	78	73	69	76	74
40°–30°N	73	78	67	71	70
30°–20°N	71	68	70	73	71
20°–10°N	74	73	78	77	76
10°– 0°	77	78	82	81	79
0°–10°S	81	81	82	80	81
10°–20°S	79	78	80	77	79
20°–30°S	79	79	80	75	77
30°–40°S	75	80	80	79	79
40°–50°S	81	81	83	79	81
50°–60°S	83	79	—	—	—

[a] Estimates by Arrhenius.
[b] Note that because of temperature variation the *absolute* moisture content decreases markedly from tropical to polar realms.

28

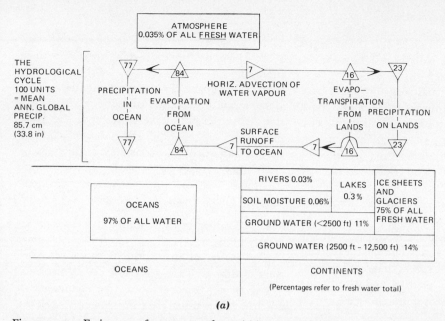

Figure 2–2.   Estimates of water transfers within the hydrologic cycle. (a) World exchanges and water storage. Units of the cycle assume a mean of 100 units of mean global precipitation. (From More, R., "Hydrologic Models and Geography", in *Models in Geography*, Chorley R.J. and P. Haggett. eds., Methuen. Used by permission.) (b) Hydrologic cycle for the United States. Typical of water utilization in a developed country, it shows that a relatively small percentage of water is withdrawn for direct use by man. The greater part is returned to the atmosphere by evaporation and transpiration (about 73%). (After Revelle, 1963.)

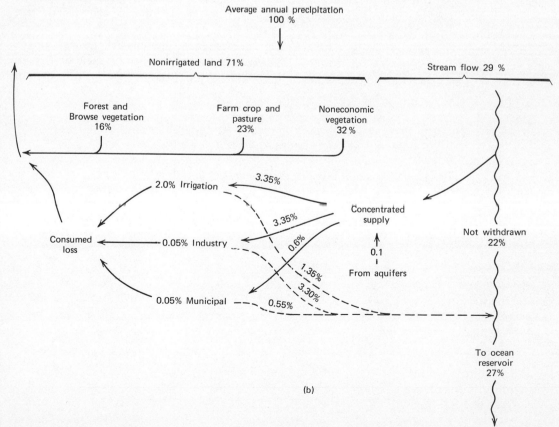

of air below its condensation level in the presence of condensation nuclei, with the resulting growth of precipitable moisture.

While dew, mist, and fog do contribute toward the total precipitation that might occur in a given region, precipitation from clouds is by far the most important process. The dominant factor in precipitation distribution is thus the study of adiabatic cooling in relation to stability and instability that occurs through cumulus-scale convection, frontal formation, or the orographic effect caused by mountain barriers. Because these processes do not occur with equal frequency over the earth, and because, as noted, the amount of moisture contained in air at different temperatures is highly variable, precipitation distribution over the earth's surface is quite uneven.

Spatial Distribution of Precipitation

It has been estimated that if the total annual precipitation that occurs over the earth's surface were evenly distributed, then the average annual total at every location would be about 34 in. (84 cm) (More, 1967). But this is clearly not the case. Figure 2–3 shows a map of the average annual distribution over the globe. Great differences are seen, with average annual amounts ranging from less than 10 in. (25 cm) to more than 100 in. (254 cm). In terms of management

of water resources by man, it is the areas of scant or excessive precipitation that merit the most attention.

While generalizations concerning the annual precipitation distribution of precipitation are not always meaningful, some concept of the cause and extent of both arid and very wet climates can be attained by considering their gross distribution. Generally, it can be shown that the arid regions of the world occur in three great realms.

1. Extending in a discontinuous belt approximately between 20 and 30°N and S are the great *tropical deserts* of the world. Such deserts owe their aridity to large-scale circulation patterns. Upper air convergence causes the dominance of adiabatically warmed descending winds, with the result that such deserts are dominated by tropical continental (cT) air masses. Occurring in similar latitudes are the *tropical littoral* deserts. Located on the eastern sides of the semipermanent high-pressure cells, they are dominated by stable maritime air masses (mT_s) with the result that their climatic characteristics differ slightly from those of the continental type (Table 2–2).

2. Located in the interiors of continents are the *continental deserts*. These owe their aridity to their distance from the sea, the major source of

Table 2–2

COMPARISON OF TEMPERATURE AND RELATIVE HUMIDITY CHARACTERISTICS OF TROPICAL AND WEST COAST DESERTS

	Average of Warmest Month			Average of Coolest Month			Mean Annual Temperature (°F)	Annual Temperature Range (°F)
		R. H.(%)			R.H. (%)			
	T(°F)	0700	1400 hr	T(°F)	0700	1400 hr		
Tropical Deserts								
Insalah (Sahara)	98	36	25	56	63	37	77	42
Riyadh (Arabian)	93	47	31	58	70	44	75.5	35
Laverton (Gt. Australian)	87	36	24	52	60	43	74.5	35
West Coast Deserts								
Arica (Atacama)	72	74	61	60	83	74	66	12
Walvis Bay (Kalahari)	66	91	73	58	83	65	62	8
Villa Cisneros (Coastal Sahara)	72	88	63	63	75	51	67	9

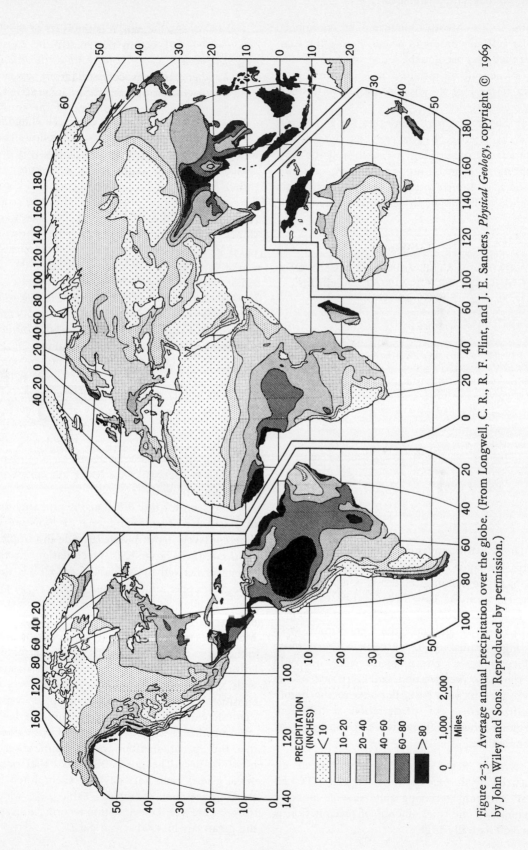

Figure 2-3. Average annual precipitation over the globe. (From Longwell, C. R., R. F. Flint, and J. E. Sanders, *Physical Geology*, copyright © 1969 by John Wiley and Sons. Reproduced by permission.)

PRECIPITATION
(INCHES)
< 10
10–20
20–40
40–60
60–80
> 80

0   1,000   2,000
Miles

water vapor. Mostly dominated by continental polar (cP) air masses in winter, during the summer they may become the source of cT air, or may be influenced by modified maritime air masses that, because of the distance they have traveled, are relatively dry.

3. The *polar deserts* constitute the third great area characterized by low precipitation. The low temperature of such regions is probably the most important contributing factor here, for the amount of moisture that a given air mass can contain depends upon its temperature (Table 2-3).

Table 2-3

MAXIMUM WATER VAPOR CAPACITY OF A CUBIC METER OF AIR AT DIFFERENT TEMPERATURES

Temperature		Grams of Water Vapor
°F	°C	
23	-5	3.261
32	0	4.847
41	5	6.792
50	10	9.401
59	15	12.832
68	20	17.300
77	25	23.094
86	30	30.371
95	35	39.599

The actual boundaries of deserts as shown on maps are quite tentative, for there are marked problems in defining the desert boundary. Many empiric formulas have been devised to express aridity, some examples being shown in Table 2-4.

In terms of generalizations concerning areas of high annual totals of rainfall, much more difficulty arises. The problems relate not only to the large precipitation variability that occurs on a year to year base, but also to the highly localized effect of precipitation distribution. Regions where high rainfall can constantly be assumed are:

1. Areas in the equatorial realm that are influenced by the *intertropical convergence* (ITC) all year. Although receiving quite variable monthly increments, the total amount of precipitation is characteristically high.

2. Coastal regions of the tropical realm where *mT air masses prevail* most of the year.

3. *The western sides of* continents in the *middle latitudes* where precipitation results from constant onshore westerlies with frequent cyclonic activity.

Because mountain barriers play such a significant role in air mass modification, it follows that topography is of consequence in analysis of the spatial distribution of precipitation. Figure 2-4 shows a topographic cross section and the distribution of precipitation in the areas differentiated. As already indicated, the windward slopes of mountains receive the greatest totals. In the lee of such areas, the precipitation decreases markedly to give a rain-shadow effect.

For the most part, there is an increase in precipitation with altitude in middle latitude areas, especially on those mountains bordering the sea. But this effect appears to be modified in the tropical and subtropical realms. Observations have shown that precipitation in such regions appears to increase to a given height and then it decreases. In Java, for example, the maximum precipitation occurs at an elevation of about 4000 ft (1219 m), thereafter it decreases. Similarly, in the Guatemalan Highlands, the maximum occurs between about 3000 (914 m) and 4000 ft (1219 m) with lower amounts both below and above these levels.

The reason for this difference is probably due to the different concentrations of moisture at different levels in the middle latitude and tropical realms. Much orographic precipitation in the tropics is derived from cumuliform clouds that have their upper limits at about 9000 ft (2743 m). Water droplets in the clouds are probably concentrated near the mean base level of the cloud. Mountain stations above that concentration level will receive less rain accordingly. In the middle latitudes, much of the precipitation is associated with stratiform clouds. These may extend to considerable heights and water droplets within them may not be concentrated at any one level.

From what has already been stated it is evident that the spatial distribution of precipitation is highly varied. The great differences that occur on a global basis can be well exemplified by studying extreme examples. Table 2-5, which shows some world precipitation records, indicates the great variability that can exist.

Table 2-4

INDICES USED FOR CHARACTERIZING THE ARIDITY OF A REGION[a]

Author	Formula	Notes
Dokuchaiev (1900)	$\dfrac{P}{E_o}$	The author gives only a qualitative evaluation of humidity by comparing the mean annual values of precipitation with those of evaporation. Latter derived using Wilde evaporimeter.
Oldekop (1911)	$\dfrac{P}{E_o}$	A calculation value of a potential evapotranspiration (?) is introduced according to the formula $E_o = \alpha d$, where d is the saturation deficit. The coefficient of proportionality, α, is derived from the comparison of evaporation from a natural lake with the course of meterorological elements. Formula may be rewritten as $P/232.d$.
Lang (1920)	$\dfrac{P}{t^\circ}$	The "rain factor." In this (and the next two methods) annual precipitation and annual average temperatures are used.
Köppen (1922)	$2(t^\circ + 7)$	See above.
de Martonne (1925)	$\dfrac{P}{t^\circ + 10}$	See above.
Meyer (1926)	$\dfrac{P}{d}$	Introduction of saturation deficit into the formula.
Reichel (1928)	$\dfrac{NP}{t^\circ + 10}$	Modification of de Martonne, where N is the number of days with precipitation.
Emberger (1939)	$\dfrac{100P}{(M + m)(M - m)}$	Evaluation using mean maximum temperature of the warmest month (M) and mean minimum temperature of the coldest month (m).
Ivanova (1941)	$\dfrac{P}{E_o}$	E = potential evapotranspiration for an open water surface using $E_o = 0.0018 (25 + t^\circ)^2 (100 - a)$, where t° = mean monthly temperature and a = mean monthly relative humidity.
Thornthwaite (1948)	$\dfrac{100s - 60d}{n}$	s = surplus moisture in humid season, d = deficit of moisture in the dry season, n = potential evapotranspiration.
Prescott (1949)	$\dfrac{P}{(S_d)}^{0.7}$	S_d is the saturation deficit and P the precipitation in inches.
Budyko (1951)	$\dfrac{R_o}{LP}$	Based upon the radiation balance (R) that exists and the latent heat of saturation (L).
Kostin (1952)	$\dfrac{P}{E_o}$	Amount of precipitation and potential evapotranspiration, derived for the same period of time. E_o is calculated using $$E_o = \frac{dn}{4}(1 + 0.004t)^2$$

[a] After UNESCO (1958).

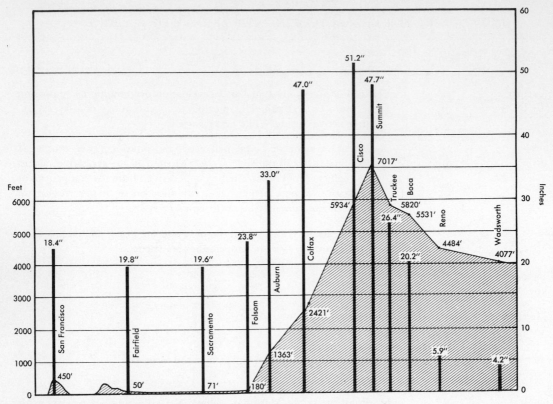

Figure 2–4. A cross section from San Francisco to Reno showing relationships between topography and precipitation. (From Powers, W.E., *Physical Geography*, 1966, Appleton, Century and Crofts. Used with permission.)

Table 2–5

WORLD RECORD PRECIPITATION DATA[a]

Variable	Station	Amount
Average annual precipitation	Mt. Waialeale, Hawaii	471.7 in.
Highest 12-month total ever recorded	Cherrapungi, India (August 1860–July 1861)	1,041.9 in.
Highest amount in a single calendar month	Cherrapungi, India (July 1861)	366.1 in.
Greatest total ever recorded in a 24-hr period	Cilaos, Reunion (March 15–16, 1952)	73.62 in.
Lowest average annual precipitation	Arica, Chile	0.02 in.
Longest period without rain being recorded	Iquique, Chile (1899–1913)	14 yr

[a]Data from Cunningham and Vernon (1968).

Spatial diversity of precipitation is not restricted to the long-term global distribution. Enormous differences occur over small areas in shorter time intervals. In many instances, intense rainfall is highly localized, a typical example of this being shown in Figure 2–5. As described later, such intense precipitation can cause havoc, with results ranging from disruption of communications and flooding in urban areas to damage of crops and accelerated erosion in rural. The study of such short-term, intense rainfall is dependent upon good instrumentation, and interesting studies in this area have been completed by researchers at Bell Laboratories. The research, originally

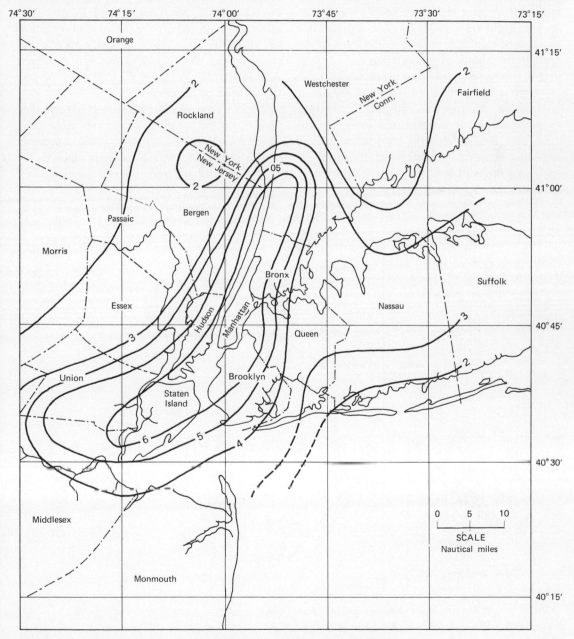

Figure 2–5.   Isohyets for a 24-hr rainfall that occurred in New York metropolitan area in September 1969. The more than 6 in. that fell on Manhattan caused extensive local flooding resulting in disruption of communications.

carried out to assess the influence of rainfall intensity upon radio communications, utilized a rain gage that allowed precipitation intensity to be measured remotely and data fed directly into computers. By generating computer maps the marked variations could be displayed. While such data may smooth themselves out over time, it most certainly indicates the significance of rainfall study at a local scale and further points to the high significance of a comprehensive network of recording stations in the study of the spatial distribution of precipitation.

It is this latter aspect that poses many of the problems arising in the study of precipitation distribution. Many maps showing distribution of precipitation are constructed using limited data. This results from the scattered nature of recording stations, and it is not infrequent to find large tracts of land represented by a single rain gage reading. How representative such data are, particularly in areas of varied topography, is questionable. This has led a number of workers to devise empiric systems for estimating precipitation received in watersheds of variable topography. The methodology varies from the formulation of regression equations to the construction of multiple correlation graphs. Figure 2–6 illustrates such an approach, showing a method for construction of a coaxial relationship between storm precipitation and runoff using other known variables. In this case they include an antecedent precipitation index (API) and storm length. As shown in the first quadrant, API versus observed runoff, a number of trial curves need be tested.

Even in areas of relatively simple topography, problems occur in estimating the area represented by a single gage reading. The most commonly used method of analysis of such data is the construction of isohyets, lines of equal rainfall. Such lines can be drawn mechanically, but frequently they reflect the relative awareness of

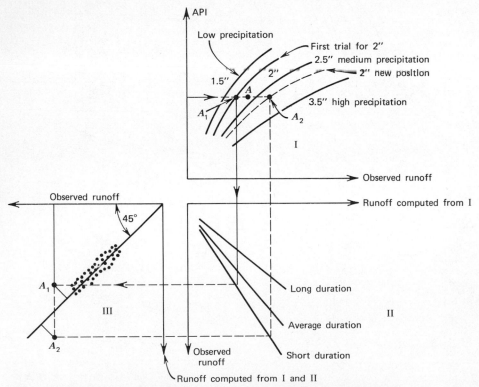

Figure 2–6. Example of method used in constructing coaxial graphs. This illustrates the coaxial relation between total precipitation and runoff for a given antecedent precipitation index (API) and storm duration. (From DeWiest, R.J.M., *Geohydrology*, Copyright © 1965 by John Wiley and Sons. Reproduced by permission.)

the compiler of the physical background of the region under study. An alternative approach is through the use of Thiessen polygons. Essentially, these consist of constructed polygons derived by weighting data according to the geometric area of the area they represent (Figure 2–7). Worked examples of the isohyetal and polygon methods are shown in Figure 2–8.

Temporal Distribution of Precipitation

Total annual precipitation alone is an insufficient measure of moisture available at the earth's surface, for it does not take into account the manner in which the precipitation is distributed throughout the year. In most realms of applied climatology that concern the use of precipitation, it is often the temporal distribution that is the dominant factor in resource management. In many areas the entire agricultural year is geared to the time at which precipitation occurs, and it is not infrequent to find that slight deviations from the seasonal pattern result in reduced crop returns.

The significance of seasonal distribution relates to the concept of precipitation effectiveness. As shown by the hydrologic equation, a considerable amount of moisture is returned to the atmosphere by evaporation and transpiration. These two factors are directly influenced by the temperature at which precipitation occurs. Similarly, the type of precipitation often differs on a seasonal scale with the result that the disposition of water falling onto a surface changes over the year.

In attempting to express precipitation effectiveness, a number of authors have derived formulas. One, proposed as early as 1869 by Linsser, expressed the factor as r/t, where r is the monthly precipitation in millimeters and t is the mean monthly temperature in degrees Centigrade. Many other expressions are available and some have been represented in map form. Bailey (1968), for example, has produced an interesting map of the world showing the distribution of effective moisture (Figure 2–9). This map should be compared with the world map showing total annual precipitation to gain some indication of the importance of precipitation effectiveness.

Apart from losses of moisture through evaporation and transpiration, the season at which

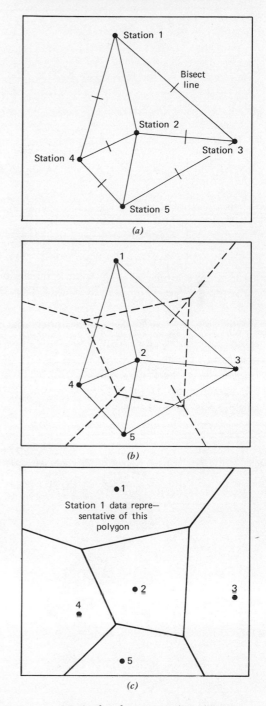

Figure 2–7. Method of constructing Thiessen polygons. (*a*) Connect nearest neighbors with a straight line. Find midpoint of line. (*b*) Draw perpendicular bisectors of lines. (*c*) Draw polygons.

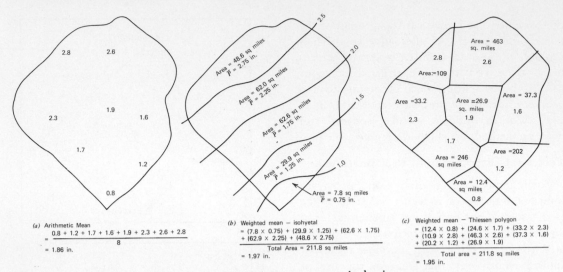

Figure 2–8. Calculation of the mean precipitation over a watershed using (a) arithmetic means, (b) isohyets, and (c) Thiessen polygons.

precipitation occurs is also significant in that it often determines the nature of precipitation occurring. The most obvious effect here is that under low-temperature conditions much of the fall accumulates as snow (Figure 2–10). Obviously, snow does not become effective moisture until melting occurs, and frequently the rapidity of melt is so great that problems of disposal occur with resulting floods.

Another effect of the time of year at which precipitation occurs concerns the size of the falling water droplets, which might vary from

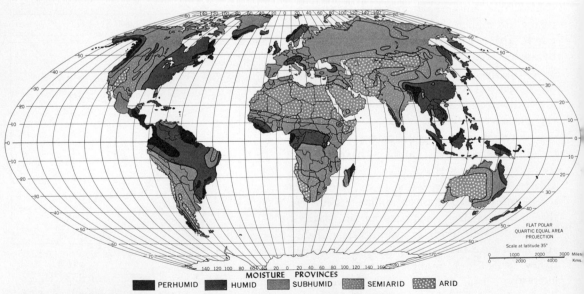

Figure 2–9. World distribution of effective precipitation (EP). The EP is calculated as a humidity index in which five humidity-index subdivisions are defined: A—Perhumid $\geq$ 16.2; B—Humid, 8.7 to 16.1; C—Subhumid, 4.7 to 8.6; D—Semiarid, 2.5 to 4.6; E—Arid < 2.5. EP is derived from the formula $EP = P/1.025T + X$, where P is annual rainfall in inches, T is the annual temperature in °F, and X is the correction supplied by a nomograph. (After Bailey, 1958, from Trewartha, G. T., *Introduction to Climate*, 4th ed., 1968. Used with permission of McGraw-Hill Book Company.)

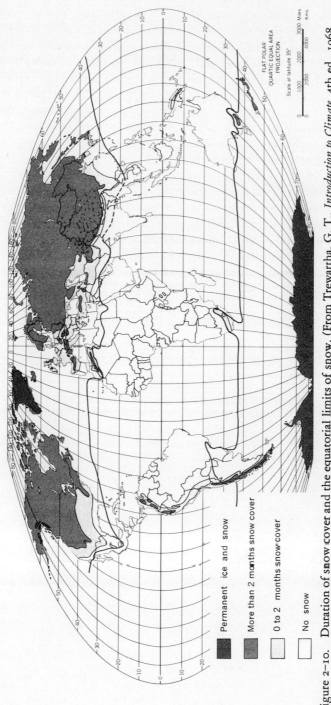

Figure 2-10. Duration of snow cover and the equatorial limits of snow. (From Trewartha, G. T., *Introduction to Climate*, 4th ed., 1968. Used with permission of McGraw-Hill Book Company.)

Permanent ice and snow

More than 2 months snow cover

0 to 2 months snow cover

No snow

FLAT POLAR
QUARTIC EQUAL AREA
PROJECTION

Scale at latitude 35°

very large droplets to minute particles of water that occur in the form of drizzle. Table 2–6 shows the various modes of precipitation that can occur and their relative sizes. The type of precipitation that occurs, with obvious low-temperature products included, often depends upon the type of air mass that prevails; in fact, a general insight to air mass dominance might be derived from gross analysis of precipitation intensity, and air mass dominance can be used as a measure of the seasonality factor of a given location. As an example, a seasonal climate that experiences cT air mass during the summer will be dry, whereas during the winter the dominant air mass might be mT, which provides moisture. The resulting climate experiences marked differences in seasonal precipitation characteristics.

The causes of the seasonal distribution of precipitation relate to the migration of the pressure belts and their associated wind systems. The results of the migrating systems can be observed on a world basis as shown in Figure 2–11. Here, the precipitation is classed as falling in winter, summer, or at all seasons. On a more localized scale, it is necessary to investigate the annual pattern on a more detailed basis. One example of a more precise analysis of seasonal variation is shown in Figure 2–12.

There are areas of the world in which the amount of precipitation received, irrespective of season, is highly variable on a year-to-year basis. Various statistical methods are available to express the amount of variability; one such approach uses the coefficient of variation (CV), which is given by the ratio between the standard deviation (σ) and the mean ($\bar{x}$), and is expressed as a percentage:

$$CV = \frac{\sigma}{\bar{x}} \times 100\%$$

As part of an analysis of rainfall distribution in the Jordon Valley, Manners (1969) used this formula to derive the pattern shown in Figure 2–13. Another method of expression of variability is mapped in Figure 2–14, where the amount of variation is shown as a percentage variation from the mean values.

Analysis of this map shows that precipitation appears to be most reliable in areas that receive a high amount of precipitation on an annual

Table 2–6

CHARACTERISTIC SIZES OF ATMOSPHERIC CONSTITUENTS

| Type | Diameter (mm) | | Terminal Velocity (cm/sec) |
	Range	Typical	
Small (Aitken) condensation nuclei	0.1×10^{-4} to 4.0×10^{-4}	—	10^{-5} to 10^{-3}
Dry haze	1×10^{-4} to 100×10^{-4}	—	10^{-3} to 10^{-4}
Fog and cloud droplets	1×10^{-3} to 200×10^{-3}	20×10^{-3}	0.01 to 70
Drizzle	2×10^{-2} to 40×10^{-2}	30×10^{-2}	1 to 170
Raindrops	0.4 to 4	1	170 to 900
Snow crystals	0.5 to 5	2	30 to 100
Snowflakes	4 to 20	10	80 to 200
Hail	5 to 75 + (largest: 140)	15	800 to 3500+

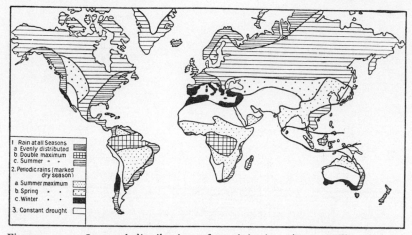

Figure 2–11. Seasonal distribution of precipitation. (From Miller, A.A., *Climatology*, 1953, Methuen. Used with permission.)

basis. The departure from normal, in, for example, equatorial, humid tropical, and middle latitude west coast areas is relatively low. The greatest amount of variability occurs in the arid and semiarid regions of the world. Such variability is, of course, to be expected. Desert precipitation often occurs in storms occurring over widely spaced intervals of time.

The variability within the arid and semiarid regions has caused enormous problems in terms of their utilization. Agricultural development has been quite successful in some years when precipitation is above "normal." But while a number of consecutive years might be moist, a drier period often follows, and agriculture becomes much more problematic, leading to the abandonment of farms. The creation of the Dust Bowl in the United States (see Chapter 4) is graphic evidence of this. The amount of variability of rainfall in an arid region is extremely

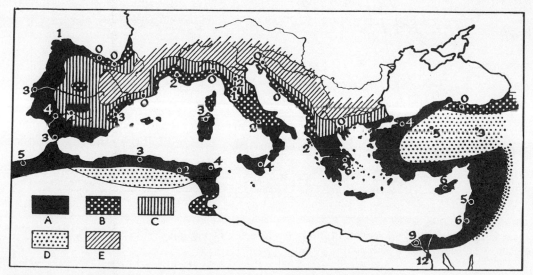

Figure 2–12. Seasonal distribution of precipitation in the Mediterranean lands. (A) Simple winter maximum. (B) Autumn maximum (winter wetter than summer). (C) Spring and autumn maxima (winter wetter than summer). (D) Spring maximum (winter wetter than summer). (E) Spring and autumn maxima (summer wetter than winter). (After Walker, 1962).

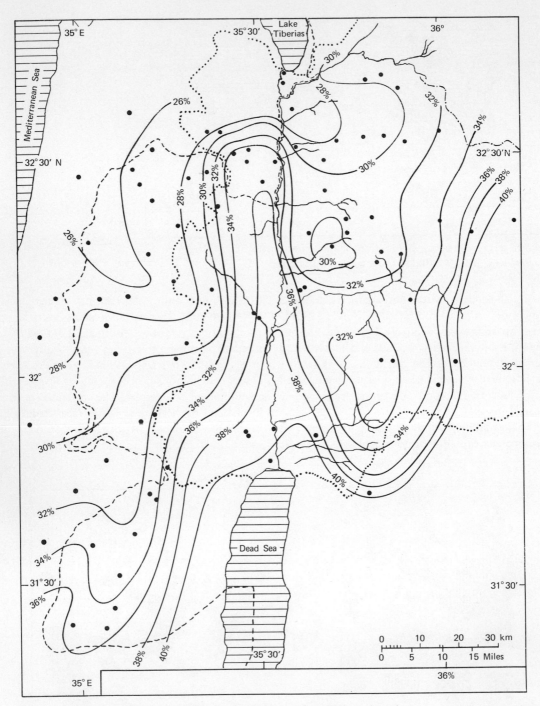

Figure 2–13. Variability of rainfall in the lower Jordan Basin using the coefficient of variation. (After Manners, 1969.)

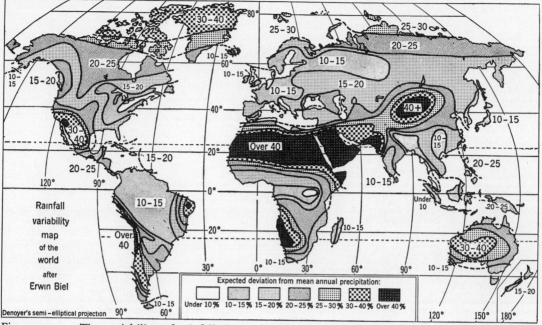

Figure 2–14. The variability of rainfall shown by percentage departure from normal .(From Strahler, A.N., *Physical Geography*, 1st ed, Copyright © 1951 by John Wiley and Sons. Reproduced by permission.)

well demonstrated in Figure 2–15, which shows the limits of the desert, according to a selected formula, over a series of years. On viewing this figure, it is little wonder that the definition of a desert area is highly problematic.

Studies of variability of precipitation are often completed by assessing the probability of a given amount of rainfall that might occur on an annual basis for a given location. The already cited study of the agriculture in Jordon by Manners (1969) uses such an approach. He derives his values by using the formula

$$d = \frac{x - \bar{x}}{\sigma}$$

where d = the required figure, x = the critical figure, $\bar{x}$ = mean value, and σ = the standard deviation. To employ the formula, tables of probability of the normal distribution curve are derived from prepared tables (e.g., Lindly and Miller, 1953). The derived values are expressed in percentage figures. Figure 2–16 shows the mapped results in relation to the probability of receiving 200 mm per year.

Obviously, statistical methods play a highly significant part in the study of rainfall, and many analytic techniques are available. Conrad and Pollak (1950) provide an excellent summary of many of the methods, giving cases concerning both probability and periodicity. Their example of the analysis of wet and dry spells shows how such methods can be used to analyze rainfall distribution over the year.

A wet spell occurs when a defined amount of rainfall falls on consecutive days. Table 2–7 shows how data are treated. The frequency distribution of days on which a given amount falls is determined and the results tabulated. In January there were 44 occasions on which the amount fell for one day. There were 33 times in which the amount occurred on a two-day consecutive period. The longest wet spell in January was 10 days. To obtain the monthly average length of wet spells, the formula

$$\bar{l} = \frac{\sum lifi}{\sum fi} \text{ days}$$

where $\bar{l}$ = average length of wet spells

li = length of spell (days)

fi = frequency of length of spell

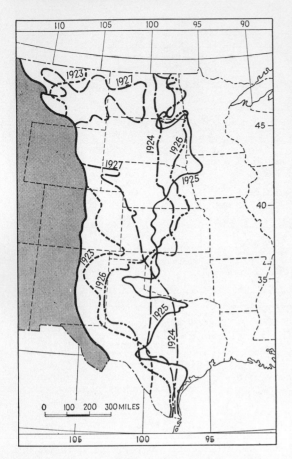

Figure 2-15. The fluctuating semiarid-subhumid boundary for the United States east of the Rockies over a five-year period. (From Trewartha, G.T., *Introduction to Climate*, 1968. Used with permission of McGraw-Hill Book Company.)

is used. For January, substitution into the formula gives

$$l_i = \frac{(1 \times 44)(2 \times 33) \ldots (10 \times 1)}{106}$$

$$= \frac{237}{106} = 2.24$$

The mean length of a January wet spell is 2.24 days. As the table shows, wet spells occur much more frequently in winter than in summer.

The year-to-year variation of precipitation is not confined to arid regions. When significant shortages or excesses occur in more densely inhabited parts of the world, the results are often far reaching. As an example, the north-eastern parts of the United States experienced precipitation well below normal for the period 1962 to 1966 (Figure 2-17) and stringent restrictions were placed upon the use of water. Such an experience was not without reward, however, for the crises led to a number of significant studies concerning the rational use of water in urban areas (e.g., Zobler and Carey, 1969). In the same way, precipitation high above normal can create massive problems. This is well illustrated by the disasters in Northern Italy in 1966 when, because of the loss of invaluable works of art, worldwide attention was focused upon the problem.

While the distribution of precipitation over both time and space requires analysis, hydrologists are often concerned with the combined effects of the two. Thus, the studies of depth-area-duration in relation to storm precipitation is a frequently used method of analysis. Most hydrology texts outline the method, good examples being given in Chow (1964), DeWiest (1965), and Ward (1967).

Modification of Precipitation Processes

The fact that water often occurs in the wrong place at the wrong time—or in some instances does not occur at all—has led to much research into methods by which water distribution can be modified to best meet the demands of man. It is possible to distinguish a number of approaches to the method of modification. There are those in which water is physically transferred from where it is in surplus to where it is needed. Aspects of this approach are considered in later sections of this book. Emphasis here is on a second approach, namely, the manner in which man attempts to modify the hydrologic cycle by altering the actual physical processes within the cycle.

Throughout history man has tried to alter the distribution of rain. The methods by which this has been attempted—ranging from the use of magic to willfull force—make exceedingly interesting reading and are well outlined in popular books by Halacy (1968) and Battan (1969). While instigators of some of the methods these authors outline professed to use "science" to obtain rain, it was not until the 1940s that the scientific method of rainfall modification really became feasible.

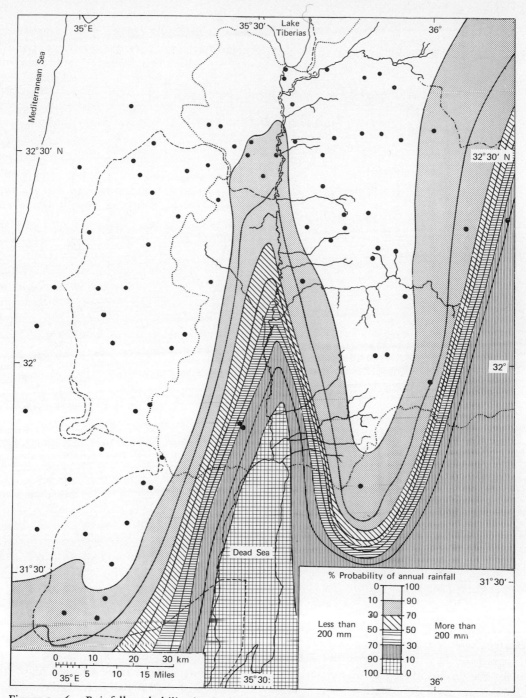

Figure 2–16. Rainfall probability in the lower Jordan Valley; probability of receiving more or less than 200 mm. (After Manners, 1969.)

Table 2-7

STATISTICS OF WET SPELLS (1901–1940) AT LUGANO, SWITZERLAND (46.0°N, 9.0°E, 902 FT)[a]

Length, l_i (Days)	Jan.	Feb.	Mar.	Apr.	May	June	July	Aug.	Sept.	Oct.	Nov.	Dec.	Year
					(a) Frequency Distribution								
1	44	45	57	58	66	84	109	85	61	49	40	69	767
2	33	27	41	42	36	43	57	58	44	35	34	37	487
3	11	11	23	44	33	33	23	26	24	24	22	13	287
4	8	7	12	14	15	21	20	16	10	8	16	12	159
5	4	6	10	14	18	7	4	3	7	13	7	6	99
6	4	3	5	7	7	7	4	1	6	9	3	2	58
7	—	4	5	5	8	3	—	2	4	5	6	1	43
8	1	1	—	2	5	1	3	—	—	3	3	3	22
9	—	—	—	1	1	4	—	—	2	1	—	1	10
10	1	—	—	1	2	—	—	—	1	1	—	—	6
11	—	—	—	—	—	2	—	—	1	2	2	—	7
12	—	—	1	—	1	—	—	—	—	—	—	—	2
13	—	—	—	—	—	—	—	—	—	—	2	—	2
14	—	—	—	—	—	—	—	—	—	—	—	—	
15	—	—	—	—	1	—	—	—	—	—	—	—	1
16	—	—	—	—	—	—	—	—	—	—	—	—	
17	—	—	—	—	—	—	—	—	—	—	—	1	1
18	—	—	1	—	—	—	—	—	—	—	—	—	1
Total	106	104	155	188	193	205	220	191	160	150	135	145	1952

(b) Average Length of Wet Spells

	Jan.	Feb.	Mar.	Apr.	May	June	July	Aug.	Sept.	Oct.	Nov.	Dec.	Year
	2.24	2.35	2.59	2.72	3.01	2.52	2.00	1.98	2.49	2.95	3.00	2.27	2.50

(c) Average Length of the Longest Wet Spells

	Jan.	Feb.	Mar.	Apr.	May	June	July	Aug.	Sept.	Oct.	Nov.	Dec.	Year
	3.25	3.50	4.68	5.10	6.00	5.18	3.92	3.42	4.68	5.15	4.92	3.98	9.42

[a] From Conrad and Pollak (1950), after H. Uttinger.

Research in cloud physics by Langmuir, Schaefer, and Vonnegut at the General Electric Laboratories in Schenectady, New York provided the practical base for the modification of clouds to give precipitation. The problem revolved around the way in which clouds form and give precipitation. A number of theories to account for the process had been postulated, but that by Bergeron (1935) provided the key to the derived methodology.

Bergeron suggested that since both ice crystals and supercooled water droplets occur together in a cloud, the ice crystals would grow at the expense of the supercooled water droplets. This would occur because the saturation vapor pressure over ice is less than that over water, that is, vapor that is saturated with respect to water is supersaturated with respect to ice. Since it can be assumed that at temperatures below −14°C some ice crystals will appear when saturated air is below freezing level, it follows that these ice crystals provide the "seeds" onto which supercooled water droplets will form. Ultimately, the ice crystals will attain sufficient size to fall to the ground. This process is illustrated schematically in Figure 2–18.

Obviously, the Bergeron effect is applicable only to rain that falls from clouds that contain

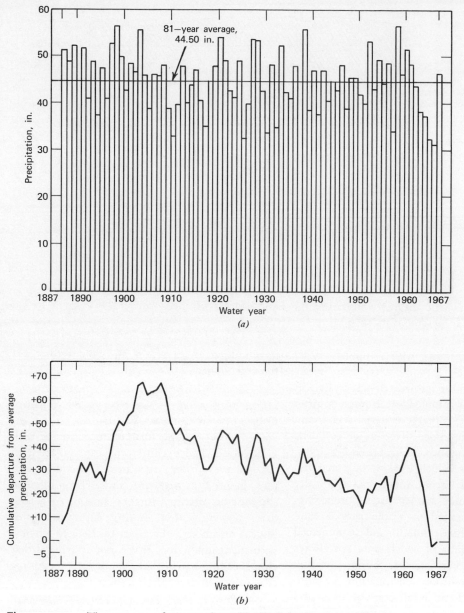

Figure 2–17.   The extent of the 1962-1966 drought in the northeast United States. (*a*) Long-term annual precipitation at Setauket, Long Island. (*b*) Cumulative departure from annual average precipitation at Setauket. (From P. Cohen et al., 1969, *Geological Water Supply Paper*, No. 1879-F.)

supercooled water and ice crystals. To explain rainfall from warm clouds, the collision hypothesis is evoked. This supposes that through collision and coalescence, water droplets can grow large enough to fall under their own weight. Research on rainmaking is almost totally concerned with cold clouds and depends upon the Bergeron explanation.

Working with supercooled moisture in the laboratory, Scheafer (1946) found that upon addition of a substance with a temperature lower than 40°, spontaneous formation of ice crystals occurred. Further experimentation showed that dry ice (solid carbon dioxide) proved a suitable medium for actual cloud seeding. At about the same time, Vonnegut (1947) found that by

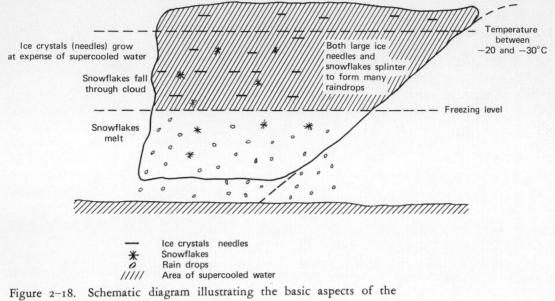

Figure 2–18. Schematic diagram illustrating the basic aspects of the Bergeron process. (After Atkinson, 1969.)

introducing a substance with a similar crystal form to that of ice, it would serve as a "seed" and thereby promote growth of ice crystals at the expense of supercooled water droplets. After much searching, a suitable material was found to be silver iodide. Since silver iodide could be introduced into clouds from ground burners, it became the most widely used seeding medium.

Results obtained from initial tests using dry ice and silver iodide seemed highly promising. The relative success led to a whole spate of cloud seeding, both for research and commercial purposes, the results of which were not always easily evaluated. Furthermore, the modification of clouds and resulting precipitation (or non-precipitation) led to legal consequences and entanglements.

Morris (1966) has surveyed legal aspects of weather modification and three cases that he cites provide evidence of problems that occur.

1. *Samples v. Irving P. Krick, Inc.*, (Civil Nos. 6212, 6223 and 6224, Western District of Oklahoma, 1954). This case arose out of cloud seeding sponsored by Oklahoma City in 1953. The plaintiff, a landowner, sued for property damages incurred in a cloudburst and flood which were coincident with the cloud seeding operations. The plaintiff failed to prove to the satisfaction of the jury that the seeding could

have influenced the storm, and their verdict was for defendant.

2. *Auvil Orchard Co., et al. v. Weather Modification, Inc., et al.* (Case No. 19268, Superior Court, Chelan County, Washington, 1956), involved cloud seeding for the prevention of hail. Flash floods had occurred on farms adjacent to the hail prevention target area. The court therefore granted a *temporary* order banning hail suppression attempts for one season. At a later date, however, after hearing expert meteorological testimony, the court refused to grant a permanent injunction. It was not convinced that cloud seeding had brought about the exceptional rainfall which caused the floods.

3. *Adams et al. v. The State of California et al.* (Docket No. 10122, Sutter County Superior Court) was decided in April, 1964, after four months of trial plus twenty-six days of pretrial motions and hearings. The defendants, Pacific Gas and Electric Company and the North American Weather Consultants, operated cloud seeding generators near Lake Almanor in the headwaters of the Feather River. A damaging flood occurred in Feather River in December, 1955, and owners of property damaged thereby sued for millions of dollars to recover their losses, claiming that the flood was caused, at least in part, by cloud seeding. The plaintiffs'

meteorological testimony asserted that the seeding material had been blown several miles outside the target area and thus it increased the rainfall and snow pack downstream of the Lake Almanor Dam. The court found that the effects of seeding were limited to Lake Almanor, which fortunately never spilled at any time before or during the flood. Accordingly, any increase produced by cloud seeding was successfully impounded in the Lake, and damages caused by the Feather River flood could not be charged to weather modification activities.

The cases above are representative of those in which the decisions were in favor of cloud seeding in that the plaintiffs failed to prove their cases against it. This is not always the case, and Morris cites other examples in which decisions went against the cloud-seeding agencies. Regardless of the court findings, the examples provide apt cases to illustrate the legal entanglements that can arise in any weather modification situation.

Subsequent to the great surge in cloud seeding programs of the 1950s, recent years have been a time for reevaluation of the results obtained. While controversy does exist in relation to the significance of cloud seeding for the inducement of precipitation, the following summation provides a reasonable guide to present thoughts on the matter (NAS-NRC, 1966).

"There is an increasing but still somewhat ambiguous statistical evidence that precipitation from some types of cloud and storm systems can be modestly increased or redistributed by seeding techniques (a) Orographic storms. Evaluations (by the panel) of 41 project-seasons of winter orographic cloud seeding by commercial operators in the western United States support the earlier conclusion (by the Advisory Committee on Weather Control in 1957) that precipitation increases of the order of 10% apparently can result from ground-based silver iodide seeding of winter orographic storms. (b) Cumulus Clouds. Experimental and operational evidence relating to the stimulation of cumulus precipitation remains highly confusing. The Panel's evaluations of 14 operational silver iodide seeding projects in the eastern United States, including *but limited to* cumulus clouds, indicate variable rainfall increases averaging about 10 to 20 percent in the nominal target area. A recent follow-on study made at the Panel's request suggests comparable increases up to 150 miles downwind of the targets.

(c) Extratropical cyclones (non-orographic). There is no clear evidence of success to date in stimulating precipitation from this type of storm in the United States; but little effort has been made except by commercial operators whose records make no distinction between cyclonic and other storm types."

A great deal of literature exists concerning statistical evaluation of rainmaking projects, the physics of clouds and precipitation processes, and the legal and economic consequences of weather modification. Excellent bibliographies concerning all these aspects are given in the NAS-NCR report mentioned previously, in Sewell (1967), and in the Report of the Advisory Committee on Weather Control (1959).

While cloud seeding is the dominant method by which rainfall is stimulated, there have been other proposals to "squeeze" water from the air. One such method, suggested by Gerard and Worzel (1967), involves condensing moisture from warm, moist, tropical maritime air. They suggest that deep, cold, offshore waters found in many tropical archipelagoes could be used as the source of cold required to chill the moisture-laden air. The chilling, of course, would cause attainment of the dew point and cause moisture in the air to be condensed.

Suggesting the use of windmill-driven generators as a cheap source of power, Gerard and Worzel point out that many other benefits might be derived from such a scheme. As shown in Figure 2–19, these would include the supply of nutrient-rich deep water for aquiculture and a natural "air-conditioning" for people living in the lee of the installation.

EVAPOTRANSPIRATION

Moisture is returned directly to the atmosphere through a number of processes. The change in state from solid or liquid form to gaseous water vapor comprises the process of evaporation and sublimation. Evaporation occurs when input of energy onto an evaporating surface causes water molecules to pass from that surface to the atmosphere; this will occur when the vapor pressure of the air is below its saturation value. The rate of evaporation is governed by the state of a number of variables including water vapor, temperature, and air motion, and several formulas

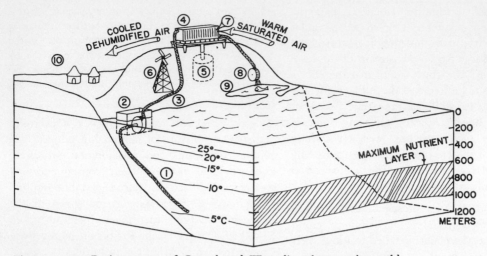

Figure 2–19. Basic aspects of Gerard and Worzel's scheme using cold ocean waters to condense atmospheric moisture. (1) Large–diameter pipe to deep water; (2) pump; (3) connecting pipe; (4) condenser; (5) fresh-water reservoir; (6) windmill electric generator; (7) baffles to direct wind; (8) small turbine to recover water power; (9) lagoon receiving nutrient-rich water for aquiculture; (10) community enjoying cooled dehumidified air. (From Gerard, R. D. and J. L. Worzel, *Science*, 157: 847-849, January 1961. Copyright © 1961 by the American Association for the Advancement of Science.)

are available to determine the rate at which it occurs. Perhaps the oldest of these is the Dalton equation given as

$$E_o = (e_s - e)\, f(u)$$

where e_s is the vapor pressure of the evaporating surface, e is the vapor pressure at some height above that surface, and $f(u)$ is a function of the horizontal wind speed. Since these parameters vary widely over the earth's surface, the rates of evaporation vary enormously. A gross estimate of the amount of evaporation that occurs on a world scale is shown in Figure 2–20.

Just as moisture is returned to the atmosphere through evaporation, it is also returned by the process of transpiration. Transpiration is the term applied to the loss of moisture by plants, such losses occurring through stomata on the exposed leaf surface, with the rate of loss of water being controlled by guard cells on the leaf. During the day these are usually open to expose the moist leaf interior, with resulting high transpiration; at night they are generally closed. Some question exists concerning the role of transpiration in plant growth and development, but it is generally agreed that transpiration

rates control the movement of moisture through plants, and this is obviously related to the transport of materials through the plant. The relative amounts of moisture lost through evaporation and transpiration obviously vary appreciably, depending upon the nature of the ground surface. But as shown in Table 2–8, transpiration rates are often significantly higher than evaporation rates over densely vegetated areas.

While the study of both evaporation and transpiration is significant in itself, it is convenient to treat them as a single process in applied climatic studies. The loss of water through the combined process is termed evapotranspiration.

The rate at which evapotranspiration occurs is dependent upon both meteorological and botanical characterisrics, but it can be assumed that under a given set of conditions, an upper limit will exist. Clearly, to satisfy this maximum rate there must be sufficient moisture available. If moisture is in limited supply, the loss of water will be lower than the maximum rate. It is therefore necessary to recognize two evapotranspiration rates. *Potential evapotranspiration* is the maximum amount of water lost, assuming

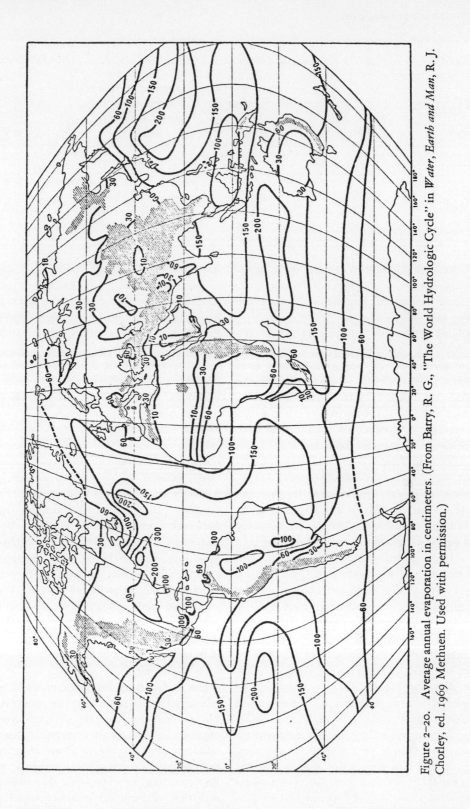

Figure 2-20. Average annual evaporation in centimeters. (From Barry, R. G., "The World Hydrologic Cycle" in *Water, Earth and Man*, R. J. Chorley, ed. 1969 Methuen. Used with permission.)

Table 2–8

EVAPORATION AND TRANSPIRATION LOSSES FOR SELECTED FOREST TYPES[a]

Forest	Forest Age (Years)	Evaporation (e) (mm/Growing Season)	Transpiration (T) + Interception (I) (mm/Growing Season)	Ratio $\dfrac{E}{(T+I)}$
Aspen	20	78	314	1:4.0
Aspen	60	84	282	1:3.4
Scots Pine	20	48	363	1:7.6
Scots Pine	60	87	340	1:3.9

[a] Data from Ward (1967).

all moisture requirements can be met; *actual evapotranspiration* is the observed amount, that, if moisture is limited, will be lower than the potential rate.

Problems exist in determining both the evaporation and transpiration rates. It follows that the estimation of evapotranspiration rates are equally problematic. In measuring the potential rate, water can be added at known quantities and its disposal can be determined more readily. A number of instruments have been designed to measure both rates, examples being shown in Figure 2–21. The evapotranspirometer, since it is measuring potential evapotranspiration, gives fairly reliable results. The example shown consists of a number of tanks that are filled with a soil of similar constituency to that of the surrounding area and covered by a continuous vegetation cover. Since all of the moisture entering the system is accounted for by evapotranspiration or by percolation into the collecting jars, then the moisture consumed by evapotranspiration can be calculated. The measurement of the actual rate is open to larger experimental error. An example of the type of instrument used, in this case the floating lysimeter, is shown in Figure 2–21.

The emplacement, location, and operation of lysimeters require great care and expenditure. Clearly, they must be large enough to provide a representative sample of a given area and deep enough so that they do not significantly alter the natural profile of soil moisture. Such instruments are rarely found outside of experimental stations and thus provide standards and supplements to evapotranspiration rates derived using other methods.

With the lack of observed data, investigators have turned to other methods of estimating evapotranspiration losses. Essentially, these concern analysis of the variables that influence moisture losses; derived formulas include such factors as solar radiation, temperature, plant cover, and wind speed. Three approaches are available: the empiric in which derived equations are based upon observation, the aerodynamic where the physics of the atmospheric processes responsible for evapotranspiration are evaluated, and the energy budget approach, which estimates the amount of energy available to cause moisture transfer back to the air. Numerous formulas have been derived using each approach. Empiric methods are represented by the work of Thornthwaite (1948), Blaney-Criddle (1950), and Makkink (1957). The aerodynamic approach is illustrated by the method of Thornthwaite and Holzman (1942), while the energy budget approach has been investigated by Bowen (1926), Penman (1963), and Budyko (1956). To demonstrate the methodology used, two methods are discussed in the following paragraphs. Other methods, particularly those pertaining to agricultural practices in defined areas, are considered in later chapters.

Penman's method uses the combined influences of turbulent transfer and the energy budget. His derived equation for the determination of evapotranspiration uses vapor pressure, net radiation, and the drying power of air at a given temperature. To derive this formula, Penman had first to obtain an expression that allowed determination of both net radiation and the drying power of air. Such a formula is necessarily quite complex, as is the expression for the evaluation of the drying power of air. Even so, and as is shown in Table 2–9, working with the system

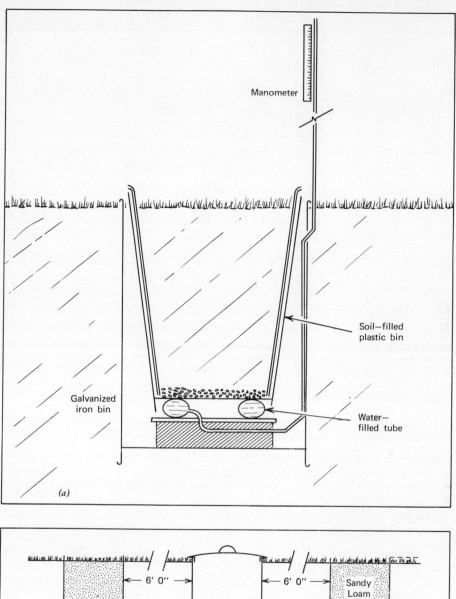

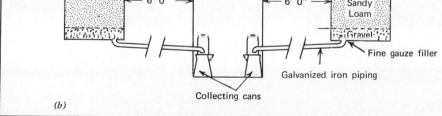

Figure 2–21. (a) Section through a simple weighing lysimeter. (b) Section through an evapotranspirometer. (After Ward, 1967.)

Table 2-9

ESTIMATION OF EVAPOTRANSPIRATION USING PENMAN'S METHOD

Formula:[a]

$$E = \frac{(\Delta H + \gamma Ea)}{(\Delta + \gamma)} \qquad (1)$$

where E = evaporation, mm per day

Δ = the slope of saturation vapor pressure versus temperature curve at air temperature T, in millibars per degree C.

H = the heat budget term (see below)

γ = the psychrometric constant

Ea = the drying power of the air (see below)

Step 1 Calculation of H

H = (incoming shortwave radiation) (reflection coefficient) − (net outward long-wave radiation)

$$H = Ra(1 - r)\,(0.18 + 0.55[n/N]) - \sigma Ta^4(0.56 - 0.092\sqrt{ed})\,(0.10 + 0.90[n/N]) \qquad (2)$$

where Ra = calculated maximum solar radiation reaching earth in absence of atmosphere. Expressed in evaporation units and available from prepared tables (e.g., Brunt)

r = reflection coefficient of surface

n/N = ratio of actual to possible hours of bright sunshine

σTa^4 = black-body function, Ta in degrees Kelvin. Available from prepared tables or $\sigma = 2.01 \times 10^{-9}$ mm/day

ed = mean vapor pressure

Step 2 Calculation of Ea

$$Ea = 0.35(1 + u/100)\,(ea - ed) \qquad (3)$$

where u = wind speed at height of 2 m in miles/day

ea = saturation vapor pressure at temperature T

Step 3 Calculation of E

Substitute values for H, Ea, Δ, and γ into Equation 1.

Note that this provides evaporation from a free water surface. For evaporation from a vegetated area an empirical constant (ratio of E to ET for that area) is introduced.

$$ET = fE$$

where f = empirical coefficient, which varies over space and time.

Example

Data for North Carolina, month of June.

Mean monthly temperature	= 75.9°F (24.4°C)
Mean monthly relative humidity	= 72%
Mean monthly sunshine	= 65%
Mean monthly wind speed at 2m	= 67.7 miles/day
Mean monthly extraterrestrial radiation (in evaporation units)	= 16.93 mm/day
Reflection coefficient	= 0.05
f value for area	= 0.7

(Continued)

Table 2-9 (*continued*)

Step 1

$$H = Ra(1 - r)(0.18 + 0.55[n/N]) - \sigma Ta^4(0.56 - 0.092\sqrt{ed})(0.10 + 0.90[n/N])$$

$Ra = 16.93$; $r = 0.05$; $n/N = 0.65$; $ed = 16.6$ mm; that is, at T of 75.9°F ea is equal to 23.0 mm, so at RH 0.65%, $ed = 16.6$.

$$H = 16.93(1 - 0.05)(0.18 + 0.55 \cdot 0.65) - 15.64(0.56 - 0.092 \cdot 4.07)(0.10 + 0.90 \cdot 0.65)$$

$$H = 6.64 \text{ mm/day}$$

Step 2

$$Ea = 0.35(1 + u/100)(ea - ed)$$
$$u = 67.7, ea = 23.0, ed = 16.6$$
$$Ea = 0.35(1 + 67.7/100)(23.0 - 16.6)$$
$$Ea = 3.71 \text{ mm/day}$$

Step 3

$$E = \frac{(\Delta H + 0.27 Ea)}{(\Delta + 0.27)}$$

$$\Delta = 0.77 \text{ mm}; H = 6.64; Ea = 3.71$$

$$E = \frac{(0.77 + 0.27 \cdot 3.71)}{(0.77 + 0.27)} = 5.88 \text{ mm/day}$$

$$ET = fE$$

where $f = 0.7$; $E = 5.88$
$ET = 0.7 \times 5.88$
$ET = 4.12$ mm/day.

---

[a] Expressed in various forms. Chang (1971) gives a method using $E_o = \Delta Qn + \gamma Ea/\Delta + \gamma$, where Qn is net radiation converted to equivalent evaporation rate.

literally involves simple arithmetic and observance of set techniques.

The formula was derived through research in the United Kingdom, but results obtained from its use appear to hold true for many other parts of the world. As shown in Table 2-10, it appears to provide fairly reliable results for the Australian region in which it was tested. Investigators looking into the agricultural potential of most regions would do well to consider the Penman approach in estimating moisture requirements.

Thornthwaite's method is probably the best known and most widely used method for estimating potential evapotranspiration in the United States. Working in the eastern part of the country, Thornthwaite devised a formula that is based essentially upon the availability of temperature data. His method (outlined by example in Table 2-11, pages 58-59) uses mean monthly temperature and an empiric heat index,

which is itself an exponential function of temperature, as inputs. The derived unadjusted potential evapotranspiration is corrected by using actual daylight hours and number of days in the month in question.

While the method is widely used, it has been criticized. Perhaps the fault most often cited is the fact that temperature, which is the major variable used in the system, is not the best indicator of evapotranspiration rates; radiation values probably provide a more precise guide. Chang (1959) gives a number of examples where, because of the time lag between incoming radiation and temperature maxima, the Thornthwaite method gives imprecise results. Furthermore, since the formula was based upon lysimeter data observed in watersheds in the eastern United States, the method does not always give good results elsewhere in the world. Criticism is also made of the fact that Thornthwaite assumes

Table 2–10

METHODS FOR ESTIMATING POTENTIAL EVAPOTRANSPIRATION FROM SHORT GRASS AT ASPENDALE, AUSTRALIA[a]

Month	$T(°C)$	ETg[b] (mm/Day)	Budyko-Penman	McIlroy	Thornthwaite	Blaney-Criddle	$0.2T$[c]
			Estimated Potential Evapotranspiration, mm/Day				
January	23.3	7.76	6.86	7.57	4.47	4.91	4.66
February	21.1	5.62	5.44	6.12	3.51	4.34	4.22
March	19.6	4.21	3.76	4.22	2.80	3.82	3.92
April	17.2	2.94	2.76	3.16	2.04	3.24	3.44
May	12.6	1.31	1.48	1.67	1.09	2.52	2.52
June	10.9	0.99	1.01	1.19	0.80	2.26	2.18
July	10.0	0.93	1.17	1.34	0.75	2.25	2.00
August	11.0	1.37	1.60	1.76	0.89	2.55	2.20
September	13.0	2.30	2.64	2.88	1.50	3.02	2.60
October	15.8	4.07	4.00	4.28	2.09	3.64	3.16
November	17.8	5.26	4.76	5.27	2.73	4.17	3.56
December	10.1	5.99	5.99	6.56	3.47	4.62	4.02
Annual total		1296	1260	1398	793	1257	1170

[a] After Sellers (1965).
[b] ETg based upon observed values of seven lysimeters.
[c] A simple empiric statement, where T is mean monthly temperature, that appears to provide a reasonably good estimate.

that evapotranspiration ceases at temperatures below 32°F (0°C).

Despite such criticism, there is little doubt that the Thornthwaite method is a useful and valuable approach, particularly when monthly data are used (shorter-term results are more questionable). Its value is enhanced since the only data required to estimate evapotranspiration is temperature, and this variable is readily available for many stations throughout the world. The numerous publications by the Thornthwaite Associates (in 1950 and following years) are also an asset to its application.

Evaporation Suppression

Large quantities of water are returned to the atmosphere through the evapotranspiration processes. In dry regions, where moisture is limited, this poses problems concerning the rational utilization of the resource. In combating the loss, it has been necessary to deal with the evaporation and transpiration components separately. The loss of moisture through plants is obviously controlled by modifying the plant cover so

that unnecessary plants (in terms of agricultural production) are eliminated, particularly those that have high transpiration rates. Loss through evaporation must be overcome by reduction of the water surface that is in contact with free air. The significance of evaporation is demonstrated by the data shown in Table 2–12, where the enormous losses from stored surface water under dry conditions are given. Clearly, one way to overcome this loss is to store water in ground water reservoirs, but this is not always practical. Alternatively, methods to reduce the losses from stored water concern covering the surface of the water to reduce evaporation.

The idea behind this is aptly demonstrated by the commonly observed rainbow colors seen when oil from automobiles falls onto a wet surface. A thin film of oil is formed on top of the water, its presence being indicated by the light interference patterns. Such a film will modify the rate at which the underlying water will evaporate.

When a pure substance is used, the film that forms on top of the water is but one molecule thick and forms a monomolecular layer (monolayer). If the covering is compact, then the transfer of water to the air will be limited and evaporation rates suppressed. While this sounds simple in theory, application in the field offers many problems. Frenkiel (1965), who has outlined the physical and chemical principles of evaporation suppression in some detail, points out that the basic experimental ideas were completed at the turn of the century. It was not until improved experimental techniques were evolved could they be tried in the field.

It is clear that a monolayer must possess basic properties to serve any useful purpose. Mansfield (1958), who was among the first to field test the method, suggests that the following are required.

1. The monolayer must have a resistance to vapor transfer at least several hundred times that of the free water surface.

2. To maintain efficiency after penetration by dust or movement by wind, the film must be "self-healing," that is, it must be fluid.

3. To resist wave action, with resulting increases and decreases of surface area, the film must have the ability to expand and contract, that is, a high degree of fluidity is needed.

4. The monolayer should exert a high surface pressure on the water greater than that exerted by the wind.

It was found that some alcohols with high molecular weights best fulfill these needs, with optimum resistance obtained using cetyl and stearyl alcohols. Recently, though, new compounds derived from long-chain alcohols have been synthesized. In the laboratory they appear superior to those in use. They have yet to be widely applied in the field.

Despite some success in the reduction of evaporation using monolayers, many problems have still to be overcome. The nature by which the material is introduced to the water surface has not been adequately solved, for, by broadcasting, much material is soon washed up on to the banks and unless the supply can be regularly maintained, little is achieved. Alternatively, by storing in a floating dispenser, the problem of disposal through the "windows" of the dispenser causes problems. Furthermore, while the materials used do exert a surface pressure on the water, a moderate wind causes dispersal of the monolayer quite easily.

There are other methods by which evaporation can be reduced without the use of monolayers. In his review of evaporation suppression, Frenkiel outlines the following alternatives.

1. Location of reservoirs at the highest possible elevation; less water is lost per unit surface area compared to those at lower altitudes.

2. Reduction of evaporation by shaping the reservoir so that it has the lowest possible area/volume ratio.

3. By rational management it might be possible to regulate the reservoir so that the least surface area is exposed during seasons of high evaporation.

4. By covering the reservoir with floating covers, such as polyethylene films or microscopic beads, evaporation can be reduced. Such materials still, however, pose many technological problems (e.g., spinning beads might promote rather than prevent evaporation).

5. Through the use of windbreaks around the reservoir. Note that while this decreases evaporation, it also contributes to water loss through transpiration.

6. In deep reservoirs, lower water levels are appreciably cooler than those at the surface. By bubbling air from the bottom, the thermal stratification is broken up. Cooler water at the surface lowers evaporation rates.

RUNOFF

Consider precipitation falling onto a relatively small land area (Figure 2–22). The disposal of precipitation fulfills the general hydrologic equation, but the amounts involved vary according to a whole host of factors. Obviously, the rate at which precipitation occurs must exceed loss through evaporation and moisture intercepted by the ground vegetation cover before any is available at the surface. Once the precipitation reaches the ground the amount that enters the surface depends upon the infiltration rate of that surface and the intensity of the rainfall. While the evaluation of these variables will give insight into the disposal of the incoming moisture, many other factors need be considered. Climatic controls will include:

The kind of precipitation that occurs—rain or snow.

Table 2-11

ESTIMATION OF EVAPOTRANSPIRATION USING THORNTHWAITE'S METHOD

Formula:

$$E = 1.6(10T/I)^a$$

where E = monthly potential evapotranspiration in cm

T = mean monthly temperature °C

I = heat index; the sum of 12 monthly i values. Constant for a given location

a = constant, a function of I

Step 1   Calculation of I

Obtained by summing prepared tables of monthly i values. The extract below provides an example.

Monthly i Values—Monthly Mean Temperature

T, °C	.0	.1	.2	.3	.4	.5	.6	.7	.8	.9
22	9.42	9.49	9.55	9.62	9.68	9.75	9.82	9.88	9.95	10.01
23	10.08	10.15	10.21	10.28	10.35	10.41	10.48	10.55	10.62	10.68
24	10.75	10.82	10.89	10.95	11.02	11.09	11.16	11.23	11.30	11.37
25	11.44	11.50	11.57	11.64	11.71	11.78	11.85	11.92	11.99	12.06

Step 2   Calculation of *Unadjusted PE*

Derived from a prepared nomogram. On the I scale plot the I value derived in Step 1. Connect this with a straight line to the point of convergence. Use the constructed line to read off unadjusted potential evapotranspiration.

Step 3   Calculation of *Adjusted PE*

Values derived from the nomogram are adjusted for day and month length. Extract of the table used for this is given below. It shows the mean possible duration of sunlight for given latitudes expressed in units of 30 days of 12 hr each. Multiply by correction factor given in the table.

Lat., °N	J	F	M	A	M	J	J	A	S	O	N	D
39	.85	.84	1.03	1.11	1.23	1.24	1.26	1.18	1.04	.96	.84	.82
40	.84	.83	1.03	1.11	1.24	1.25	1.27	1.18	1.04	.96	.83	.81
41	.83	.83	1.03	1.11	1.25	1.26	1.27	1.19	1.04	.96	.82	.80

Example

Monthly Temperatures (°C) for Seabrook, N.J.

J	F	M	A	M	J	J	A	S	O	N	D
0.9	1.2	5.9	11.3	17.5	22.3	24.7	23.7	20.2	14.0	7.6	2.3

Step 1

For each month refer to table for i value. For example, June 22.3 = 9.62; July 24.7 = 11.23

J	F	M	A	M	J	J	A	S	O	N	D
0.7	1.2	1.29	3.44	6.66	9.62	11.23	10.55	8.28	4.75	1.89	.31

$$I = \Sigma i = 58.21$$

(Continued)

Table 2-11 (*continued*)

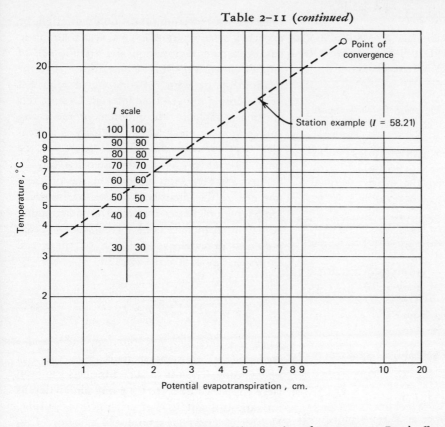

Potential evapotranspiration , cm.

Step 2 Locate $I = 58.21$ on nomogram. Join to point of convergence. Read off unadjusted *ET* values.

J	F	M	A	M	J	J	A	S	O	N	D
.1	.2	1.6	4.1	7.5	10.6	12.2	11.5	9.3	5.5	2.3	.4

Step 3 Calculate monthly (adjusted) potential evapotranspiration. Using latitude of station, multiply each month by correction factor.

J	F	M	A	M	J	J	A	S	O	N	D
(.1)	(.2)	1.6	4.6	9.2	13.1	15.4	13.6	9.7	5.3	1.9	.3

Each gives the potential transpiration for each month. The sum $= 75.0$, gives the annual potential evapotranspiration in centimeters.

The size of the raindrops—the intensity of the storm.

The duration of the precipitation.

The intermittency of precipitation.

The prevailing temperature—in relation to *ET* rates.

The season of the year—that is, frozen ground.

The study of the surface needs to include:

Surface permeability data.

The moisture-holding capacity of the soil.

The antecedent moisture conditions, and so on.

One classification of runoff characteristics, which includes some of the above variables, is given in Table 2-13. The amount of water that becomes runoff is thus highly variable and, in effect, represents surplus moisture available after all other hydrologic requirements have been met.

Table 2-12

SELECTED EXAMPLES OF EVAPORATION
RECORDS REDUCED TO RESERVOIR
SURFACE EVAPORATION[a]

Station	Elevation (ft) and Dates	Annual Evaporation (in.)
Columbus, Ohio	763 (1918–1930)	26.81
Lincoln, Nebraska	1250 (1917–1930)	42.04
El Paso, Texas	3700 (189 –1993)	71.16
Elephant Butte, N. Mexico	4265 (1916–1930)	66.99
Tucson, Arizona	2400 (1929–1930)	60.26
San Juan, Puerto Rico	82 (1919–1930)	55.29
Alexandria, Egypt	— (1920–1929)	28.76
Atbara, Sudan	— (1905–1929)	123.67

[a] After Wisler and Brater (1959).

To express this, the hydrologic equation may be written

$$RO = P - (ET + I + In \pm \Delta St)$$

where RO is runoff (that is, surplus moisture), P is precipitation, ET is evapotranspiration, I is interception, In is infiltration, and ΔSt is change in storage. It is quite evident that, depending upon local conditions, any one of the factors might be dominant to the exclusion of the others.

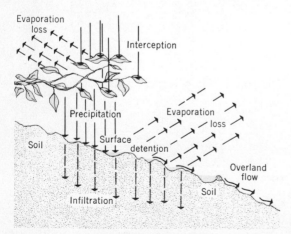

Figure 2-22. Disposal of precipitation falling on a small area. (From Strahler, A.N., *Physical Geography*, Copyright © 1969 by John Wiley and Sons. Reproduced by permission.)

In studies concerning the runoff that might be expected in any particular area, much attention has been paid to variation and significance of rainfall intensity. This is expressed by the relationship between the amount of rain that occurs during a given time interval. Despite this apparent simplicity, the problem of predicting and estimating precipitation intensity is a complex one. Partially, the problem relates to the time interval involved, for intensity can be defined for very short periods, in the order of minutes, to the intensity expressed over a much longer period. Many formulas for determining rainfall intensity have been devised. Chow (1962) provides an extensive coverage of these and shows that most take the form of

$$I = \frac{a}{t + b}$$

where I is intensity, t is duration, and a and b are empirical constants pertaining to the area under study. The use of such formulas has been largely replaced by maps for point rainfall depths for various durations over given periods of time. Hershfield (1961) has produced a series of 54 maps providing such data for the United States. Another example shown in Figure 2-23, gives the 6-hr rainfall that might be expected once every 10 yr. For other durations, tables are available for relevant locations (Table 2-14). Such maps are concerned with probabilities of expected intensities and are used for planning purposes. For actual intensities the United States Weather Bureau has published maps showing actual conditions.

Just as rainfall intensity is highly variable, so is the infiltration capacity of different surfaces. Figure 2-24 shows infiltration differences that occur on surfaces of different textural classes and land utilization. Presupposing that initial soil moisture is below capacity, then infiltration is initially similar irrespective of the type of land use or surface texture. But the diagrams clearly show that infiltration soon begins to vary considerably. On the fine-textured surface a large amount of runoff might occur. In the same way, the poorly managed land also allows much runoff to result, and the result often leads to accelerated erosion so typical of improper use of land on steep slopes or in semiarid regions.

Table 2-13

CLASSIFICATION OF RUNOFF-PRODUCING CHARACTERISTICS[a]

Designation of Watershed Characteristics	Runoff-Producing Characteristics			
	100 Extreme	75 High	50 Normal	25 Low
Relief	(40) Steep, rugged terrain; average slopes generally above 30%	(30) Hilly; average slopes of 10–30%	(20) Rolling; average slopes of 5–10%	(10) Relatively flat land; average slopes of 5%
Soil infiltration	(20) No effective soil cover; either rock or thin soil mantle of negligible infiltration capacity	(15) Slow to take up water; clay or other soil of low infiltration capacity, such as heavy gumbo	(10) Normal, deep loam; infiltration about equal to that of typical prairie soil	(5) High, deep sand or other soil that takes up water readily and rapidly
Vegetal cover	(20) No effective plant cover; bare or very sparse cover	(15) Poor to fair; clean-cultivated crops or poor natural cover; less than 10% of drainage area under good cover	(10) Fair to good; about 50% of drainage area in good grassland, woodland, or equivalent cover; not more than 50% of area in clean-cultivated crops	(5) Good to excellent; about 90% of drainage area in good grassland, woodland, or equivalent cover
Surface storage	(20) Negligible; surface depressions few and shallow; drainage ways steep and small; no ponds or marshes	(15) Low, well-defined system of small drainage ways; no ponds or marshes	(10) Normal; considerable surface-depression storage; drainage system similar to that of typical prairie lands; lakes, ponds, and marshes less than 2% of drainage area	(5) High; surface-depression storage high; drainage system not sharply defined; large flood-plain storage or a large number of lakes, ponds, or marshes

Each column shows the contribution of the different watershed characteristics, relief, soil infiltration, vegetation, and surface storage to a particular amount of runoff. For example, under extreme conditions of 100% runoff, relief accounts for 40% of the runoff, while the other three elements account for 20% each. The right-hand column shows that a low proportion of runoff is considered to be 25%, to which relief contributes twice as much as either soil infiltration, vegetation, or surface storage.

[a] From *Farm Planners' Engineering Handbook for the Upper Mississippi Watershed*, U.S. Soil Conservation Service, Milwaukee, Wis., 1953.

Table 2–14

RAINFALL AMOUNTS FOR DURATIONS OTHER THAN 6 hr

Percent of 6-hr Rainfall[a]

Duration	Zone A	Zone B	Zone C
5 min	9	14	18
15 min	19	30	38
30 min	25	40	50
1 hr	35	55	62
2 hr	55	70	76
6 hr	100	100	100
12 hr	130	124	118
24 hr	165	150	140

[a] Zone A is the area in states of Calif., Oreg. and Wash. west of Sierra Nevada and Cascade Mountains. Zone B, between Zones A and C. Zone C, east of the Continental Divide.

Once a surplus is available it will flow in a number of forms. Initially it may occur as sheet-flow runoff, where excess moisture flows downslope as a continuous film of water. Depending upon the amount of water available, the degree of slope, and the length of flow, this laminar flow will become turbulent overland flow that appears as small rivulets of water. Ultimately, most of this surface water will find its way into a water system flowing in well-defined channels.

The discharge of any stream (Q), which is given by the formula

$$Q = AV$$

where A is the cross-sectional area of the stream and V its cross-channel mean velocity, depends not only upon the climate of an area but also upon the geomorphic characteristics of the drainage basin in which it occurs. Such factors are considered in some detail in Chapter 3.

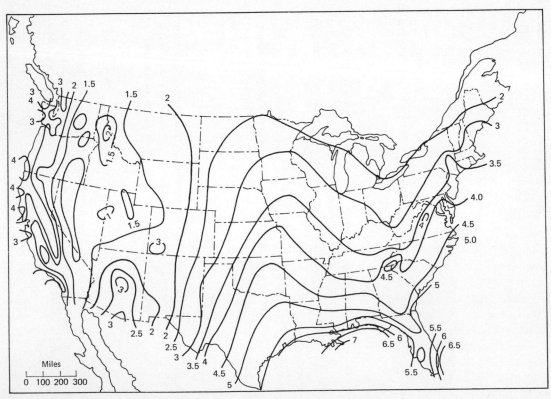

Figure 2-23. Six-hr rainfall (inches) to be expected once every 10 years in the United States. (After Kirpich and Williams, 1969.)

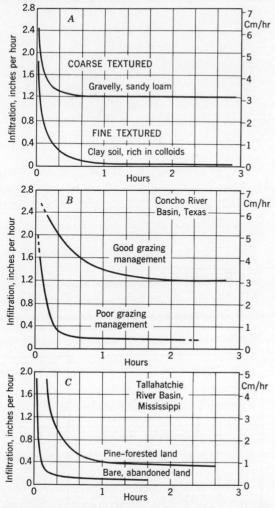

Figure 2–24. Infiltration is influenced by texture, land management, and vegetation cover. (From Strahler, A.N., *Physical Geography*, Copyright © 1969 by John Wiley and Sons. Reproduced by permission.)

Despite the significance of the geomorphic parameters, the role of climate in relation to the seasonal discharge of streams is of consequence. Several European hydrologists have classified rivers according to their discharge characteristics in relation to climatic regime. One example of note is that described, with appropriate regime examples, by Guilcher (1965). He divides the continents into hydrologic regions each characterized by symbols derived from the Köppen climate classification. Another interesting representation is that devised by Pardé (1955), who

divides northern hemisphere rivers into four categories.

1. Rivers that receive much of their moisture from melting snow and ice so that their maximum discharge occurs in summer, with a minimum in winter. These comprise rivers of the *glacial regime*.

2. Rivers under a *rainy maritime regime* receive a similar input of moisture throughout the year, but because of low evaporation losses in winter, show a maximum winter discharge.

3. River regimes that have their maximum discharge at a time coincidental with the period of maximum solar radiation, with a minimum during the low-sun season, form the *tropical rainy regime*.

4. The *snowy plains regime* is characteristic of rivers in which the largest discharge occurs in spring subsequent to the melt of snow that has accumulated during the winter. Frequently, the thawing period is followed by intense convectional rainfall with the result that the maximum discharge occurs from March to June.

The actual discharge of streams experiencing these regimes are well shown by stream hydrographs. On a hydrograph, the amount of runoff that occurs is plotted against time and an analysis of the resulting graph reveals much about the characteristics of the area, or the drainage basin, from which the water discharges. Examples illustrative of the previously outlined regimes are shown in Figure 2–25.

The use of the hydrograph is not restricted to long-term discharge patterns, and short-term hydrographs are equally illustrative of variations that occur on a watershed. In the analysis of short-term stream hydrographs, it is necessary to take into account the amount of water that enters the stream from ground water discharge, for this plays a significant role in the pattern that the hydrograph exhibits. Ward (1967) outlines four basic types of hydrographs, which are based upon those first identified by Horton (1935). The four are illustrated in Figure 2–26.

Type O (so-called because nothing happens so far as the stream hydrograph is concerned) shows what might result when precipitation is so light that it never exceeds infiltration. No surplus moisture is available, with the result that there

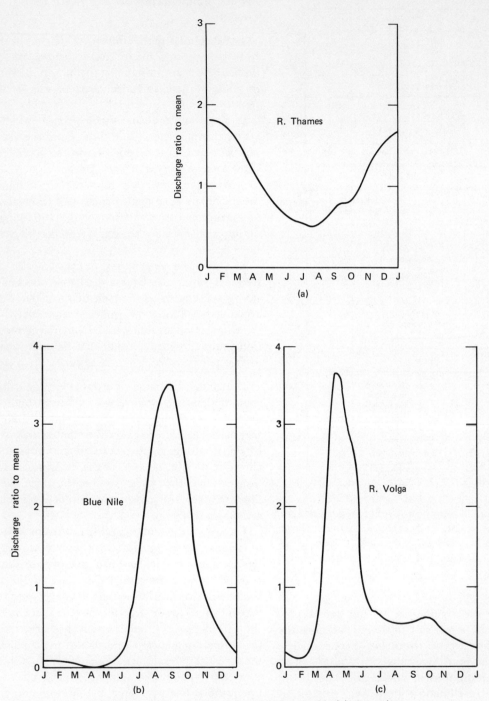

Figure 2–25. Examples of river regimes identified by Pardé. (*a*) Oceanic, (*b*) Tropical, (*c*) Plains snowmelt. Note that the vertical axis shows discharge as a ratio to the mean discharge.

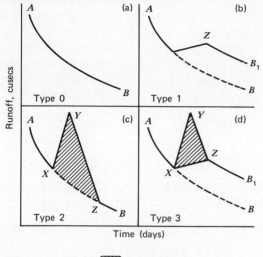

Figure 2–26. Stream hydrographs showing various responses to inputs of precipitation. See text for discussion. (After Ward, 1967.)

is no addition to the runoff. The graph remains unchanged, for it is assumed that even though precipitation does occur and result in infiltration, there is not sufficient moisture to bring soil moisture up to capacity, thus no water is added to the ground water reservoir and its output is unchanged.

In Type 1 a similar situation is assumed, but here the stream hydrograph is modified because it is assumed that infiltration exceeds the soil moisture deficit and goes to ground water. This causes a change in the ground water flow and additional water is discharged into the stream. Type 2 might be considered as the reciprocal of Type 1. Here it is assumed that the infiltration rate is exceeded and runoff occurs. The peak in the hydrograph is a result of this, but note that there is no addition to the stream discharge from ground water. After the peak due to the actual runoff, the graph reverts back to the original depletion curve. In Type 3, the pattern of the hydrograph shows that both surface runoff and ground water contribute toward the discharge. The first peak of the curve is a result of additional water added by runoff, and it is seen that this does not revert to the original depletion level. Supply of ground water maintains the level of discharge above that resulting from direct runoff alone.

Modification of Runoff

The high variation in the patterns of stream discharge creates enormous problems in the management of water and often results in large financial losses. To combat the problem of too much discharge, many flood control practices have been devised. There is an extensive body of literature concerning the causes, problems, and control of flood water, with comprehensive accounts being given by Hoyt and Langbein (1955), Lane (1965), and Burton (1965).

Essentially though, from the many methods used, two main approaches can be identified. Firstly, there are those that rely upon the actual control of flood water itself. Emphasis here is on large-scale modification of actual stream flow with the creation of man-made levees, flood spillways, storage reservoirs, and the like, forming the defense mechanism. A second approach, which is of more concern from the climatological standpoint, is through the modification of the variables included in the surplus water equation

$$[RO = S = P - (ET + I + In + \Delta St)]$$

defined earlier. It will be appreciated that the modification of any one of the variables will alter the disposition of the others, to result in an increase in surplus moisture available for runoff.

The modification of precipitation—the input factor—has already been touched upon in the discussion of effects of cloud seeding. It should be noted, however, that cloud seeding is largely concerned with the necessity of increasing input rather than control of its disposal. Even so, there are methods by which the input factor can be modified to control runoff, with one of the most significant aspects concerning the control of snow melt.

As indicated in the discussion of river regimes, snow melt often forms an appreciable portion of the discharge of some rivers. Snow that has accumulated over the winter season thaws with the onset of warmer weather, and a great influx of melt water into surface streams often causes floods. Thereafter, once the thaw is over, the amount of water carried by the streams may be minimal. Such flashy flow of water is extremely difficult to manage, and much can be gained if the flow is regularized.

Methods for regulating the flow resulting from snow melt depend upon a number of principles. One way concerns the development of alternating vegetation cover, with grass, for example, alternating with forests. The different vegetation cover will result in microclimatological differences which modify both the amount of snow that accumulates and the rate at which it melts. Such modification has proven quite successful in the western parts of the United States where the U.S. Department of Agriculture has developed the concept of *barometer watersheds* covering broad climato-physiographic regions. Intensive study of such watersheds, which vary in extent from 50,000 to 150,000 acres, provide data that allow criteria to be applied to other areas (U.S.D.A., 1967). As part of the same study, a second method of modifying snow melt is being examined. This depends upon modification of the reflectivity of the snow surface.

When radiant energy strikes a surface, it is either absorbed or reflected. The amount that is reflected depends upon the albedo of the surface. As shown in Table 1–1, albedoes are quite variable for different types of surface. By modifying the albedo of a surface it is possible to alter the amount of energy that it absorbs and change the energy balance of that particular surface. Thus, by covering a surface of snow with some form of blacking (for example, soot, ashes, or tar), the amount of heat energy available for melting the snow is increased considerably. It follows that the rate of snow melt can be controlled by selective covering of given surfaces.

Experiments in this field are not confined to the western United States. Researchers at the University of Chile, for example, treated large areas of the Coton Glacier in the Andes with absorbent black dust. The treatment caused a melting increase of up to 300%, and it is estimated that in 1969 an additional 10,000 m³ of melt water were produced each day (Science Services, 1970). The sponsor of the program, a public utility company, obtained marked economic benefits from the sustained runoff that resulted.

The significance of vegetation cover in relation to runoff is not restricted to that concerning snow accumulation and melt. Vegetation plays a very important part in determining the rates of interception, evapotranspiration, and infiltration. Many studies assessing the influence of vegetation upon runoff have been completed.

A fine discussion of such effects is given by Penman (1963) and the following examples are drawn from his account.

In Ota, Japan, a catchment area consisted of broad-leaf forest and, subsequently, was effectively replaced by grassland. The effects of this change are indicated in the following data.

Event	Year	Rain (mm) R	Runoff (mm) r	(R − r)
Undisturbed broad-leaf forest	1908	1509	869	640
	1909	1503	942	561
	1910	1856	?	?
	1911	1768	1008	760
	1912	1521	881	640
	1913	1535	884	651
August 1914 to July 1915 clearing	1914	1390	713	677
	1915	1717	901	816
Grassland cover, later small trees	1916	1619	1118	501
	1917	1481	984	497
	1918	1348	784	564
	1919	1456	806	650

The value $(R - r)$ provides a measure of the disposition of precipitation. A high value indicates that a relatively low proportion of precipitation remains to pass as runoff. In the preceding table, it can be seen that subsequent to clearing the forest $(R - r)$ decreased appreciably. The mean for the years 1908 to 1913 is 671 mm, while for the post-clearing period, it is 603 mm. Notice the marked increase in runoff immediately following the clearing and how, with increase of other surface cover, it decreased within the next few years.

Another series of observations from Japan were obtained at Kamabuchi. One of two catchment areas, both originally under mixed coniferous and deciduous trees, was cleared and subsequent growth of grass and bush was cut and removed. The cleared catchment showed a marked increase in the amount of runoff.

	Mean Annual Data, 1947 to 1950	
	Forest Area (2½ Hectares)	Bare Area (3 Hectares)
Rain (mm)	2355	2355
Runoff (mm)	1790	1955
Evaporation (mm)	565	400

Of a number of examples in the United States, the results obtained from the Coweeta Experimental Forest in North Carolina are perhaps the best known. The data given below were derived from a 33 acre catchment that was cleared of forest between January and March, 1941. Trees were left where they fell and limbs and small branches were cut off and scattered to protect the soil; given such conditions, the vegetation sprouted vigorously. At first (August 1941, and between June and August 1942) growth was cut back; thereafter it was left to grow.

Catchment No. 17
(33 Acres)

Year	Rain (in.) R	Runoff (in.) r	$(R - r)$ (in.)	
1936–1937	74.1	32.3	Mean (1936 to 1940)	
1937–1938	65.2	25.4	= 41.7	
1938–1939	76.8	36.9		
1939–1940	62.5	16.9		
1940–1941	51.4[a]	22.4	29.0	
1941–1942	72.0	42.6	29.4	mean
1942–1943	83.8	47.9	35.9	= 35.0
1943–1944	74.6	35.2	39.4	

[a] To that time, the driest year on record.

While the mean $(R - r)$ value is lower for the post-cutting period, it is evident that under the conditions that existed (litter left and new growth uninterrupted) that by 1943, little difference was seen in the disposal of precipitation over the catchment area.

Grandiose schemes altering river courses are also part of the modification of surface water. A good example of such a scheme is the Soviet plan to reverse the flow of rivers in what is known as the Kama-Vychegda-Pechora project. The purpose of this is to divert massive quantities of water from the Vychegda and Pechora, which flow northward into the Arctic Ocean, into the southward-flowing Kama. As shown in Figure 2-27, it is thought that the diversion would meet a variety of needs in the central and southern parts of the country.

Aspects of the scheme have been lucidly analysed by Micklin (1969). The most pertinent in relation to the area under discussion concern the climatic aspects of the proposed modification.

Micklin makes this comment:

"Research on possible climatic changes was carried out by means of analogues, that is, by examining such changes as have occurred due to the construction of smaller reservoirs in similar environmental zones and then extrapolated to the probable effects of the proposed Pechora-Vychedga-Kama reservoir. The conclusions reached were (1) the area whose climate would be affected by the reservoir to one degree or another is approximately 60,000 square kilometers, or nearly four times the area of the water body itself; (2) the affected zone would be cooled by the water at the beginning of the warm season but warmed by it at the end; however, the influence of the reservoir on plant growth conditions, and consequently on agriculture, would be negative because of a net loss of heat available to plants during the growing season; and (3) the reservoir would reduce the continentality of the climate with winters becoming less severe and summers becoming both shorter and more humid."

Extended effects of these theorized climatic changes would be reflected in vegetation, soil, and ground water conditions. Just what the results might be are difficult to ascertain with any exactness, but the extended effects might be marked. Such "side effects" of any great surface modification scheme requires careful study of all aspects prior to any initiation of the scheme. It is obvious that in the case of a huge program, such as that of the North American Water and Power Alliance (NAWAPA), in-depth research is needed in many areas. This plan was prepared to ". . . utilize the excess water of the northwestern part of the North American continent and distribute it to the water deficient areas of Canada, the United States and Mexico." (Wagner, 1971). The plan would take water flowing to the Arctic, store it in high elevation reservoirs, and then distribute it by canals and waterways as needed.

Perhaps the most significant modification of runoff rates are the result of gross alteration of the surface by man, particularly through urbanization. Writing on the significance of the conservation of wetlands in relation to building development, Niering (1970) notes: "Wetlands are of major importance in the nation's hydrologic regime. Because of their water-holding capacity they act as storage basins assisting in minimizing erosion and serving to reduce the destruction of floods. In cities, this is especially

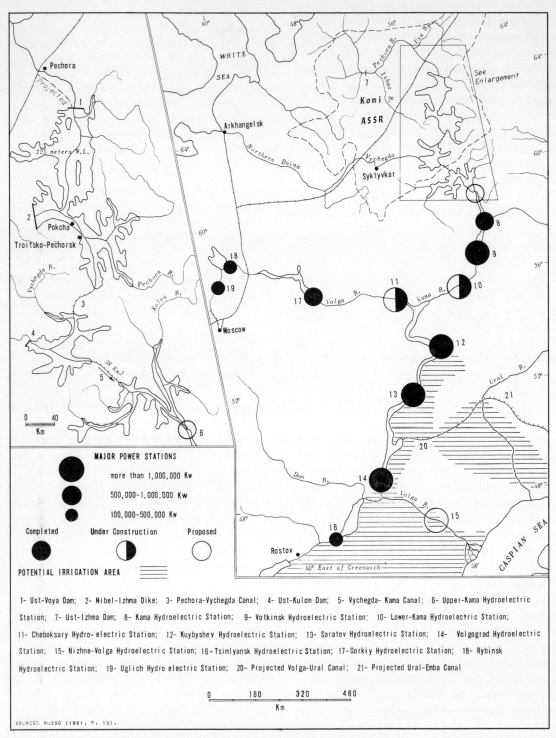

Figure 2–27. The Kama-Vychegda-Pechora project and its area of influence. (From Micklin, P.P., *Canadian Geographer*, 13: 199-215, 1969. Used with permission of the author.)

The labels and legend within the figure read:

MAJOR POWER STATIONS

more than 1,000,000 Kw

500,000-1,000,000 Kw

100,000-500,000 Kw

Completed   Under Construction   Proposed

POTENTIAL IRRIGATION AREA

1- Ust-Voya Dam; 2- Nibel-Izhma Dike; 3- Pechora-Vychegda Canal; 4- Ust-Kulom Dam; 5- Vychegda- Kama Canal; 6- Upper-Kama Hydroelectric Station; 7- Ust-Izhma Dam; 8- Kama Hydroelectric Station; 9- Votkinsk Hydroelectric Station; 10- Lower-Kama Hydroelectric Station; 11- Cheboksary Hydro- electric Station; 12- Kuybyshev Hydroelectric Station; 13- Saratov Hydroelectric Station; 14- Volgograd Hydroelectric Station; 15- Nizhne-Volga Hydroelectric Station; 16- Tsimlyansk Hydroelectric Station; 17- Gorkiy Hydroelectric Station; 18- Rybinsk Hydroelectric Station; 19- Uglich Hydro electric Station; 20- Projected Volga-Ural Canal; 21- Projected Ural-Emba Canal

SOURCE: RUSSO (1961, P. 13).

important, because urbanization intensifies the rate of runoff as buildings, concrete and asphalt tend to concentrate large volumes of precipitation. Since cities are deficient in soak-in areas, the runoff is usually rapid and in excessive volumes. Wetlands, including floodplains, act as catchment basins and tend to slow the speed of flow, thus minmizing flood damage. In 1955 when the severe floods struck eastern Pennsylvania, hundreds of bridges were washed out along the stream courses. However, two bridges of the type destroyed were left standing below the Cranberry Bog, a natural area preserved by the Nature Conservancy."

The hydrologic effects of urbanization have been explored in some detail by Leopold (1968). As might be expected, his survey shows that urbanization causes runoff to become peaked and that flood frequency is appreciably higher than in nonurbanized areas. Two examples of his results are shown in Figure 2–28. Figure 2–28*a* indicates how the ratio between discharge before and after urbanization increases markedly and Figure 2–28*b* gives the flood-frequency curves for a square mile basin in varying degrees of urbanization.

GROUND WATER

The movement of water through a surface hydrologic system resolves itself into input (precipitation), flowage, and discharge. Precisely the same situation occurs in ground water systems, with the essential difference being the limitation of water that can enter and its much slower motion through the ground water system. The permeability, the size, and the slope of an aquifer—the water-bearing strata—depend upon geologic composition and structure. The significance of climatology in ground water studies is therefore essentially concerned with the rate at which the aquifer is recharged from surface water supplies. It has already been stressed that the disposition of precipitation among the variables of the hydrologic equation depends upon many factors, so that study of ground water supply depends upon evaluation of other components of the equation. Generally, however, shallow ground water (i.e., unconfined ground water where the water table is close to the surface) react to precipitation regime in a

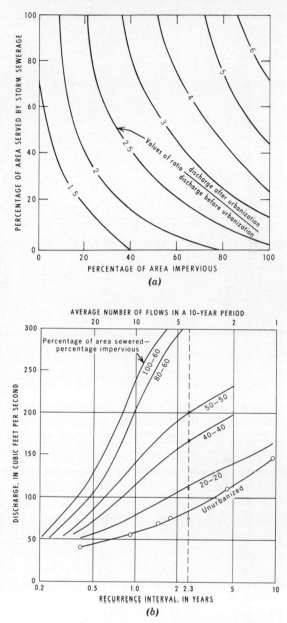

Figure 2–28. (*a*) Effect of urbanization on mean annual flood for a 1 sq mile drainage area. (*b*) Flood-frequency curves for a 1 sq mile basin in various stages of urbanization. (From Leopold, L.B., *Geologic Survey*, Circular No. 554.)

similar way to runoff, with peaks occurring after copious inputs of precipitation. Figure 2–29 shows a typical example of where the level of the water table is correlated to storm precipitation.

Since the water table does vary after an individual storm, it follows that marked differences

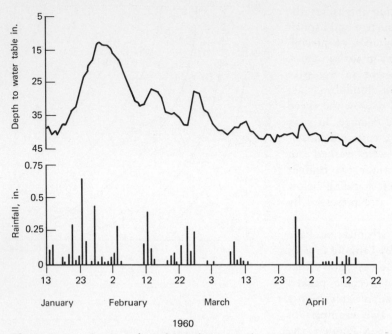

Figure 2–29. Graphs of rainfall and height of a shallow water table. Note the slight delay in rise of the water table after precipitation and also the fact that the amount of rise differs with similar inputs. This is because precipitation is not always budgeted in the same way and often less is available for ground water. (After Ward, 1967.)

will occur when precipitation regime varies from wet to dry in the course of a year. Figure 2–30 shows the effect. During the wet season the water table is high and ground water provides ground flow to the stream, springs occur, and shallow wells contain water. In the dry season, the resulting drop in the level of the water table causes flow of water from the river to the ground, dry springs, and dry wells.

The preceding outline presupposes that ground water is constantly replenished by precipitation. In fact, of the three types of ground water that occur, only meteoric water is recharged by present-day precipitation. Juvenile water occurs deep in the ground and is a by-product of igneous activity. Quantitatively, the addition of new water to the hydrologic cycle from this source is minimal. Connate water is water that has been trapped in sedimentary or volcanic rocks at the time of their formation. Water derived from such a source is nonrenewable and if used by man, must ultimately be depleted.

Climatic changes, particularly in relation to pluvial periods, have influenced ground water availability. In a study of the subsurface water of the Sahara Desert, Ambroggi (1966) has shown that seven well-defined ground water basins exist together possess a capacity of some 15,000,000 m³ of ground water. In effect, a tremendous reservoir of ground water occurs in the aquifers of the desert. But in terms of exploitation many problems exist; while some recharge of the aquifers is occurring today, the water is mostly a relic of past climates. Flow of water through the system is slow, perhaps attaining half a mile each year, with the result that in the areas of present discharge "Water coming out . . . today is rain that fell between the last Saharan pluvial period and the Roman Empire." Carbon 14 has been used to date the water and the oldest found thus far, in the western desert of Egypt, is estimated to be 25,000 years old.

Ground Water Recharge

Overuse of ground water ultimately leads to a lowering of the water table. If pumping is continued at an excessive rate, the effect is a temporary—in relative terms—loss of the resource. Rational use of ground must therefore

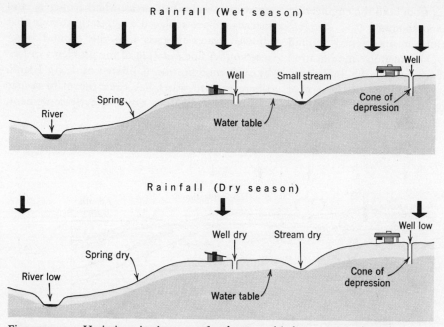

Figure 2–30.   Variations in the unconfined water table between wet (*upper*) and dry (*lower*) seasons. (From Longwell, C.R., R.F. Flint, and J.E. Sanders, *Physical Geology*, Copyright © 1969 by John Wiley and Sons. Reproduced by permission.)

be considered in relation to recharge of the aquifer through precipitation. The variable nature of precipitation distribution however, often means that ground water is needed at a time when input is low, for water needs, which often reflect a seasonal trend, are sometimes out of phase with river discharge and levels of reservoir storage. The concept of conjunctive management is an attempt to use excess water that occurs during wet periods to recharge ground water; the ground water supply is then used at a time when river flow is minimal. The concept is schematically illustrated in Figure 2–31. Obviously, such utilization of water needs

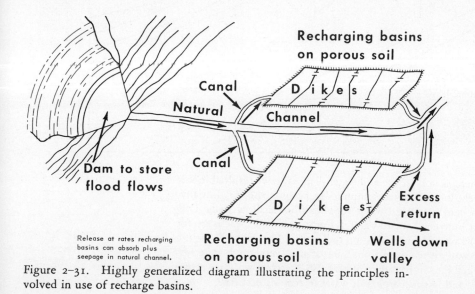

Figure 2–31.   Highly generalized diagram illustrating the principles involved in use of recharge basins.

considerable planning from both the engineering and the managerial standpoints.

While the variation in unconfined ground water is a natural phenomenon, modification of the surface that comprises the infiltration area has caused many problems in the use of ground water. This is particularly evident in urban and suburban areas where concrete and macadam have replaced a vegetation cover that allowed water to pass into the ground water reservoir. A fine example of the problems caused by such modification is offered by Long Island, where ground water has been the main source of water supply since times of early settlement.

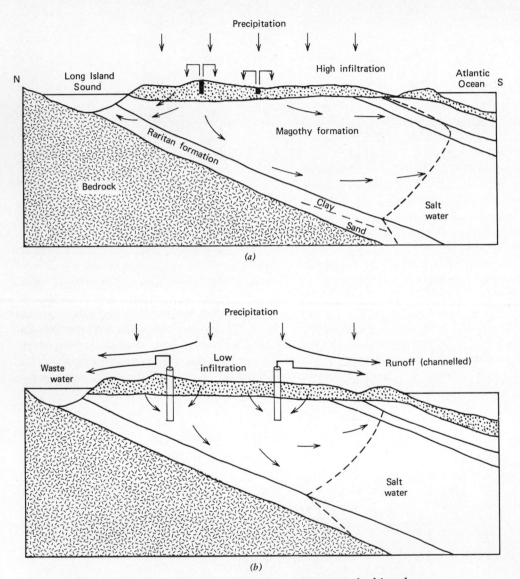

Figure 2–32. Changes in land use and water disposal have resulted in salt water encroachment in ground water supplies in Long Island. (*a*) Initially, a low population density derived water supplies from shallow wells. Disposal of water in individual waste-water systems provided a recharge for ground water. (*b*) Increasing urbanization reduced infiltration while, at the same time, waste water is removed from the system in modern sewage systems and discharged into surrounding waters. With greater withdrawal, reduced infiltration, and less recharge, salt water intrusions occurred.

The actual evolution of the problem is well illustrated in Figure 2-32, where it can be seen that not only is the level of the water table lowered, but the increased withdrawal with the reduced infiltration has led to the intrusion of salt water below the fresh water reservoir. At the present time the U.S. Geological Survey has established areas in which treated sewage water is being recharged into the ground water system (USGS, 1966).

THE WATER BALANCE

It will appear evident, from all that has been said thus far, that a balance is attained between incoming and outgoing moisture and that the balance will reflect the climatic regime that exists. A number of writers have made use of the budget approach to evaluate the climatic and hydrologic differences that exist over the earth's surface. Probably the best known is the Thornthwaite method, and Table 2-15 illustrates his bookkeeping method.

The PE, shown in row 1, is the adjusted potential evapotranspiration that is derived by substituting monthly temperature (°C) into an empiric formula derived by Thornthwaite. The heat index, i, is calculated at the same time. The results obtained provide the unadjusted PE, which is corrected through the use of a prepared nomograph. (See Table 2-11.)

Row 2 gives the monthly precipitation data, P. The difference between P and PE, row 3,

provides the amount of moisture available after evapotranspiration requirements have been satisfied. The excess above the difference will go either to soil water or will occur as runoff. The amount of water stored by the soil is highly variable and will depend upon the nature of the soil. In his initial work, Thornthwaite used a value of 4 in. (10 cm) as a general value to be applied to all water balance studies. While this does enable comparison of the balance for different stations, it is not realistic in terms of precise water balance studies. One of the most significant modifications of the Thornthwaite budget is the introduction of the varying soil moisture-holding capacities. However, for demonstration purposes, the 4 in. value is retained here.

Rows 4 and 5 give the amount of moisture stored in the soil and the change in storage (ΔSt) since the previous month, using the 4 in. value. As long as P is greater than PE, the value remains 4 in. However, as soon as the relationship is reversed, PE is greater than P, plants draw upon the available soil moisture, and the soil storage falls below capacity. As soon as the cumulative excess of PE over P is greater than 4 in., then soil moisture is assumed totally utilized. A deficit period then exists. As row 6 shows, actual evaporation is lower than the potential at this time. Obviously, at the end of the deficit period, when P becomes greater than PE, the 4 in. must be restored to the soil before any surplus occurs. Rows 7 and 8 provide the balance of

Table 2-15[a]

		J	F	M	A	M	J	J	A	S	O	N	D	Yr.
1	PE[b]	0.5	0.7	1.2	2.0	3.1	3.9	4.8	4.4	3.4	2.0	0.9	0.4	27.3
2	P	5.6	4.4	4.0	2.3	2.0	1.9	0.5	0.7	1.6	3.6	5.6	6.7	38.9
3	P–PE	5.1	3.7	2.8	0.3	−1.1	−2.0	−4.3	−3.7	−1.8	1.6	4.7	6.3	
4	ΔST	0	0	0	0	−1.1	−2.0	−0.9	0	0	+1.6	+2.4	0	
5	ST	4.0	4.0	4.0	4.0	2.9	0.9	0	0	0	1.6	4.0	4.0	
6	AE	0.5	0.7	1.2	2.0	3.1	3.9	1.4	0.7	1.6	2.0	0.9	0.4	18.4
7	D	0	0	0	0	0	0	3.4	3.7	1.8	0	0	0	8.9
8	S	5.1	3.7	2.8	0.3	0	0	0	0	0	0	2.3	6.3	20.5

[a] From Carter (1965).
[b] PE—potential evapotranspiration; P—precipitation; P–PE—precipitation minus potential evapotranspiration; ΔST—change in soil moisture storage since previous month; ST—soil moisture storage at the end of the month; AE—actual evapotranspiration; D—water deficit; S—water surplus. Data in inches.

water in terms of deficit and surplus, and it will be noted that the deficit does not occur as soon as PE is greater than P because of the period of soil moisture utilization.

The preceding data can be shown graphically as indicated in Figure 2-33. Notice that the graph is divided into four areas, the surplus and deficit periods, the period of utilization, and the period of recharge.

Representative graphs for selected climatic regions are shown in Figure 2-34. As can be seen, the water balance approach provides considerable insight into both the climatic and hydrologic regime of any given region.

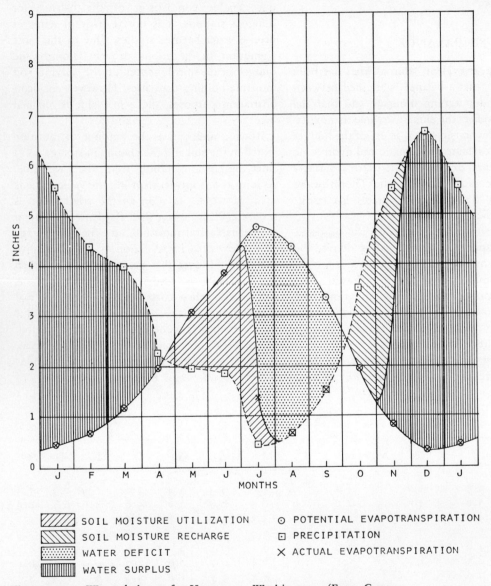

Figure 2-33. Water balance for Vancouver, Washington. (From Carter, D.B., *Fresh Water Resources*, Association of American Geographers, 1965. Used with permission.)

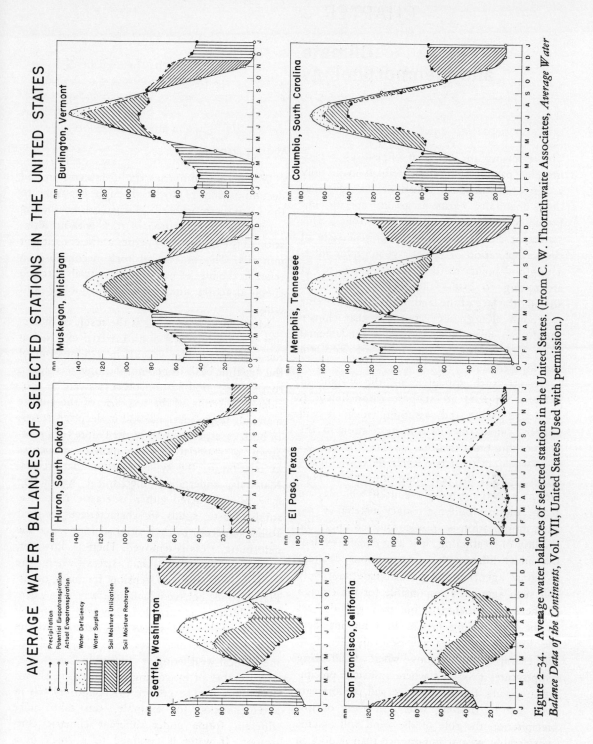

Figure 2-34. Average water balances of selected stations in the United States. (From C. W. Thornthwaite Associates, *Average Water Balance Data of the Continents*, Vol. VII, United States. Used with permission.)

CHAPTER 3

Climate
and Geomorphology

"In reviewing the place of climate as a geomorphic parameter one can perhaps make only a single uncontested statement, namely, that little unanimity exists as to its specific importance." (Holzner and Weaver, 1965.) This statement clearly indicates the present state of the interpretation of climatic-geomorphic interaction and points to the difficulties involved in attempting to assess the role of climate in shaping the face of the land. Indeed, one extreme viewpoint (King, 1953) suggests that climate should not even be considered a fundamental controlling factor in the study of landform development.

Despite such an opinion, the weight of evidence appears to suggest that climate is an active factor and, assuming such, it will appear evident that its role in relation to the geomorphic processes that dominate in any region will operate at two levels. First, a climatic regime reflects the prevailing atmospheric conditions and the atmosphere itself will play an active role in modifying surface material *in situ*. Such modification is aptly considered under the heading of weathering. It is also equally evident that under different climatic conditions, different erosional agencies will predominate; not only will such agents be responsible for the erosion of the surface materials, but they will also act as the transporting media. In some instances the different processes will be easily identified; for example, in regions where subfreezing temperatures exist and where sufficient precipitation occurs, the work of ice will predominate over that of running water. In areas of low precipitation the role of the wind will assume considerably greater importance than it does in humid regions, for in humid regions the effect of the wind is subsidiary to other, more active, forms of erosion. While both erosion and weathering occur simultaneously, it is convenient to treat them separately when assessing their roles in the denudation process.

WEATHERING

Weathering comprises the disintegration and decomposition of rocks through exposure to the earth's atmosphere. Mechanical weathering causes the disintegration of rock without alteration of the chemical properties of the constituent minerals. The process of rock decomposition refers to the breakdown of rock minerals through chemical action, and is termed chemical weathering.

Mechanical weathering is the result of molecular stresses imposed upon material exposed at the surface. Strahler (1952), in a discussion of the dynamic basis of geomorphology, has shown that surface modification processes can be considered in terms of the response of the surface materials to gravitational and molecular stresses. In both cases, the nature of the failure (or breakdown) of material depends upon its properties in relation to the type of stress that occurs. Materials undergoing mechanical weathering may be considered either as elastic or plastic solids. Elastic solids are characterized by their ability to return to their original form once the deforming stress is removed. There is, however, an upper limit—the yield stress—where the elastic solid fails and tension fractures occur. Plastic solids behave in a similar way, but where the yield stress is exceeded, they will deform by flowing. Table 3-1 shows the nature and the results of molecular stresses as they apply to mechanical weathering.

A number of interesting responses are shown in Table 3-1. The growth of foreign crystals in surface rocks, for example, can take quite different forms under different climatic conditions. If water is confined to a crack in a rock and caused to freeze, under normal temperature and pressure conditions, expansion will increase its volume about 9%. If the expansion is confined and the freezing point depressed, the pressure exerted by the frozen water becomes very large and tremendous pressure is exerted

Table 3–1[a]

Molecular Stresses

Materials Involved	Properties of Material	Stress and Cause	Kind of Failure	Weathering Process and Form
1. Rock: strong, hard crystalline, glassy, or crystal aggregate	Elastic solid, non-homogeneous	Shear stress due to nonuniform expansion-contraction in cyclic temperature changes	Rupture by shear or tension fractures between grains, along cleavages, joints, bedding planes	Granular or blocky disintegration of rocks, esp. coarse-grained crystalline rocks.
	Elastic solid, homogeneous	Shear stress set up by thermal gradient from surface heating	Rupture between layers paralleling rock surface	Exfoliation of rock by fire, lightning; solar or atmospheric heating-cooling.
2. Permeable rock and water	Elastic solid	Shear stress set up by interstitial ice crystal growth	Rupture between grains, cleavage pieces, joint blocks, beds	Frost disintegration of rocks. Felsenmeer.
Clay soils	Plastic solid	Stress from growth of ice lenses, wedges	Plastic deformation of clays adjacent to ice	Heaving of clay soils, frost mounds, polygons
3. Permeable rock or soil and water and salts	Elastic solid or elastic continuum	Shear stress set up by interstitial growth of salt crystals	Rupture between grains, cleavage pieces, joint blocks or beds	Efflorescence, granular disintegration in dry climates. Caliche heaving.
4. Rock or soil and colloids and water	Elastic or plastic solid	Shear stress set up by dilatation accompanying water adsorption and drying	Rupture between grains. Plastic deformation of clays during swelling	Exfoliation of basaltic, granitic rock upon alteration of silicates. Slaking of shales.
5. Rock or soil and capillary water	Elastic or plastic solid	Shear stress set up by dilatation accompanying changes in capillary film tension	Rupture between grains or masses of soil	Disintegration of granular permeable rocks. Heaving or subsidence of clays, silts
6. Rock or soil and plant roots	Elastic or plastic solid	Shear stress set up by swelling of rootlets under osmotic pressure	Rupture between grains, cleavage pieces, joint blocks, beds	Disintegration of rock by prying of roots. Deformation of soils
7. Strong, hard, monolithic bedrock	Elastic solid	Shear stresses of tectonic origin stored as elastic strain at depth	Rupture of rock on planes paralleling surfaces after release of confining pressure	Exfoliation of domes, slabs, shells, Quarry rupture, rock-burst

[a] From Strahler (1952).

on the surrounding rocks. The freezing of water at night, with melting during the day, is a common feature in upland areas of middle latitudes.

This freeze-thaw process, which may extend over considerable periods of time, causes alternating pressure loads on the surrounding rocks and ultimately leads to their shattering. The detritus that results from the weathered cliff face often falls to the foot of the cliff to form extensive screes and talus slopes (Figure 3–1a).

Just as the growth of ice crystals can cause disintegration of surface rocks, so can the growth of crystals of mineral salts. When ground water reaches the surface in arid conditions, in a situation such as is shown in Figure 3–1b, the water evaporates rapidly. The ground water that

occurs is frequently heavily charged with mineral salts that remain after the water evaporates. Salt crystals then grow in the pore spaces of the rock, to cause disintegration of the rock particles. As shown in the figure, elongated hollows may occur along the line at which the water table appears at the surface.

Further examination of Table 3–1 shows that the term exfoliation appears in several places. The interpretation of exfoliation, the surface peeling of rocks, has caused a number of problems. Because the process is found most often in areas experiencing large diurnal temperature changes, it was originally attributed to differential thermal expansion and contraction of rocks and rock constituents. But experiments by Blackwelder (1933) showed that rocks could

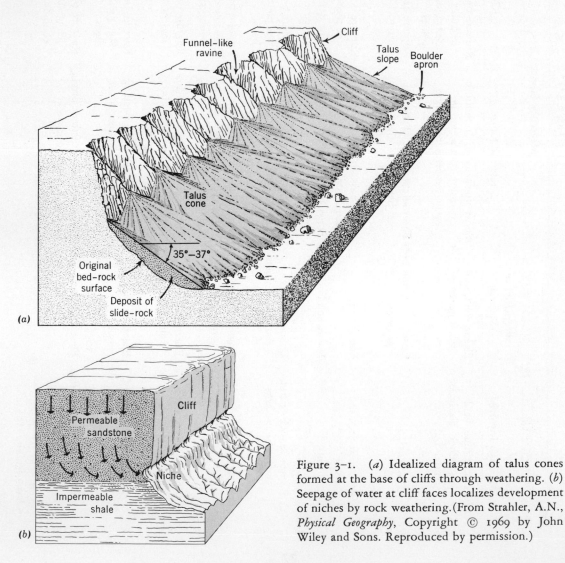

Figure 3–1. (a) Idealized diagram of talus cones formed at the base of cliffs through weathering. (b) Seepage of water at cliff faces localizes development of niches by rock weathering.(From Strahler, A.N., *Physical Geography*, Copyright © 1969 by John Wiley and Sons. Reproduced by permission.)

be subjected to great extremes of sudden heating and cooling without any visible results of disintegration. Griggs (1936) found a similar result but took the experiment one step further. By introducing water during the cooling process he found evidence of exfoliation occurring after the experimental equivalent of two and a half years. Formerly, without the introduction of water, no change had occurred for the equivalent of 244 years. He concluded that, with the introduction of moisture, chemical weathering was of more significance than mechanical in the exfoliation process.

Evidence of exfoliation is not limited to desert areas however, and another atmospheric factor also contributes toward the process. Rocks, particularly igneous formations, are formed under pressures much higher than that found under normal atmospheric conditions. When denudational processes ultimately result in the exposure of the rock at the surface, minerals that were stable at the higher pressures are less stable at the surface pressure. This instability, together with chemical weathering, causes *unloading* to occur, and the rock breaks along lines of sheeting structures and the surface appears to "peel off" from the rock below.

Recently, Ollier (1969) has attempted to reconcile the varying interpretations of the exfoliation process and suggest a terminology that allows the various types to be distinguished. Thus, large-scale exfoliation might be considered as spalling, while the exfoliation of smaller rock areas is considered as flaking. Ollier also notes that it is difficult to assess the process in terms of any single climatic element.

The relative significance of mechanical weathering under different climatic regimes has been assessed by a number of authors. Figure 3–2 shows a climograph on which is superimposed the relative preponderence of mechanical weathering. This qualitative assessment shows that it is strongest in cool areas that receive ample precipitation.

Chemical weathering is an extremely important process in surface modification. In some ways, chemical weathering can be viewed as the modification of minerals to meet a condition of equilibrium in relation to prevailing atmospheric conditions. As already noted, many minerals are formed under high temperature and pressure

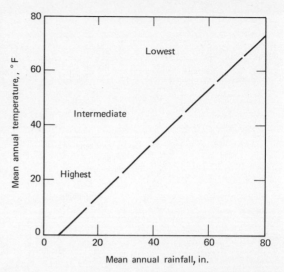

Figure 3–2. Relative intensity of physical weathering under different temperature-precipitation regimes. (After Leopold et al., 1964.)

conditions, with the result that they are most stable under those conditions. Upon exposure to the atmosphere, or when subjected to circulating meteoric water, a new set of conditions exist and the minerals may react accordingly. It is found that the order of crystallization of minerals in a magma, as indicated by the Bowen Series (Figure 3–3), is the order in which minerals react to chemical weathering processes. Minerals formed at high temperatures and pressures (e.g., plagioclase feldspars) are more highly susceptible to chemical weathering, with resulting chemical change, than those formed at a lower temperature and pressure (e.g., quartz) in the original magma.

The various forms of chemical weathering—oxidation, hydration, hydrolysis, and carbonation—all require water or water vapor as part of the chemical change involved. Furthermore, the rates at which chemical decomposition occurs often depends upon prevailing temperature conditions with the results that the rates of chemical activity vary considerably from one set of climatic conditions to another. Figure 3–4 provides a guide to the differences that occur.

The humid tropics, with high annual temperatures and precipitation, provide the best examples of extreme chemical weathering. Consider, for example, the weathering of feldspar, a common constituent of granite. Under less extreme climatic conditions hydrogen ions

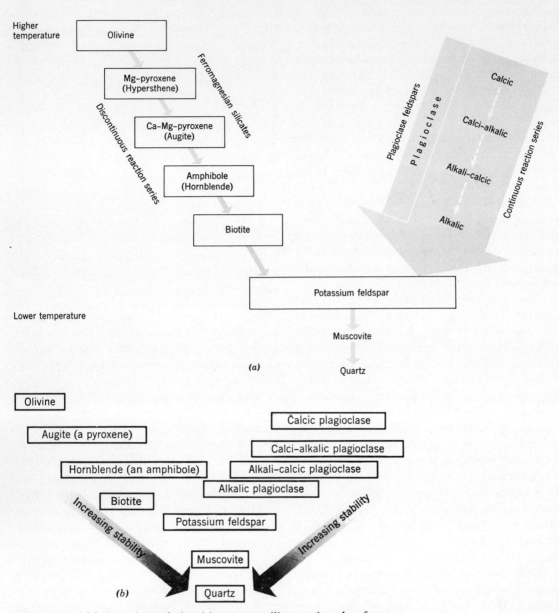

Figure 3-3. (*a*) Reaction relationship among silicate minerals of many igneous rocks. The diagram shows relative temperatures at which crystallization occurs. (*b*) Mineral stability in weathering shows that most stable minerals are those that crystallize at lower temperatures. In some ways the lower diagram is the reciprocal of the upper diagram. (From Longwell C. R., R. F., Flint, and J. E. Sanders, *Physical Geology*, Copyright © 1969 by John Wiley and Sons. Reproduced by permission.)

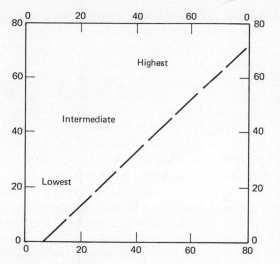

Figure 3–4. Relative intensity of chemical weathering under different temperature-precipitation regimes. (After Leopold et al., 1964.)

(which are derived from the reaction between rainwater and the carbon dioxide of the air or, more significantly, from rainwater percolating through decayed vegetation), cause the following reaction:

$$2KAlSi_3O_8 + 2H^+ + H_2O$$

potassium feldspar   hydrogen   water
                      ions

$$= 2K^+ + Al_2Si_2O_5(OH)_4 + 4SiO_2$$

potassium   kaolinite   silica
ions

The potassium is washed out in solution with, perhaps, some of the silica, leaving the secondary mineral, kaolinite. This is a clay mineral and, in comparison to the other minerals, is relatively insoluble. If, however, the weathering process is continued under warm temperatures and moist conditions, the kaolinite itself might react with water.

$$Al_2Si_2O_5(OH)_4 + 5H_2O$$

kaolinite   water

$$= Al_2O_3.3H_2O + 2H_2SiO_4$$

aluminum oxide   silicic acid

The result is the production of an impure earthy aluminum oxide, which forms the main ore for the mineral bauxite. The silica content is leached out through the process of desilication, a characteristic of chemical weathering under moist tropical conditions. In the same way, iron rich minerals are modified so that iron oxides also form a characteristic residue of extreme chemical weathering. As is noted later, the lateritic soils of the tropical realm are a result of this chemical weathering, and they are one of the major problems to be overcome in the development of agriculture in such regions.

The secondary clay minerals also illustrate another form of chemical weathering. Hydration refers to the combination of water with a chemical compound, but a combination that is reversible because the water can be driven off by increasing the temperature. Some clay minerals have the property of becoming hydrated when moisture is available, revert to a dehydrated state when water supply is limited. The addition of water causes the clays to swell; on release of water they shrink or crack. Such swelling clays can cause great problems in terms of utilization, particularly from the engineering standpoint.

Chemical weathering through carbonation also illustrates the importance of different climatic conditions in the rate at which chemical activity occurs. The most obvious effect of carbonation is seen in the weathering of carbonate rocks such as limestone.

$$CaCO_3 + H_2CO_3 = Ca^{++} + 2HCO_3^-$$

calcium   carbonic   calcium   bicarbonate
carbonate acid   ions   ions

Because the bicarbonate is about 30 times more soluble than the carbonate, the rate of solution is considerably increased. But the availability of water is not the only criterion in the rate at which limestone is weathered. Temperature and pressure also play significant roles. Carbon dioxide dissolves more readily in cold water than in warm, while the rate is also increased with increasing pressure. The significance of this has led to important postulations concerning the formation of caverns in limestone formations (Moore, 1960) as well as the interpretation of rates of exchange of carbon dioxide in the biosphere (Bolin, 1970). It follows, too, that the rate of carbonation will vary considerably from one climatic regime to another. In respect to this, Bloom (1969) writes "Limestone in the tropics is carbonated and dissolved intensely both by carbonic acid solutions and by acidic nitrogenous and organic compounds from the decaying vegetation. Tropical weathering of limestone may reduce a region to a series of sponge-like hills and cavern-ridden lowlands. Few streams flow across tropical karst landscapes, for most of the drainage is underground. The strange, needlelike mountains of classical Chinese art originated in the tropical karst landscape of the southern provinces of China. The scenes seem exotic and dream-like to Western eyes, but they are actually reasonable geomorphic sketches."

As can be seen, the process of chemical weathering essentially concerns the reaction between soil, rock, and acidic water, with the resulting formation of new minerals. Since such reactions vary considerably from one climatic realm to another, the role of climate in the dominant process assumes a factor of considerable importance. Further, and as is outlined in the discussion of pedogenic regimes in Chapter 4, the nature of soil-forming processes rely heavily upon the dominant climate of the region involved.

THE CYCLE OF EROSION

The modification of a land surface may be described in terms of the geomorphic cycle of erosion, which itself forms part of the geologic cycle (Figure 3–5). Essentially, the geomorphic cycle concerns the erosion, transport, and deposition of material at the earth's surface. The analysis of the geomorphic (or geographic) cycle was, for many years, dominated by the concepts of one man. In a series of essays extending over some thirty years, W. M. Davis postulated a process of landform analysis that provided the base for most of the work in landform interpretation. His deductive reasoning method has been strongly criticized in recent years because it lacks a rigorous quantitative base through which meaningful scientific analysis could occur. Nonetheless, Davis provided a method of analysis that allowed the comprehension of landscape and gave meaning to seemingly incomprehensible complexities of the physical environment.

At the base of Davis' approach is the assumption that any landscape is a function of stage, structure, and process. Stage refers to a point in time at which a given landform exhibits topographic features indicative of a defined sequence. It was this aspect of the cycle that most concerned Davis and his followers, and the inter-

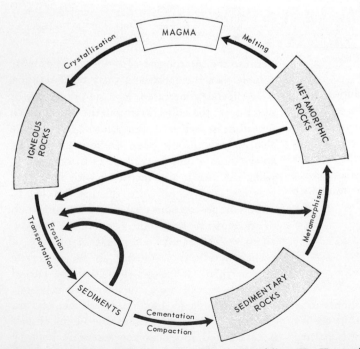

Figure 3–5. The geologic cycle. (From Hamblin, W. K., and J. D. Howard, *Physical Geology Laboratory Manual*, 1st ed., 1964. Reprinted by permission of Burgess Publishing Company.)

pretation was highly anthropomorphic. Subsequent to an initial emergence of a landmass it passes through a sequence identified as youth, maturity, and old age, after which the entire cycle may be repeated through the process of rejuvenation.

The structure part of the formula concerns the basic geologic nature of the landform; in some instance, of course, this is of prime importance. Process refers to the manner by which a landform is modified through the youth-old age sequence. It is at this level that the role of climate becomes significant. Under humid climatic conditions the normal cycle of erosion is supposed to exist, the sequence of its evolution being shown in Figure 3-6. It is assumed that river valley development is the most important single factor influencing the topography, with valley creep and slope development occurring concurrently. In contrast to the normal cycle, Davis and subsequent workers (e.g., Cotton, 1942) assumed that other erosion cycles were "climatic accidents." Thus, the cycle under arid conditions (Figure 3-7) and glacial conditions must be considered "nonnormal." This consideration has been met with skepticism in recent years, as has the assumption that the initial landform uplift took place rapidly over a relatively short period. A good critique of the Davis methodology is given by Dury (1969).

The analysis of the process aspect of the Davisian cycles is only one way, and indeed a rather gross way, in which the impact of climate on landforms can be assessed. But the logical simplicity of the system caused it to be the major method of analysis, particularly in the United States, for many years. While Davis did work in Germany, European geomorphologists were not, for the most part, influenced by his work. Other workers, particularly Penck and de Martonne, produced alternative interpretations, but their work was not accorded the attention given to Davis. More recently, however, with the development of quantitative geomorphology, movement is away from the Davisian system and the relationship between climate and geomorphology is treated in more rigorous terms. While, as discussed in the following paragraphs, the organizational relationship between climate and geomorphology has been given much attention, it should not be over-

looked that quantitative assessment is also highly applicable at the more detailed level. The quantitative analysis of drainage basin characteristics provides an apt example of such analysis.

When precipitation occurs on a watershed, the surplus moisture will ultimately find its way into stream channels. The amount of water in the stream, its discharge, is of utmost importance in the evaluation of such diverse factors as the amount of debris it is carrying and its flood potential. Insight into the amount of water in a river may be attained through application of stream ordering and subsequent drainage analysis. Figure 3-8 shows a watershed with its divide clearly marked. The surplus water falling into the area defined by the divide will ultimately pass Point X. But it will be appreciated that the large basin shown consists of many smaller drainage areas. Each of these may be defined by characterizing the stream that flows within the basin. The basin of a first-order stream is smaller than that of a second-order stream, which is smaller than that of a third-order stream, and so on. It is found (Figure 3-8b) that when plotted on semilog paper, the relationship is shown as a straight line. Clearly there will be a relationship between the amount of precipitation that will fall on a given watershed and the amount that is discharged by the stream. Studies have shown that the relationship is expressed as

$$Q = jA^m$$

where Q is discharge, A is the watershed area, and the constants j and m are derived by fitting a regression line to the available data. As an example, in a study of 12 streams in New Mexico, Leopold and Miller (1956) found the relationship best described by the equation

$$Q_{2.3} = 12 \, A^{0.79}$$

where $Q_{2.3}$ is the flood discharge (in cfs) equaled or exceeded every 2 to 3 years and A is the drainage area of the basin in square miles.

Other aspects of drainage basin morphology, as they relate to precipitation disposal, have also been investigated. These range from analysis of shape of basin as it influences storm precipitation to analysis of slope and surface cover as parameters in assessing the nature of the discharge. Strahler (1964) and Leopold et al. (1964) provide lucid accounts of such aspects.

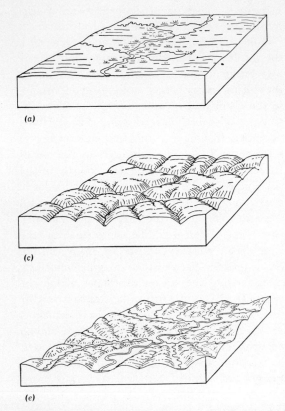

(a)

(b)

(c)

(d)

(e)

(f)

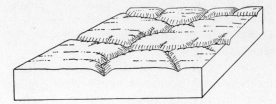

(g)

Figure 3–6. Cycle of land-mass denudation in a humid climate. (*a*) In the initial stage, relief is slight, drainage poor. (*b*) In early youth, stream valleys are narrow, uplands broad and flat. (*c*) In late youth, slopes predominate but some interstream uplands remain. (*d*) In maturity, the region consists of valley slopes and narrow divides. (*e*) In late maturity, relief is subdued, valley floors broad. (*f*) In old age, a peneplain with monadnocks is formed. (*g*) Uplift of the region on a rejuvenation, or second cycle of denudation, shown here to have reached early maturity. (From Strahler, A. N., *Physical Geography*, Copyright © 1969 by John Wiley and Sons. Reproduced by permission.)

Figure 3–7. (right) Cycle of land-mass denudation in an arid climate. (*a*) In the initial stage, relief made by crustal deformation is at the maximum. (*b*) In the mature stage, the mountains are completely dissected and the basins are filled with alluvial fan material and playa deposits. (*c*) In the old stage, relief is low and alluvial deposits have largely buried the eroded mountain masses, whose remnants project here and there as islandlike groups. (From Strahler, A.N., *Physical Geography*, Copyright © 1969 by John Wiley and Sons. Reproduced by permission.)

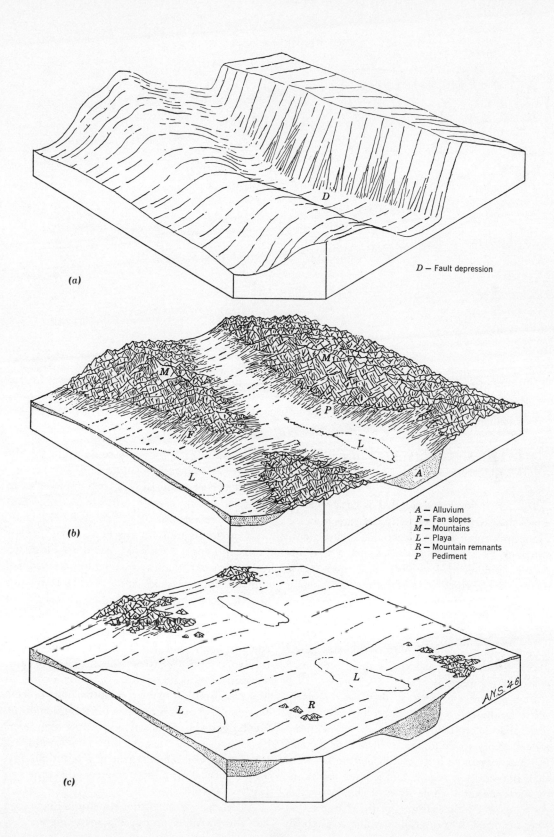

(a)

D — Fault depression

(b)

A — Alluvium
F — Fan slopes
M — Mountains
L — Playa
R — Mountain remnants
P   Pediment

(c)

A.N.S. '46

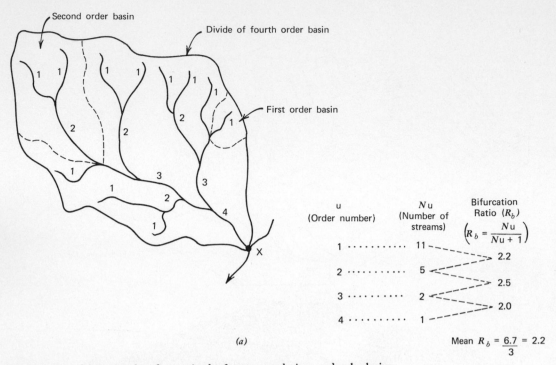

Figure 3–8. (a) Example of a method of stream ordering and calculation of the bifurcation ratio.

CLIMATIC AND CLIMATOGENETIC GEOMORPHOLOGY

The correlation between gross climatic regions, established through an empiric classification, and geomorphic processes would, at first sight, appear to be a simple task. But such a relationship is difficult to establish for a number of reasons. First, climatic classifications are usually too complex to be related to selected geomorphic processes, for such processes often transcend identified climatic boundaries. It is generally found, in fact, that grouping of climates based upon geomorphic aspects usually comprise relatively few regions.

The second problem is that the currently active geomorphic processes are not necessarily those that determined the landscape as it appears today. Many morphological characteristics and features may have been established under more than one climatic regime and a single landform might represent a whole series of climatic cycles. As a result of the great climatic changes that have occurred, it is often impossible to establish

which dominant process is responsible for the present appearance of a landform.

As a partial result of this, two approaches to the relationship between climate and geomorphology are available; these comprise the study of climatic geomorphology and climatogenetic geomorphology. "Climatic geomorphology refers to separate studies of contemporary or fossil climatic influences on landforms, and climatogenetic geomorphology . . . to the study of different generations of surface forms within a single landscape which have resulted from both contemporary and fossil climatic influences." (Holzner and Weaver, 1965.) Despite this apparent simplicity, much confusion occurs in interpretation of the two approaches. Indeed, Büdel (1969) even provides a different definition. He suggests that climatic geomorphology refers to the currently active processes modifying landforms, while climatogenetic geomorphology, a term that he introduced, merely adds time as a further variable.

As an expression of the climatic influence on landforms, Büdel produced a map (Figure 3–9)

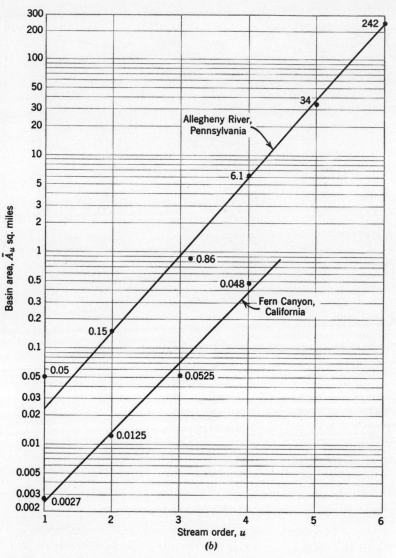

Figure 3–8. (*b*) Actual examples showing the relationship between size of basin area and stream order. Note the logarithmic axis. (From Strahler, A.N., *Physical Geography*, Copyright © 1969 by John Wiley and Sons. Reproduced by permission.)

showing the climatomorphological regions of the world. These clearly correspond with the large-scale climatic regions and are distinguished as follows:

I. Zone of glaciated areas—the polar realms and high mountains.

II. Zone of pronounced valley formation—those areas of the subpolar region that experience permafrost but that at present are ice free.

III. Temperate-extratropical-tropical zone of valley formation characterized by moderately active processes and usually strongly marked by features derived from the cold glacial periods.

IV. Tropical zone of planation formation surfaces—most marked in the humid, but particularly in the seasonally humid tropical zones.

It will be noted that the realms shown by Büdel are essentially temperature zones, as related to both present and past climates, and that no differentiation is made between humid and arid climates; the present great deserts of the world are not distinguished. In his review of climatic

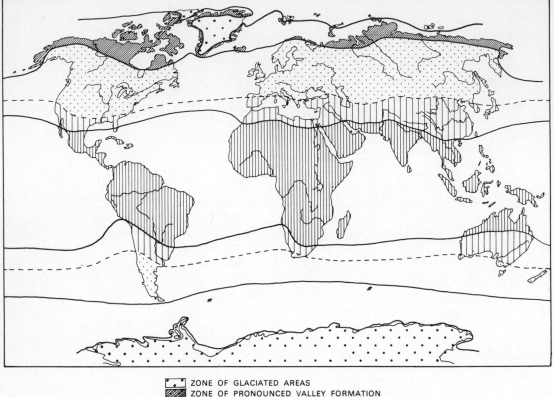

ZONE OF GLACIATED AREAS
ZONE OF PRONOUNCED VALLEY FORMATION
EXTRA-TROPICAL ZONE OF VALLEY FORMATION
SUBTROPICAL ZONE OF PEDIMENT AND VALLEY FORMATION
TROPICAL ZONE OF PLANATION SURFACE FORMATION

Figure 3-9. Climatomorphological zones of the world. (After Büdel, from C. Board et al., *Progress in Geography*, Vol. 1, 1970. Reprinted with permission of Edward Arnold Publishers.)

geomorphology, Stoddart (1969) compares Büdel's scheme with that of Tricart and Cailleux (1965). As shown in Figure 3-10, these authors show many more regions and their approach includes the role played by both climate and vegetation cover.

Further insight into the world distribution of currently or recently active geomorphic processes can be derived through study of the worldwide system of landform classification devised by Murphy (1968). This system has three categories of information to identify landform distribution. The first level shows structural regions by differentiating mountain systems (e.g., Alpine or Caledonian), sedimentary, and volcanic areas. The second level is based upon topographic regions, and Murphy devises a system for identification of plains, hills, mountains, and the like. The third level of organization

concerns erosional and depositional forms, and it is here that the role of climate becomes evident. Figure 3-11 shows the distribution of these forms according to the following definitions:

h. Humid landform areas. Areas in which the pattern of permanent streams has a density of at least one stream every 10 miles traverse distance, and which have not been subject to glaciation since the beginning of the Pleistocene epoch.

d. Dry landform areas. Areas in which the stream density is less than one stream in every 10 miles and which have not been subject to glaciation since the beginning of the Pleistocene epoch.

g. Glaciated areas. Areas covered by glacial ice at some time since the beginning of the Pleistocene epoch, but earlier than the Wisconsin and Würm glaciations (see Chapter 11).

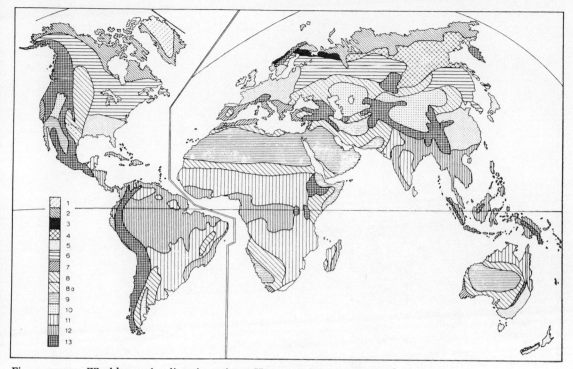

Figure 3-10. World morphoclimatic regions. Key to regions: 1 - glaciated regions; 2 - periglacial regions with permafrost; 3 - periglacial conditions without permafrost; 4 - forests on Quaternary permafrost; 5 - midlatitude forests with maritime climate or lacking severe winter; 6 - mid latitude forests with severe winter; 7 - midlatitude forests, Mediterranean type; 8 - semiarid steppes and grasslands; 8a - semiarid steppes and grasslands with severe winters; 9 - deserts and degraded steppes without severe winters; 10 - deserts and degraded steppes with severe winters; 11 - savannas; 12 - intertropical forests; 13 - azonal mountainous regions. (From D. R. Stoddart, *Water, Earth and Man*, R. J. Chorley, ed., Methuen, 1969. Reprinted with permission.)

w. Wisconsin and Würm glaciations. Areas covered by glacial ice during or since the Wisconsin or Würm glaciations but now free of glacial ice.

i. Icecaps. Areas covered by glacial ice at present.

The classification devised by Murphy follows the general trend among geomorphologists in that it considers only the gross aspects of climatic differences. These are essentially dependent upon ground cover and follow the observation by Garner (1968) that "The only consistent lithospheric response to atmospheric conditions is . . . ground cover." Garner suggests that the most critical forms are:

a. Effectively continuous plant cover (corresponding to the *h* regions of Murphy).

b. Ice cover (the *i* and formerly *g* and *w* regions of Murphy).

c. Exposed land (equated to the *d* regions of the Murphy system).

In the study of the evolution of landforms, climatogenetic geomorphology becomes the aspect of most importance. Many landforms are the result of complete cycles of climatic change, with active processes varying enormously under the different prevailing climatic regimes. For example, studies of landforms extensively modified during Pleistocene times must concern themselves with both the glacial conditions that existed during the period of ice advance as well as the periglacial or humid conditions that prevailed during interglacial periods. While the effects of such changes are dealt with more

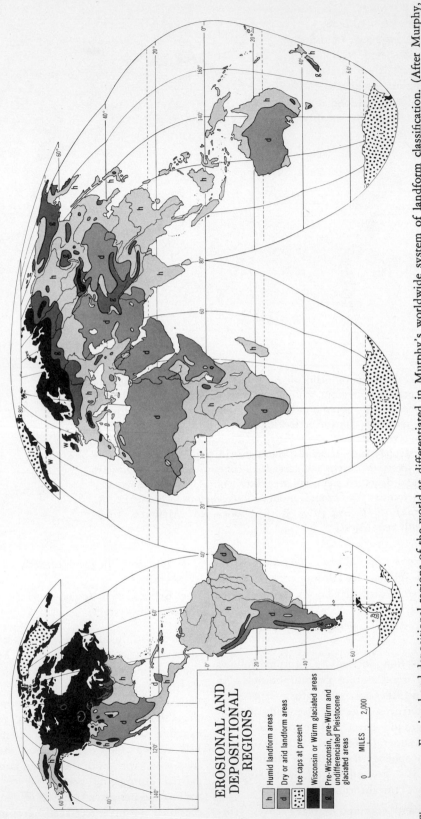

Figure 3-11. Erosional and depositional regions of the world as differentiated in Murphy's worldwide system of landform classification. (After Murphy, 1968. From Strahler, A. N., *Physical Geography*, Copyright © 1969 by John Wiley and Sons. Reproduced by permission. Based on Goode Base Map. Copyright by the University of Chicago Press. Used with permission.)

EROSIONAL AND DEPOSITIONAL REGIONS

Humid landform areas

Dry or arid landform areas

Ice caps at present

Wisconsin or Würm glaciated areas

Pre-Wisconsin, pre-Würm and undifferenciated Pleistocene glaciated areas

MILES

0   2,000

extensively elsewhere, it is fitting that one or two examples be cited here.

While the influence of different geomorphic processes associated with the Pleistocene have been most widely investigated in the temperate and cool climate regions, recent work has shown that the different climatic regimes have played a very important role in the development of tropical landforms. For the most part, such differences are associated with arid-humid differences rather than glacial-nonglacial effects. One such example is shown in Figure 3–12, which outlines the sequence of geomorphic events active in shaping the lowlands of northeastern South America. It is cyclic events such as this that gives rise to fossil landscapes, which

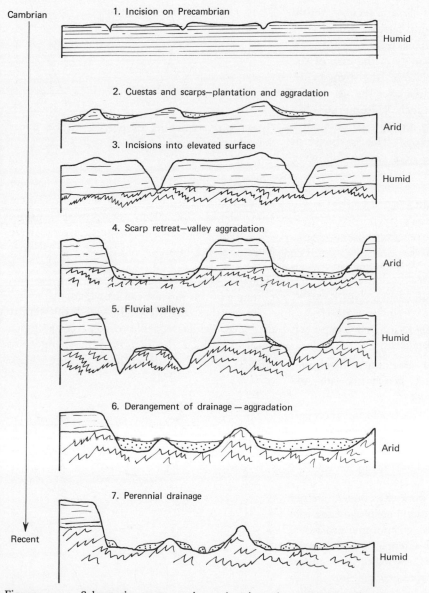

Figure 3–12. Schematic cross sections showing the sequence of geomorphic development under humid/arid conditions in the northeastern South American lowlands. Events depicted range from incision of Precambrian deposits to the development of the most recent event, a perennial drainage system. (After Garner, 1968.)

seem to bear little relationship to the geomorphic processes presently active in the areas where they occur.

Morphogenetic Regions and Climate-Process Systems

An interesting development in the treatment of climate relationships has been the formulation of the concept of morphogenetic regions. Such regions are defined by Peltier (1950) as comprising ". . . a series of climatic regimes within which the intensity and relative significance of the various geomorphic processes are . . . essentially uniform." But, as in the case of the general relationship between the two fields, some confusion has arisen over the use of the term morphogenetic; as such, Wilson (1968) has proposed a scheme that attempts to rationalize the use of the terminology. This is shown in Table 3–2 and will be used in the following analysis.

Table 3–2
DEFINITION OF TERMINOLOGY FOR MORPHOGENETIC CLASSIFICATION[a]

Term	Definition
Climate-process system	Concept relating climatic factors to geomorphic processes
Morphogenetic system	Concept relating climate process to landforms
Morphogenetic region	Actual area where landforms reflect present climate and processes
Paleomorphogenetic region	Area where landforms reflect past climate and processes.

[a] After Wilson (1968).

Following pioneer work by Büdel, Peltier (1950) has produced a useful analysis of the role of climate in relation to geomorphic processes. Table 3–3 outlines the essential aspects of the regions that he identifies; the regions recognized are based upon temperature-precipitation characteristics of the various realms. Peltier further analyzes the relationship by assessing the various active processes that occur at different temperature-precipitation values. Representing his data on climographs, he divides the field of the graph

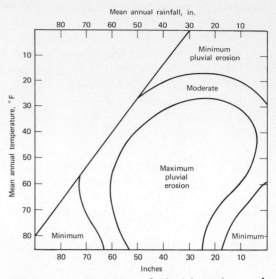

Figure 3–13. Intensity of pluvial erosion under different climates. (After Peltier, 1950.)

into areas, depending upon the relative activity of each erosional process. For example, Figure 3–13 shows the effects of pluvial erosion. Maximum effect is seen to occur over a wide range of the graph, and corresponding climatic limits of the effects of pluvial erosion are similarly broad. Nonetheless, the relative importance of this process can be gaged.

By synthesizing data illustrated on similar diagrams, Peltier was able to construct a climograph in which the identified morphogenetic regions could be differentiated (Figure 3–14).

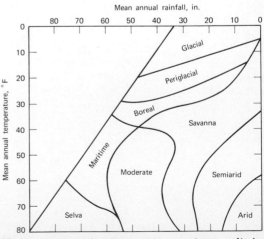

Figure 3–14. Morphogenetic regions distinguished by Peltier. (After Peltier, 1950.)

Table 3-3

MORPHOGENETIC REGIONS[a]

Morphogenetic Region	Estimated Range of Average Annual Temperature, °F	Estimated Range of Average Annual Rainfall, in.	Morphologic Characteristics
Glacial	0–20	0–45	Glacial erosion Nivation Wind action
Periglacial	5–30	5–55	Strong mass movement Moderate to strong wind action Weak effect of running water
Boreal	15–38	10–60	Moderate frost action Moderate to slight wind action Moderate effect of running water
Maritime	35–70	50–75	Strong mass movement Moderate to strong action of running water
Selva	60–85	55–90	Strong mass movement Slight effect of slope wash No wind action
Moderate	38–85	35–60	Maximum effect of running water Moderate mass movement Frost action slight in colder part of the region No significant wind action except on coasts
Savanna	10–85	25–50	Strong to weak action of running water Moderate wind action
Semiarid	35–85	10–25	Strong wind action Moderate to strong action of running water
Arid	55–85	0–15	Strong wind action Slight action of running water and mass movement

[a] From Peltier (1950).

Note that the climatic data he uses are expressed as mean annual figures. Perhaps this is a disadvantage of the analysis, for it has been shown that seasonality is of prime importance in many geomorphic processes, particularly as it pertains to rates of erosion. A severe summer rainstorm over an area of sparse vegetation can remove much more debris than the same amount of precipitation falling over a longer period on a forested area. Furthermore, and as indicated in Figure 3–15, the temperature at which the precipitation occurs—a measure of seasonality—causes large differences in the amount of sediment removed.

A somewhat similar approach to the identification of morphogenetic regions has been sug-

gested by Tanner (1961), who, like Peltier, uses a climograph to represent his evaluation. But, as shown in Figure 3–16, he used potential evapotranspiration instead of temperature as one variable, while the scale of his climograph is logarithmic. Such treatment expresses erosion rates as a function of effective precipitation. In contrast to the number of morphogenetic regions identified by Peltier, only four classes—Arid, Selva, Moderate, and Tundra-Glacial—are present.

Wilson (1969) introduced the term climate-process systems (CPS) in an attempt to clarify the abundance and somewhat confusing terminology used in climatic geomorphology (Table 3–2). Defined as ". . . sets of climatic conditions

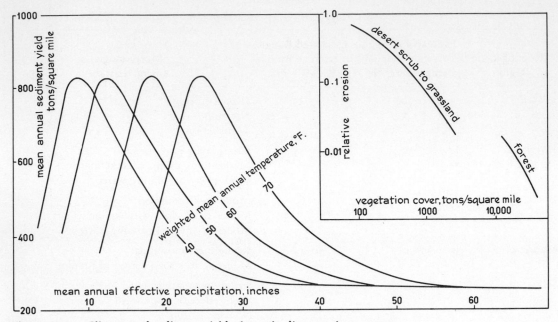

Figure 3–15. Climate and sediment yield: the main diagram shows curves peaking in areas that range from 10 to 25 in., depending upon the mean annual temperature. Inset shows relative rates of erosion under different vegetation covers, based upon weight of vegetation per unit area. (From Dury, G. H., *Perspectives on Geomorphic Processes*, Association of American Geographers, Resource Paper No. 3, 1969. Used with permission.)

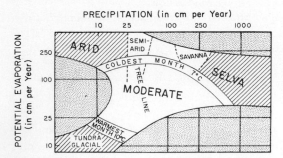

Figure 3–16. Tanner's morphogenetic regions. (After Tanner, 1961.)

under which particular suites of geomorphic processes are dominant," he uses the concept to outline a relationship between climate and geomorphology that is significant in two respects.

a. Climate is considered as a dynamic process and treatment is in terms of air mass dominance.

b. The important effects of climatic seasonality are given weight in evaluation of dominant processes.

Through analysis of the dominant processes under different climatic regimes, Wilson delineates six CPS (Figure 3–17). This graph does resemble that of Tanner, but in fact only represents the first step in the approach used. A second step is the consideration of seasonality, which is attained by equating the recognized CPS to the simple air mass system devised by Strahler. This identifies seven basic air mass regimes, three of which correlate directly to three of the CPS as follows:

Equatorial regime (mT, mE)	Selva system
Polar regime (cP, cA)	Glacial system
Desert regime (cT, mT$_s$)	Arid system

These are areas in which a similar air mass dominates all year. The remaining area of single air mass dominance recognized by Strahler (that of mP dominance) does correspond to the humid

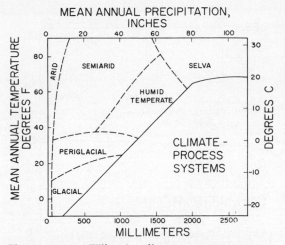

Figure 3-17. Wilson's climate-process systems. (After Wilson, 1968.)

temperate system of Wilson but since ". . . few areas have been subjected to this climate system long enough for equilibrium landforms to develop," the correlation is not easy to establish.

The seasonal regimes of the air mass system, continental (mT — cP seasonally dominant), Mediterranean (mP — cT seasonally dominant), and tropical wet-dry (mT — cT seasonally dominant) involve the interaction of more than one CPS. As such, the landscape characteristics either show a combination of those found under the regimes influenced by a single air mass or features developed only under seasonal regimes. Examples of the climatic regimes and associated CPS are best represented using the thermohyet diagram, details of which are given in Chapter 6. Actual examples are shown in Figure 3-18.

A further merit of the CPS analysis is its applicability to palaeoclimatology. If landscape characteristics can be identified and associated with a defined CPS—other than that currently active—it becomes possible to infer the climatic regime and hence the dominant air masses that formerly existed. One such example is suggested by Wilson: ". . . consider a buried channel deposit which shows both frost-wedged particles and granitic grus, modified of course by stream transplant. The deposit in question shows evidence of periglacial conditions, yet also indicates process characteristics of one of several warmer regimes . . . the dominant climatic regime during the time of formation of the de-

posit was continental. This specifically indicates process characteristics of one of several warmer regimes . . . the dominant climatic regime during the time of formation of the deposit was continental. This specifically indicates that the area in question was dominated by cP air masses during the winter season and by mT during the summer." Such a method of analysis holds much promise for investigation of the dynamics of past climates.

CLIMATE AND THE REGIONAL DIVERSITY OF LANDFORMS

From the foregoing analysis it is evident that processes actively modifying landscape vary over the surface of the earth. As such, it is possible to analyze regional diversity of landforms resulting from the varied processes that occur. Obviously, a complete connection between actual landscape and climate process cannot be assumed, for the enormous differences between mountain belts and ancient shields are the result of structural differences. Nonetheless, the modification of such landforms can be related to the distribution of process, which itself relates to climate. Not only will the processes differ spatially (e.g., ice will dominate in some areas, running water in others) but the rates of erosion by the agencies will also differ. Figure 3-19 gives Corbel's (1964) estimates of the rates of regional erosion under different climatic conditions for lowland and upland areas. Rates of erosion in humid areas are appreciably higher than those in arid areas, while mountains experience more rapid erosion than lowlands in similar climatic regimes. Such data can be used to estimate total world erosion; one estimate is given in Table 3-4.

The preceding data are, of course, gross estimates, and variations will occur. Note that while the rate varies, the nature of the resulting landform through erosion of a given rock type can also vary, for the different intensities of chemical and physical weathering, or the role of water in its various phases, can cause similar rock types to assume quite different forms. Figure 3-20 illustrates how limestone—here bordered by granite and schist formations—will give rise to different landform types under different climatic conditions.

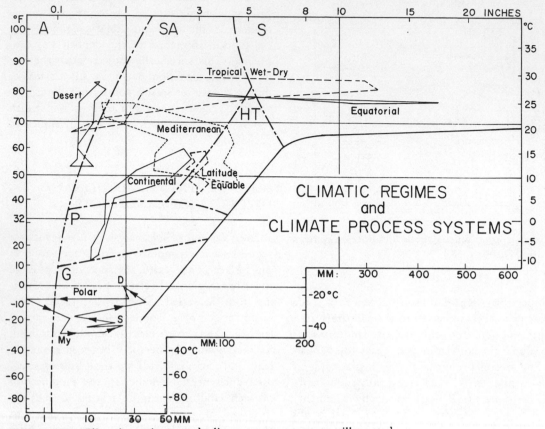

Figure 3-18. Climatic regimes and climate-process systems illustrated using the thermohyet diagram. Boundaries of climate-process systems shown by dash-dot line and each system is labelled by one of two letters where A = arid, S = selva, HT = humid temperate, SA = semiarid, P = periglacial, G = glacial. Climatic data from stations typical of seven climatic regimes are plotted. Representative stations: Equatorial—Padang, Sumatra; tropical wet-dry—Calcutta, India; Desert—William Creek, Australia; Mediterranean—Izmir, Turkey; Continental—Winnipeg, Canada; Middle-latitude equable—Dunedin, New Zealand; Polar—McMurdo Sound, Antarctica. (Climate-process systems from Wilson, 1969; regime diagrams from Oliver, 1969.)

To facilitate analysis of regional diversity resulting from different intensity of geomorphic processes, it is useful to employ the concept of air mass dominance over given regions. Figure 3-21 shows a conceptual arrangement wherein four distinctive realms are considered. Within each of these, as a result of the dominant effect of a single air mass, the geomorphic processes will be relatively uniform and sustained. Between each of the realms distinguished, processes will reflect those of the neighboring regimes either as a subdominant or seasonal form.

The cP realm. Dominated by cold, polar air masses for much of the year, the climatic characteristics will, nonetheless, vary on a seasonal basis. In winter, with short or non-existent daylight and low insolation, temperatures are invariably low. Precipitation, in the form of snow is light, a reflection of the limited moisture capacity of cold air. In summer, the long daylight hours often cause temperatures to rise above freezing and some precipitation results either from local controls or through incursions of mP air masses. Depending upon

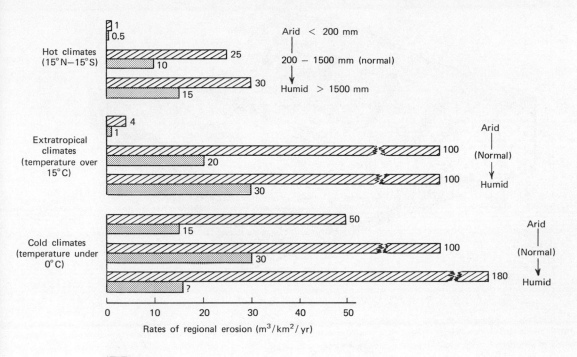

Figure 3–19. Relative rates of regional erosion under different climatic regimes. As indicated, arid areas are designated as having less than 200 mm precipitation per annum, humid regions more than 1500 mm. Plotted from data given by Stoddart (1969).

Table 3-4

TOTAL WORLD EROSION (M³ × 10⁶) PER ANNUM[a]

	Arid	Normal	Humid	Total
Warm				
Equatorial	1.0	300.0	52.5	353.5
Intertropical	3.7	225.0	20.0	248.7
Extratropical	19.0	360.0	80.0	459.0
Total	23.7	885.0	152.5	1061.0
Temperate	175.0	1050.0	325.0	1550.0
Cold				
Extrapolar	70.0	550.0	90.0	710.0
Polar	77.5	350.0	75.0	502.5
Total	147.5	900.0	165.0	1212.5
Total unglaciated	346.2	2835.0	642.5	3823.7
Glaciated lands	550.0	4000.0	200.0	4750.0

[a] From Corbel (1964).

the length of time that temperatures rise above freezing, and the relationship between precipitation-ablation rates, two main realms may be distinguished.

In areas where ice accumulation exceeds ablation (the melting and evaporation of accumulated snow and ice), ice will collect and ice sheets may result. This is evident in the present-day ice sheets of Antarctica and Greenland. Landform modification through such conditions, while locally evident in, for example, the form of nunataks, is largely inferred from regions that experienced similar conditions during the Pleistocene. While glacial ice is most widespread in the continental ice sheets, a similar erosional-depositional set of characteristics occurs in mountain areas where suitable climatic conditions prevail (see Chapter 12).

Where ice does not accumulate, periglacial conditions prevail. Well-known features associated with such conditions are shown in Figure 3-22. While the major active process results

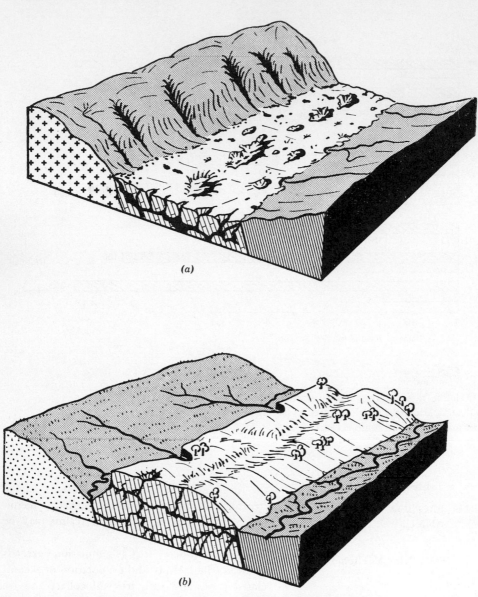

(a)

(b)

Figure 3-20. Landscape types according to Corbel. (*a*) *Humid subarctic.*
From left to right: gneiss-granite, limestone, schist. The limestone is in
depressions and shows intense karstification. Example taken from Norwegian Lapland (65°N), with swamps and marshes. (*b*) *Humid temperate*
(same structure as in *a* but with sandstone for granite-schist). The sandstone
is more resistant than the limestone. Moderate deep karstification. Example
taken from Belgium Ardennes (about 50°N). Pastures and croplands on
the sandstone and schist, meadows and woods on limestone.

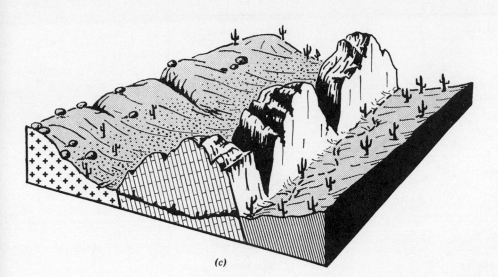

(c)

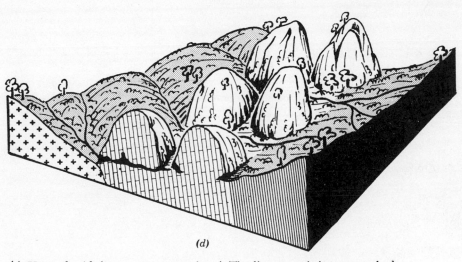

(d)

(c) *Hot and arid.* (same structure as in *a.*) The limestone is in very marked relief, a few cave tunnels and canyons inherited from colder Quaternary times. The granite rock decomposes and forms pediments. Streams are temporary wadis. Example taken from Arizona-Sonora deserts (about 32°N). Sporadic vegetation. (*d*) *Hot and humid* (same structure as in *a.*) The top of the limestone rocks rise above the granite and schist. The limestone is dissected by exogenous streams, it is resistant to erosion by indigenous streams. Subterranean karstification is weak. Example taken from Caribbean between 15 and 20°N. Forests and croplands. (From Corbel, J., *Annales de Géographie*, 73, 385–409, 1964. Used with permission.)

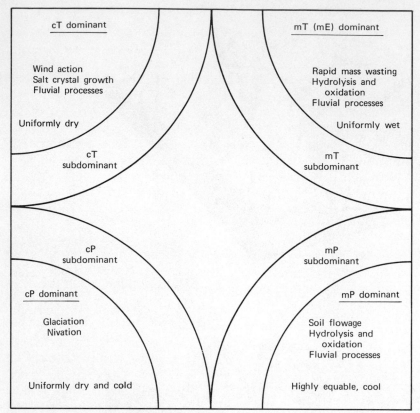

cT dominant		mT (mE) dominant

Wind action
Salt crystal growth
Fluvial processes

Rapid mass wasting
Hydrolysis and
  oxidation
Fluvial processes

Uniformly dry

Uniformly wet

cT
subdominant

mT
subdominant

cP
subdominant

mP
subdominant

cP dominant

mP dominant

Glaciation
Nivation

Soil flowage
Hydrolysis and
  oxidation
Fluvial processes

Uniformly dry and cold

Highly equable, cool

Figure 3-21. Schematic relationship between realms of dominant air masses and geomorphic (climate–process) systems.

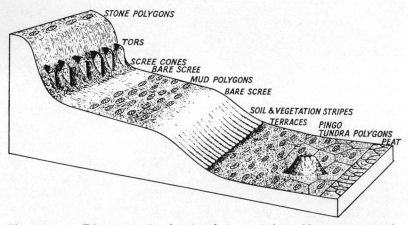

STONE POLYGONS
TORS
SCREE CONES
BARE SCREE
MUD POLYGONS
BARE SCREE
SOIL & VEGETATION STRIPES
TERRACES   PINGO
TUNDRA POLYGONS
PEAT

Figure 3-22. Diagrammatic sketch of the varied landforms associated with slope in the tundra. On the steep slope, stone polygons pass to stone stripes while screes are common at cliff base. On river banks, polygons pass to tetragons, which are parallel and perpendicular to the bank. (From Baird, P. D., *The Polar World*, 1964. Redrawn from FitzPatrick. Used with permission of Longman Publishers.)

from physical weathering, frost action, solifluction, and the like, these regions also experience the highly pervasive problem of frozen
ground, the permafrost.

Frozen ground may be divided into two
groups. First, the seasonally frozen ground that
thaws annually, and second, ground that is
permanently frozen, the permafrost. If it is
assumed that permafrost exists where temperature
is constantly below freezing (actually, because
of dissolved solids, surface forces between
grains, etc., the temperature for permafrost is
usually below that given by the freezing point
of water, 32°F), then it is possible to envisage
permafrost under a number of conditions. If
the surface temperature is low enough, then it
may extend to the surface; if, however, the
surface temperature is above freezing for a
limited period, then surficial thawing will occur.
A thawed layer at the surface, termed the "active
layer," will overlie permafrost below.

Under climatic conditions where thawing does
not extend far below the surface, the depth of
the permafrost is determined by the relationship
between the surface temperature and the amount
of heat passing up from the earth's interior. It
is generally assumed that because of flow of
heat from the earth, in the upper portions of
the crust there is a temperature increase of about
1°C for every 100 to 200 ft of depth. This means
that the permafrost will have a lower limit, or
equilibrium depth, where the flow of internal
heat offsets the effect of atmospheric freezing.
This is illustrated in Figure 3–23, which shows
that if we assume a mean surface temperature
of −20°C (−4.0°F) and an increase of temperature with depth of 1°C every 100 ft, then the
base of the permafrost is at 2000 ft. As noted,
such an assumption allows the formulation of a
rule of thumb concerning the lower limit of
permafrost, ". . . the depth of the bottom of
permafrost would be 100–200 feet multiplied
by the negative mean of the ground surface in
degrees Centigrade" (Lachenbruch, 1968).

Such a rule of thumb assumes that the surface
conditions have been in equilibrium for a considerable period. This is obviously not true for
many areas. Because the conduction of heat is
relatively slow in the earth's layers, permafrost
may continue to exist long after the mean surface

temperatures have risen. Figure 3–23 (lower)
shows how the lag is related to surface temperature.

The present distribution of permafrost has
been assessed by Black (1954) and his results
summarized in a map (Figure 3–24). Three types
of permafrost are shown and described as continuous, discontinuous, and sporadic. Obviously,
represented as such a scale, the boundaries
cannot be well defined. Many variations in depth
and extent are caused by local variations that
modify the gross effect. Freshwater bodies,
shallow ponds, and the effect of topography are
responsible to numerous variations.

The presence of permafrost presents enormous
difficulties in terms of development of the cP
realm. It is not difficult to visualize mass movement of the active layer, a supersaturated
admixture overlying the rock-hard permafrost.
Nor is it difficult to foresee the enormous
engineering difficulties posed by the nature of
the permafrost and active layer. Buildings, both
surface and underground, require special building
techniques, while transport facilities, ranging
from roads to pipelines, pose great engineering
problems. As described elsewhere, the modification of permafrost through construction has
significant ecological effects.

The cT realm. Areas dominated by cT air masses
throughout the year correspond to the great
tropical deserts. These areas are noted both for
the monotony and magnificence of their scenery,
the lack of vegetation cover often giving a
stark appearance and frequently unmistakable
topographic effects.

The nature of the desert climate is dominated
by lack of moisture, large diurnal variation in
temperature, and frequent high winds of varying
periodicity. In comparison to humid realms,
the frequency of climatic influences is quite
different. Under arid conditions, on the whole,
gradation is slower than in humid areas, and
the relative significance of running water and
mass movement—in comparison to the effect
of the wind—is appreciably less. That is not to
say, however, that running water does not play
a highly significant role in the development of
desert topography.

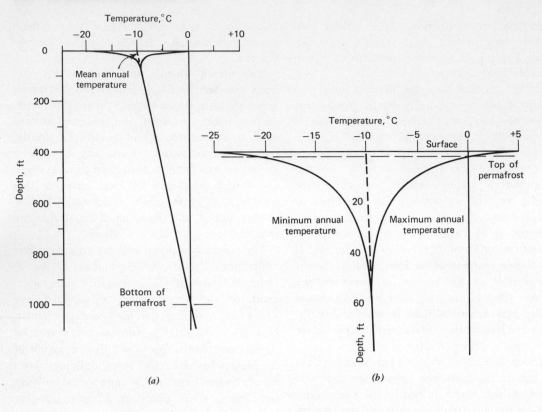

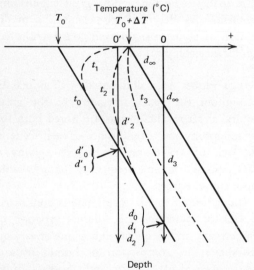

(a)

(b)

Depth

Figure 3-23. (*Upper*) Relationship between temperature and depth showing how (*a*) the lower limit and (*b*) the upper limit of permafrost is established. (*Lower*) Ground temperatures at successive times (T_0, T_1, . . . T_∞) after an increase in mean ground surface temperature from T_0 to $T_0 + \Delta T$. If the final temperature is below o°C (the case in which o°C lies at point O), the base of the permafrost rises successively from d_0 to d_∞. If the final temperature is above o°C (o°C lies at point O'), the permafrost degrades first from the top (curve t_1) and then from both top and bottom (curve t_2) and finally vanishes. (From Lachenbruch, A. H., 1968. Used with permission of Reinhold Publishing Company.)

While wind action does occur in moister areas, its influence is obscured by the more rapid and clearly evident agents of erosion. Thus, in the desert realm the action of the wind through deflation, attrition, and deposition are most easily distinguished. In the same way, the various features resulting from such activity can be analyzed more readily (Figure 3–25).

Desert rains, although infrequent, are usually of high intensity and this, together with the fact that the desert floor may be sun-baked or highly impermeable, leads to high initial runoff and flash floods. Obviously, resulting flood water will not remain evident for a long time; high evaporation rates plus ultimate infiltration cause it to disappear. When runoff does occur, how-

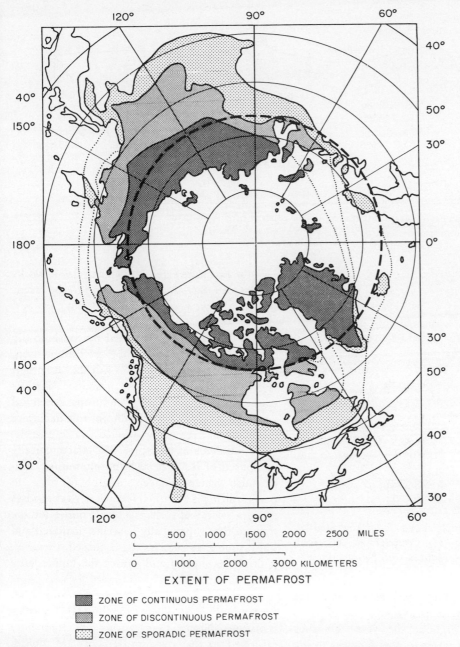

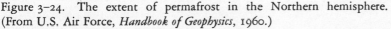

EXTENT OF PERMAFROST

▓ ZONE OF CONTINUOUS PERMAFROST

░ ZONE OF DISCONTINUOUS PERMAFROST

▒ ZONE OF SPORADIC PERMAFROST

Figure 3–24. The extent of permafrost in the Northern hemisphere.
(From U.S. Air Force, *Handbook of Geophysics*, 1960.)

ever, it often finds its way into preexisting channels that occur in desert realms. These channels—called arroyos in Latin America and the American Southwest, or wadis in Arabic countries—may represent former channels that were initiated in wetter climatic periods or the channels of intermittent streams that occur.

During flash floods such channels might be entirely filled with water and the steep banks rapidly eroded.

Running water also plays a role in the formation of alluvial fans (Figure 3–26) and the evolution of large ravines and canyons. The latter are often cut by exotic streams that derive

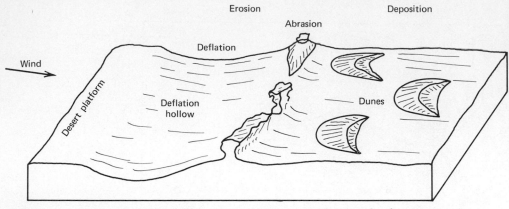

Figure 3-25. Schematic diagram illustrating major aspects of the work of the wind in desert regions.

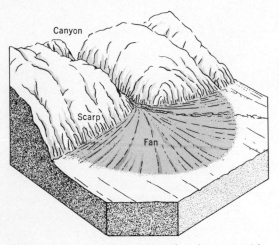

Figure 3-26. A simple alluvial fan. (From Strahler, A. N., *Physical Geography*, Copyright © 1969 John Wiley and Sons. Reproduced with permission.)

their water from humid areas remote from the desert. Enough water is available for them to sustain flow throughout the year and they provide important water sources for man in the arid environment.

The evolution of the desert landscape as suggested by Davis (Figure 3-7), does give insight into the nature of the geomorphic processes under arid conditions. Note, though, that there is considerable disagreement about the mode of formation of many of the features shown. Such differences are well illustrated in the diverse opinions concerning the formation of pediments. Extensive coverage—and dispute

—in the literature extends from the classic works of McGee (1897) and Paige (1912) to the recent study and review given by Rahn (1967).

The mT realm. Distinguished by continuously high temperatures and copious precipitation, this realm is characterized by the luxuriant vegetation response to the sustained climatic conditions. Under such conditions, chemical weathering plays a very significant role in surface modification. A well-weathered waste mantle, often hundreds of feet thick, is characteristically composed of residual iron and aluminum compounds. Where surface outcrops of rock do occur, rotting is a marked feature. This comprises the rapid decay of perhaps a single mineral in the rock so that superficially it seems unaltered and quite sound; as soon as it is exposed, however, it crumbles rapidly and causes the rapid decay of the formation.

Despite the prolific vegetation and high evaporation rate, there is abundant surplus moisture (Figure 3-27). Soils are frequently saturated so that mass movements are typical. So marked is this feature that Machatschek (1969) refers to its movement as "porridgelike," while Sapper (1935) terms it "Subsilvan soil flow." In areas where steep slopes are visible, a knife-edged ridge effect results from numerous landslides, hill slips, and mud flows.

The large amount of surplus moisture available in the mT realm causes it to be the source of enormous quantities of runoff. As shown in Table 3-5, rivers of this realm discharge a high

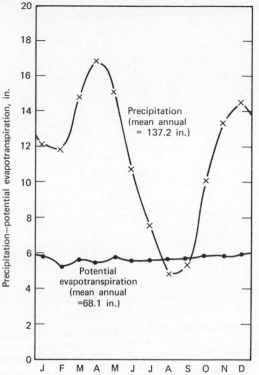

Figure 3–27. Average monthly precipitation and potential evapotranspiration for Madang, New Guinea (5°S., 146°E., elevation 20 ft.). Note the minimal monthly variation in potential evapotranspiration rates resulting from the equatorial location. Despite marked monthly variations in precipitation, a large surplus exists for most of the year.

proportion of the total world runoff that reaches the oceans.

It should be appreciated that the geomorphic interpretation in the densely vegetated areas of the mT realm is made exceedingly difficult. There are many controversial topics (the role and significance of planation, for example) that have yet to be resolved. But, clearly, the role of climate under such extreme conditions must be given considerable weight in the geomorphic assessments of such regions.

The mP realm. Equated to the moist regions of middle-latitude areas, the geomorphic characteristics of this realm have been widely investigated. Despite the intensity of research, many landforms are difficult to explain in terms of a dominant geomorphic process and many diverse opinions are expressed. Part of the problem is

Table 3–5

DISCHARGE OF MAJOR WORLD RIVERS (NOTE CONTRIBUTIONS OF TROPICAL REALM)[a]

	Discharge (Thousands of cfs)
Amazon	3500[b]
Congo	1400
Yangtse	770
Bramaputra	700
Ganges	660
Yenisei	614
Mississippi	611
Orinoco	600
Lena	547
Parana	526

[a]Note Holeman (Water Resources Research: 4,737-747,1968) gives somewhat different values.
[b]Variously estimated at between 3 and 4 million cfs.

the fact that such regions have experienced considerable changes of climate in recent geologic history and "equilibrium" states have not been established. This point was stressed earlier as a criticism of the Davisian method, which assumed a normal cycle of erosion.

At the present time the dominant erosional agency consists of fluvial systems that carve landscapes into intricate patterns. The nature of valley formation, with the modification of valley sides and slopes by physical weathering, is the topic for much research at the present. Since the realm lies in the critical zone that experienced both glacial and nonglacial conditions during the Pleistocene, much of the interpretation of present-day topography must relate to the past events. Thus, drainage systems (Figure 3–28), valley forms (Figure 3–29), and coastal formations require both insight into currently active and past climatic conditions.

In this realm, as in the others, important variations in climate occur because of the existing topographic conditions. Local modifications of climate in this way are covered in the study of topoclimatology. In outlining the role of topoclimates, Geiger (1965) notes:

"If the surface of the ground is sloping, the heat balance will be modified owing to the different angle of incidence of solar radiation. In moun-

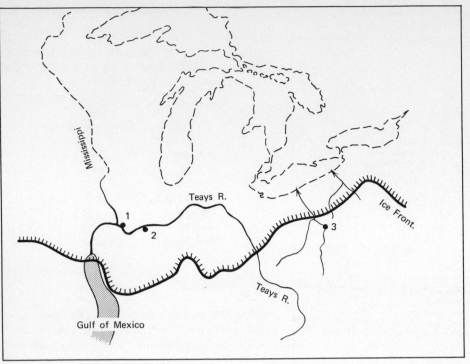

(a)

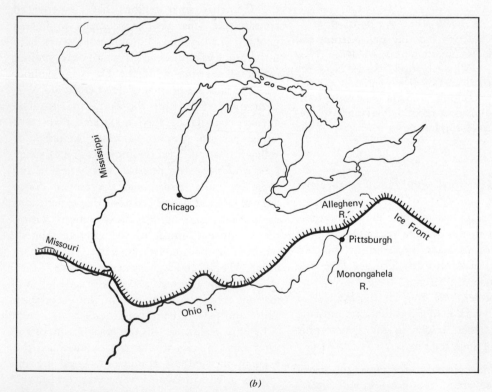

(b)

Figure 3–28. Evolution of the Ohio River drainage system. (*a*) Drainage system prior to glaciation shows a major river (River Teays) rising in what is now North Carolina and flowing to a much more extensive Gulf of Mexico. The Mississippi was a tributary of the Teays. A divide separated

(a)

(b)

(c)

(d)

Figure 3–29. The development of the Yosemite Valley illustrates the effects of both fluvial and glacial erosion. (*a*) Broad valley stage prior to Sierra Nevada uplift. (*b*) Subsequent to uplift river deepens it valley some 700 ft. (*c*) A second uplift causes formation of deep valley of Merced River some 1300 ft lower than original level. (*d*) The valley subsequent to glacial retreat when it contained a lake dammed by a glacial moraine. *Legend:* B - Mt. Broderick; C - Clouds Rest; CR - Cathedral Rocks; EP - Eagle Creek; LC - Liberty Cap; MR - Merced River; ND - North Dome; SD - Sentinal Dome; W - Washington Column; BV - Bridal Veil Creek; CC - Cascade Cliffs; EC - El Capitan; HD - Half Dome; LY - Little Yosemite Valley; MW - Mt. Watkins; R - Royal Arches; TC - Tenaya Creek; YC - Yosemite Creek. (From Foster, R. J., *General Geology*, 1969, Merrill. Used with permission.)

the drainage basin of the River Teays from that of northward flowing rivers. Locations 1, 2, and 3 shows present towns of Lincoln, Champaign and Pittsburgh, respectively. (*b*) The present-day drainage system shows how the Monongahela and Allegheny Rivers now flow into the Ohio, a tributary of the Mississippi. (After Janssen, 1952.)

tainous regions, direct solar radiaton may be cut off due to the effect of shadows, in which case only diffuse sky radiation is effective. The flow of water down sloping surfaces and also in the ground near the surface, modifies the water balance of the ground. The influence of hills, mountains, and valleys on winds leads to a modification of the amount of precipitation, since raindrops, and even more so, snowflakes, are carried along by the wind. These modifications in the heat and water balance of the ground surface result in a topoclimate which, under otherwise similar conditions, differs widely from the climate on a horizontal surface."

Geiger goes on to note the influence of other factors. Local winds are generated and have important effects on the circulation pattern, this in turn modifies the heat characteristics, and obviously this influence is impressed upon the vegetation cover that exists. It is noted, too, that topoclimates can occur under specialized geologic conditions, with the topoclimate of caves providing an apt example.

The study of topoclimates becomes extremely important in the investigation of micro- and mescoclimatic differences in the realms described above. Beyond that, its importance relates to vegetation cover and to meet this, in relation to man-made vegetation regions, the highly specialized field of agrotopoclimatology has developed.

CHAPTER 4

Climate and Soils

Considered for many years as a simple mixture of weathered rock and organic debris, soil is now known to be a dynamic body in which complex chemical, physical, and biologic processes take place. Characterized by specific properties (Table 4-1, Figure 4-1), the nature of soil is the result of the interaction of many environmental and biologic processes.

Much of the groundwork in distinguishing the processes responsible for soil formation must be credited to the Russian scientist Dokuchaiev who, in 1897, published the first meaningful classification of soils. At the base of his system was the identification of a number of soil-forming factors such as recognition of the roles played by local climate, parent material, plant and animal life, relief, elevation, and the lay of the land. More recently, Jenny (1940) set down such factors in the form of a soil equation:

$$\text{Soil} = f(cl, o, r, p, t),$$

which expresses the formation of soil as a function of climate (cl), all organisms (o), relief form (r), parent materials (p), and time (t). Three of the factors included in the equation (r, p, and t) might be considered as passive, and two (cl, o) as active soil-forming agents. Climate, as one of the active factors is of prime importance not only for its direct influence but also because of its role in determining the nature of the second active factor—plant and animal life.

While conceptually most adequate, this soil equation has caused problems in interpretation. With defined terms, it would seem that quantitative evaluation of the variables expressed would allow the relative impact of each to be determined. This is not the case because, as Crocker (1952) has pointed out, "The fact that it is only possible to apply the functional, factorial approach with absolute certainty to monogenetic soils means that functions and reliable sequences can only be developed for a very limited part of the soils of the world." It has been noted too

that analysis of soils using the soil equation may be of limited value in other respects. Bunting (1965), for example, notes that a result of ". . . using the factorial approach is that, though we may gain great insight into the factors, we learn very little about the soil itself. . . ."

Clearly, separation of any environmental complex into compartmentalized categories leaves much to be desired because reciprocal effects between variables are often overlooked. However, bearing such disadvantages in mind, use of the soil equation does allow a qualitative assessment of the input factors responsible for soil formation.

CLIMATE, WEATHERING AND SOIL FORMATION

The main principles of climate-weathering relationships have been discussed in Chapter 3. At this point, emphasis is given to the fate of the weathered products and the rate at which weathering occurs in the soil-forming process.

The Soil Profile

Russell (1968) has pointed out that chemical analysis of spring and river water provides much information into the nature of the materials that are removed from soils. Analysis shows that such water often contains high quantities of simple cations (sodium, potassium, magnesium, calcium) and anions (sulfates, nitrates, bicarbonates) together with silicic acids. On the other hand, it contains low concentrations of aluminum and iron compounds (Table 4-2). This points to the fact that selective removal of soil constituents occurs and that, where adequate moisture and good drainage conditions exist, a prime process in the modification of surficial formations is the removal of bases and silica, leaving a residue of iron and aluminum together with cations of insoluble hydroxides. This effect is well demonstrated in the formation of soil profiles.

Table 4-1[a]

a. Classification of Soil Structural Units[b]

	Type (Shape and Arrangement of Peds)						
		Prismlike; Horizontal Dimension Much Less Than Vertical. Vertical Faces		Blocklike; Polyhedronlike, or Spheroidal; 3 Dimensions Fairly Even			
				Blocklike; Blocks or Polyhedrons Having Surfaces That are Casts of the Molds Formed by Ped Faces		Spheroids or Polyhedrons Having Plane or Curved Surfaces not Related to Faces of Nearby Peds	
	Platelike; Vertical Dimension Much Less Than Length or Width	Without Rounded Caps	With Rounded Caps	Flat Faces; Most Vertices Angular	Mixed Rounded and Flat Faces; Rounded Vertices	Nonporous Peds	Porous Peds
	Platy, mm	Prismatic, mm	Columnar, mm	Blocky or angular, mm	Subangular blocky, mm	Granular, mm	Crumb, mm
Very fine or very thin	1	−10	−10	−5	−5	−1	−1
Fine or thin	1− 2	10− 20	10− 20	5−10	5−10	1− 2	1−2
Medium	2− 5	20− 50	20− 50	10−20	10−20	2− 5	2−5
Coarse or thick	5−10	50−100	20−100	20−50	20−50	5−10	
Very coarse or very thick	+10	+100	+100	+50	+50	+10	

b. Origin of Soil Colors

Color	Constituent
Black	Carbonate ions, usually Ca^{++} or Mg^{++}, plus highly decomposed organic material; other cations ($Na^{+} \cdot K^{-}$), plus highly decomposed organic material; sulfur compounds;[c] manganese oxide[c]
Red	Ferric iron oxides: *hematite* (Fe_2O_3); *turgite* $2(Fe_2O_3) \cdot H_2O$; *goethite* ($Fe_2O_3) \cdot H_2O$
Yellow (ocherous or light yellowish-brown)	Hydrous ferric oxides: *limonite* $2(Fe_2O_3) \cdot 3(H_2O)$
Brown	Partially decomposed, acid, organic material; combinations of iron oxides, plus organic material
White to colorless	Aluminum oxides and silicates (*kaolinite*, *gibbsite*, *bauxite*); silica (SiO_2); alkaline earths ($CaCO_3$, $MgCO_3$); gypsum ($CaSO_4 \cdot 2H_2O$); highly soluble salts (chlorides, nitrates, borates of sodium and potassium); certain organic colloids[c]
Bluish	*Alloysite*, a hydrous aluminum oxide;[c] and *vivianite*, a hydrous ferrous phosphate[c]
Green	Ferrous (incompletely oxidized) iron oxides[c]

[a] From Van Riper (1970).
[b] Source: Soil Survey Staff, *Soil Survey Manual*, U.S. Dept. of Agriculture, Handbook No. 18, p. 228, U.S. Govt. Printing Office, Washington, D.C., August, 1951.
[c] Relatively minor soil constituents that require unusual conditions for their accumulation.

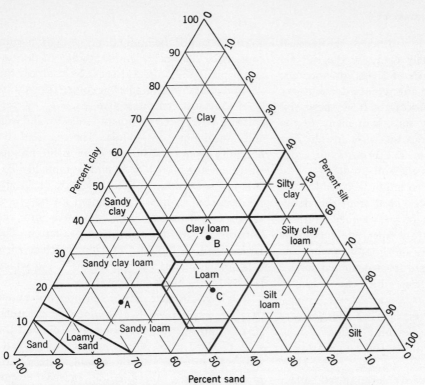

Figure 4–1. Soil texture diagram. Texture classes are bound by heavy lines. Soil sample A (65% sand, 20% silt, 15% clay) is classed as a sandy loam. Sample B (33% each of sand, silt and clay) is a clay loam and Sample C (40% sand, 42% silt, 18% clay) is a loam.

Table 4–2

AVERAGE COMPOSITION OF RIVER WATER[a]

	North America	South America	Europe	Asia	Africa	Average for the Five Continents
CO_3	33.40	32.48	39.98	36.61	32.75	35.15
SO_4	15.31	8.04	11.97	13.03	8.67	12.14
Cl	7.44	5.75	3.44	5.30	5.66	5.68
NO_3	1.15	0.62	0.90	0.98	0.58	0.90
Ca	19.36	18.92	23.19	21.23	19.00	20.39
Mg	4.87	2.59	2.35	3.42	2.68	3.41
Na	7.46	5.03	4.32	5.98	4.90	5.79
K	1.77	1.95	2.75	1.98	2.35	2.12
$(FeAl)_2O_3$	0.64	5.74	2.40	1.96	5.52	2.75
SiO_2	8.60	18.88	8.70	9.51	17.89	11.67

[a] From Thompson (1957). Expressed as percent of the ions in the water.

As parent material is weathered, the residual materials that ultimately comprise the soil are characterized by a series of layers, or horizons, each of which has been affected in a different way by the weathering processes. It is these soil horizons that provide soils with many of the distinctive characteristics that give insight into its mode of formation. The horizons are considered collectively as the soil profile.

Figure 4-2 shows what might be considered as the "ideal" soil profile—ideal from the standpoint that it consists of well-defined horizons formed under moderate temperature conditions with adequate moisture and good drainage. The profile grades downward from unaltered surface litter to the bedrock from which the weathered material was derived.

The characteristics of each of the horizons result largely from the leaching action of rainwater although, of course, the temperature at which reaction occurs assumes importance. Below the fresh and partly decomposed surface litter, the A horizon is seen to comprise a number of distinctive layers. The A_1 is often very dark in color, a result of the high humus content; humus consists of the finely divided residue of decomposed plant and animal tissues. By contrast, the A_2 horizon is much lighter in color. This results from eluviation, the selective leaching by percolating water of clays, organic colloids, and iron and aluminum compounds. The A_3 and B_1 horizons form a transitional zone in the profile, with A_3 being more like A_2 and B_1 more like B_2. It is in the B_2 horizon that the effects of illuviation are seen. Iron and aluminum leached from above are deposited and cemented into the horizon to give it a red-yellow coloration. The B_3 is transitional to the C horizon, which consists of a similar material to that of the B but which lacks B's structural characteristics.

It is obvious that such a soil horizon cannot develop where there is a shortage of moisture. In recognizing this, Marbut (1935) divided the soils of the United States into two major categories (Figure 4-3). In the humid, eastern parts of the country, precipitation exceeds evapotranspiration on an annual base. Profiles formed in such regions will be marked by leaching and

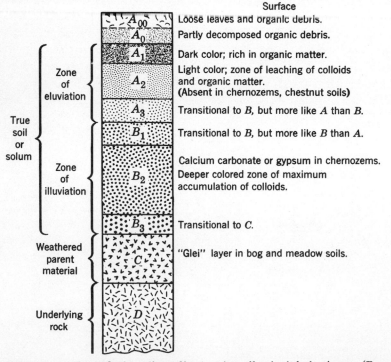

Figure 4-2.   An "ideal" soil profile showing all principle horizons. (From Strahler, A.N., *Physical Geography*, Copyright © 1969 by John Wiley and Sons. Reproduced with permission.)

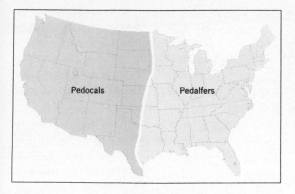

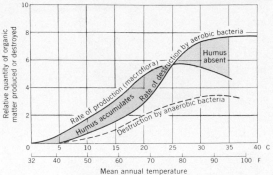

Figure 4-3. Division of the United States showing location of pedocal and pedalfer soils. (After Marbut.)

Figure 4-4. Relative rates of production and destruction of organic matter at given mean annual temperatures. (After Senstius, from Strahler, A. N., *Physical Geography*, Copyright © 1969 by John Wiley and Sons. Reproduced with permission.)

will tend to develop profiles in which iron and aluminum form a high percentage of residual materials. Marbut termed such soils pedalfers—the *al* and *fer*—referring to aluminum and iron, respectively.

Where, on an annual average, evaporation exceeds precipitation, leaching will not occur. The main response will be the upward motion of water through the soil profile. Water carries salts in solution and, with the evaporation or use of the water, these will be deposited in the upper levels of the soil. Soils resulting from this process were termed pedocals by Marbut, where *cal* refers to the predominance of calcium salts that are deposited.

It is evident from the pedalfer–pedocal classifications that variations in precipitation lead to different soils characterized by different profiles. In addition, since temperature plays such an important role in chemical weathering, and since it affects the rates of decay of organic matter (Figure 4-4), marked variations in soil profiles will occur under different climatic regimes. To assess the significance of this, it proves convenient to consider soils in terms of pedogenic regimes.

Pedogenic Regimes

Recognition of the concept of pedogenic regimes allows basic trends leading to major soil groups to be unified in terms of the climatic regimes under which they form. Prior to evaluation of identified regimes, however, it is necessary to clarify a few of the factors that are inherent in the discussion of pedogenesis.

Soils consist of both organic and inorganic constituents. The inorganic components consist of two types. First, there are silts and sands, not appreciably altered by chemical weathering, that have been derived from the parent material. These unaltered fragments consist of resistant minerals with quartz and mica often represented. Second, there are the clay fractions that have been derived through alteration of original materials (often from easily weathered minerals such as feldspar). These clay colloids are an important part of the soil because they influence the cation-exchange capacity of the soil. A cation is a positively charged atom, and both metallic and nonmetallic cations are identified. The metallic cations, K^+, Ca^{++}, and Mg^{++}, for example, are termed bases to differentiate them from the nonmetallic, such as H^+.

A high cation-exchange capacity means that the clay colloids are well supplied with cations that can be exchanged for different cations in the soil solution. It is from the soil solution that plants acquire nutrients and adsorbed cations (cations attracted to the surface of the colloid—not those that have been absorbed) act as a "bank" of nutrients that are partly available. Also, as weathering takes place, the soluble nutrients tend to be washed out of the soil by leaching and adsorption retards the rate at which this occurs.

The capacity of the colloids to exchange cations is a function of their chemical composition and their crystal lattice. Figure 4-5 shows

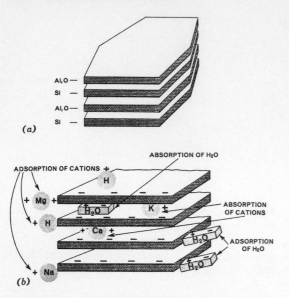

Figure 4–5. (*a*) The sandwichlike arrangement of individual clay-colloid particles. (*b*) Example illustrating adsorption and absorption within a clay colloid. (From Van Riper, J. E., *Man's Physical World*, 1962. Used with permission of McGraw-Hill Book Company.)

the layered "sandwichlike" lattices of clay minerals derived from the weathering of primary silicates. It is evident that the exchange capacity will not only depend on the amount of clay that occurs but will also depend upon the type of clay mineral that is dominant. The type that exists is a function of the amount of weathering that the soil has experienced.

Illite and vermiculite are associated with slightly weathered soils, with illite being more common in those that are acid. Montmorillonite is an important component of moderately weathered soils, especially those with a high pH value and containing large amounts of humus. Kaolinite is characteristic of the clay minerals associated with highly weathered, strongly leached soils.

Clay minerals make up the bulk of the clay fraction found in cool, temperate climates. In hot, humid regions, kaolinite becomes predominant. This, however, may be weathered even further into iron and aluminum hydroxides as follows:

$$Al_2Si_2O_5(OH)_4 + 5H_2O$$
$$\rightarrow Al_2O_3 \cdot 3H_2O + 2H_4SiO_4$$

These sesquioxides may, ultimately, make up the bulk of the clay fraction of soils in hot, humid regions.

When the cation-exchange capacity of the different clay minerals and sesquioxides is compared, it is evident that as the weathering process progresses, the cation-exchange capacity becomes lower (Table 4–3) and fewer nutrients become available for plants. It can also be assumed that the type of clay minerals in the soil is indicative of the amount of weathering that the soil has undergone. In recognizing this, Thompson (1957) identifies five phases of soil weathering based upon mineral content and secondary clay minerals that are present (Table 4–4). The least-weathered (youngest) soils have a high cation-exchange capacity, with illite and vermiculite as the main clay minerals. These occur together with easily weathered materials that are still in the soil. By the middle stage of his spectrum (Group 3), the easily weathered minerals are no longer found, and clays with a lower base saturation occur. At the advanced stage of weathering (old age), the soil might consist almost entirely of free alumina, iron oxides, and some kaolinite. Such a spectrum of weathering events is graphically shown in Figure 4–6. This illustrates that with rapid chemical weathering, the process leading to "old soils" occurs much more quickly under some climatic realms and not others.

The basic differences that occur can best be dealt with through consideration of the pedogenic regimes. The major regimes identified are laterization, podzolization, calcification, salinization, and gleization. Some authors also identify ferralization and cryogenesis as distinctive types. The processes active in each of the regimes are shown in Figure 4–7.

Laterization is a pedogenic regime associated with climates that experience high average annual temperatures and copious precipitation. With precipitation occurring all year, the resulting vegetation forms the highly specialized equatorial forest biome. Despite the great biomass of these forests, there is little humus in the soil. This is because organic matter undergoes rapid bacterial decay at the high prevailing temperature, and is rapidly removed by the large amounts of moisture available. The lack of humus means

Table 4–3 [a]

a. Percentage Base Saturation of Different Soils at Like pH Values

Great Soil Group	Soil Type	Location	Total Exchange Capacity (meq./100 g)	Percentage Base Saturation at pH Values of:				
				4.8	5.0	5.5	6.0	6.5
Yellow podzolic	Norfolk sandy loam	Alabama	1.83	9	16	32	44	60
Red podzolic	Cecil clay loam	Alabama	4.85	6	23	41	58	74
Gray-brown podzolic	Miami silt loam	Wisconsin	9.79	43	50	63	72	82
Planosol-grass	Grundy silt loam	Illinois	26.33	57	60	69	80	91

b. Cation-Exchange Capacity

Meq.	Per 1% of
2	Humus
1.5	Vermiculite
1.0	Montmorillonite
0.3	Illite
0.1	Kaolinite

[a] From Thompson (1957).

that the soil is low in humic acid and soils are often mildly basic of alkali, although variations from pH 5 to pH 7 are found.

The high amount of precipitation results in rapid leaching while, with the high temperatures, it also promotes rapid chemical weathering. This usually leads to the process of desilication—the removal of combined silica from the soil—which means that the entire profile may ultimately consist of iron and aluminum sesquioxides.

The results of laterization means that soils lack well-defined horizons, and are typically red or

Table 4–4

A GROUPING OF SOILS IN TERMS OF WEATHERING [a]

I	Young	High content of easily weathered calcium and magnesium minerals. High percentage of base saturation of clays of 2:1 crystal lattice
II		Low content of easily weathered calcium and magnesium minerals. High content of slowly decomposing sodium and potassium minerals. Medium percent of base saturation. Mixed clays of 2:1 and 1:1 crystal lattice with dominance of 2:1 type
III		Some slowly weathering sodium and potassium minerals. Very little unweathered calcium and magnesium minerals. Predominance of clays of 1:1 lattice, but some 2:1 remain. Low percent base saturation
IV		High content of quartz sand and little or no silt-size fraction in the top soil. Clays of 1:1 crystal lattice with high percent of hydrogen saturation. Large amount of free alumina and iron oxide either hydrated or dehydrated. Little if any unweathered silicate minerals
V	Old	Some quartz sand and a high proportion of free alumina and iron oxide. Some clays of 1:1 crystal lattice. Forms hard crust on desiccation

[a] From Thompson (1957).

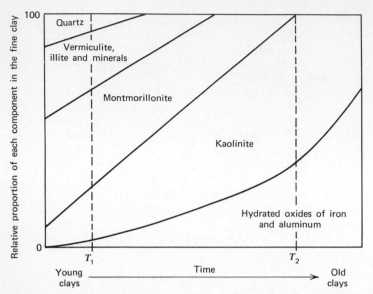

Figure 4–6. Relative proportions of the main clay components in the fine-clay fraction of soils of different ages. A fairly young soil (at T_1) has a low proportion of kaolinite and iron/aluminum oxides. These increase with time so that at T_2 they comprise the main clay components. Thereafter, kaolinite decreases and the hydrated oxides increase. (After Thompson, 1957.)

yellow colored as a result of the predominance of the iron and aluminum present. Cation-exchange capacity is low because the advanced stage of weathering precludes the presence of bases and, as such, the soil is considered infertile. Thus arises the strange situation in which one of the most luxuriant of the world biomes develops in a region of limited soil fertility. Figure 4–8 offers an explanation of how this apparent anomaly can be explained

Podzolization produces a soil that has well-defined horizons. In fact, the idealized soil profile discussed earlier is one that might occur through podzolization. The process occurs where sufficient moisture permits a forest cover to grow but where temperatures are low enough to inhibit rapid bacterial decay of organic matter. It is thus often associated with the coniferous forest biome.

The slow rate of decomposition causes percolating rainwater to be acid. This then dissolves out free and absorbed basic ions that are leached from the soil. The acidity of the soil solution also causes solution of the sesquioxides of iron and aluminum that are precipitation in the B horizon as a result of the reduced acidity. The A horizon is thus greatly leached and is largely composed of silica that is insoluble in the acid solution. The B horizon is a zone of accumulation of the removed materials and differs markedly from the A. At times, deposition of the aluminum and iron sesquioxides gives rise to a well-cemented, impermeable hardpan that obstructs soil drainage.

Podzolization produces well-marked horizons and a soil of limited fertility. Clay fractions that do exist exchange metallic for hydrogen ions, and the soil is highly acidic. Perhaps the most characteristic feature is in the ash-gray A2 horizon; the podzols derive their name from Russian terms for this color.

Calcification is frequently associated with the temperate grassland biome; it results from a seasonal climatic regime when, during one time of the year, evaporation exceeds precipitation while at the other time of the year the reverse holds true. Such marked seasonal variations, given the temperature conditions that exist, causes the role of water to vary appreciably in the soil-forming process. Under the forest biomes, with a surplus

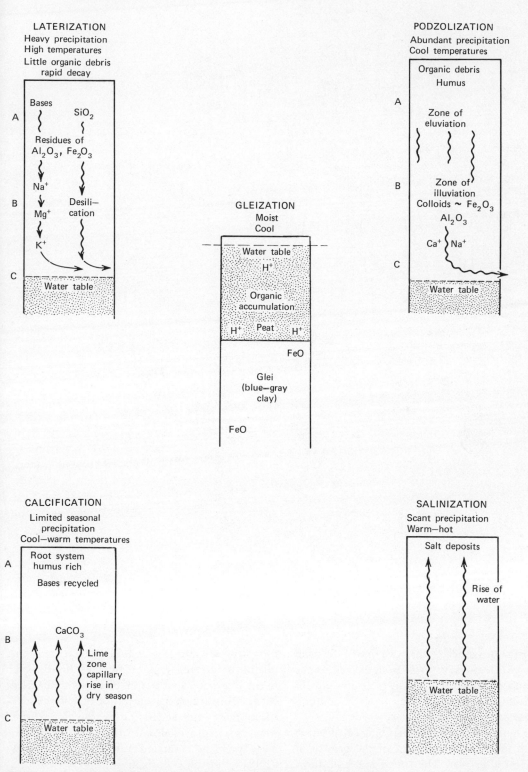

Figure 4–7. Highly schematic diagrams illustrating the processes associated with the major pedogenic regimes.

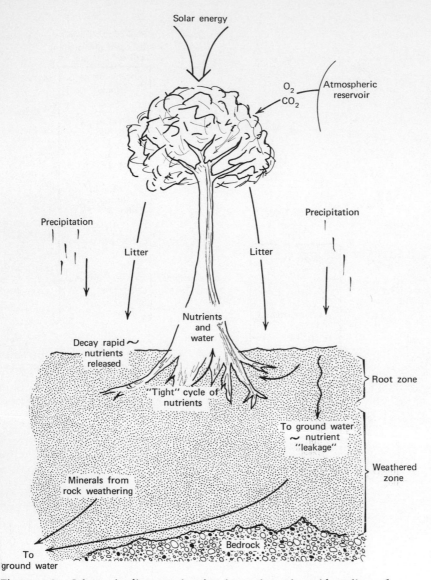

Figure 4-8. Schematic diagram showing how, through rapid cycling of nutrients, tropical forests occur on soils of limited fertility. (After Carter, 1968.)

of moisture for much of the year, one of the main processes is removal of bases from the soil. In the grassland biome, precipitation is insufficient for this leaching process and any bases in solution are soon precipitated out.

The A horizon of soils undergoing calcification are rich in humus as a result of the growth of grass year after year. The soil is not acid, however, for bases are recycled through the extensive root structure and clay colloids, which are often calcium saturated, are not removed from the soil. In the B horizon a saturation of calcium salts may occur. This accumulation might be enriched by addition of salts added during the dry season when water moves upward through the profile. Such water is saturated with calcium that is deposited in the B horizon.

The neutral pH, the ample supply of nutrients

and the presence of colloids of high-base saturation have caused soils formed under this process to be among the most fertile used by man.

Salinization may be considered as one step beyond the calcification process in that a marked water deficit occurs throughout the year. Vegetation cover will be much sparser than the grass of the grassland biome and the A horizon associated with calcification will not occur. In fact, under advanced conditions, the major process may become the upward movement of water through the soil and the deposit of salts at or close to the surface. This assumes that subsurface water is available and obviously this is not true of all semiarid locations. Unfortunately, in some arid regions irrigation water provides the mean by which salinization can occur where it would not normally and, as is noted later, this poses great problems in terms of use of water in the soils of arid regions.

Gleization occurs where low temperatures combine with poor drainage conditions. The low temperatures permit the accumulation of much organic debris to provide a high hydrogen ion concentration and resulting acidic soils. The poor drainage conditions cause the lower part of the soil profile to become waterlogged and it often consists of a structureless clay—the glei horizon. This is usually blue-gray in color because of the presence of reduced iron oxide, caused by the poorly aerated conditions of the lower horizons.

Gleization is a marked feature of the tundra realm, and a circumpolar belt of soils with a glei horizon exists. Because of the importance of freezing and thawing in conditions in which it forms, the process of cryogenesis is often used to describe such regimes.

While found over wide areas of the tundra biome, a glei horizon can occur where local conditions are such that drainage through the soil is limited, and glei is often found in association with soils of the temperate realm.

THE SOIL CLIMATE

While the atmospheric climate is characterized by physical processes active within a gaseous environment, soil climate occurs in a milieu that is a biologic-mineral system governed, by the most part, by its own laws of development. As Shul'gin states, "Soil climatology . . . has much in common with atmospheric climate. It is similarly characterized by temperature, humidity, daily and annual fluctuations of both indices, their spatial distribution, their connection with the environment and the determining phenomena of nature. Nevertheless, soil climate has its own specific traits as compared with the climate of the atmosphere and that of the air layer near the ground. . . . One of the distinguishing features of soil climate consists of the environment where it forms." The importance of this can be readily demonstrated by considering factors influencing the thermal and hygric environments within the soil.

To study soil temperature, it is necessary to turn to laws governing heat transfer in solids. If, in a uniform material, a temperature gradient exists a heat flux will occur. The rate of heat flow will depend not only upon the size of the gradient but also upon the ability of the material to conduct heat. This is measured through the thermal conductivity of the body (k). Marked differences are found in k values for different materials (Table 4–5); in the case of soils, evaluation of k is complicated by the fact that soil is not a homogenerous body but consists of particles of different size and composition. Thus, to evaluate the thermal conductivity of soils, it is necessary to take into account:

a. The thermal conductivity of the individual particles comprising the soil.

b. The manner in which the particles are sorted and the porosity (volume of air space to total volume of the soil sample) of the soil.

c. The amount of soil moisture contained in the soil, water having a higher conductivity than air.

Generally, it is found that soils with a high silica content have high thermal conductivity, a fact of some significance in the silica-rich upper horizons of some podzols, while those rich in organic matter have a low k value. Because of the different thermal properties of air and moisture, wet soils have a higher conductivity than dry.

An important characteristic of soils is thermal diffusivity for, as Lowry (1969) points out, "It is a measure of the time required for a thermal im-

Table 4-5

THERMAL PROPERTIES OF REPRESENTATIVE MATERIALS COMMONLY FOUND IN THE PHYSICAL ENVIRONMENT[a]

Natural Material	Thermal Conductivity, k_m (cal deg^{-1} cm^{-1} sec^{-1})	Density ρ_m (gm cm^{-3})	Specific Heat c_m (cal gm^{-1} deg^{-1})	Thermal Capacity $(\rho c)_m$ (cal deg^{-1} cm^{-3})
Granite	0.011	2.6	0.2	0.52
Ice	0.0055	0.9	0.51	0.45
Wet sand	0.004	1.6	0.3	0.48
Wet marsh soil	0.002	0.9	0.8	0.7
Still water	0.0015	1.0	1.0	1.0
Old snow	0.0007	0.5	0.51	0.22
Dry sand	0.0004	1.4	0.2	0.3
Wood (typical)	0.00035	0.6	0.3	0.18
New snow	0.0002	0.1	0.5	0.05
Peat soil	0.00015	0.3	0.44	0.1
Still air	0.00005	0.001	0.24	0.00024

[a] From Lowry (1969).

pulse to travel over a given distance. In terms of soil temperature . . . it is a measure of the time for a certain kind of temperature change—for instance the rapid heating of the surface at sunrise—to be felt at a certain depth in the soil." Thermal diffusivity is derived by dividing k by the thermal capacity of the soil sample.

To obtain the thermal capacity, it is first necessary to derive the specific heat of the soil. Specific heat is defined as the amount of heat energy required to raise the temperature of 1 gram of substance 1°C. In relation to soils, derivation of this value is complicated for soil is a non-homogeneous body. Thus a weighted mean is derived. Consider a soil that consists of just two constituents. Each constituent has its own density (ρ_1 and ρ_2), its own specific heat (C_1 and C_2), while each comprises a given volume of the soil sample (V_1 and V_2). To obtain the weighted mean the equation must read

$$C_m = \frac{(V_1\rho_1C_1) + (V_2\rho_2C_2)}{V_1\rho_1 + V_2\rho_2}$$

Lowry shows that this expression can be simplified to read

$$\rho_mC_m = (\rho C)_m = (V_1\rho_1C_1) + (V_2\rho_2C_2)$$

The resulting factor $(\rho C)_m$ gives the thermal capacity of the soil sample. It is this, together with the thermal conductivity that provides

thermal diffusivity, $k/\rho C$. The inverse relationship of the two variables means that soils with a high k and low heat capacity react most rapidly to temperature changes in the atmosphere. For a fuller treatment of these factors, the reader is referred to the excellent account given by Lowry. Not only does he analyze the factors already outlined, he also provides a number of examples of application of the soil heat-budget equation.

The preceding thermal characteristics are used in many ways to determine the temperature characteristics of soils. For example, the formula derived by the United States Corps of Engineers draws upon thermal conductivity and heat capacity in calculation of the depth to which soil freezes.

$$X = \left[\frac{24\,KF}{L + C(V_0 - 32 + F/2T)} \right]^{1/2}$$

where X = depth of frost penetration in ft
K = thermal conductivity in BTU. ft^{-1} °F^{-1} hr^{-1}
F = freezing index (days $\times$ frost °F)
L = average latent heat in BTU. ft^{-3}
V_0 = mean annual air temperature in °F
T = duration of freezing period in days
C = average volumetric heat capacity in BTU. ft^{-3} °F^{-1}

Flow of heat through the soil, its direction and value, depends upon the initial assumption that a temperature gradient exists. Thus, the temperature at the soil-air boundary layer is of prime importance in determining soil temperature characteristics. Temperature ultimately depends upon input of solar radiation so that it follows that many soil characteristics are derived from evaluation of the energy budget of the area in which the soil occurs. Indeed, Bunting (1965), quoting Yaarlov, writes ". . . for many soil processes the supply of incoming solar radiation sets the upper limit to rates of change in the soil."

Such an effect is well demonstrated when considering the way in which soil temperatures vary under different surface conditions. Table 4-6 shows how snow cover modifies soil temperature: notice, for example, how the greater depth of snow results in higher soil temperatures despite similar prevailing air temperature. A variation is also found when different cropping patterns are used (Table 4-7a), while aspect, as it relates to slope exposure, also results in soil temperature variations (Table 4-7b).

Moisture in the soil also plays a very important role in determining the nature of the soil climate. As already outlined, moisture influences

Table 4-6

INFLUENCE OF SNOW CARPET UPON AVERAGE SOIL TEMPERATURE (°C) IN WINTER (NOV–MARCH) AT BARNAUL[a]

Winter	Average Air Temperature (°C)	Average Thickness of Snow (cm)	Average Soil Temperature at 40 cm	80 cm
1929–1930	−13.8	20	−1.9	0.8
1926–1927	−13.9	40	−0.1	1.7
1927–1928	−13.9	61	1.0	2.1
1931–1932	−12.0	24	−0.8	1.4
1921–1922	−12.7	43	0.4	—

[a] From Shul'gin (1965).

both the movement of nutrients and the entire process of soil formation. In respect to the soil climate, its role has already been mentioned because the amount of water in the soil modified its thermal characteristics.

The study of the hygric environment of the soil relates to the amount of water in the soil, the form in which it occurs, and the phase under

Table 4-7

SOIL TEMPERATURES IN RELATION TO a. DIFFERENT FARMING PRACTICES AND b. DIFFERENT SLOPE EXPOSURE

a. Soil temperatues (°C) at depths 3 and 10 cm under winter wheat with different farming practices

Date/ June	Hours of Sunshine	3 cm			10 cm		
		Continuous Sowing	Wide-Row Spacing	Fallow	Continuous Sowing	Wide-Row Spacing	Fallow
13	1.8	17.5	21.0	22.3	16.5	17.7	18.0
14	6.5	23.8	29.4	32.5	20.0	24.0	25.0
15	9.6	24.0	29.0	32.5	19.5	22.5	25.0
16	13.2	25.0	29.5	33.5	20.0	23.7	26.5

b. Mean soil temperature (°C) in the Inn Valley related to slope exposure

Exposure	Winter	Summer	Year
North	4.2	15.3	9.5
East	4.0	18.6	11.3
South	5.3	19.3	12.6
West	5.5	18.5	12.2

[a] Data from Shul'gin (1965).

which it operates. Like moisture in the air, soil moisture can occur in gaseous, liquid, and solid forms, and each modifies the properties of the soil appreciably.

The gaseous phase, represented by water vapor, will always be present in soils that have capillary water. The humidity of air in soil pores is nearly always close to 100% while the carbon dioxide content, because of the addition from decomposition of organic matter, is much higher than that in the atmosphere. Water in its solid phase creates special problems in the soil climate. As explained in Chapter 3, the frozen soils of northern climates are a special study in themselves.

In its liquid phase, soil water can occur in a number of forms. After effective precipitation occurs, water will move through the soil in response to gravity. This gravitational water will pass to the ground water reservoir and, ultimately, to stream flow. As it passes through the soil, it leaves moisture deposited in small pores, suspended in larger ones, or as water films surrounding soil particles. This moisture represents capillary water—the water that supplies the needs of plants.

Through evaporation and plant use, capillary water, unless replenished, diminishes. As the depletion occurs, the force that binds the water to soil particles increases. Ultimately, a point is reached where the moisture is no longer available to plants. It is now chemically and biologically inactive and is described as hygroscopic water.

Hygroscopic water can be driven off by heating the soil in which it occurs to 105°F (40.6°C). After its removal, the only moisture that remains is that in the hydrated oxides of soil minerals. This combined water can also be driven off by heating to much higher temperatures, but for practical purposes it is a form of soil moisture that can be overlooked.

The relationship between the various kinds of soil moisture is shown in Figure 4-9. As indicated in the diagram, other terms used to describe various soil moisture conditions are used and are appropriately defined. It must be noted that the spatial distribution of soil water on both macro- and microscales is highly variable. Macrodifferences reflect such major factors as precipitation regime and gross soil characteristics. On the microlevel, and as illustrated in Figure 4-10,

plant roots and plant activity are highly significant in the distribution.

CLIMATE AND SOIL CLASSIFICATION

The classification of soils, like that of climate, is problematic. As Flawn (1970) notes, "Where items being classified fall into natural groups on the basis of well-defined properties, the job of classification is infinitely more simple than in cases where the classification must deal with a continuum in which there are gradations rather than clear cut boundaries. Soil is, of course, a continuum of materials of varied composition and size; it thus falls into the latter category and is very difficult to classify in a meaningful way because divisions must be drawn arbitrarily."

Historical perspective often plays an important part in the nature of classification of natural phenomena, and this is most evident in soil classification. As noted, Dokuchaiev placed great emphasis on the role of climate in soil formation and, indeed, classifications that resulted from his pioneering work placed great emphasis upon the role of variables (notably climate) in soil formation. Dokuchaiev's classification, as modified by Sibircelf, is shown in Table 4-8a. It shows that the system is based upon the assumption that "normal" soils are the result of a long, sustained set of conditions and that where such conditions have not occurred the soils are either transitional or "abnormal." The correlation between soils and climate is well established in the system.

Following Dokuchaiev, Glinka (1927) devised a classification (Table 4-8b) that is again dependent upon climate—in this case, moisture conditions becoming most important. The first order shows that the ektodynamomorphic soils are those that are well established and will only alter if external conditions—climate—are modified. The endodynamomorphic soils in which internal changes take place are considered transitional. Glinka's scheme formed the base of that introduced into the United States by Marbut, who modified them accordingly. As already noted, his classification (Table 4-9) identifies the pedalfers and pedocals, each of these being divided into six categories. This classification was further modified, particularly by Kellogg, to become the base for the Great Soil Groups

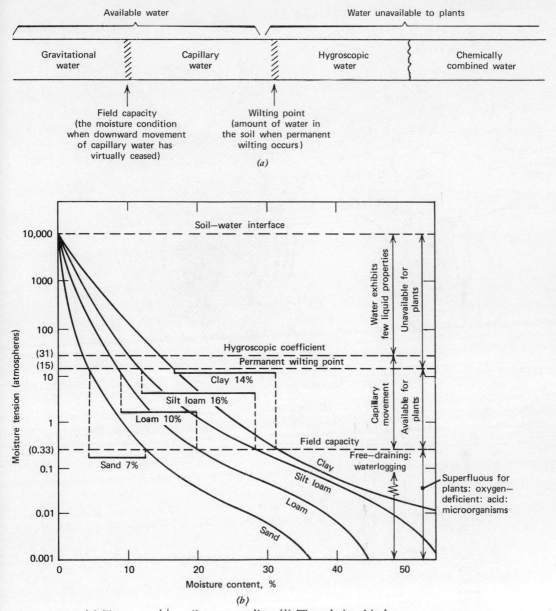

Figure 4-9. (*a*) Terms used in soil water studies. (*b*) The relationship between moisture content and moisture tension in four different types of soil. Note the range of moisture available for plants. (From More, R. J., Water and Crops in *Water, Earth and Man*, R. J. Chorley, ed., Methuen, 1969. Used with permission.)

classification published in the 1938 Yearbook of Agriculture. It is this 1938 system that has formed the base for most soil studies in recent times.

The system is based upon soil genesis, and climate obviously provides the basic genetic input. The first orders recognized are the zonal,

azonal, and intrazonal soils. The zonal soils are those formed in well-drained areas where the interaction between climate, vegetation, and the soil-forming process has been sustained. The intrazonal soils are those formed in poorly drained conditions, or where a special rock type—such as limestone—gives rise to exceptional

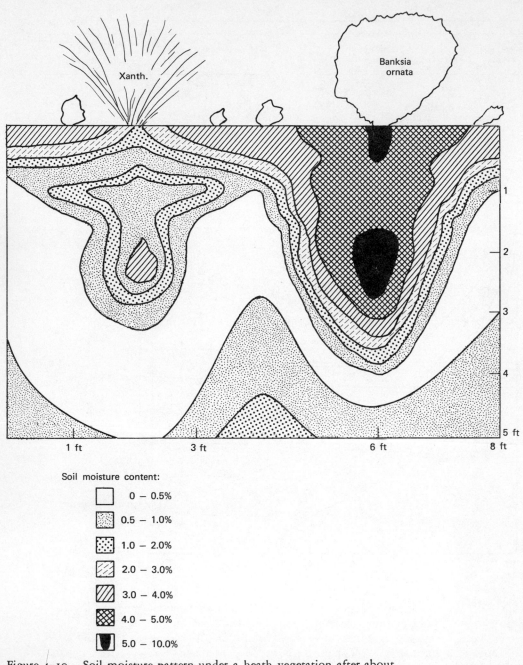

Soil moisture content:

- [ ] 0 – 0.5%
- 0.5 – 1.0%
- 1.0 – 2.0%
- 2.0 – 3.0%
- 3.0 – 4.0%
- 4.0 – 5.0%
- 5.0 – 10.0%

Figure 4–10. Soil moisture pattern under a heath vegetation after about 1 in. of rain had fallen on a profile near its wilting point. Values of soil moisture given in percent, where wilting point is less than 1% and field capacity above 6%. (After Specht, 1958.)

Table 4-8a
DOKUCHAIEV'S CLASSIFICATION

Class	Zone	Soil Type
A. Normal	I. Boreal	Tundra (dark brown) soils
	II. Taiga	Light grey podzolized soils
	III. Forest—steppe	Gray and dark gray soils
	IV. Steppe	Chernozem
	V. Desert—steppe	Chestnut and brown soils
	VI. Aerial or desert	Aerial soils, white soils, yellow soils
	VII. Subtropical and zone of tropical forests	Laterite or red soil

	Zone or Soil	
B. Transitional soils	VIII. Dry land moor soils or moor meadow soils	
	IX. Carbonate containing soils (rendzina)	
	X. Secondary alkali soils	
C. Abnormal soils	XI. Moor soils	
	XII. Alluvial soils	
	XIII. Aeolian soils	

Table 4-8b
GLINKA'S CLASSIFICATION

I. Ektodynamomorphic soils
1. Soils developed under optimum moisture conditions
   (a) Laterite
   (b) Terra-rossa
   (c) Yellow soils
2. Soils developed under average moisture conditions
   (a) Podzol soils
   (b) Gray forest soils
   (c) Degraded Chernozem
3. Soils developed under moderate moisture conditions
   (a) Chernozem
4. Soils developed under insufficient moisture conditions

Group A	Group B
(a) Chestnut soils	(a) Brown crusts
(b) Brown soils	(b) Lime crusts
(c) Gray soils	(c) Gypsum crusts
(d) Red soils	

5. Soils developed under excessive moisture conditions

Group A	Group B
(a) Moor soils (peat and muck soils)	(a) Soils of the mountain meadows
	(b) Peat soils of the dry tundras and mountain peaks

6. Soils developed under temporarily excessive moisture conditions
   (a) Solonetz soils
   (b) Solontshak soils
   (c) Transition forms of (a)
   (d) Transition forms of (b)

II. Endodynamormorphic soils
   (a) Rendzina
   (b) Various skelatal soils

Table 4–9

MARBUT'S SOIL CATEGORIES

	Pedalfers (VI–1)	Pedocals (VI–2)
Category VI	Pedalfers (VI–1)	Pedocals (VI–2)
Category V	Soils from mechanically comminuted materials Soils from siallitic decomposition products Soils from allitic decomposition products	Soils from mechanically comminuted materials
Category IV	Tundra Podzols Gray-brown podzolic soils Red soils Yellow soils Prairie soils Lateritic soils Laterite soils	Chernozems Dark-brown soils Brown soils Gray soils Pedocalic soils of Arctic and tropical regions
Category III	Groups of mature but related soil series Swamp soils Glei soils Rendzinas Alluvial soils Immature soils on slopes Salty soils Alkali soils Peat soils	Groups of mature but related soil series Swamp soils Glei soils Rendzinas Alluvial soils Immature soils on slopes Salty soils Alkali soils Peat soils
Category II	Soil series	Soil series
Category I	Soil units, or types	Soil units, or types

conditions. Azonal soils lack a well-defined profile through immaturity, a lack of time for complete profile development.

Suborders are related to conditions under which the soil forms and essentially indicate the climate, vegetation, or drainage conditions that exist. Thereafter, the Great Soil Groups are identified (Table 4–10). The description of these groups depends upon climatic conditions and, as indicated in Figure 4–11, the relationship between climate and the groups can be readily shown.

The basic concepts of the 1938 classification have been severely criticized. For instance, it is assumed that the zonal soils are mature systems in equilibrium with their environment. By this, it must be assumed that intrazonal and azonal soils are immature and do not reflect the conditions under which they are found. In some cases this is not true because these soils must also be considered as a reflection of the environment in which they occur. The azonal soils comprise a "hazy" category, and it is suggested that the group is often used as a dumping ground for soils that are difficult to classify. It must be noted,

too, that to explain the mature soil characteristics using the system, one must assume that they have experienced long-term, sustained climatic conditions. In view of climatic change over the past million years, this is difficult to assume.

It is the soil scientists, however, that have been most critical of the system. Their primary concern is that the 1938 system is an applied climatic system rather than a classification of soil itself. To meet this shortcoming, a new soil classification system has been devised. Developed by American pedologists and presented at the Seventh International Congress of Soil Science in 1960, the scheme has been carried through a number of stages; the system finally presented is known as the Seventh Approximate (hereafter referred to as the 7th Approx.). While some modifications are likely to occur as the system is field tested, it is believed that the 7th Approx. will remain the system to be substantially adopted.

According to Simonson (1962) the 7th Approx. ". . . reflects evolution in the concept of the soil itself. Basic to the scheme is the concept that soil

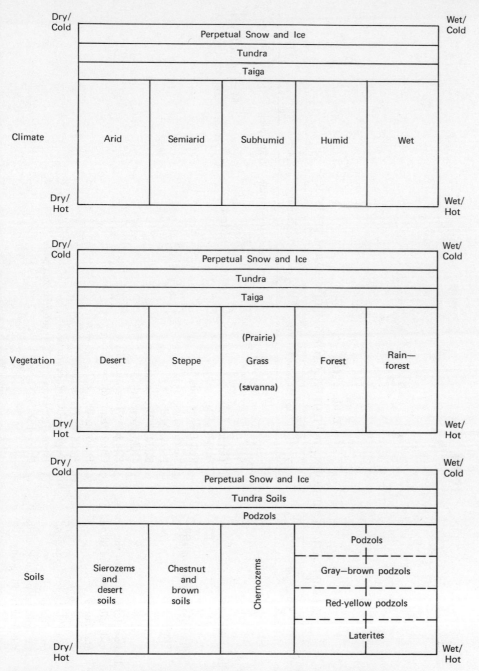

Figure 4-11. Conceptual relationships between distribution of climate, natural vegetation, and soils. (After Blumenstock and Thornthwaite, 1941.)

comprises a continuum on the land surface, one that can be subdivided into classes in a variety of ways. Also basic to the scheme is an effort to achieve more quantitative definitions than have been devised heretofore. Definitions of classes at every categoric level are expressed in terms that can be observed or measured." Here, then, is a

classification that breaks away from the genetic tradition and that, as Smith (1963) notes, fulfills the objective of an attempt to classify soils rather than soil-forming processes.

Like other classifications, the 7th Approx. is a multiple category system (Table 4-11). The six categories used are orders, suborders, great

Table 4–10

CLASSIFICATION OF SOILS ON THE BASIS OF THEIR CHARACTERISTICS[a]

Order	Suborder	Great Soil Group	Family	Series	Type
			[example]	[example]	[example]
ZONAL SOILS					
Pedocals	Soils of the cold zone	1. Tundra soils	Mesa	Mesa	Mesa gravelly loam
	1. Light, colored soils of arid regions	2. Desert soils 3. Red Desert soils 4. Sierozem 5. Brown soils 6. Reddish Brown soils		Chipeta	Chipeta silty clay loam
	2. Dark-colored soils of the semi-arid, subhumid and humid grasslands	7. Chestnut soils 8. Reddish Chestnut soils 9. Chernozem soils 10. Prairie soils 11. Reddish Prairie soils			
Pedalfers	3. Soils of the forest-grassland transition	12. Degraded Chernozem soils 13. Noncalcic Brown or Shantung Brown soils			
	4. Light-colored podzolic soils of the timbered regions	14. Podzol soils			

ZONAL SOILS

Pedalfers

4. Light-colored podzolic soils of the timbered regions (con't.)
   - 15. Brown Podzolic soils
   - 16. Gray-Brown Podzolic soils
   - 17. Yellow Podzolic soils

5. Lateritic soils of forested warm-temperate and tropical regions
   - 18. Red Podzolic soils (and Terra Rossa)
   - 19. Yellowish-Brown Lateritic soils
   - 20. Reddish-Brown Lateritic soils
   - 21. Laterite soils........ Nipe...... { Nipe...... Nipe clay / Rosario.... Rosario clay

INTRAZONAL SOILS

1. Halomorphic (saline and alkali soils of imperfectly drained arid regions and littoral deposits).
   - 1. Solonchak or saline soils
   - 2. Solonetz soils
   - 3. Soloth soils

2. Hydromorphic soils of marshes, swamps, seep areas and flats
   - 4. Meadow soils................ Clyde..... { Clyde..... Clyde silty clay loam / Webster.... Webster silty clay loam
   - 5. Alpine Meadow soils
   - 6. Bog soils
   - 7. Half-Bog soils
   - 8. Planosols
   - 9. Ground-water Podzol soils
   - 10. Ground-water Laterite soils

3. Calomorphic
   - 11. Brown Forest soils
   - 12. Rendzina soils

AZONAL SOILS................
   - 1. Lithosols............... { Houston.. Aguilita.. } { Houston... Houston clay / Soller..... Soller clay loam / Bell...... Bell clay
   - 2. Alluvial soils
   - 3. Sands (dry)

a U.S.D.A. (1938).

Table 4-11[a]

ALFISOLS . . . Soils with gray to brown surface horizons, medium to high base supply, and subsurface horizons of clay accumulation; usually moist but may be dry during warm season

A1   AQUALES (seasonally saturated with water) gently sloping; general crops if drained, pasture and woodland if undrained (some low-humic gley soils and planosols)

A2   BORALES (cool or cold) gently sloping; mostly woodland, pasture, and some small grain (gray wooded soils)

A2S   BORALES steep; mostly woodland

A3   UDALFS (temperate or warm, and moist) gently or moderately sloping; mostly farmed, corn, soybeans, small grain, and pasture (gray brown podzolic soils)

A4   USTALFS (warm and intermittently dry for long periods) gently or moderately sloping; range, small grain, and irrigated crops (some reddish chestnut and red-yellow podzolic soils)

A5S   XERALES (warm and continuously dry in summer for long periods, moist in winter) gently sloping to steep; mostly range, small grain, and irrigated crops (noncalcic brown soils)

ARIDISOLS . . . Soils with pedogenic horizons, low in organic matter, and dry more than 6 months of the year in all horizons

D1   ARGIDS (with horizon of clay accumulation) gently or moderately sloping; mostly range, some irrigated crops (some desert, reddish desert, reddish-brown, and brown soils and associated solonetz soils)

D1S   ARGIDS gently sloping to steep

D2   ORTHIDS (without horizon of clay accumulation) gently or moderately sloping, mostly range and some irrigated crops (some desert, reddish desert, sierozem, and brown soils, and some calcisols and solonchak soils)

D2S   ORTHIDS gently sloping to steep

ENTISOLS . . . Soils without pedogenic horizons

E1   AQUENTS (seasonally saturated with water) gently sloping; some grazing

E2   ORTHENTS (loamy or clayey textures) deep to hard rock; gently to moderately sloping; range or irrigated farming (regosols)

E3   ORTHENTS shallow to hard rock; gently to moderately sloping; mostly range (lithosols)

E3S   ORTHENTS shallow to hard rock; steep; mostly range

E4   PSAMMENTS (sand or loamy sand textures) gently to moderately sloping; mostly range in dry climates, woodland or cropland in humid climates (regosols)

HISTOSOLS . . . Organic soils

H1   FIBRISTS (fibrous or woody peats, largely undercomposed) mostly wooded or idle (peats)

H2   SAPRISTS (decomposed mucks) truck crops if drained, idle if undrained (mucks)

INCEPTISOLS . . . Soils that are usually moist, with pedogenic horizons of alteration of patent materials but not of accumulation

11S   ANDEPTS (with amorphous clay or vitric volcanic ash and pumice) gently sloping to steep; mostly woodland, in Hawaii mostly sugar cane, pineapple, and range (ando soils, some tundra soils)

12   AQUEPTS (seasonally saturated with water) gently sloping, if drained, mostly row crops, corn, soybeans, and cotton; if undrained, mostly woodland or pasture (some low-humic gley soils and alluvial soils)

12P   AQUEPTS (with continuous or sporadic permafrost) gently sloping to steep, woodland or idle (tundra soils)

13   OCHREPTS (with thin or light-colored surface horizons and little organic matter) gently to moderately sloping, mostly pasture, small grain, and hay (sols bruns acides and some alluvial soils)

13S   OCHREPTS gently sloping to steep; woodland, pasture, small grains

14S   UMBREPTS (with thick dark-colored surface horizons rich in organic matter) moderately sloping to steep, mostly woodland (some regosols)

MOLLISOLS . . . Soils with nearly black, organic-rich surface horizons and high base supply

M1   AQUOLLS (seasonally saturated with water) gently sloping, mostly drained and farmed (humic gley soils)

M2   BOROLLS (cool or cold) gently or moderately sloping, some steep slopes in Utah, mostly small grain in North Central States, range and woodland in Western States (some chernozems)

(Continued)

Table 4-11 (*continued*)

M3   UDOLLS (temperate or warm, and moist) gently or moderately sloping; mostly corn, soybeans, and small grains (some brunizems)

M4   USTOLLS (intermittently dry for long periods during summer) gently to moderately sloping; mostly wheat and range in western part, wheat and corn or sorghum in eastern part, some irrigated crops (chestnut soils and some chernozems and brown soils)

M4S   USTOLLS moderately sloping to steep, mostly range or woodland

M5   XEROLLS (continuously dry in summer for long periods, moist in winter) gently to moderately sloping, mostly wheat, range, and irrigated crops (some brunizems, chestnut, and brown soils)

M5S   XEROLLS moderately sloping to steep, mostly range

SPODOSOLS . . . Soils with accumulations of amorphous materials in subsurface horizons

S1   AQUODS (seasonally saturated with water) gently sloping; mostly range or woodland, where drained in Florida, citrus and special crops (ground-water podzols)

S2   ORTHODS (with subsurface accumulations of iron, aluminum, and organic matter) gently to moderately sloping, woodland, pasture, small grains, special crops (podzols, brown podzolic soils)

S2S   ORTHODS steep; mostly woodland

ULTISOLS . . . Soils that are usually moist with horizon of clay accumulation and a low base supply

U1   AQUULTS (seasonally saturated with water) gently sloping, woodland and pasture if undrained, feed and truck crops if drained (some low-humic gley soils)

U2S   HUMULTS (with high or very high organic-matter content) moderately sloping to steep; woodland and pasture if steep, sugar cane and pineapple in Hawaii, truck and seed crops in Western States (some reddish-brown lateritic soils)

U3   UDULTS (with low organic-matter content; temperate or warm, and moist) gently to moderately sloping, woodland, pasture, feed crops, tobacco, and cotton (red-yellow podzolic soils, some reddish-brown Lateritic soils)

U3S   UDULTS moderately sloping to steep; woodland, pasture

U4S   XERULTS (with low to moderate organic-matter content, continuously dry for long periods in summer) range and woodland (some reddish-brown latentic soils)

VERTISOLS . . . Soils with high content of swelling clays and wide deep cracks at some season

V1   UDERTS (cracks open for only short periods, less than 3 months in a year) gently sloping; cotton, corn, pasture, and some rice (some grumusols)

V2   USTERTS (cracks open and close twice a year and remain open more than 3 months), general crops, range, and some irrigated crops (some grumusols)

AREAS with little soil . . .

X1   Salt flats

X2   Rockland, ice fields

NOMENCLATURE

The nomenclature is systematic. Names of soil orders end in *sol* (L. *solum*, soil), e.g., ALFISOL, and contain a formative element and as the final syllable in names of taxa in suborders, great groups and subgroups.

Names of suborders consist of two syllables, e.g., AQUALF. Formative elements in the legend for this map and their connotations are as follows:

and	—Modified from Ando soils; soils from vitreous patent materials
aqu	—L. *aqua*, water; soils that are wet for long periods
arg	—Modified from L. *argilla*, clay, soils with a horizon of clay accumulation
bor	—Gr. *boreas*, northern; cool
fibr	—L. *fibra*, fiber; least decomposed
hum	—L. *humus*, earth; presence of organic matter
ochr	—Gr. base of ochros, pale; soils with little organic matter
orth	—Gr. *orthos*, true; the common or typical
psamm	—Gr. *psammos*, sand, sandy soils
sapr	—Gr. *sapros*, rotten; most decomposed
ud	—L. *udus*, humid; of humid climates
umbr	—L. *umbra*, shade, dark colors reflecting muck organic matter
ust	—L. *ustus*, burnt, of dry climates with summer rains
xer	—Gr. *xeros*, dry; of dry climates with winter rains

---

a From U.S. Department of Agriculture, Soil Conservation Service.

groups, subgroups, and series. The series level has remained unchanged from the 1938 system so that at the field level no new nomenclature is required. In the higher categories, however, new names were derived. Their selection was based on a number of criteria, including such diverse factors as the avoidance of existing terms, the use of Greek and Latin roots, and the selection of terms that were as short and euphonic as possible. Importantly, the nomenclature should give some insight into the properties of the soil in question.

The hierarchy established appears quite logical; orders consist of words of three or four syllables ending in *sol*. Suborders are named by removing *sol* from the order term and substituting a new prefix. Thus an order is termed Histosol, derived from the Greek *histos*, tissue, which refers to the high organic content of the soils and from *sol*, soil. A suborder of this is the Fibrists—the peats; this is derived by dropping *sol* and retaining *ist* of histosol; the *fibra* is derived from the Latin word for fibrous. A similar schema is used in derivation of remaining categories. Great groups are derived by adding a prefix to the suborder term while an adjective is added to the suborder to give a subgroup.

Despite the obvious care taken in the derivation of terms it has not been above rather strong criticism. Heller (1963) has described it as "bizarre and incomprehensible . . . barbarous . . . lacking in euphony." Interesting differences of opinion have also been expressed in the literature. For example, Van Royan (1970), in response to an article by Bunting (1970), writes:

"The article . . . entitled 'Concept, Class and Terminology in Studies of Tropical Soils' is a useful contribution, if only because it shows the Babylonian confusion that still reigns in the pedological nomenclature and horribilia to which scientific classification can give birth. Plinthaquults! Haplustalfs! Chromoxererts! The terms sound as though Jonathan Swift brought them straight back from the land of the Houyhnhnms."

Needless to say, such comment was countered by the 7th Approx. advocates. Mausel (1971), for example, wrote "The nomenclature of the new soil classification is no more of a monstrosity than botanical or zoological classifications . . . once a few score of formative word elements are known, (it) provides more insight into the char-

acter of a specific soil than other classifications provide for their particular type of phenomena."

With such factors in mind, the question of importance here is how does the 7th Approx. relate to climate? The answer is, of course, that it does not. It is a generic system based upon actual chemical, biological, and morphological characteristics as opposed to the genetic system that stress formation characteristics. Analysis of a soil in terms of composition, degree of horizination, the presence or absence of various horizons, and combined measure of the weathering and weatherability of minerals present can lead to the grouping of soils of quite diverse origin. Entisols, for example, are mineral soils with little horizination and are identified as lithosols, regosols, and some alluvial soils of the older system.

That is not to say, however, that a genetic thread cannot be traced through the 7th Approx. The spodosols, for instance, are characterized by a spodic horizon that corresponds to the well-formed B2 horizon found under podzolization. The horizons of the mollisols are characteristically formed under conditions found under prairie vegetation in semiarid to subhumid conditions. Thus, while genesis *per se* is not a factor in the 7th Approx., it can be related to some categories merely because the identified characteristics are formed under given pedogenic regimes. A number of writers have correlated the Great Soil Groups of the 1938 system to the 7th Approx., examples being shown in Table 4-12.

It is evident that the 7th Approx. is admirably suited for investigation of soils by soil scientists. Whether it is of more value than the 1938 system in related disciplines is a topic of controversy. This is demonstrated in the correspondence between Van Royan and Mausel, cited earlier. Van Royan, in stressing the need for a conceptual approach to environmental relations, feels that the genetic system best serves the need. Mausel counters this by saying that the "climate" approach to soils is an incomplete one and further notes that development of the 7th Approx. will ultimately lead to a time when it will become the only system used in communication of soil science. The argument will undoubtedly go on for some time because it is evident that the 7th Approx. is not always the system used in introductory soil study. The introduction to, and often the only contact with, soil science often

Table 4-12

SUMMARY OF THE GLOBAL SOIL ORDERS[a]

Soil Order	Diagnostic Properties	Equivalent Soils
Entisols	No diagnostic horizons or profile development	Azonal soils; tundra, lithosols, regolith, alluvium, sands and gravels
Vertisols	Dark topsoils that crack badly when dry; slanted columnar structure	Grumusols, tropical black clays, regur, tirs
Inceptisols	Young soils, with profile features just beginning; no clear-cut diagnostic horizons, usually dark epipedons	Subarctic brown forest, brown forest, ando, lithosols, regosols, some humic gleys
Aridisols	Desert soils that have one or more diagnostic horizons but no spodic or oxic horizons: light-colored epipedons	Desert, red desert, gray desert (sierozems), reddish brown, lithosols, regosols, solonetz, solod, solonchak
Mollisols	Dark brown to black, mellow epipedons: one or more diagnostic horizons, except spodic or oxic horizons: often a *ca* horizon	Chernozems prairie (brunizem), chestnut, red prairie, humic gleys, planosols, black rendzinas, some brown forest, reddish chestnut soils, splonchak, solonetz.
Spodosols	Spodic horizon; usually also an albic horizon; usually coniferous forests over sandy soils	Podzols, brown podzolic, groundwater podzols
Alfisols	Pronounced argillic horizon without a spodic or oxic horizon; may have a natric horizon; no pronounced color changes vertically; high base exchange	Gray-brown podzolic, gray forest, noncalcic brown soils, planosols, half-bogs, solods, and terra rossas
Ultisols	Red or yellow argillic horizon; no oxic horizon: low base-exchange throughout	Red-yellow podzolic, reddish-brown lateritic, ruburzem, humic gleys, low humic gleys, groundwater laterites
Oxisols	Oxic horizon, with many free sesquioxides: little horizon differentiation; deep soils	Latosols, ferrallites, ground water laterites, laterites
Histosols	A surface organic layer at least 12–18 in. thick	Peat, muck

[a] From Van Riper (1970).

occurs in basic environmental courses given at the undergraduate level. At this level, a survey of typical texts used shows that the older classification scheme is still in favor. Some ignore the 7th Approx. completely (e.g., Patton et al., 1970), others use the old system but add the 7th Approx. as an appendix (Strahler, 1969), while it is sometimes used in conjunction with the 1938 system by adding equivalent soil types in parenthesis after the Great Soil Groups (e.g., Trewartha et al., 1967). Despite its obvious shortcomings, the 1938 system is still in favor. This is to be expected, of course, for surveys of environmental

processes require interaction between them to be assessed. A genetic system certainly is of value in this.

CLIMATE AND SOIL EROSION

The soil-forming process is a long-term one. In most cases soil that is formed, or is in the process of undergoing active modification, forms part of an environmental system in which the variables interact one with another. By modifying any one of the variables, the original characteristics are altered. In his quest for food, man has long been a modifying influence. Through experience and forethought, many relatively infertile soils have been rendered productive and, by addition of required inputs, soils lacking the necessary nutrients for cultivation have been made usable. At the same time, however, many thousands of acres of soils have been destroyed through lack of knowledge or sheer carelessness.

Surface erosion of soils is a natural process, it goes on regardless of land utilization. What man has done, in many instances, is to accelerate the natural processes. Interruption of the equilibrium that may have occurred under the natural state leads to accelerated erosion. This interruption is most likely to occur when the surface vegetation cover is modified. Such an effect is well seen in the creation of the Dust Bowl in the United States (Figure 4-12).

In the semiarid, brown soil areas of the United States, precipitation is highly variable on an annual basis. Some years, usually occurring in concurrence, are moist. Inevitably these wet years are followed by years of severe drought. Under a natural grass cover, drought had rela-

tively little impact upon soil losses. As settlers moved west, however cultivation of the land became widespread, and the grass cover was ploughed under for the use of agriculture. Initial use of the land usually occurred during a series of wet years and good crop yields were obtained. These were invariably followed by dry periods and cropped areas became bare soils, devoid of vegetation. The exposed soil was loose and dry, and erosion on a massive scale followed.

The wet 1880s were followed by the dry 1890s. At the turn of the century, wet years again prevailed, and more land was plowed. By 1910 another drought set in. Thus it continued until 1931, the time of the great economic depression, when there began a drought that culminated in great dust storms. In describing these, Dasmann (1968) writes, "Dust blew across the continent, darkened the sky, reddened the sunsets, and made the plain region almost uninhabitable for man or livestock. Millions of acres of farms were damaged, with an estimated loss of topsoil ranging between 2 and 12 inches in places. Drifting dunes moved over farms, burying

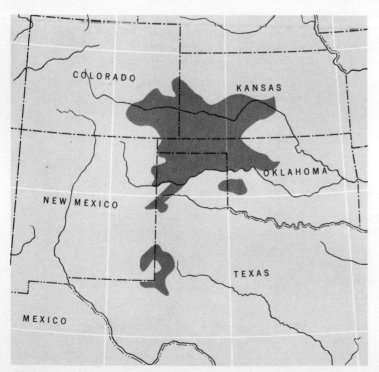

Figure 4-12.   Location of the Dust Bowl, an area of serious soil erosion in the United States. (From Dasmann, R., *Environmental Conservation*, Copyright © 1968 by John Wiley and Sons. Used with permission.)

roads, fences and even dwellings. . . . It was a period of misery and privation difficult to match in American history."

While the disaster of the Dust Bowl prompted Federal action—the Soil Conservation Service was started in 1935—and rational conservation practices were instituted, farming on the marginally wet lands continued. Even in the 1950s, severe drought led to widespread damage.

Accelerated erosion by wind in marginally moist regions is, of course, only one aspect of the way in which climate can influence soil losses. The effects of sheet erosion and gullying are equally problematic. One concern of the problem is the determination of how much soil is eroded under different land use conditions. To gain some idea of the amount that might occur, the United States Department of Agriculture has used a soil-loss equation (Wischmeier and Smith, 1965), which is given as

$$A = R. K. L. S. C. P$$

where A = computed soil loss per unit area
  R = rainfall factor
  K = soil-erodibility factor
  L = slope length factor
  S = slope-gradient factor
  C = crop management factor
  P = erosion-control practice factor

Of interest from the climatic aspect is the rainfall factor R. This is derived empirically and determined from two rainfall characteristics, namely the total kinetic energy of the storm (E) and its maximum 30 min intensity (I). The product EI relates the effects of rainfall impact on the ground and the relative ability of resulting runoff to transport soil particles. The role of precipitation in this respect has been outlined by Ellison (1948), who shows that erosion by rain drops can reach very high levels when large drops occur on bare soil. The significance of this is seen in moist tropical areas where the backspatter of rain-loosened soil stains the walls of houses up to two or three ft.

In use of the soil-erosion equation, R is derived from the sum of EI values for a given time period and is expressed numerically as an average yearly total of storm EI values. To facilitate its use, isoerodent maps have been prepared (Figure 4–13).

Clearly, the great differences that are shown in Figure 4–13 can be offset by the nature of the ground onto which the precipitation is falling. The soil-loss equation takes these into consideration by expressing the nature of the surface material in respect to its relative erodibility (K), the way in which runoff will be facilitated (L and S), and the role of the surface vegetation cover in inhibiting or accelerating erosion (C and P). Values have been derived for each of these factors and it is possible to determine a numerical value for soil loss. For example, consider the following facts:

Area: Fountain County, Indiana (Location A on Figure 4–13)
  Soil: Russell silt loam
  Slope: 8%, 200 ft long
  Cropping system: 4-yr rotation (wheat—meadow—corn—corn).

By interpolation of Figure 4–13, the R factor is determined to be 185. K, the soil erodibility is equated to values of similar silt loams for which values have been derived; = 0.83. L and S, the slope length and gradient factors are derived simultaneously and read from a slope-effect chart (Figure 4–14); = 1.41. The cropping management factor, based upon the crops crown and the rotation, is determined as 0.119. Finally, the introduction of erosion control practices—contour plowing on steep slopes, for example—depends upon derived coefficients ranging from 0.6 to 0.9. In this case it is = 0.6.
Substituting these values into the soil-loss equation:

$$A = R. K. L. S. C. P.$$
$$= 185. \ 0.38. \ 1.41. \ 0.6. \ 0.119$$
$$= 7.1 \text{ tons of soil loss per acre per year}$$

The importance of the climatic factor in this result can be assessed by assuming that the same farm be transferred to an area where the R factor alters. (This, of course, is purely hypothetical and, in some ways, is misleading.) Consider the relocation of the Indiana farm to Madison County, Tennessee. As shown in Figure 4–13 (Location B), the value of R in this location is 300. Substituting this into the soil-loss equation, we get:

$$A = 300 \times 0.38 \times 1.41 \times 0.6 \times 0.119$$
$$= 11.47 \text{ tons of soil loss per acre per year}$$

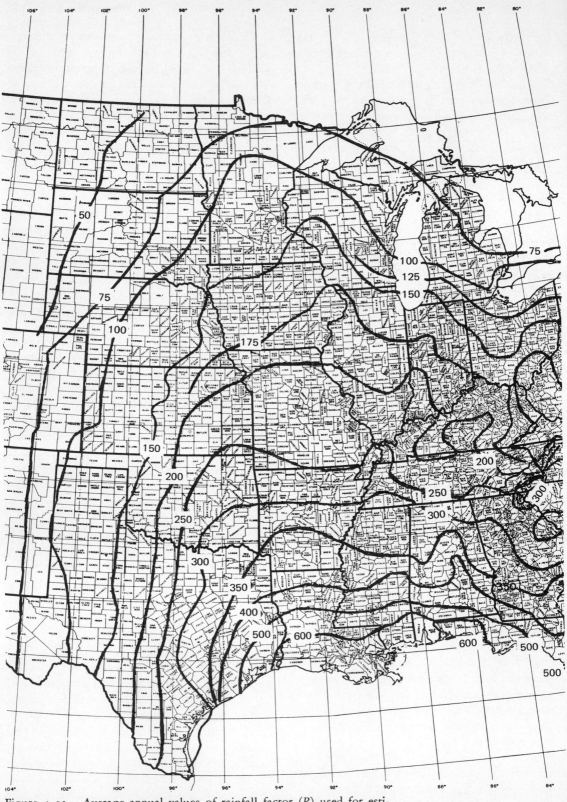

Figure 4-13. Average annual values of rainfall factor (R) used for estimating rainfall-erosion losses from cropland. See text for discussion. (From
U.S. Department of Agriculture, 1965, *Agricultural Handbook*, No. 282.)

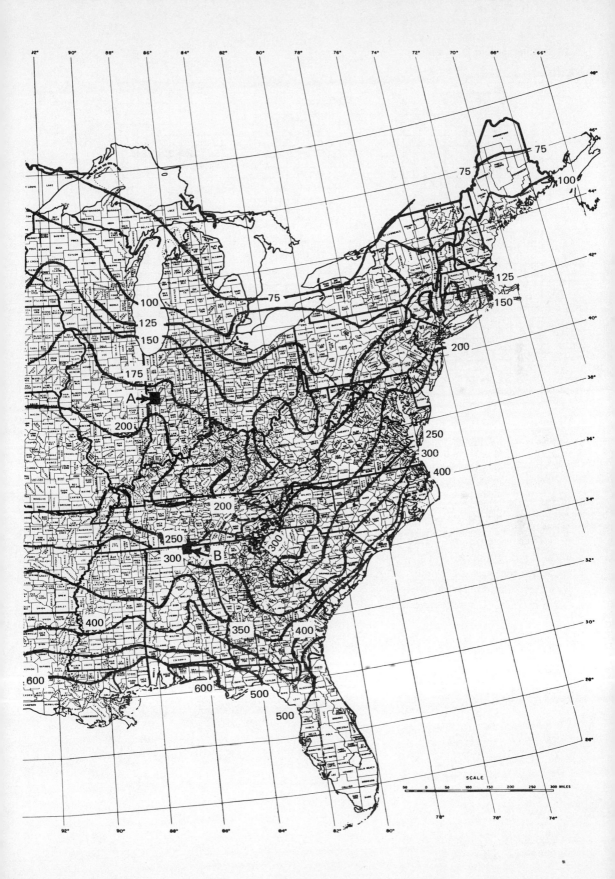

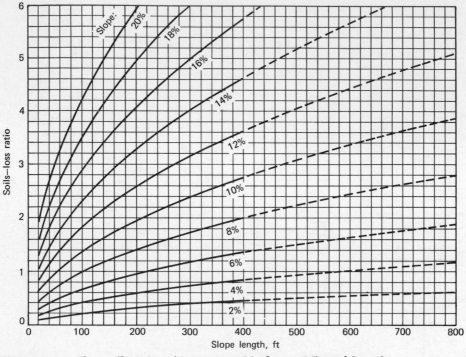

Figure 4–14.   Slope-effect chart (the topographic factor, *LS*) used in estimating rainfall-erosion losses from cropland. See text for discussion. (From
U.S. Department of Agriculture, *Agricultural Handbook*, No. 282.)

Climate obviously plays a most important role in determining soil loss.

Soil erosion can be measured through other means. In gullying, it is possible to actually measure the rate of erosion and estimate the volume of material used over time. In a similar way, the siltation of reservoirs provides a good estimate of the amount of material removed from the watershed that feeds it. In evaluation by these two means it is found, as in the case of estimation through the soil-loss equation, that many variables enter the picture. Climate, particularly the effect and nature of rainfall that occurs, invariably assumed importance.

CHAPTER 5

Climate, Plants, and Natural Vegetation

The relationship between plant life and climate is an intimate one. The fact that early climatic classifications used vegetation as an index of climate and that it has been suggested that ". . . the easiest way to recognize a climatic region . . . is through its effect upon the predominant grouping of plants that grow on the earth's surface" (Rumney, 1968) is testimony to this relationship. Indeed, a rare note of agreement is found between disciplines for, as Polunin (1960), a plant geographer, comments, ". . . climate is the most far-reaching of the natural 'elements' controlling plant life."

To assess the significance of the relationship between climate and plant life, it is useful to deal with it from two standpoints. First, the relationship can be viewed from that aspect in which individual plants are considered in terms of optimum climatic conditions required for their growth and development (assuming that growth denotes the increase of plant size over time while development refers to the passage of the plant through given morphological stages irrespective of rate of growth). Such an approach allows evaluation of the role played by individual climatic elements as they affect plant life.

Second, the relationship might be viewed as the interconnection between the total climatic environment and plant communities. At this level, where the plant cover of a region is considered *in toto*, the relationship concerns the prevalence of a given form of vegetation in relation to all environmental controls, one of which is climate.

Note at the outset that any discussion concerning the natural plant cover needs some qualification. The natural vegetation of any region is the end product of a long process of succession, adaptation, and acclimatization; the resulting climax vegetation represents a surface covering that is in dynamic equilibrium with its environment. Changes in this equilibrium through most of geologic time have resulted

from natural forces; so marked have been the changes that some contend that a climax vegetation community is rarely attained. In more recent times (in geological terms), man has become an active agent of modification and has induced enormous changes in plant cover and plant distribution over the globe. It is often exceedingly difficult to reconstruct the natural vegetation of a region so effective has been man's modification. Even so, in this chapter the concern is with "natural" processes, and the role of man as an innovator and modifier of the plant world is considered in a later chapter.

CLIMATE AND PLANTS

Just as the study of solar radiation is basic to the understanding of the climatic environment, it is basic to the comprehension of plant life. The conversion of radiant energy to chemical energy by plants represents the way in which solar radiation can directly enter the biological system of earth. The conversion to chemical energy, which resides in the bonds and structure of sugar in plants, is achieved through photosynthesis. In this process, atmospheric carbon dioxide diffuses through stomata into leaves and dissolves in water saturating the cell walls. Both of these components are devoid of usable energy; this is derived through a conversion process, the key to which is the green pigment, chlorophyll. The overall reaction might be simplistically represented by

$$6CO_2 + 6H_2O + \text{solar energy} = C_6H_{12}O_6 + 6O_2$$

carbon dioxide / water / sugar / oxygen

Energy from the sun is converted into the free energy of hexose sugar. Note here that advanced work concerning the reaction between light and plants is often considered in terms of quantum efficiency. When light interacts with matter, it operates as though it were composed of "packets" of energy, termed quanta or photons.

Quanta are considered ultimate particles, such as protons or electrons, but lack an electrical charge. The work unit used in this approach is the *einstein*, where one einstein equals 2.854×10^7 gram calories divided by the wavelength of the quanta expressed in millimicrons. This inverse relationship shows that long-wave (high-frequency) radiation has less energy per quantum than short-wave radiation. Several workers have used quantum efficiency in estimating the variable rates of photosynthesis over the earth's surface (see pp. 260–261).

Clearly, sunlight is a basic requirement for the process, and the rate at which photosynthesis proceeds depends upon the amount of light that is available. There is, however, an upper limit beyond which increase in intensity has no effect upon oxygen production; at such a limit, a light-saturation condition is said to exist. For most species, the photosynthetic response of a single leaf to sunlight increases in almost a linear form until light saturation occurs, after which the rate remains almost constant (Figure 5–1).

Since both sunlight and carbon dioxide enter into the photosynthetic process, it follows that the rate of production will also depend upon the availability of carbon dioxide. At low concentrations, carbon dioxide might become the limiting factor. Similarly, other factors might limit the process; in some cases these might be enzymic rather than photochemical.

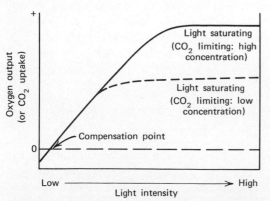

Figure 5–1.   Relationship between rate of photosynthesis and light intensity at high and low levels of carbon dioxide concentration. (After Fogg, 1966.)

The reliance of photosynthesis upon solar radiation, assuming an adequate carbon dioxide and water supply, means that marked differences in rates of photosynthesis will be found over the globe. As part of an evaluation of potential photosynthesis on a world basis, Chang (1970) derived the distribution shown in Figure 5–2. Net photosynthesis shown is derived from the difference between gross photosynthesis and respiration, so that the graph gives insight into the net production of organic matter. A number of interesting features are seen; for example, potential photosynthesis does not, as might be expected, attain its maximum value at the equator. The significance of the distribution is commented on more fully in Chapter 9, where it is discussed in terms of agricultural production.

There are plant responses, other than photosynthesis, that relate directly to the amount and duration of incoming solar radiation. For example, plants may be classified according to their relative requirements of direct sunlight or shade. Those that thrive best in full sunlight are termed the heliophytes, while those that develop in shade are termed sciophytes. The reasons for the existence of these two types are complex. As Daubenmire (1959) points out, heliophytes may do well in full sunlight for a variety of reasons, including:

1. High heat requirements.
2. To escape destruction by fungi, which are not found under the low humidity environment associated with bright sunlight.
3. To stimulate flowering.
4. To open guard cells and thereby promote intake of carbon dioxide.
5. To overcome the destructive effect of shade in which, as a result of slow rates of decay, a nitrogen deficiency may occur.

The explanation of why sciophyllous plants do well in shade is equally complex, with one major difference being that they have a much lower compensation point than heliophytes. The compensation point is defined as the point at which carbon dioxide is neither absorbed nor evolved by a plant; this will occur when the amount of light required for photosynthesis equals the respiratory use of carbon compounds (Figure 5–1). The compensation point for some sciophytes might be as high as 4200 L while for some

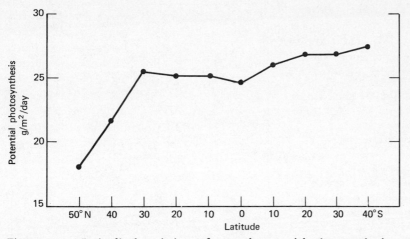

Figure 5–2. Latitudinal variation of annual potential photosynthesis. (After Chang, 1970.)

shade plants it is as low as 27 L.[1] The relatively slow rate at which the sciophytes manufacture chlorophyll is also significant, for light continuously decomposes chlorophyll and a plant will remain green only if enough chlorophyll is manufactured. In shaded conditions, decomposition is slower.

Much research has also gone on into the way in which plants respond to the varying lengths of day and night that occur over the globe. Much of the understanding of this aspect of plant growth and development, photoperiodism, is the result of work by Garner and Allard in the United States. In their investigation of why the growth of a variety of tobacco was delayed when planted north of its usual growing area, they found that at a critical stage of plant development, the daylight hours were too long (or the nights too short). Further work allowed them to classify plants according to the following criteria:

Long-day plants	Flower only when daylight is greater than 14 hr
Short-day plants	Flower only when daylight is less than 14 hr
Day neutral plants	Bud under any period of illumination

| Intermediate plants | Flower with 12 to 14 hr daylight but will not outside of these limits |

It would appear to follow that photoperiodism must play a significant role in explaining why plants flourish at some latitudes and not at others. Obviously, it cannot be cited as the sole reason, for other climatic events must be considered. In the same way, some plants that extend over wide latitudinal areas do not have the same response to varying lengths of day and night. While physiologically similar, such plants have different photoperiods, having adjusted genetically to the day length period[2] of the area in which they exist.

Temperature and Moisture Requirements

The thermal and moisture conditions under which plants exist probably represent the most closely studied aspect of climate-plant relationships. The two factors are, of course, intimately related and ultimately both depend upon the energy budget that exists in any given climatic regime.

In the discussion of the energy exchange at the earth's surface, it was shown that the atmosphere is heated from below and, as a result, a thermal gradient exists upward in the troposphere. Furthermore, because of differences in the distri-

[1] L = lux; where 1 L is the amount of light received at a distance of 1 m from a standard candle. Light intensity at 1 ft is 1 ft-c, where 1 ft-c = 10.764 L. Such units are rarely used in climatology; they are presented here as an example of the various illumination units available.

[2] Note that it is the length of night instead of the length of day that is the most important factor in photoperiodism.

bution of energy over the globe, horizontal gradients will also exist. Early plant geographers recognized this gradient and equated it to plant distribution. A typical representation was that devised by de Candolle who identified the following:

Physiologic Plant Groups	Temperature Limits, °F
Megistotherms[a]	over 86 (30.0°C)
Megatherms	68 to 86 (20.0 to 30.0°C)
Xerophiles[b]	59 to 68 (15.0 to 20.0°C)
Mesotherms	59 to 68 (15.0 to 20.0°C)
Microtherms	32 to 59 (0 to 15.0°C)
Hekistotherms	below 32 (0°C)

[a] Gentilli (1958) notes that this group was common during geologic time (e.g., during the Carboniferous) but that descendants are restricted to such locations as thermal springs.

[b] The Xerophiles are plants differentiated according to moisture conditions.

Whether or not a given plant is found within any one of these differentiated zones presupposes that all plants are limited by a critical temperature boundary; there will be an upper and a lower limit as well as a value at which plants develop most rapidly. While it is true that in some cases such limits can be defined with accuracy, the derivation of such critical (or cardinal) temperatures is not a simple matter because:

1. Different physiological processes within a plant function most efficiently at different temperatures.

2. The value of the critical temperature varies over time with the development of the plant.

Despite these difficulties, it is possible to relate the distribution of some plants to a given set of temperature conditions. Figure 5–3 provides an example where the areal limits of a plant can be equated to the location of a given isotherm or, as in the lower map, how the limits of a plant can be related to a number of climatic variables.

Since most plants grow and develop within given temperature limits, any variation above or below a critical temperature causes plant injury. Excessive heat can result in the destruction of protoplasm at a temperature of 54°C (130°F), while even at slightly lower temperatures injury

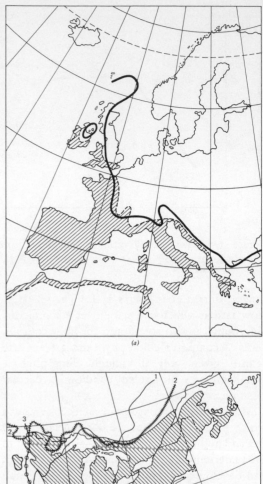

(a)

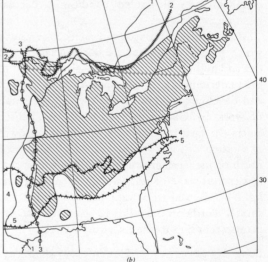

(b)

Figure 5–3. (a) The 45°F January isotherm, a climatic limit for madder (*Rubia peregrina*) in Europe. (b) Bioclimatic limits of sugar maple (*Acer saccharophorum*) showing coincidence with meteorological elements. The shaded area shows distribution of sugar maple. Key: (1) 30-in. annual rainfall; (2) –40°C mean annual minimum; (3) B/H (dry-humid) boundary; (4) 10-in. mean annual snowfall; (5) –10°C mean annual minimum. (From Dansereau, P., *Biogeography: An ecological perspective*, copyright © 1957 by the Ronald Press. Used with permission.)

can occur. Sunscald, for example, often occurs in trees with bark when the cambium (a layer of actively dividing cells) is killed. Such injury hastens the destruction of trees, for it provides entry to the tree for disease-causing organisms. Excessive heat also has a desiccating effect on plants, and rapid moisture loss may lead to wilting. Under such conditions, as the plant gets hotter, respiration will exceed photosynthesis, and a metabolic imbalance will occur. Such an imbalance occurs over short periods quite frequently; if it is sustained, however, the plant will die.

Low temperatures can also result in destruction of plant life. Rapid temperature reduction with freezing causes the living matter of cells to freeze; slow freezing results in the freezing of intercellular water and ultimately in cell dehydration. Frost damage can also occur in plants that usually survive under very cold conditions. A sudden warm spell during a cold period can result in transpiration—but water that is used is not replenished and as the temperature falls again, winter burn may occur.

There are many other effects of extreme cold and warmth on plants. Ranging from mechanical damage of frost glaze and frost heaving to the effects of fires on vegetation cover, such effects are extensively covered in the available literature (e.g. Daubenmire, 1959, Polunin, 1960).

While temperature differences, with resulting plant differences, occur latitudinally, important modifications of temperature also occur vertically. In fact, one of the best-known variations of plant type is that associated with elevation. Many maps and diagrams (e.g., Figure 5-4) have been produced showing plant zonation as a function of altitude. As the example indicates, there is a superficial resemblance between the plant grouping found on the side of a mountain with that of an equatorial-pole traverse. This resemblance has caused wide misconceptions pertaining to mountain climates and the plants associated with them. Preston James (1959) outlines the climatic case clearly

"A very common and often repeated error is to think of high altitude climates of the tropics as similar to the climates with the same average temperatures found at sea level in middle latitudes. . . . As one ascends the mountains (of Colombia) the range of temperature becomes less. At Bogota, 8660 feet above the sea level, the average annual temperature

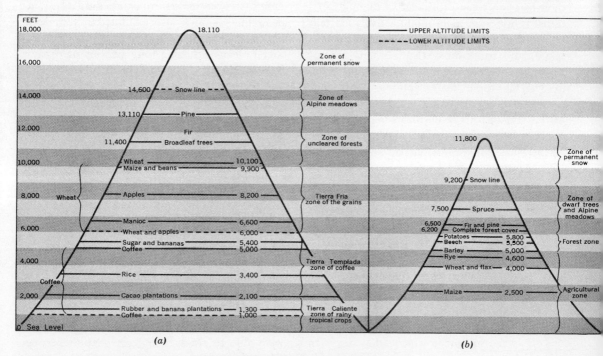

Figure 5-4. Plant zonation as a function of altitude. (a) A tropical mountain. (b) A middle-latitude mountain. (From Trewartha, G.T., *Introduction to Climate*, 1968. Used with permission of McGraw-Hill Book Company.)

is 58.1°F (approximately 14.4°C)—exactly the same as the average annual temperature of Knoxville, Tennessee; but in Bogota the difference between the averages of the warmest and coldest month is only 1.8°, while the difference at Knoxville is 38.1°. To describe Bogota as having a "perpetual spring climate" as is so frequently done is to create a very false impression, for there is none of the weather variety characteristics of a middle latitude spring."

It follows from this that while—on both a latitudinal and altitudinal basis—similar vegetation changes will occur, that is, forests give way to grasslands, the individual plant members of such vegetation groups might be quite dissimilar. Note, too, that temperature is not the only variable that will influence the plant form. It has already been noted that precipitation varies with altitude while the pressure decrease that occurs will mean that less carbon dioxide is available for plants per unit volume of air.

Temperature, then, places limits on the distribution of plants as a result of its influence upon the rate of chemical activity that occurs within the plant. Because of the varying metabolic rates and responses, plants vary enormously in their reaction to and tolerance of thermal conditions. While some algae might actively exist near hot springs where temperatures are in excess of 200°F, and some mosses and lichens continue to exist in temperatures as low as −90°F (−67.8°C), these are exceptions. Most plants cease their activities when soil temperature falls below 42°F (5.0°C), while plants in extremely hot regions are usually limited in growth not by temperature conditions, but by associated lack of moisture.

Moisture is essential for plant growth and development, for it provides the medium by which chemicals and nutrients are dispersed through the plant. In outlining the role of water in plants, Chang (1968) points out:

1. Water is the main constituent of the physiological plant tissue.
2. It is a reagent in photosynthesis and in hydrolytic processes such as starch digestion.
3. It is the solvent in which salts, sugar, and other solutes move from cell to cell and organ to organ.

4. It forms an essential element in the maintenance of plant turgidity.

The actual amounts of water consumed and transpired by plants has been the focus of intensive research. Much of the research is focused upon water needs of domesticated plants, an aspect dealt with in some detail in Chapter 9.

While plants need moisture, it is not always available in the same amounts at different locations. As a result, an enormous variety of ways exist in which plants are adapted to water availability. Indeed, an entire spectrum exists, with some plants thriving in conditions of surplus moisture while others maintain themselves under conditions of extreme drought.

Plants existing in the perpetual drought of arid regions are the xerophytes. These have the ability to exist and reproduce in areas where soil moisture is invariably below field capacity. The methods by which plants overcome this shortage is highly variable, with some acting as drought evaders and some as drought resisters. Some annuals evade drought by remaining in seed form through extensive dry periods and pass through their life cycles when the occasional desert shower makes moisture available. Such drought evasion is also exhibited by succulents, which store moisture in their stems or leaves and thereby can withstand long periods of dryness. True xerophytes, the drought resisters, have the ability to obtain moisture from relatively dry soil. Phreatophytes, for example, overcome the paucity of soil moisture by drawing upon ground water. Figure 5-5 illustrates examples of the variable plant forms that exist under arid conditions.

While both moisture availability and temperature influence plants, the two variables do not operate in a vacuum. There is a close interrelation between the two and other environmental variables. In assessing the combined influence of temperature and moisture on plant distribution, Mather and Yoshioka (1968) make use of the Thornthwaite water balance analysis. By plotting data representative of vegetation areas of the coterminous United States on a climograph, they were able to show a high correlation between selected climatic variables and plant distribution. The axes of the climograph (Figure 5-6) uses

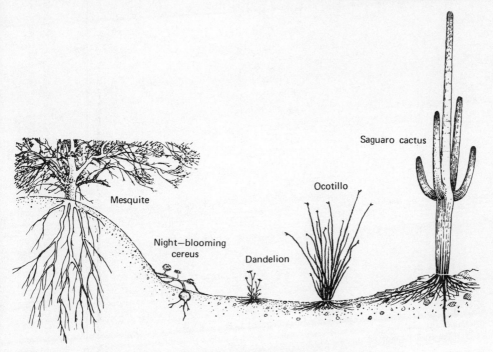

Figure 5–5. Adaptation of desert plants are typified (left to right) by the deep roots of the mesquite; the storage roots of night-blooming cereus; the drought-evading desert dandelion; the leaf-dropping habit of the ocotillo; and the collecting roots and storage capability of the saguaro cactus. (From Farb, P., *Face of North America*, 1963, Harper and Row, Used with permission.)

Thornthwaite's index (*Im*) on the abscissa and potential evapotranspiration on the ordinate.

The moisture index is derived as follows:

$$Im = 100(S - D)/PE$$

where *S* is surplus moisture, *D* is deficit, and *PE* is potential evapotranspiration. (See Chapter 6.) It can be modified because

$$S = P - AE$$

and

$$D = PE - AE$$

(*P* is precipitation and *AE* is actual evapotranspiration.) These can be substituted to simplify calculation of *Im*:

$$Im = (P/PE - 1) 100$$

By locating different vegetation types on the climograph, a distinctive pattern resulted. The results were so marked that the authors concluded that ". . . it is possible to speak not only of a forest climate, as opposed to a grassland or a desert climate, but that it may be possible to identify a birch-maple as opposed to a spruce or oak-chestnut forest climate in a region such as the United States." It might be that the analysis does go beyond that of "vegetation" distribution and provides guidelines to actual plant types that dominate under specific climatic regimes.

While, thus far, emphasis has been placed upon the growth and development of plants on a macroscale, it is evident that microscale conditions are also of prime importance. Unfortunately the highly variable nature of the soil-air interface make generalizations about microconditions extremely difficult. The number of variables involved—ranging from soil condition and thermal conductivity to the nature and amount of vegetation cover—practically makes each study unique. At this point, merely a few examples of the role of microclimate are given (Figure 5–7).

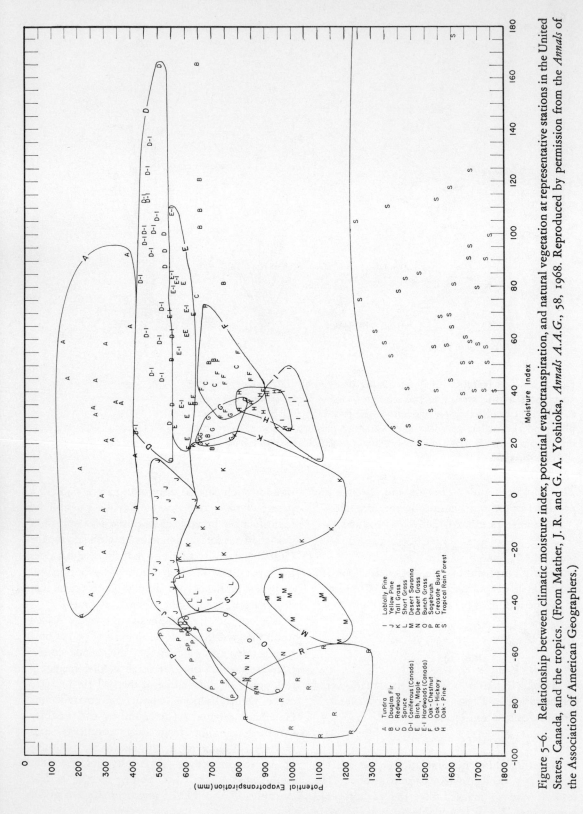

Figure 5–6. Relationship between climatic moisture index, potential evapotranspiration, and natural vegetation at representative stations in the United States, Canada, and the tropics. (From Mather, J. R. and G. A. Yoshioka, *Annals A.A.G.*, 58, 1968. Reproduced by permission from the *Annals of the Association of American Geographers*.)

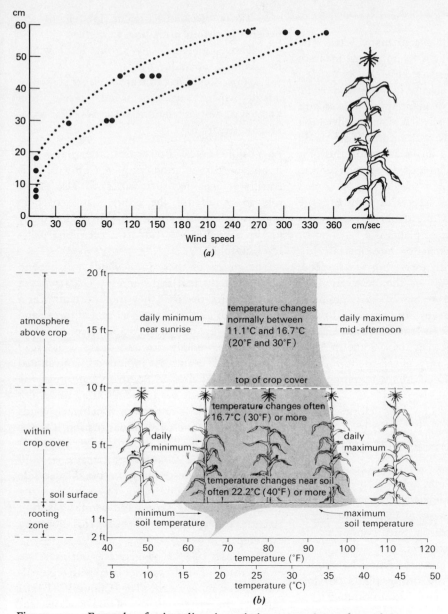

Figure 5-7. Example of microclimatic variations near the surface of the earth. (*a*) The wind profile in a silage crop. The broken lines denote the limits of fluctuations. (After Munn, 1966.) (*b*) Temperature variations around crop plants vary with altitude and soil depth. (After Williamson, from Janick. J., R.W. Schery, F.W. Woods, and V.W. Ruttan, *Plant Science: An Introduction to World Crops.* Copyright © 1969 by W.H. Freeman and Company. Used with permission.)

The Role of the Wind

The wind affects plants in many ways. The most obvious is that it is an important medium by which plants are dispersed thereby comprising a factor of some significance in evaluation of the origin and dispersal of plant species. A second obvious effect is the physical damage that high winds can cause. The Beaufort Wind Scale (Table 5-1) provides an example of this, for wind speeds can be judged from their effects upon trees.

In relation to other aspects of plant life, the role of the wind has already been mentioned. Carbon dioxide intake of plants is a function of wind speed, and transpiration rates increase (up to a given value) with increasing wind speed. Winds can thus result in the desiccation of plants; one extreme result of this produces dwarfing of trees. Because of the rapid loss of water, some trees lack the turgidity that enables them to expand to normal sizes. Some trees, as much as 100-yr old, may never become larger than a shrub for the rate of dry matter production is extremely low.

The impact of wind upon the deformation of trees has been investigated by Yoshino (1967), who has devised a classification of wind-shaped trees. He identifies:

1. Those whose trunks are vertical but which have branches that are bent drastically to the leeward side by prevailing winds that occur during the growing season. Such conditions are found exclusively in conifers.

2. Those whose trunks are vertical but whose branches are severed by strong winds carrying snow or frozen rain. This "storm-pruning" results from

a. Snowflakes, frozen rain, and so on, adhering to branches on the windward side of the tree to cause mechanical injury.

b. Low temperatures, low humidity, and the icy blast of strong winds on the windward side of the tree; buds are killed on that side every year with the result that branches are absent.

3. Those whose trunks and branches are deformed drastically by prevailing winds that occur during the growing season. These will include

a. Those trees that occur in regions experiencing strong prevailing westerlies.

b. Those influenced by strong local winds, for example, up-valley winds.

c. Those found in other areas, either coastal or inland, where persistent winds result from dominant air circulation patterns plus topographic situation.

As Yoshino notes, local winds often play a very important part in modifying tree growth. It is equally evident that such winds will also play a role in modifying the entire plant cover in regions where they prevail. There are many locally named winds, Table 5-2 providing just a few examples.

The effects of these local winds will depend upon their thermal and humidity characteristics and the velocities that they might attain. Warm winds, such as the Santa Ana of California and the Sukhovey of the U.S.S.R. (see Lydolph, 1959) are extremely dry and cause widespread desiccation of plants. As indicated by the annual event of forest fires in Southern California, the prevalence of such winds and resulting tinder-dry plant cover, makes regions in which such winds occur highly prone to such destruction. On the other hand, such warm winds can be of appreciable benefit. The Chinook of Canada, a wind associated with descending air on the lee-side of mountains (Figure 5-8), often hastens the melting of snow in spring and permits planting activities to begin at an earlier date than normally expected.

The causes of local winds vary, but most are often a result of air flows associated with regional-scale pressure systems. The Chinook (Föhn) wind of the eastern slopes of the Rockies, for example, often occurs under circulation conditions similar to those shown in Figure 5-9. In the Mediterranean, the passage of a low-pressure system often gives rise to some of the winds already described and, as is clearly evident from Figure 5-10, many others besides.

CLIMATE AND NATURAL VEGETATION

So far, treatment of climate-plant relationships has been in terms of the response of plants to individual climatic elements; but just as the

Table 5-1
THE BEAUFORT SCALE

| Effect Caused by the Wind | | Beaufort Number | Description | Speed | |
On Land	At Sea			(m/sec)	(miles/hr)
Still; smoke rises vertically	Surface mirror-like	0	Calm	0–0.2	0–1
Smoke drifts but vanes remain still	Only ripples form	1	Light air	0.3–1.5	1–3
Wind felt on face, *leaves rustle*, vane moves	Small, short wavelets, distinct but not breaking	2	Light breeze	1.6–3.3	4–7
Leaves and small twigs move constantly, streamer or pennant extended	Larger wavelets beginning to break, glassy foam, perhaps scattered white horses	3	Gentle breeze	3.4–5.4	8–12
Raises dust and loose paper, *moves twigs and thin branches*	Small waves still but longer, fairly frequent white horses	4	Moderate breeze	5.5–7.9	13–18
Small trees in leaf begin to sway	Moderate waves, distinctly elongated, many white horses, perhaps isolated spray	5	Fresh breeze	8.0–10.7	19–24
Large branches move, telegraph wires whistle, umbrellas hard to control	Large waves begin with extensive white foam crests breaking; spray probable	6	Strong wind	10.8–13.8	25–31
Whole trees move; offers some resistance to walkers	Sea heaps up, lines of white foam begin to be blown downward	7	Stiff wind or moderate gale	13.9–17.1	32–38
Breaks twigs off trees; impedes progress	Moderately high waves with crests of considerable length; foam blown in well-marked streaks; spray blown from crests	8	Stormy wind or fresh gale	17.2–20.7	39–46
Blows off roof tiles and chimney pots	High waves, rolling sea, dense streaks of foam; spray may already reduce visibility	9	Storm or strong gale	20.8–24.4	47–54
Trees uprooted, much structural damage	Heavy rolling sea, white with great foam patches and dense streaks, very high waves with overhanging crests; much spray reduces visibility	10	Heavy storm or whole gale	24.5–28.4	55–63
Widespread damage (very rare inland)	Extraordinarily high waves, spray impedes visibility	11	Hurricane-like storm	28.5–32.6	64–72
	Air full of foam and spray, sea entirely white	12	Hurricane	32.7–36.9	73–82

Table 5-2

CHARACTERISTICS OF SELECTED LOCAL WINDS

Name	Location	Characteristics	Season	Origin[a]
Bora (L, boreas, north)	Adriatic coast	Cold, gusty northeasterly wind. Frequency at Trieste, 360 days in 10 yr. Mean winter speed 52 km/hr, summer 38 km/hr	Most violent in winter (may reach 100 km/hr)	1
Chinook (from Chinook Indian territory)	Eastern slope of Rockies	A warm wind that may, at times, result in sudden and drastic rise in temperature. May attain 60 or 70 °F in spring with R.H. 10%	Most notable in spring	1
Etesian (Gr, etesiai, annual)	Eastern Mediterranean	Cool, dry northeasterly wind that recurs annually	Summer and early autumn	2
Föhn (German, possibly from L, favoniun = growth, i.e., favoring wind)	Alpine lands	Similar to Chinook. Characterized by warmth and dryness	Most frequent in early spring	1
Haboob (Arabic)	Southern margins of Sahara (Sudan)	Hot, damp wind often containing sand. Of relative short duration (3 hr) average frequency of 24/yr	Early summer	Associated with northward advance of ITC
Harmattan (Arabic)	West Africa	Hot, dry wind characteristically dust laden	All year, but most effective in low-sun season	Circulation associated with N.E. trades

150

Wind	Location	Description	Season	Type[a]
Khamsin (Arabic)	N. Africa and Arabia	Hot, dry southeasterly wind. Regularly blows at a 50-day period (Khamsin = 50). Temperatures often 100–120°F. Same wind with adiabatic modifications include Ghibli (Libya), Sirocco (Mediterranean), Leveche (Spain)	Late winter, early spring	2
Levanter (from Levant, eastern Mediterranean)	Western Mediterranean	Strong easterly wind often felt in Straits of Gibraltar and Spain. Damp, moist, sometimes giving foggy weather for perhaps 2 days	Fall, early winter to late winter, spring	2
Mistral (maestrale of Italy = master wind)	Rhone Valley below Valence	Strong, cold wind channeled down Rhone Valley. May reach 100 km/hr in north. Can cause sudden chilling in coastal regions. (Note also the Bise, an equivalent cold north wind in other parts of France)	Most frequent in winter	2
Norther	Texas, Gulf of Mexico to W. Carribean	Cold, strong, northerly wind whose rapid onset may suddenly drastically lower temperatures (also Tehuantepecer of C America)	Winter	Related to circulation pattern over United States
Pampero	Pampas of S. America	Southern hemisphere equivalent of the Norther	Winter	Related to large-scale pressure patterns
Zonda	Argentina	A warm, dry wind on lee of the Andes. Can attain 120 km/hr. Comparable to Chinook and Föhn. In dry weather carries much dust	Winter	1

[a] 1 refers to winds associated with adiabatically warmed descending air. 2 refers to Mediterranean/North African winds that result from prevailing pressure conditions and frontal situations.

Figure 5-8. The Chinook or Föhn wind. (From Kuenen, P.H., *Realms of Water*, Copyright © 1955 by John Wiley and Sons. Used with permission.)

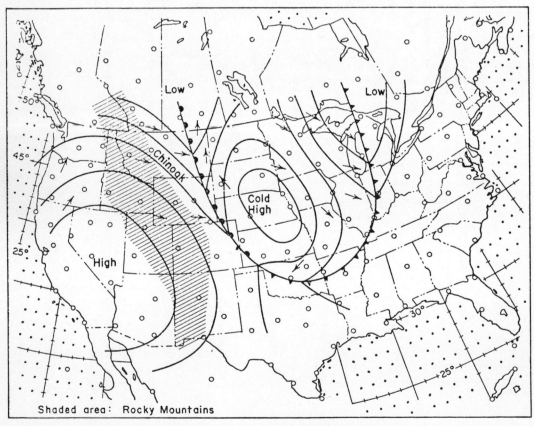

Figure 5-9. Diagram of a synoptic situation inducing the Chinook (Föhn) wind down the east slope of the Rockies. (From Rumney, G.R., *Climatology and the World's Climates*, 1968, The Macmillan Company. Used with permission.)

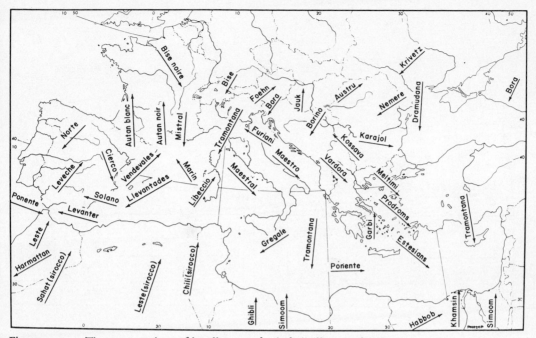

Figure 5–10. The great variety of locally named winds is illustrated in local winds of the Mediterranean Basin. (From Rumney, G.R., *Climatology and the World's Climates*, 1968, The Macmillan Company. Used with permission.)

climatic elements are interrelated so are plants, for they do not exist in isolation but rather as part of a larger plant community. Plants will thus respond not only to external physical elements but will also experience limitations placed upon them by other, often competitive, species. To describe the plant cover of an area, irrespective of the species making up the group, the term vegetation is used. Gleason and Cronquist (1964) have aptly described the concept of vegetation:

"The term vegetation . . . refers to the general aspects of the plants of an area taken collectively and regardless of the kinds of plants which produce that aspect. It is based upon the impression which the plant makes through our eyesight, not individually but en masse. . . . Why do a forest and a meadow look unalike? Because each is characterized by the prevalence of plants of a certain vegetative form. . . . The great bulk of plants in a forest consist of trees . . . under the trees are shrubs and herbs and even grasses, but the chief impression on the mind is made by the trees. In a meadow the bulk of plant life consists of grasses; various other plants of different growth form are

also in the meadow, but it is the grasses which give the impression which we call meadow."

It is this collective concept of plant life that has most occupied plant geographers and ecologists. The floristic approach of the botanist is quite secondary in such studies. The equatorial forest, the savanna, and the tundra are identified, not by the individual plants of which they are comprised, but by the collective impression of the entire vegetation cover.

The relationship between the world distribution of climate and vegetation has been investigated by many writers. Figure 5–11, for example, shows a schematic relationship between temperature and precipitation extremes and the way in which they are related to both climate and vegetation regions. Another example (Figure 5–12) shows how Miller deduced thermal boundaries for his climatic classification using vegetation distribution. By plotting distinctive vegetation types on a climograph, he derived equations used as boundary values. Note that the vegetation used on the climograph is repre-

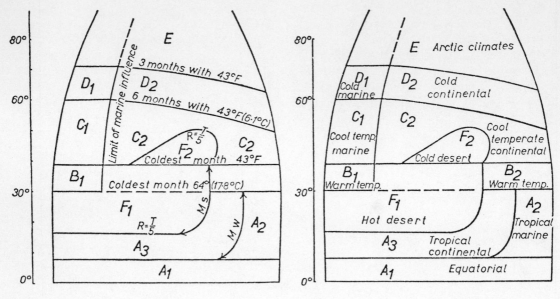

(a)

(b)

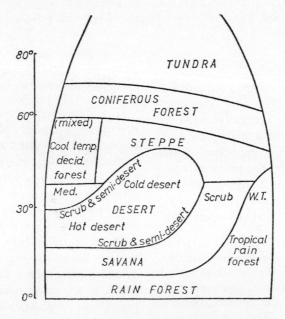

(c)

Figure 5–11. Schematic arrangement and relationships between (*a*) climatic limits of a classification system, in this case, A. A. Miller's scheme, (*b*) the identified climatic regions, and (*c*) the corresponding natural vegetation distribution of an idealized continent. (After Bucknell, 1966.)

sented in terms of major groups and does not reflect any particular plant species.

To gain some insight into the role played by climate in determining major vegetation groups, it is useful to assess the impact of climate upon the grouping found in an orderly vegetation classification scheme. A useful system has been devised by Dansereau (1957) who, following Schimper (1903), uses the following approach:

BIOSPHERE
(World: that part of the earth's crust and atmosphere which is favorable to some form of life)

↓

BIOCYCLE
(Major divisions of the Biosphere)

| Land | Freshwater | Saltwater |
| biocycle | biocycle | biocycle |

↓

BIOCHORE
(Divisions of the Biocycle e.g., Land Biocycle)

| Forest | Savanna | Grassland | Desert |
| biochore | biochore | biochore | biochore |

↓

CLIMAX AREA

↓

HABITAT
For example, Moving dune, bottomland, cliff

↓

LAYER
(Occurrence in space at a definite layer)

↓

BIOTYPE
(Subdivision of lowermost plants)

Examination of this shows different environmental variables operating at various levels. The biosphere is, of course, the total earth system, and its subdivision into biocycles represents the dominant physical media that comprise its near-surface characteristics, the land and water in both sea and land. Climate has its first impact in the derivation of the biochores. It is clear that the four groups named (forest, savanna, grassland, and desert) are differentiated on the basis of moisture availability. But within these biochores,

Dansereau recognizes vegetation associations that, under conditions of environmental equilibrium, might produce climax communities. The basis of the differentiation of these vegetation associations (Table 5–3) depends upon thermal characteristics. Within each group there is a spectrum of vegetation association that depends upon prevailing temperatures. For example, within the desert biochore, the following exists:

Hot dry← ─────────── →*Cold dry*
Tropical desert← ─────── →Arctic fell field

Passing to the lower orders of the scheme, topographic, geomorphic, and pedogenic factors become of prime concern. Thus, while a climax community of any biochore might exist, variations will be found within it as a result of such differences. It is in these lower orders that microclimatology becomes significant, with modifica-

Table 5–3

FORMATION CLASSES OF DANSEREAU'S VEGETATION SYSTEM

I. Forest Biochore
 1. Equatorial rainforest
 2. Tropical rainforest
 3. Monsoon forest
 4. Temperate rainforest
 5. Summergreen deciduous forest
 6. Needleleaf forest
 7. Evergreen-hardwood forest (Sclerophyll forest)

II. Savanna Biochore
 8. Savanna woodland
 9. Thornbush and tropical scrub
 10. Savanna
 11. Semidesert
 12. Heath
 13. Cold woodland

III. Grassland Biochore
 14. Prairie
 15. Steppe
 16. Grassy tundra

IV. Desert Biochore
 17. Dry desert
 18. Arctic fell field

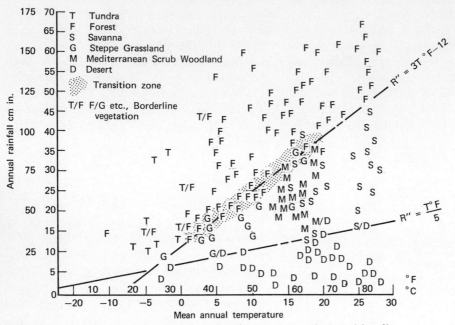

Figure 5-12. The use of vegetation distribution to attain empiric climatic limits. (After Miller, 1965.)

tions due to slope and surface moisture conditions playing an important role. A schematic representation of the scheme and the dimensions of the environment to which it applies is shown in Figure 5-13.

Some other climate-vegetation organizational schemes show a similar approach to that outlined in the vegetation associations of Danseareau. While boundaries alter from scheme to scheme, the nature of the grouping is similar, a fact that is well illustrated in Table 5-4, which lists world biomes according to another author. A biome, a term frequently used in ecologic texts, is essentially the same as a vegetation association. The similarity between biomes and identified climatic regions points to the significance of climate in ecologic studies. Vegetation response in a given area reflects the utilization of energy in that realm and, specifically, the rate at which radiant energy is stored by photosynthetic activity of producer organisms—chiefly the green plants—in the form of organic substances used as food materials. The gross primary production of a system (as compared to the net primary production, see Chapter 9) represents the total rate of photosynthesis, including organic matter

used in respiration. As Figure 5-14 indicates, the gross primary production varies appreciably, with the result that gross productivity within the identified biomes will vary accordingly.

Most of the biomes represented experience vulnerabilities, often as a result of climate. In the tundra, for example, the slow regenerative growth of vegetation poses great problems in terms of utilization by man. Problems of the construction of the trans-Alaskan pipeline system, already outlined, provide a guide to such difficulties. A good example is also shown in the case cited by Belous (1970). He tells of an instance in which a survey team gouged the initials of the company for which they worked, in the surface of the tundra. After a few years the depth of the letters had deepened from a few inches to many feet. The slow growth of vegetation allowed accelerated erosion to occur and it is estimated that it will take several hundred years before the eyesore disappears.

The vulnerability of both the hot desert and equatorial forest biomes are also related to climate. In the deserts the rapid evaporation rates may well result in the upward movement of water with eventual salinization of the surface.

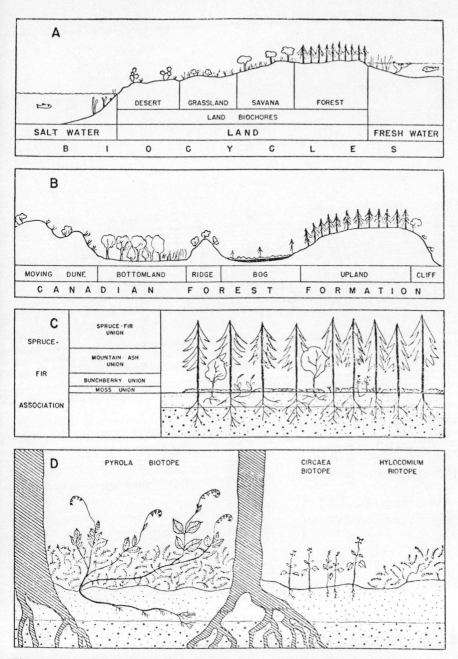

Figure 5–13. Dimensions of the environment. (A) The three biocycles and related land biochores. (B) An example of a formation within the forest biochore and of the several habitats determined by topography. (C) A single upland association with its four layers. (D) The lowermost layers of a spruce-fir association with their ultimate subdivisions, the biotopes. (From Dansereau, P., *Biogeography: An Ecological Perspective*, Copyright © 1957 by The Ronald Press. Used with permission.)

157

Table 5-4

LOCATIONS AND GENERAL ENVIRONMENTAL CONDITIONS FOR CERTAIN TYPES OF TERRESTRIAL ECOSYSTEMS[a]

Climax Ecosystem Type	Principal Locations	Precipitation Range (in./yr)	Temperature Range (°F) (Daily Maximum and Minimum)	Soils
Tropical rain forest	Central America (Atlantic coast) Amazon Basin Brazilian coast West African coast Congo Basin Malaya East Indies Philippines New Guinea N.E. Australia Pacific islands	50–500 Equatorial type: frequent torrential thunderstorms Tradewind type: steady, almost daily rains No dry period	Little annual variation Max. 85–95 Min. 65–80 No cold period	Mainly reddish laterites
Tropical savanna	Central America (Pacific coast) Orinoco Basin Brazil S. of Amazon Basin N. Central Africa East Africa S. Central Africa Madagascar India S.E. Asia Northern Australia	10–75 Warm season thunderstorms Almost no rain in cool season Long dry period during low sun	Considerable annual variation; no really cold period Rainy season (high sun) Max. 75–90 Min. 65–80 Dry season (low sun) Max. 70–90 Min. 55–65 Dry season (higher sun) Max. 85–105 Min. 70–80	Some laterites: considerable variety

Type	Distribution	Rainfall	Temperature	Soils
The atoll	Principally in tropical Pacific and Indian oceans	15–150 Convectional, but some tropical cyclones Droughts common	Little annual variation or range Max. 80–100 Min. 65–75 No cold period	Calcareous sand, gravel, and rubble Some atolls with phosphate "Jemo" hardpan soils
Broad-sclerophyll vegetation	Mediterranean region California Cape of Good Hope region Central Chile S.W. Australia	10–35 Almost all rainfall in cool season Summer very dry	Winter Max. 50–75 Min. 35–50 Summer Max. 65–105 Min. 55–80	Terra rossa, noncalcic red soils; considerable variation
Temperate grasslands	Central North America Eastern Europe Central and Western Asia Argentina New Zealand	12–80 Evenly distributed through the year or with a peak in summer Snow in winter	Winter Max. 0–65 Min. −50–50 Summer Max. 70–120 Min. 30–60	Black prairie soils Chernozems Chestnut and brown soils Almost all have a lime layer
Warm deserts	S.W. North America Peru and N. Chile North Africa Arabia S.W. Asia East Africa S.W. Africa Central Australia	0–10 Great irregularity Long dry season, up to several years in most severe deserts	Great diurnal variation Max. 80–135 Min. 35–75 Frosts rare	Reddish desert soils, often sandy or rocky Some saline soils

159

(Continued)

Table 5-4 (*continued*)

Climax Ecosystem Type	Principal Locations	Precipitation Range (in./yr)	Temperature Range (°F) (Daily Maximum and Minimum)	Soils
Cold deserts	Intermountain W. North America Patagonia Transcaspian Asia Central Asia	2–8 Great irregularity Long dry season Most precipitation in winter; some snow	Great diurnal variation Winter Max. 20–60 Min. −40–25 Frosts common ½–¾ of year Summer Max. 75–110 Min. 40–70	Gray desert soils, often sandy or rocky Some saline soils
Temperate deciduous forest	Eastern N. America Western Europe Eastern Asia	25–90 Evenly distributed through year Droughts rare Some snow	Winter Max. 10–70 Min. −20–45 Summer Max. 75–100 Min. 60–80	Gray-brown podzolic Red and yellow podzolic
Temperate rain forest	N.W. Pacific coast, North America W. coast, New Zealand Southern Chile Tasmania and S.E. Australia	50–350 Evenly distributed through year; wetter in winter Some snow	Winter Max. 35–50 Min. 25–45 Summer Max. 55–70 Min. 50–65	Podzolic, deep humus

			Temperature		Soil
Montane coniferous forests	Western North America Appalachian N. America European mts. Asian mts.	15–100 Evenly distributed or with summer dry season Snow may be very deep in winter	Winter Max. −20–60 Min. −55–35 Summer Max. 45–80 Min. 20–60		Various, podzolic, often shallow, rocky
Boreal coniferous forest	Northern N. America Northern Europe Northern Asia	15–40 Evenly distributed Much snow	Winter Max. −35–30 Min. −65–15 Summer Max. 50–70 Min. 20–55		True podzols Bog soils Some permafrost at depth, in places
Alpine tundras	Western N. America N. Appalachian N. America European mts. Asian mts. Andes African volcanoes New Zealand	30–80 Much winter snow; long-persisting snowbanks	Winter Max. −35–30 Min. −60–10 Summer Max. 40–70 Min. 15–35		Usually rocky Some turf and bog soils Polygons and stone nets Some permafrost
Arctic tundra	Northern N. America Greenland Northern Eurasia	4–20 Shallow snowdrifts, but many bare and dry areas in "High Arctic"	Winter Max. −40–20 Min. −70– 0 Summer Max. 35–60 Min. 30–45		Rocky or boggy Much patterned ground Permafrost

161

a From Billings (1970).

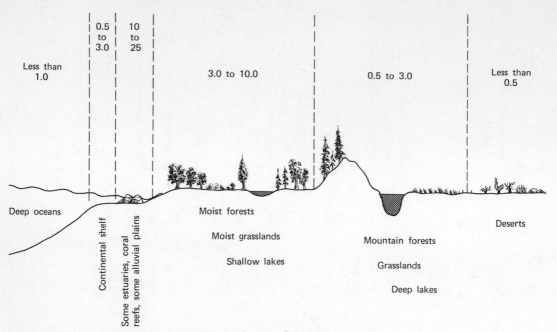

Figure 5-14. Estimates of annual gross production for major ecosystems. Values in 10^3kcal/m²/yr. (After Odum. 1971.)

This fragility is well demonstrated by Jacobson and Adams (1958) who relate the decline of Mesopotamian agriculture to increased soil salinity. A similar problem has been shown to exist in irrigated areas of dry regions today. Just as climate can cause excessive soil salinity in arid regions so can it cause rapid depletion of soil nutrients in the equatorial rainforest regions. High temperatures and rapid bacterial decay, together with copious water supply, causes excessive leaching under such climates. In agricultural development of the tropics this is one aspect of applied climatology that needs much consideration.

All of the other biomes face similar climatic problems: the desiccation of plants and resulting forest fires in the Mediterranean realm; the variability of precipitation in monsoon regions; and the periodic droughts that may occur even in "well-watered" biomes of the world. To explain, and possibly mitigate such problems, requires much groundwork in climatology.

In attempting to understand the distribution and potential utilization of vegetation zones, Holdridge (1947) introduced an interesting model. His original hypothesis was that vegetation should evolve selectively and survive only in rather limited sectors of a broad climatic environment. He suggests that the structure, form, and growth habits of plants could be grouped into well-marked plant communities of natural vegetation and that their grouping would reflect the climate of a given location. This is an assumption made by many; Holdridge's main contribution to understanding the distribution comes from the variables he uses and the model he constructs.

As variables, he uses mean annual precipitation, a potential evapotranspiration ratio, and a mean annual biotemperature given by the formula

$$\frac{\sum \text{mean monthly temperature} > 0°C}{12}$$

The axes of his model (Figure 5-15) are logarithmic and the areas defined are shown by hexagons, each representing a *natural life zone*. A number of workers have applied the Holdridge model to climate-vegetation studies, particularly in relation to Central American studies. Tosi (1964) has also outlined its applicability in terms of economic development.

The dominant life form that occurs within a given area can also be assessed using the vegeta-

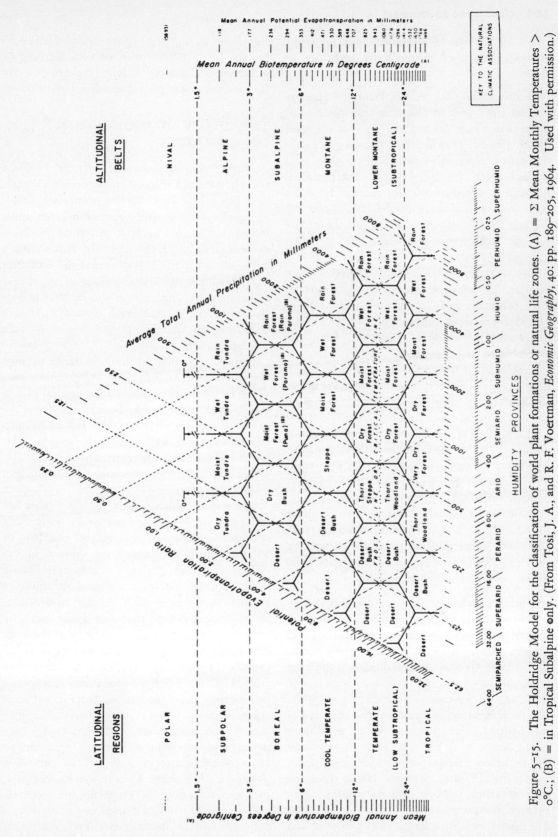

Figure 5-15. The Holdridge Model for the classification of world plant formations or natural life zones. (A) = Σ Mean Monthly Temperatures > 0°C.; (B) = in Tropical Subalpine only. (From Tosi, J. A., and R. F. Voertman, *Economic Geography*, 40: pp. 189–205, 1964. Used with permission.)

tion classification of Raunkaier (1934). In this scheme, groups of plant life are identified as follows:

Phanerophytes (Tall Aereal Plants).   The perennial trees and shrubs that bear their buds more than 25 cm (about 10 in.) above the level of the ground. They are thus exposed to all conditions of unfavorable weather.

Chamaephytes (Surface Plants).   The perennial herbs and low shrubs whose renewal buds are close to, or not more than 6 cm above the surface. They are found most frequently in cold regions (both Alpine and Arctic) and are often protected by snow cover in winter.

Hemicryptophytes (Half-Earth Plants).   Characterized by herbs and grasses, the aereal parts of these plants die down to soil level where budding occurs. By their position at or in the topmost soil layers they are afforded protection from unfavorable weather elements.

Geophytes (Earth Plants).   Plants whose perennating organs (bulbs, tubers, rhizomes) are below the surface so they are protected from such hazards as surface frost or desiccating winds.

Hydrophytes (Water Plants).   Plants that thrive within a water environment. The aqueous equivalent of the geophytes in that the reproductive parts of the plant are shielded from direct climatic variation.

Therophytes (Annuals).   Plants that complete their life cycles within a time span when conditions are most favorable to their budding activities. Seeds produced remain inactive until unfavorable conditions are modified.

The significance of this classification is seen when the composition of identified life forms is examined in relation to different climatic regimes (Figure 5-16). Under the hostile climatic conditions of the Arctic realm, the dominant life form is provided by the hemicryptophytes, comprising some 61% of the total population. As might be expected, hemicryptophytes are also common in the temperate realm where winter conditions might be severe. In this zone, though, tall aereal plants become more prevalent, forming about 15% of the life forms that exist. It is in the moist tropical realms, however, that the phanerophytes become dominant, comprising 61% of the total. The specialized plant forms of the tropical arid regions have already been discussed and it is not

surprising to find that phanerophytes make up 50% of the total life forms of such regions. As indicated in Figure 5-16, hemicryptophytes also form an important part of the desert plant cover.

THE EFFECT OF VEGETATION ON CLIMATE

From all that has been said so far, it is evident that climate has a marked influence upon vegetation. But this is not a one-way relationship and it is equally evident that vegetation cover must influence climate. Such a reciprocal effect is evident at various levels of study. It has already been pointed out (Chapter 1) that the evolution of atmosphere and the evolution of vegetation are intimately connected. It can be shown, too, that at a more local level, vegetation has a marked impact upon the climatic regime on both a meso- and microscale level.

Perhaps the most striking example of how climate can be modified by vegetation is provided by forests. Because of their vertical extent, forests modify the energy budget well above the earth-atmosphere interface. The results of such modification are well known to anyone who walked into a forest and experienced the modification that occurs in relation to the surrounding nonforested areas.

The way in which forests modify climate has intrigued workers for many years. One early controversy centered around the problem of whether the forest caused the climate or the climate caused the forest. The reasoning behind this controversy relates to the modified moisture balance of a forest, with graphic evidence of moisture contribution to the air being shown by "smoking forests." It was assumed that the large amount of moisture transpired by forests would provide the necessary moisture to be returned as rain. While there is little probability that forests do, in effect, cause themselves, there is still some argument about whether or not forests do induce higher precipitation values. Geiger (1965) cites many examples of the argument. One applies to India where, subsequent to a new forest law passed in 1875, large areas in the south-central part were reforested. Investigating the precipitation regime of the area caused one researcher to conclude that an increase in rainfall did occur. Another worker suggested, however, that the

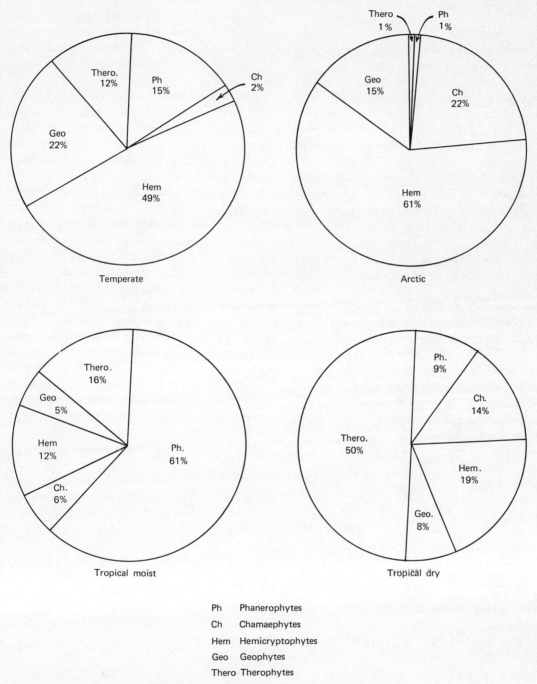

Ph   Phanerophytes
Ch   Chamaephytes
Hem   Hemicryptophytes
Geo   Geophytes
Thero Therophytes

Figure 5–16. The relative distribution of identified life forms under different climatic conditions. (Data from Polunin, 1960.)

change detected was merely part of a wide climatic variation that affected the whole country.

Aside from the controversial precipitation problem, forest climates do differ from those of the neighboring nonforested areas. Attainment of specific data has been slow, however, for the vertical extent of forests requires atmospheric data at all levels, and these can only be attained with ease by the construction of towers. Furthermore, interpretation of climatic modification by forests is made exceedingly difficult, because completed studies can only be used for forests of given ages and structures. The wide diversity of forest types and their relationship to local topographic conditions often exclude applicaton of derived data to other areas. One has only to think of the difference in structure between tropical and deciduous forests to appreciate the difficulty.

The boundary layer of the forest is its canopy, for it is at this level that energy exchanges will occur. Some insolation is returned directly to space, the amount depending upon the albedo or reflectivity of the canopy layer. In some forests this will vary enormously from season to season. Some energy will be trapped within the canopy layer while some will penetrate to the floor of the

forest. As illustrated by the data in Table 5–5, the amount of penetration is characteristically low. The evaluation of the data is further complicated in that the amounts involved in the dispostion of radiant energy varies both with the state of the sky and the amount of foliage that exists (Table 5–5*b*).

Compared to the flow of air over open areas, wind inside the forest is slight. Again, the amount of decrease will depend upon the type and structure of the forest concerned. Barry and Chorley (1970) cite the example of a European forest in which the wind velocity is reduced by as much as 60 to 80% at a distance of 100 ft inside the forest. Similarly, in a Brazilian forest, the wind speed was found to decrease from about 5 mph to 1 mph with the same penetration distance. The flow of air is, of course, highly complex and varies in the vertical as well as the horizontal dimensions. Figure 5–17 shows an example where winds of 5 to 15 mph above the canopy were less than 2 mph at the surface. Modification of wind speed in this way has been put to good use in the construction of shelter belts in areas that lack a natural tree cover (Chapter 9).

The forest environment also modifies local moisture conditions. Evaporation from the forest

Table 5–5

VARIATIONS IN RADIATION RECEIPTS WITHIN FORESTS[a]

a. Daily totals of net radiation in and above a young pine forest (ly/day)

Height (m)	10.0	5.0	4.1	3.3	2.1	0.2
July total 7th	566	555	223	36	—	35
1952 percent	100	98	39	6	—	6
November total 9th	291	—	104	—	14	—
1951 percent	100	—	35	—	5	—

b. Solar radiation received on a horizontal surface at the 1-m level in an oak forest as a percentage of incident radiation above the canopy

	Clear Sky	Overcast Sky
Foliaged	9%	11%
Defoliaged	27%	56%

[a] From Munn (1966).

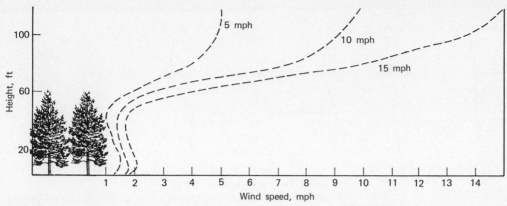

Figure 5-17.   Wind velocity profiles showing the influence of a 45-yr-old stand of ponderosa pine (*Pinus ponderosa*). (After Fons and Kittredge, 1948.)

floor is relatively low because of reduced insolation. This is counterbalanced by the fact that with the profuse vegetation, high transpiration occurs. As such, the humidity within a forest depends upon the density of the forest and the rates of transpiration that occur. It is found generally that the relative humidity in the forest may be from 2% to 10% higher than that of non-forested areas, with the highest humidities occurring during the high-sun season. Note that comparison of humidity using relative humidity values is not always meaningful, for the modified thermal environment directly influences the water-holding capability expressed using relative humidity.

The thermal differences that do occur result from a combination of the factors already outlined: shelter from direct rays of the sun, heat modification through water transfer, and the "blanket" effect caused by the canopy. The essential result of the interaction of these factors is that temperatures inside a forest are moderated, the maximum is lower, and the minimum higher than those in nonforested areas experiencing a

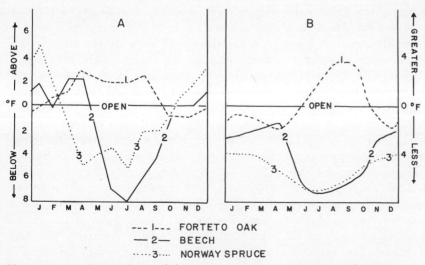

Figure 5-18.   Comparison of forest temperatures with those of neighboring open areas. (A) Mean monthly temperatures. (B) Mean monthly temperature ranges. Note the anomalous conditions of the forteto oak forests. (After FAO, 1962.)

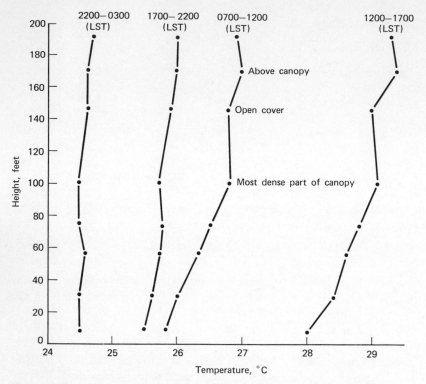

Figure 5-19. Forest temperature profiles at different times of the day. Data derived from a 200-ft tower in a Colombian forest. (Data from Munn, 1966.)

similar climatic regime. The amount of variation is seasonal, the mean difference in summer being as much as 5°F (2.8°C) within a low-altitude, middle-latitude forest (Figure 5-18). Exceptions to such moderating influences do occur; the Forteto Oak forests of the Mediterranean, for example, experience higher temperatures than neighboring nonforested areas. Such trees transpire slowly with the result that the "usual" hygric and thermal conditions of the forest are modified.

Forest temperatures vary in a vertical as well as a horizontal direction. For the most part, temperature increases with height during the day while lapse conditions occur at night. This reflects the role of the canopy in the energy exchange. Figure 5-19 shows the results obtained from a tower (one of two erected 1100 m apart) in a Colombian forest. The expected responses in vertical temperature distribution

over a 24-hr period are seen to occur. The data derived from the second tower were not significantly different and indicated that the vertical temperatures shown in the diagram are typical for wide areas.

While the forest, with its extensive vertical cover, provides the most marked influence of vegetation upon climate, almost all other vegetation types result in modified climatic regimes. Clearly, many such modifications will occur at the microlevel for, as has been shown, the energy exchange that occurs at any given location depends, in part, upon the nature of that surface. Differences caused by varying albedos, transpiration rates, soil characteristics, and so on are aspects of plant growth being actively investigated. For the most part, much of the derived data applies to agricultural systems and such problems are covered later within the realm of agricultural climatology.

CHAPTER 6

Organization of
the Climatic Environment

Devising a suitable classification scheme is difficult in most disciplines because, inevitably, a decision must be made concerning the selection of representative variables used as inputs into the system. In the formulation of a climatic classification the problem is compounded considerably. Climate is an abstract concept, it represents the summation of all interacting atmospheric processes and, as such, does not exist at any given moment. It is a man-designed concept that must be superimposed upon an environment in which climate per se has no meaning. Thus, while it is necessary to systematize the long-term effect of interacting atmospheric processes, the manner in which they are systematized is a field of controversy and contradiction.

That climatology is, in part, concerned with spatial aspects of the environment means that the grouping of similar climates revolves around the problem of where to draw the line. The solution of this problem, together with the selection of appropriate input variables, provides the reason for the many different interpretations of the grouping and location of climatic types. Much of the literature dealing with climatic classification appears more concerned with the limits of a climate group than with the climate itself. The reason for this is suggested by Carter (1966), who writes ". . . the limits or boundaries of the climate types, not the cores of the climatic regions, are the intellectual fields where toil has produced the most elegant results." Whether such an emphasis is valid is debatable, for some might feel the desert itself is more important than the desert boundary.

Regardless of the problem of emphasis on core or boundary, it remains a fact that any demarcation line showing the limit of a climatic region is often quite arbitrary and usually most subjective. Changes in nature, except in highly specialized circumstances, are transitional and a well-defined boundary shown on a map usually represents a transition. Where the line is drawn depends upon the criteria used in the completed system and the purpose to which it is to be put. Johnson (1968), in a sophisticated discussion of choice in classification, suggests that it is a subjective process and that, although analysis may be based upon numerical techniques, an investigator can make his data show what he wants it to show, whether it be the existence or the nonexistence of boundaries of transitional zones.

Because different criteria are used, and because classifications are usually devised for a particular purpose, climatic regions will vary. The amount of variation depends partially on the "fineness" of the system, so that the *number* of regions differentiated will vary enormously. It is possible to recognize a complete spectrum of identified regions ranging from the assumption that each individual climatic station represents a climatic region (so that for every station there is an equivalent climatic region) to an assumption that the earth, as a planet at a given distance from the sun, experiences (in comparison to other planets of the solar system) a unit climate. Actual classification schemes do not vary to such great extremes but, as shown in Figure 6–1, the number of climatic regions used to designate a single area can be quite variable.

As a result of the many variables pertaining to the creation of a climatic classification system, the selection of a classification scheme for use in a particular problem must be treated with care. At the present time, two or three climatic schemes dominate the field, to the exclusion of most others. Often, an investigator in a field other than climatology might be called upon to use a classification system. The natural response is to look into general texts on climate and utilize the system shown because it is clearly "acceptable." Unfortunately, and all too frequently, the scheme is based upon principles and biases other than those required by the investigator. Subsequent difficulty is usually

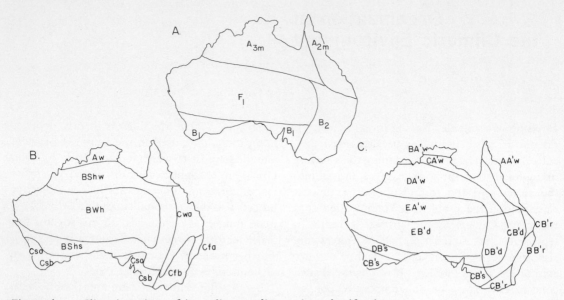

Figure 6–1. Climatic regions of Australia according to three classifications.
(A) Miller differentiates 6 climatic regions. (B) Köppen differentiates 10
climatic regions. (C) Thornthwaite differentiates 12 climatic regions.

encountered in attempting to correlate climatic influences and distribution with the investigator's own needs. Many good, if rather specialized, climate systems have been devised; often they are overlooked because of unawareness on the investigator's part.

APPROACHES TO CLIMATIC CLASSIFICATION

That different climatic environments exist over the earth's surface has been noted since early historic times. The classical Greeks made great advances in astronomy, and one result of such endeavours was the division of the earth into zones of illumination. Such zones, using the significant parallels 0°, 23½°, and 66½° are related to the distribution of zonal temperatures. Ptolemy used the base to divide the world into seven climatic types essentially dependent on length of day. The relative shortcomings of a climatic division based upon solar illumination-temperature zones were recognized by Arabic scholars between the ninth and twelfth centuries, and they modified the system. Europeans, however, inherited the Greek scheme and used it as a base for many years. The division of the world into torrid, temperate, and frigid zones (an aspect still found in some elementary texts) is an outcome of this inheritance.

It is significant that this early classification used only one element of climate. It is quite apparent that the easiest way to divide the earth into climatic regions is to select a single variable and plot its distribution on a map. While such divisions are simplistic, and obviously not representative of the complete climatic complex, they are useful. For example, Figure 6–2 shows the thermal divisions of the earth, following the work of Herbertson (1905). The division, used to facilitate organization of natural regions in relation to crop growth, is an appreciable advance on the three-part zonal concept. Similar distributional maps could be presented showing the world distribution of other climatic elements. Each, in its way, would form a single-element classification and would be useful for a given purpose; but beyond that very limiting purpose, the map would have little significance.

To formulate a meaningful climate system, it is therefore necessary to incorporate more than one climatic variable. Most of the classifications currently available use temperature and precipitation as their variables. For the most part, the choice of these two has been determined historically. Reliable data have been available

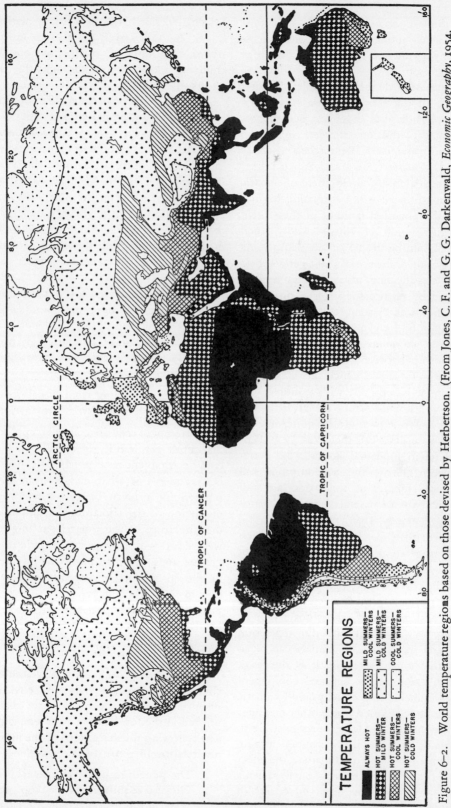

Figure 6–2. World temperature regions based on those devised by Herbertson. (From Jones, C. F. and G. G. Darkenwald, *Economic Geography*, 1954. Used with permission of the Macmillan Company.)

TEMPERATURE REGIONS

- ALWAYS HOT
- HOT SUMMERS— MILD WINTER
- HOT SUMMERS— COOL WINTERS
- HOT SUMMERS— COLD WINTERS
- MILD SUMMERS— COOL WINTERS
- MILD SUMMERS— COLD WINTERS
- COOL SUMMERS— COLD WINTERS
- HOT SUMMERS— COLD WINTERS

ARCTIC CIRCLE

TROPIC OF CANCER

TROPIC OF CAPRICORN

only over the last 100 years (much less for most climatic stations) and many early records consist only of temperature and precipitation. Much of the early trendsetting work in classification was limited to the use of these two variables. Furthermore, much of the early work was associated with plant physiologists and plant geographers who found a correlation between vegetation and the two parameters. Researchers such as Dove (1864), Linsser (1865), and Supan (1877) were all influential in the development of climatic classification so that it is not surprising to find, in view of the botanical training of these men, that the distribution of climate and natural vegetation should be treated simultaneously. As an example of the influence of plant geographers on climatic classification, many climatic regions are identified by plant association, a procedure that is still followed by some writers today; it is not unusual to find a climate type described as *savanna*, *taiga*, or *tundra*, all vegetation terms. The relationship between climate and biology is illustrated in Table 6–1, which shows the 1900 classification of Köppen. It will be noted that both plant and animal names are given to climatic regions. While it would be quite unusual to find a climate described the "Yak" or "Penguin" climate in modern literature, the correlation between climate and vegetation is still prevalent and climates of the world are still described in terms of natural vegetation distribution (e.g., Rumney, 1968).

It is evident that, in relating the distribution of natural vegetation to the distribution of climate, the effect of climate is being measured, instead of the climate itself. It is assumed that a given climate gives rise to a distinctive vegetation association and, to identify the climate type, it is first necessary to determine the vegetation and then infer the climate. If climate can be so identified, that is by expressing it as the result of the distribution of one selected component of the environment, then equally good climatic regions may be recognized using other measures of the effects of climate. It becomes possible to devise climatic schemes using factors ranging from the human response to climate to the effect of climate on rock weathering. Such systems would be based upon the observed effects of climate and the criteria used to delimit their boundaries established by best-fit properties.

Systems derived through such methodology might be collectively termed *empiric classifications*. The use of qualifying term empiric connotes identification through the observed effects of climate.

The empiric systems essentially concern themselves with *where* similar climate types occur. An equally valid approach to climatic distribution concerns the study of *why* climatic types occur in distinctive locations. Systems that attempt to deal with the question "why" must concern the cause of climate. As such, they may be termed collectively the *genetic classifications*. Like the empiric systems, there are a number of methods of examining the cause of climates so that the basis of genetic systems can vary appreciably.

Much controversy exists concerning the relative merits of the genetic and empiric approaches to classification. The view presented here is a reiteration of a summary by Mortensen (1934), who states that two great groups of climatic divisions exist:

i. Those proceeding from actual observation.
ii. Those proceeding from explanation.

If it is accepted that there are two approches and that both are valid, then the genetic versus empiric argument need not be of concern. What is of concern is what classification should be selected for a given distributional problem. A discussion of the climatological implications of the migration of the polar front might not be well dealt with using an empiric system; a genetic approach may not prove of great value in the discussion of specific temperature requirements for plant types. In effect, the approach used and the classification selected depends on the purpose to which it is to be put.

The Empiric Classifications

Because it is possible to observe the effects of climate on a whole range of environmental phenomena, there are many bases that can be utilized in the formulation of an empiric system. The following illustrate some of the innumerable interrelationships that can be examined.

1. The human response to climate.
2. Climatic requirements for crop growth.
3. Water needs and precipitation effectiveness related to vegetation.

4. Study and identification of climatic analogs, for example, agricultural analogs.

5. Vegetation distribution related to climatic controls.

6. Geomorphic processes acting under different climatic conditions.

7. Climate and soil-forming processes.

8. Continentality and oceanicity as climatic determinants.

A little thought could provide many other relationships, and for each there is probably a climatic classification available.

With so many approaches it is little wonder that there are many empiric systems. Some of the

Table 6–1
CLIMATIC REGIONS IDENTIFIED IN KÖPPEN'S (1900) CLASSIFICATION

A. Megathermal or tropical lowland climates. Coldest months 18°C or above
1. Liana
2. Baobab

B. Xerophytic climates—arid and semiarid. Continuous scarcity of precipitation
  I. Coastal deserts of low latitudes
   3. Garua (fog)
  II. Lowland desert and steppe experiencing great summer heat
   4. Date palm
   5. Mesquite
   6. Tragacanth
   7. East Patagonian
  III. Lowland desert and steppe with cold winters and short, hot summers
   8. Buran
   9. Prairie

C. Mesothermal or temperate climates, Coldest month below 18°C, warmest month over 22°C.
  I. Subtropical climates with moist, hot summers
   10. Camelia
   11. Hickory
   12. Maize
  II. Subtropical climates with mild, wet winters and dry summers
   13. Olive
   14. Heather
  III. Tropical mountain climates and maritime climates of middle latitudes
   15. Fuchsia
   16. Upland savanna

D. Microthermal or Cool Climates. Warmest months between 10°C and 22°C.
   17. Oak
   18. Spruce
   19. Southern beech

E. Hekistotherm of Cold Climates. Warmest month between 0°C and 10°C
   20. Tundra
   21. Penguin or Antarctic
   22. Yak (Pamir)
   23. Rhododendron
   24. Ice cap

Note: For each of the climate regions, Köppen supplied quantitative precipitation and temperature boundaries. For details see Köppen (1900), Knoch, and Schulze (1954), for Gentilli (1958).

diverse approaches are listed in Table 6–2. For each of the systems listed, its purpose and method of climatic differentiation is given. While the list is lengthy, it is by no means complete. Clearly, it is impossible to discuss all of the systems given, so to gain some insight into the different methodologies, selected systems are outlined in the Appendix. Those given are not necessarily the best available; they merely represent a cross section of available empiric systems.

Of the many systems that have been devised, it is inevitable that one or two develop into what might be termed "standard systems." Such classifications become standard as a function of their wide usage; since contact with climatic classification occurs mostly in introductory courses, it follows that the most widely used systems are those that facilitate an orderly description of world climates. To achieve any prominence such a system must, of course, be acceptable conceptually.

One system that has developed along these lines is that formulated by Köppen. In its various forms, this is probably the most widely used of all climatic systems. A second scheme that, largely as a result of its innovative methodology, is also widely used is that devised by C. W. Thornthwaite.

The Köppen System (See Appendix).

Wladimir Köppen has made one of the most lasting and important contributions in the field of climatic classification. He was trained as a botanist and, in the early stages of his work he was strongly influenced by the work of de Candolle and Supan. The systems he has formulated range from a highly descriptive vegetation zonal scheme to a classification in which boundaries are defined in relatively precise mathematical terms. The Köppen system has been considerably modified over time and a recent version is shown in Figure 6–3.

Beginning with his doctoral dissertation (Leipzig, 1870) and continuing up to his death in 1940, Köppen proposed, modified, and re-modified a system of which Hare (1951) has said some regard "... as an international standard, to depart from which is scientific heresy." Such rigorous interpretation was not probably intended by Köppen, to whom the scheme was

never completely satisfactory. The evolution of the system, so admirably reviewed by Wilcock (1968), shows that Köppen was not so concerned with the precise boundaries as he was with attempting to use simple observations of selected climatic elements to provide a first-order world pattern of climates.

Köppen's early work was completed at a time when plant geographers were first compiling vegetation maps of the world. His early publications (1870, 1884) were concerned with temperature distribution in relation to plant growth, and it was not until 1900 that any of his publications were really concerned with world climatic classification. The 1900 system (Table 6–1), which did not obtain much notice, is a highly descriptive scheme making use of plant and animal names to characterize climate. In 1918, Köppen produced a system which is substantially that in use at the present time. Boundary values have changed, new symbols introduced, but the framework of the present system was clearly evident. The scheme demonstrates Köppen's major contribution to the systematic treatment of the climates of the world. He recognizes a pattern underlying world climatic regions and introduces a quantification method that allows any set of data to be located within the system. The classification is considerably enchanced by the introduction of a unique set of letter symbols that obviate the necessity of long descriptive terms.

If the success of a climatic classification can be judged by the amount of literature concerning it, then the Köppen system is highly successful Most texts in general climatology provide, at the very least, an outline of the system. Similarly, innumerable papers have been published that have dealt with the Köppen classification in one form or another. Wilcock (1968) suggests that publications concerning the system are of three main types. First, there are those in which the system is faithfully used to describe and analyse world climates; second, there are papers that are concerned with the modification and clarification of its methodology; and third, there are those publications that call for its rejection.

Faithful reproduction of the Köppen system is most evident in papers published in the 1920s and 1930s. During the 1930 period, practically every major region was analysed, the coverage

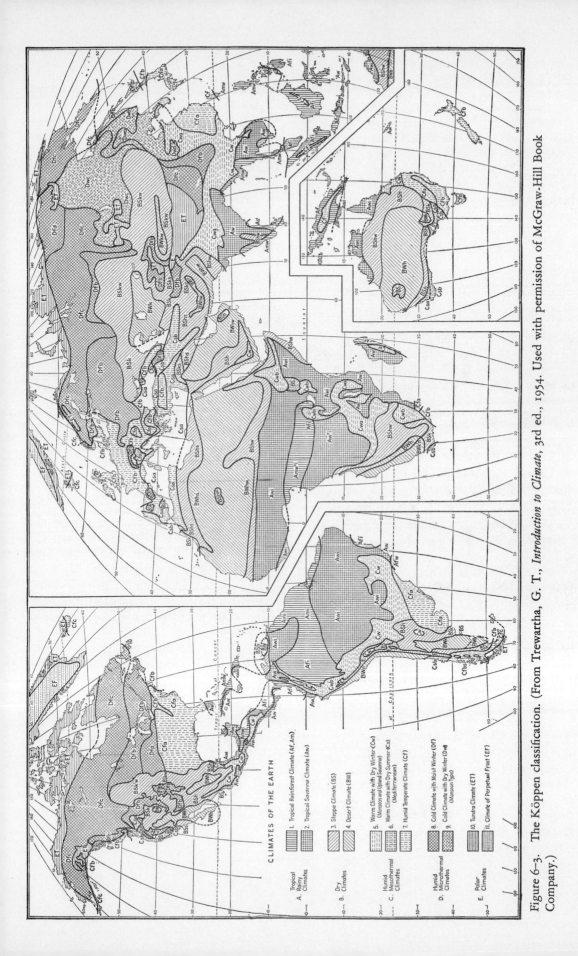

CLIMATES OF THE EARTH

A. Tropical Rainy Climates

1. Tropical Rainforest Climate (Af, Am)

2. Tropical Savanna Climate (Aw)

B. Dry Climates

3. Steppe Climate (BS)

4. Desert Climate (BW)

C. Humid Mesothermal Climates

5. Warm Climate with Dry Winter (Cw) (Monsoon and Upland Savanna)

6. Warm Climate with Dry Summer (Cs) (Mediterranean)

7. Humid Temperate Climate (Cf)

D. Humid Microthermal Climates

8. Cold Climate with Moist Winter (Df)

9. Cold Climate with Dry Winter (Dw) (Monsoon Type)

E. Polar Climates

10. Tundra Climate (ET)

11. Climate of Perpetual Frost (EF)

Figure 6-3. The Köppen classification. (From Trewartha, G. T., *Introduction to Climate*, 3rd ed., 1954. Used with permission of McGraw-Hill Book Company.)

Table 6–2
SOME EMPIRICAL APPROACHES TO CLIMATE CLASSIFICATION

Author(s)	Purpose	Base of System
Bagnouls and Gaussen (1957)	Biological climates	Duration of the dry season based upon the derived Xerothermal Index (X). Twelve major regions identified according to X values, temperature of coldest month, and frost/snow data
Blair (1942)	An orderly description of world climates	Five main zonal climates distinguished: tropical (T), subtropical (ST), intermediate (I), subpolar (SP), and polar (P). Fourteen types and 6 subtypes distinguished using letter notation. Based upon precipitation and temperature data and related to vegetation types
Brazol (1954)	Human comfort zones	Use of wet and dry bulb temperatures to establish comfort months. Twelve ranked classes ranked from No. 12—lethal heat—to No. 1—glacial cold
Budyko (1958)	Distribution of energy in relation to water budget	Use of rational index of dryness to relate ratio of net radiation to energy required for vaporizing moisture. Moisture regions form basic unit
Carter and Mather (1966)	Environmental biology	A modification of the 1948 Thornthwaite system
Creutzberg (1950)	Climate-vegetation relationships	The annual rhythm of climate based upon identification of "Isohygromen" (lines of equal duration of humid months) and "Tag-Isochione" (lines of equal daily snow cover duration). Four major zones differentiated, subdivided by monthly moisture values
de Martonne (1909 and following years)	World regional (land-form) studies	Nine first-order divisions based upon temperature and precipitation criteria. Numerous subdivisions named for local areas in Europe. Considerable attention given to desert limits, but most boundaries derived nonquantitatively
Emberger (1955)	Biologic (ecologic)–climate relationships	Two main climates differentiated, deserts and non-deserts. Differentiated and subdivided in terms of annual range of temperature and duration of light periods
Federov[a]	The "complex" method, utilization of day-to-day weather observations	An incomplete system that relies upon codification of daily weather events. For example, the first letter indicates character of prevailing wind, second letter character of temperature, third letter character of precipitation, cloudiness, humidity, fourth letter character of various phenomena of atmosphere and state-of-ground surface
Gorsczynski (1945)	The Decimal System	Ten "decimal" types associated with five main zonal climates. Emphasis on continental versus marine climates and definition of aridity
Geiger-Pohl (1953)	World map of climate types	Modification of the Köppen system
Köppen (1918 and following years)	See description in text	

(Continued)

Table 6–2 (*continued*)

Author(s)	Purpose	Base of System
Malmstrom (1969)	Precipitation effectiveness as a teaching scheme	Retention of the basic concepts of the Thornthwaite (1948) system but with arbitrary threshold values to express water needs of plants. Warmth index $(N - 38m/100)$ also used
Miller (1931 and following years)	See description in Appendix	
Papadakis (1966)	Agricultural potential of climatic regions	Use of "crop-ecologic" characteristics of a climate based upon empirically derived threshold values. Ten main climate groups recognized, each divided into subgroups which are themselves divided
Passarge (1924)	Climate-vegetation relationships	Recognition of five climatic zones and their subdivision into ten regions emphasizing vegetation distribution
Penck (1910)	World climates in relation to studies in physical geography and physiography	Recognition of three main types of climates significant in determining weathering and erosion. Humid, Arid, and Nival. Each subdivided into two
Peguy (1961)	See description in Appendix	
Philippson (1933)	Climatic regionalization on both world, continental and regional level	Based upon temperature of warmest and coldest months and upon precipitation characteristics. Five climatic zones with 21 climatic types and 63 climatic provinces
Putnam et al. (1960)	Coastal environments of the world, climate-vegetation characteristics	Fourteen types recognized. Climatic characteristics expressed as those occurring between the 25th and 75th percentile of the frequency distribution appropriate to each climatic types. Variables include mean maximum and minimum temperatures, mean annual precipitation and monthly precipitation frequency
Terjung (1968)	See description in Chapter 7	
Thornthwaite (1948)	See description in text	
Trewartha (1954)	An orderly description of world climates	Modification of the Köppen system (See Appendix)
Troll (1963)	See description in Appendix	
U.S. Army, Natick Laboratories (1969)	See description in Appendix	
Vahl (1919)	World climates related to vegetation	Five zones appraised by temperature limits which are a function of the data for warmest and coldest months. Subdivision by precipitation expressed as a percentage of number of wet days in a given humid month
von Wissman (1948)	World distribution of climate related to vegetation	Related to the Köppen approach. Five temperature zones subdivided by precipitation distribution and temperature regimes

[a] For an account of Federov's system, see Lydolph (1959).

gaining much impetus from the Köppen-Geiger Handbuch (1936) where the system was used almost exclusively. While many authors followed rules, others suggested methods by which the system might be improved in its application to different parts of the world. Modifications proposed by Ackerman (1941) and, more recently, Shear (1964) are provided as examples in the Appendix.

A number of climatologists have proposed systems that are, for the most part, simplifications of the Köppen classification. Not all are successful and the notation "after Köppen" found at the foot of maps resulting from such simplification frequently means that the depiction is a Köppen map with many important features omitted. Some modifications are, however, more meaningful—those proposed by Trewartha provide a good example.

In simplifying Köppen's system, Trewartha (1954) states that his purpose is to produce ". . . a general scheme of climatic classification for use in a first course in climatology where the goal of instruction is the development of the world pattern of climates." In effect, Trewartha is attempting to make the Köppen system a better teaching device. That the modification is successful is attested by its wide use; it appears, for example, in successive editions of *Goode's World Atlas* and is available as a demonstration wall map.

In a recent edition of his work (1968), Trewartha has modified the system to such a degree that its departure from the original Köppen system is quite radical. The new system stands as a classification in its own right and can no longer be merely considered a "modified Köppen." The new classification (Figure 6–4), makes an interesting comparison between the original Köppen map (Figure 6–3) and the early Trewartha modification. (See Appendix for discussion).

The Thornthwaite Classification. In a paper published in 1931, C. W. Thornthwaite proposed a climatic classification that appeared as a marked departure from preexisting systems. Unlike most classifications available at the time (based upon temperature zonation for first-order grouping), Thornthwaite based his system on the concept of precipitation and thermal efficiency. Many authors prior to Thornthwaite had suggested that the relationship between precipitation and evaporation could provide a useful measure of precipitation effectiveness but few had utilized the concept because of lack of evaporation data. Faced by the same shortage, Thornthwaite produced a precipitation-evaporation index that could be determined empirically from available data. Using this index, Thornthwaite determined humidity provinces, and these formed the first-order division of his classification scheme. Unlike the Köppen system, boundaries between provinces are not related to any practical vegetation or soil criteria; instead, they are based upon regular arithmetic intervals of derived values. The 1931 system (Table 6–3) has been adequately described and analyzed by a number of authors (Hare, 1951, Carter and Mather, 1966, Hidore, 1968) and, despite some adverse criticism, it may be considered as a major contribution to the process of climatic classification.

Of more significance at present (because it has superseded the earlier system), is Thornthwaite's 1948 classification. This shows a radical departure from the 1931 system because it makes use of the important concept of evapotranspiration. While the earlier system had been concerned with the loss of moisture through evaporation, the new approach considers loss through the combined process of evaporation and transpiration. Plants are considered as physical mechanisms by which moisture is returned to the air. The combined loss is termed evapotranspiration, and when the amount of moisture available is non-limiting, the term potential evapotranspiration is used.

As with any widely used system, the 1948 classification has been subject to criticism. Many of the criticisms relate to the empiric formula used to express evapotranspiration and to the way in which the water budget of a station is manipulated. Such criticisms are dealt with in a later section because, at this point, concern is with the actual classification system.

The two major aspects of the system are the use of a moisture index and a thermal efficiency index. The moisture index is obtained by comparing amounts of moisture received (precipitation) with the amount of moisture needed to satisfy plant requirements under a given set of

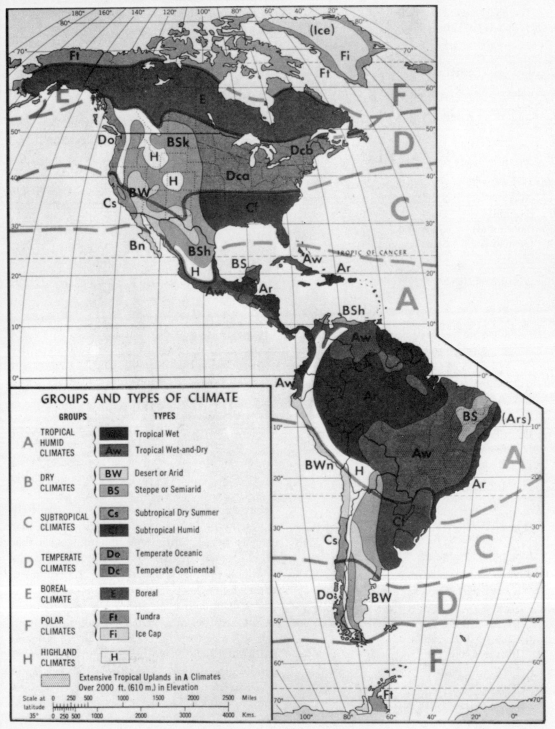

Figure 6–4. Trewartha's 1968 classification as applied to the western hemisphere. (From Trewartha, G. T., *Introduction to Climate*, 4th ed., 1968. Used with permission of McGraw-Hill Book Company.)

Table 6-3
STRUCTURE OF THE FIRST THORNTHWAITE CLASSIFICATION OF CLIMATE[a]

Precipitation effectiveness

P-E index:

$$I = \text{sum of 12 monthly values of } 115\left(\frac{P}{T-10}\right)^{\frac{10}{9}}$$

P = mean monthly precipitation in inches
T = mean monthly temperature in °F

Humidity Province	Vegetation	P-E Index
A (Wet)	Rain forest	128 or above
B (Humid)	Forest	64–127
C (Subhumid)	Grassland	32–63
D (Semiarid)	Steppe	16–31
E (Arid)	Desert	Less than 16

Temperature efficiency

T-E Index:

$$I' = \text{sum of 12 monthly values of } \left(\frac{T-32}{4}\right)$$

T = mean monthly temperature in °F

Temperature Province	T-E Index
A' (Tropical)	128 or above
B' (Mesothermal)	64–127
C' (Microthermal)	32–63
D' (Taiga)	16–31
E' (Tundra)	1–15
F' (Frost)	0

Seasonal distribution of precipitation

r	Rainfall adequate in all seasons
s	Rainfall deficient in summer
w	Rainfall deficient in winter
d	Rainfall deficient in all seasons

Climatic types

AA'r	BA'r	CA'r	DA'w	EA'd	D'	E'	F'
AB'r	BA'w	CA'w	DA'd	EB'd			
AC'r	BB'r	CA'd	DB'w	EC'd			
	BB'w	CB'r	DB's				
	BB's	CB'w	DB'd				
	BC'r	CB's	DC'd				
	BC's	CB'd					
		CC'r					
		CC's					
		CC'd					

[a] From Hidore (1969), adapted from C. W. Thornthwaite, "The Climates of North America According to a New Classification," *Geog. Rev.*, 21, 4 (October 1931), 633–55.

temperature conditions (the need = potential evapotranspiration). If, in a given month of the year, the amount of moisture required for evapotranspiration is exceeded by precipitation, then a moisture surplus occurs:

$$\text{surplus} = \text{rainfall} - \text{need}$$
$$s = r - n$$

Such conditions will occur under humid conditions and a humid index may be expressed:

$$I_h = 100s/n$$

Under dry conditions, the amount of precipitation is less than that required to support active plant moisture needs. A deficit occurs:

$$\text{deficit} = \text{need} - \text{rainfall}$$
$$d = n - r$$

and an aridity index may be derived:

$$I_a = 100d/n$$

For humid months of the year, the I_h values will be derived, for the arid months the IA. By evaluating the two over the course of a year, it is possible to derive a moisture index:

$$I_m = I_h - I_a$$

Clearly, under extremely moist conditions, when every month of the year experiences a surplus, the I_m value will be a high positive number. In arid climates, with a deficit throughout the year, the I_m will be negative.

But the relationship shown is not completely representative of the sequence of events that actually occur. Consider a climatic station that experiences a number of consecutive months in which there is a moisture surplus. Suppose, at the end of this period, that a number of deficit months occur. In the transition from the humid to the arid periods, plants do not suffer from lack of moisture immediately. They draw on moisture in the soil derived from the surplus period. Thus, there is a period of soil water utilization prior to the onset of the effect of a deficit period. In a similar way, at the end of a dry period, soil moisture must be recharged prior to a period of surplus moisture.

To allow for such exchanges, Thornthwaite weighted his equation by considering that a surplus of 6 cm in one season compensates for a deficit of 10 cm in another. Thus, the aridity index (I_a) carries less weight than the humid (I_h) by a factor of 0.6.[1] The I_m is amended to read

$$I_m = I_h - 0.6 \, I_a$$

By substituting the earlier equations for I_h $(= 100s/n)$ and I_a $(= 100d/n)$ we get

$$I_m = \frac{100s}{n} - 0.6 \, \frac{100d}{n}$$

$$= \frac{100s - 60d}{n}$$

that, in verbal form, reads

moisture index

$$= \frac{(100 \times \text{water surplus}) - (60 \times \text{water deficit})}{\text{potential evapotranspiration}}$$

The procedure for determining the moisture index, together with graphical illustrations of the results, are shown in the Appendix.

After determination of the moisture index, the second major consideration concerns the thermal index, a measure of thermal efficiency. Since thermal efficiency is a function of temperature and length of daylight, such factors are already considered in Thornthwaite's equation for the evaluation of potential evapotranspiration. As shown in Chapter 2, this is:

$$e = 1.6 \, (10t/I)^a$$

where e = 30-day ET in cm
  t = mean monthly temperature in °C
  a = an arbitrary constant depending upon location $(= f(I))$
  $I = \sum(t/5)^{1.514}$, that is, the sum of 12 monthly heat indices

Values derived from the formula provide a quantitative measure of thermal efficiency and actual numerical values are available for many stations (Thornthwaite Associates, 1956).

Both the moisture index and the thermal index are subdivided into groups that depend upon the seasonality of the climate. Such a breakdown is clearly evident in Table 6–4 which shows the complete classification scheme.

In relation to its use, the Thornthwaite system has been more widely applied in analysis of

water budgets than as an actual classification scheme. This is unfortunate because while much can be criticised in the empiric equations used, the classification procedure is most logical. Perhaps its relative lack of use in classification results from the rather complicated classification methodology, a fact that can be easily overcome by application of one of the classification techniques given by a number of authors. One such technique, proposed by Basile and Corbin, is outlined in the Appendix.

Of course, in light of the necessity of rather complicated classification procedures, the complaint that the Thornthwaite system is too complex might be true. The case for this is aptly given by Jen-Hu Chang (1959) who writes:

"The Thornthwaite classification is indeed so complicated that it has not been possible to combine all four symbols on one map. The identities of Köppen's Cs with the Mediterranean type of climate, ET with the tundra, and Cfa with the corn belt, are all to (sic) familiar. Few, if any, could name the corresponding symbols in the Thornthwaite system. . . . Furthermore, such salient features as the prevalence of fog, small annual temperature range, and the Ganges type of temperature curve are simply disregarded in the Thornthwaite system. It is therefore extremely difficult to use the Thornthwaite classification as a framework to expound climatic patterns of the world."

Perhaps here the case against the system is overstated. It is not surprising that, after much classroom drill, many students do identify the Köppen symbols. But had the Thornthwaite system been given the same amount of coverage, then the reverse might hold true. Further, while it is true that some climatic features are omitted in the system, the fact that one can analyse climates on both a macro- and microlevel probably outweighs such a consideration.

The case for the Thornthwaite system has been stated strongly by Carter and Mather (1966). Not only do these authors attempt to place the Thornthwaite system into a natural evolution of climatic classification, they also provide some amendments to meet problems arising from use of the system.

The Genetic Systems

Just as climates may be classified according to the effect of climate upon selected environ-

---

[1] In a later work, this factor was dropped because it became possible to employ a variable evapotranspiration rate (Mather, 1966).

Table 6-4

THORNTHWAITE'S (1948, pp. 55–94) SYSTEM FOR CLIMATE CLASSIFICATION

There are nine climatic types as determined by moisture index.

	Climatic Type	Moisture Index
A	Perhumid	100 and above
B_4	Humid	80 to 100
B_3	Humid	60 to 80
B_2	Humid	40 to 60
B_1	Humid	20 to 40
C_2	Moist subhumid	0 to 20
C_1	Dry subhumid	−20 to 0
D	Semiarid	−40 to −20
E	Arid	−60 to −40

These nine types are modified by letters indicating seasonality of precipitation. (Aridity index is: water deficit/water need. Humidity index is: water surplus/water need.)

	Moist Climates (A, B, C_2)	Aridity Index
r	Little or no water deficiency	0–16.7
s	Moderate summer water deficiency	16.7–33.3
w	Moderate winter water deficiency	16.7–33.3
s_2	Large summer water deficiency	33.3+
w_2	Large winter water deficiency	33.3+

	Dry Climates (C_1, D, E)	Humidity Index
d	Little or no water surplus	0–10
s	Moderate winter water surplus	10–20
w	Moderate summer water surplus	10–20
s_2	Large winter water surplus	20+
w_2	Large summer water surplus	20+

Nine divisions are made based on temperature efficiency.

TE Index (cm)	(in.)	Climatic Type	
		E'	Frost
14.2	5.61		
		D'	Tundra
28.5	11.22		
		C'_1	
42.7	16.83		Microthermal
		C'_2	
57.0	22.44		
		B'_1	
71.2	28.05		
		B'_2	
85.5	33.66		Mesothermal
		B'_3	
99.7	39.27		
		B'_4	
114.0	44.88		
		A'	Megathermal

(Continued)

Table 6-4 (continued)

Temperature distribution provides eight further subdivisions.

			Summer Concentration Type
	A'		a'
44.88	114.0	48.0	
	B'_4		b'_4
39.27	99.7	5.19	
	B'_3		b'_3
33.66	85.5	56.3	
	B'_2		b'_2
28.05	71.2	61.6	
	B'_1		b'_1
22.44	57.0	68.0	
	C'_2		c'_2
16.83	42.7	76.3	
	C'_1		c'_1
11.22	28.5	88.0	
	D'		d'
5.61	14.2		
	E'		

mental systems, so may they be grouped according to factors that contribute toward their cause. These causative effects relate to the origin (or genesis) of the climates concerned and resulting systems may be termed the *genetic classifications*. It has already been noted that controversy exists over the relative merits of the genetic versus empiric approaches to classification. It is certainly true that there is no single widely used genetic system at the present time; that this is the case relates not to the relative merits of the approach but instead to difficulties involved in creating a meaningful genetic system. Such difficulties will become apparent in the discussion that follows.

The ultimate cause of the climate upon earth is the sun, and a study of the earth-sun relationships is basic to the study of physical climatology. While some exceedingly meaningful work relating solar radiation to climate has been completed (e.g., Milankovitch, 1930, Budyko, 1956, Miller, 1968), classification systems based upon the earth's energy balance are still in their formative stages. Perhaps the closest workable system is that recently outlined by Terjung (1970).

Terjung bases his classification upon the variation of net radiation (R) at various points over the earth's surface. By graphing the seasonal march of values for R at 1123 points on the earth, Terjung analyses the maximum input, deviation from the maximum, the number of months in which R is less than zero, and the shape of the curve. Using these variables, he derives a matrix to form the base of his classification. Details of this are given in the Appendix.

Apart from the use of the energy budget approach, there are two other ways in which genetic classifications can be approached. First, there is the method that considers the cause of climate in relation to the actual physical determinants acting upon it. Such determinants include prevailing winds, effect of continentality, orographic influences, and so on. For the most part, such an approach resolves itself into recognition of the zonal wind patterns and superimposes other determinants upon it. A second approach is through the analysis of relative air mass dominance and the influence and incidence of frontal conflict. These two approaches are considered separately in the following discussion.

Genetic systems based upon identified physical determinants. One of the earliest workable systems of classifying climate by cause was introduced by Hettner in 1930. Unlike the Köppen system, which became popular at about this time, Hettner's system never obtained the notice that it merited. The basic controls that he utilized to differentiate climates can be inferred through analysis of regions shown on his world map (Figure 6-5). Table 6-5 lists the climates and their inferred causes, and it is plainly seen that many different climatic controls are used. Unfortunately, the lack of a quantitative base, plus the rather complex subtypes differentiated were deterrents against the use of the system. Nonetheless, the introduction of the classification was significant and factors expressed within it have been incorporated into subsequent systems of this type. The more recent classifications of Hendl (1960) and Kupfer (1954) illustrate this. In Hendl's system, for example, factors such as leeward versus windward side, continental versus maritime, and other influences can be seen. Kupfer's system shows

Table 6-5

HETTNER'S CLIMATIC REGIONS

Climate (Numbers refer to Figure 6-5)	Inferred Cause
1. Polar and subpolar	Effect of *reduced insolation in high altitudes*
2. Temperate maritime	Effect of latitude and *oceanicity* on equability
3. Cool temperate lowland	*Latitudinal variation* of temperature zones; effect of *mountains*, or lack thereof
4. Warm temperate lowland	As above
5. Prairie climates	Subhumid climates resulting from *continental location*
6. High-latitude dry	Desert climates in the interior of continents differ in *origin* from those low latitudes
7. Etesian climate	*Seasonal effects* of different wind systems—the Mediterranean
8. Dry trade wind	Deserts *differentiated by cause*, not by formula
9. Subtropical east coast	Significance of *east coast versus west coast* location in relation to climatic regimes
10. Tropical continental and monsoon	Location in relation to *a. land-sea distribution* and *b. seasonal wind* shifts
10*b*. Wet windward climates of the tropics	Influence of location (i.e., *windward/leeward* sides) in relation to prevailing winds
11. Wet equatorial	Climates dominated by *low pressure* system—the doldrums
12. Tropical and subtropical highland climates	Role of *mountains*
13. High latitude monsoon	Modification of seasonal climates by *latitude*

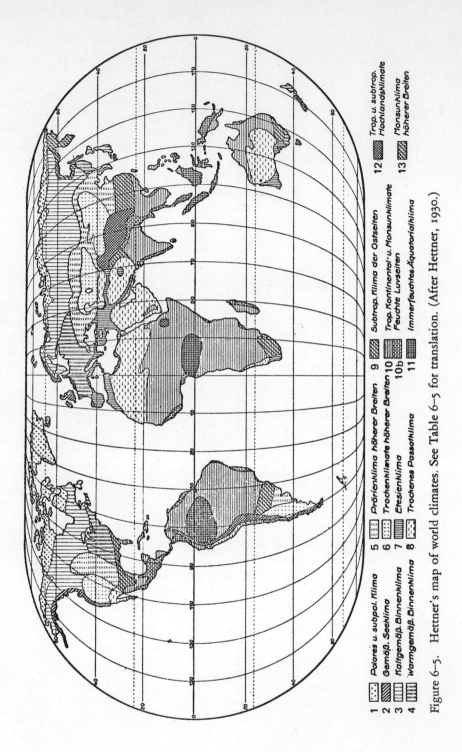

1 Polares u. subpol. Klima
2 Gemäß. Seeklima
3 Kaltgemäß. Binnenklima
4 Warmgemäß. Binnenklima
5 Prärienklima höherer Breiten
6 Trockenklimate höherer Breiten
7 Etesienklima
8 Trockenes Passatklima
9 Subtrop. Klima der Ostseiten
10 Trop. Kontinental-u. Monsunklimate
10b Feuchte Luvseiten
11 Immerfeuchtes Äquatorialklima
12 Trop. u. subtrop. Hochlandsklimate
13 Monsunklima höherer Breiten

Figure 6–5. Hettner's map of world climates. See Table 6–5 for translation. (After Hettner, 1930.)

that similar influences are considered. Obviously both of these systems contain factors not considered in the earlier classification and the organizational approach is more acceptable in terms of present-day knowledge.

Both Hendl's and Kupfer's systems can also be related to the genetic system proposed by Flohn (1950). Flohn's classification represented a distinct step forward in the genetic approach to climate because it incorporated a dynamic viewpoint. The system is based upon a model continent—the Idealkontinent—with extensions over the oceans. Seven types of climate are differentiated in a zonal pattern. Four of these, termed the "Homogene Klimates," represent regions that experience a climate caused by a single dominant effect that occurs throughout the year. Examples include locations dominated by the intertropical convergence or the subtropical trade winds all year. Seasonal shifts in world pressure systems cause migration of the zones identified with the "Homogene Klimates." Thus, between each zone is a climate that results from the influence of adjacent climates on a seasonal base. The resulting climates are a mix of the two neighboring zones and are termed the "Heterogene Klimates."

While Flohn's system attempts to overcome the static picture by which climate is represented, it is highly qualitative so that it is not possible to identify, with any degree of certainty, into which region a given climatic station might occur. Further, few regions are identified with the result that enormous areas of the earth are grouped under a single climatic heading. It seems that this system, like others taking a similar approach, provide excellent illustrative examples of climatic controls on a world basis. But, for rigorous study of world climates, it would appear that such systems supplement—instead of replace—the quantitative and more rigorous empiric systems.

The air mass system. The development of synoptic meterology and the concept of air masses led to enormous improvements in comprehending weather. It seemed that a similar methodology could provide a suitable base for the study of genetic climatology and, during the formative years, a number of significant climatological studies based upon air masses were published.

Despite these, no comprehensive climatic classification based upon air masses was formulated and, even at present, such systems are still only first approximations of the climatic complex of the earth.

That this is so probable relates to the difficulties involved in assessing the influence of air masses on an annual basis. Basic to the problem are:

1. The fact that the early development of nomenclature and air mass classification used, as its base, air mass characteristics of the North Atlantic. Subsequent work showed that the conditions in this area are not easily transferred to other parts of the world.

2. The problem of identification of air masses that, although given the same code designation, have properties that vary enormously from core to periphery. Similarly, the same notation is used for seasonal description of air masses despite their very different characteristics.

Despite these problems some workers have attempted to devise air mass classifications. To facilitate his description of world regional climates, Strahler (1951) introduced an air mass system based upon a global model. Figure 6–6 shows the model and points to the basis of his classification. Essentially, three main groups of climates are differentiated:

Group I. Climates dominated by equatorial and tropical air masses all the year.

Group II. Climates that occur between Groups I and III and that are influenced by the interaction between tropical air masses (Group I) and polar air masses (Group III).

Group III. Climates controlled by polar air masses.

The complete system dependent upon this approach is shown in the Appendix.

Strahler has further contributed toward the development of air mass classifications through introduction of a method by which climatic regimes can be expressed in terms of interacting air masses. To analyze such regimes, Strahler introduced the thermohyet diagram (Figure 6–7). This graph is the most recent refinement of a series of climatic representations collectively termed climographs.

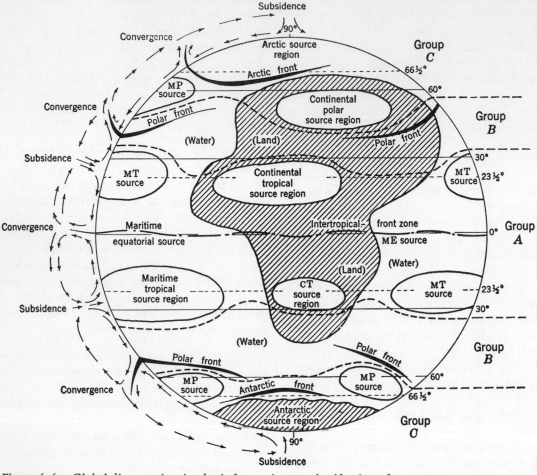

Figure 6–6.   Global diagram showing basis for an air mass classification of climate. (From Strahler, A. N., *Physical Geography*, 1st ed., copyright © 1951 by John Wiley and Sons. Used with permission.)

The essential feature of any climograph is the plotting of two elements of climate against one another. For example, the mean monthly temperature of a station may be plotted against the mean monthly precipitation of the same station. Each calendar month is represented by a point and the points can be connected to form a closed circuit representative of the annual cycle. The utility of such a graph lies in the fact that a particular climate is represented by a figure of distinctive shape. Stations having the same shapes and located on the same portion of the climograph can be assigned to a particular climate type. The invention of the climograph has been attributed to Ball (1910) but its dissemination is largely the result of its use by Griffith Taylor.

While climographs were frequently used for showing such things as comfort frames (Taylor, 1946) and the agricultural response to climate (Visher, 1944), it was never widely used as a tool to analyze climate on a monthly basis. A possible deterrent to its use might have been the fact that on the conventional climograph, with its arithmetic axis, the station figures tend to become overcrowded in one sector of the graph. The extremes are barely used at all. Moreover, arid regions are difficult to differentiate from the semiarid on the arithmetic scale for the all-important fraction of an inch occupies a small space on the axis.

By use of a nonarithmetic axis, the thermohyet diagram overcomes the difficulty of crowding.

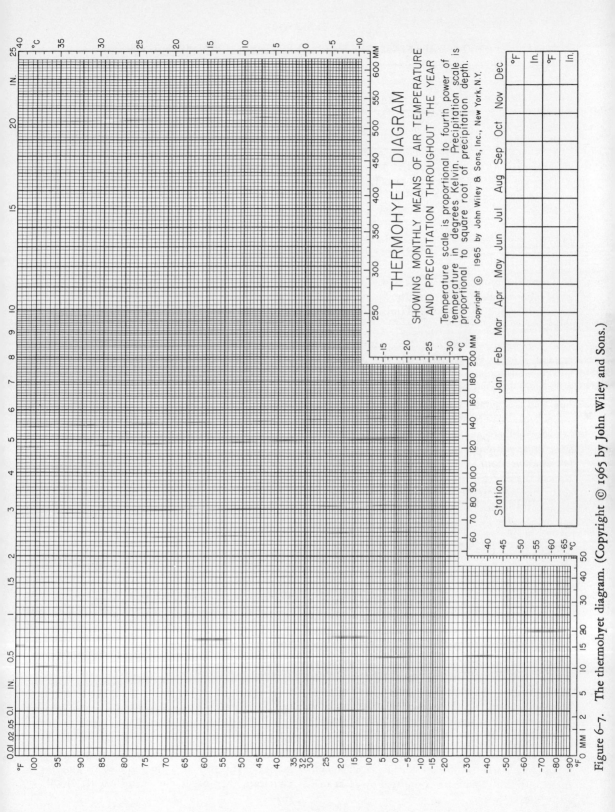

Figure 6-7. The thermohyet diagram. (Copyright © 1965 by John Wiley and Sons.)

Table 6–6

SEVEN AIR MASS REGIMES AND THEIR CHARACTERISTICS[a]

Regime	Air Mass	Dominant Feature	Representative Station (Figure 6-8)
a. Regimes under a single dominant air mass			
1. Equatorial	mT or mE	Thermal equability	Padang, Sumatra
3. Desert	cT or mT$_s$	Limited precipitation	William Creek, Australia
5. Middle-latitude equable	mP	Equability	Dunedin, New Zealand
7. Polar	cP or cA	Limited precipitation	Barrow Point, Alaska
b. Regimes under two dominant air mass influences			
2. Tropical wet-dry	mT and cT	Extremes of precipitation	Calcutta, India
4. Mediterranean	mP(mT) and cT	Extremes of precipitation	Izmir, Turkey
6. Continental	mT(mP) and cP(cA)	Thermal extremes	Winnipeg, Canada

1. The Equatorial Regime. Exceptional thermal equability is shown by stations in this group. Low latitude and the dominant influence of mT and mE air masses give uniformly high monthly temperatures and precipitation. In some instances equinoctal rainfall maxima are noted showing the relationship between climate and the passage of the overhead sun. Some stations show a large difference of rainfall amount from season to season, but in all instances the amount is large. On the T. H. Diagram such climates are shown by a horizontal linear trend and a location in the upper right-hand field of the graph.

2. The Tropical Wet-Dry Regime. This regime is typified by very high rainfall during the high sun season and by drought during the season of low sun. The seasonal contrast is caused by the dominance of mT and cT air masses during the respective seasons. Temperatures reach a peak prior to the onset of the rainy season, after which cloud cover reduces insolation. The temperature range, particularly during the dry season, is considerably greater than that of the equatorial regime. On the T. H. Diagram the regime is characterized by great horizontal extent. In the low-precipitation area the figure often shows a greater temperature range and a figure-eight loop may appear.

3. The Desert Regime. The low monthly rainfall totals are clearly shown on the T. H. Diagram. Areas with this regime are dominated by cT or mT$_s$ air masses all year.

4. The Mediterranean Regime. Characterized by high-sun drought, caused by the dominance of cT air masses, and low-sun precipitation, the result of maritime air masses, this regime yields a figure axis strongly sloping down from left to right on the T. H. Diagram.

5. The Middle Latitude Equable Regime. Dominance of mP air masses all the year, together with marked oceanic influences, causes this regime to experience a small range of both temperature and precipitation. On the T. H. Diagram stations within this regime are represented by a small cluster of points.

6. The Continental Regime. Continentality is expressed by the great annual range of temperature. Significant precipitation may occur in all months, although a summer maximum often exists. The temperature range and seasonal precipitation are shown on the T. H. Diagram by a figure with the axis rising steeply from left to right. In the drier months cA or cP air masses prevail. In summer mT or mP are dominant.

7. The Polar Regime. The climate of this regime is largely influenced by cA or cP air masses all year and is characterized by low temperatures and small amounts of precipitation. The figures on the T. H. Diagram occupy positions in the lower left-hand corner of the grid. A slight summer maximum precipitation may occur due to incursions of mP air masses.

[a] From Oliver (1968).

The abscissa, on which is represented the mean monthly precipitation, has a scale that is proportional to the square root of the depth of precipitation. This treatment expands the scale in the lower precipitation region where small differences are infinitely more important than with very high totals of rainfall. The temperature scale on the ordinate is drawn proportional to the fourth power of temperature in degrees Kelvin. This expands the scale with increasing temperature so that more space is available for temperatures between 32°F (0°C) and 100°F (37.8°C).

Distribution of plotted stations on the thermohyet diagram allows climates to be classified by climatic regimes. Strahler (1968), following Dansereau (1957), recognizes seven well-defined regimes. These are listed in Table 6–6, with examples of each type plotted on the thermohyet diagram in Figure 6–8.

Use of these climatic regimes provides a good approach to the introduction of climatic distribution and climate-vegetation relationships (Oliver, 1968). The limited number of regions differentiated, however, does not allow a meaningful world distributional map to be constructed.

Other authors have outlined air mass systems using varying approaches. Allisow (1954), Hidore (1969), and Critchfield (1966) use qualitative approaches to facilitate a description of regional climatology. Approaches using a quantitative base have been proposed by Brunnschweiler (1957) and Oliver (1970). These are described in the Appendix.

The Arbitrary Systems

There remains a third approach to climatic grouping that cannot be considered either genetic or empiric. Figure 6–9 shows a map of India illustrating climatic regions according to Kendrew (1965). In a classic work, Kendrew has described the climates of the world on a continental scale and has divided each continent into what he considers suitable climatic regions. The map of India illustrates that boundaries between identified regions are given no values; they merely divide the subcontinent into regions that are characterized by the verbal description given in each unit. Kendrew has devised a system of classification that best fits his own needs for a discussion of regional climatology. Such an approach cannot be considered as a rigorous attempt to classify climates and the

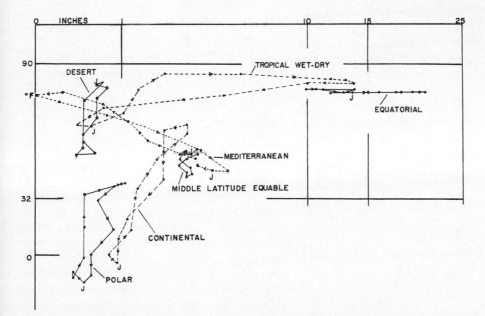

Figure 6–8. Plotted station data on the thermohyet diagram illustrating stations typical of the seven major climatic regimes given in Table 6–6. (From Oliver, 1968.)

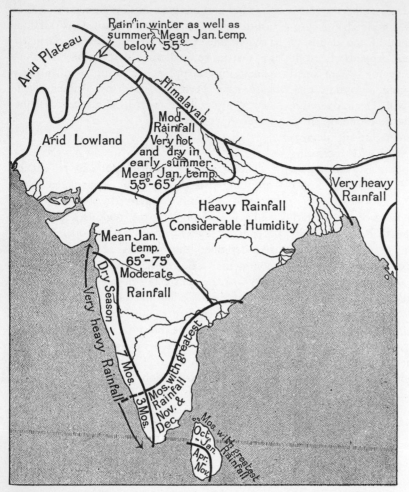

Figure 6–9.   Kendrew's climatic divisions of India. (From Kendrew, W. G., *The Climates of the Continents*, 1922. Used with permission of Oxford University Press.)

divisions shown are quite arbitrary. A number of authors have used similar analyzes and resulting depictions may be classed together and termed the *arbitrary* classifications.

As illustrated by Kendrew's work, the arbitrary systems can prove extremely useful and other authors have used the type of approach to good effect. Borisov (1965), for example, has described the climates of the U.S.S.R. in a similar fashion. After a discussion of various classification schemes pertaining to the U.S.S.R., he uses the following outline for regional analysis:

A. Climates of low-lying areas

Climates of the seas (three differentiated)

Climates of the land (six differentiated)

B. Climates of mountainous areas (nine differentiated)

Using such a pattern of organization, Borisov succeeds in characterizing the diverse climates of the U.S.S.R.

Unfortunately, not all arbitrary systems are as useful as the two mentioned. Many maps of climate types for individual regions, notably continental areas, have been produced. Many such maps appear to indicate little more than the whims of their authors and little merit can be found in the mode of differentiation. Problems of such depiction are often compounded by inappropriate use of climatic terms.

The difference between the arbitrary and other methods of classification is determined by the logic of the conceptual framework of the underlying methodology and by the rigor with variables used in formulating the method are defined. To meet the demands of a rigorous classification a number of requirements must be met. Critchfield (1966) suggests that the system must first attempt to organize facts in an orderly arrangement of categories. Second, the classification must establish a language for communication in terms of the material classified. Third, it must establish a means of identifying boundaries so that facts can be mapped. Only in this way can the evolved principles of the classification be used to establish analogs for various places in the world so that climatic distribution can be meaningfully compared.

PART II

Climate, Man, and Man's Activities

CHAPTER 7

Climate and Man

Man, as an animal, exists in a climatic environment that directly affects his body and body functions. The influence of this environment is determined by the microclimate that surrounds the body, a microclimate that Lee (1958) has termed the proprioclimate. The nature of proprio-climate-body interaction comprises the field of human bioclimatology,[1] the study of which draws upon both medical and physical sciences.

The varied aspects of human biometeorology are shown by Tromp (1967) to comprise a number of research areas. Physiologic biometeorology deals with the direct effects of climate and weather upon the body and comprises evaluation of the "normal" physiologic response to both acceptable and extreme conditions. Social biometeorology relates the effects of weather and climate to the social activities of man, with study ranging from the psychological effects on the individual to the nature of group response and perception. The relationship between man, climate, and disease is covered in pathological biometeorology, in which the role of climate in both disease and therapy are considered.

Apart from these three main areas, Tromp also distinguishes the rather specialized aspects of urban and nautical meteorology with, of course, the former becoming an important area of interest in view of present-day, urban-industrial ills.

THE PHYSIOLOGICAL RESPONSE

The successful functioning of the human body depends upon conditions of both internal and external environments. The external environment supplies the energy and nutrients needed for the organs, tissues, and cells of the body, while their healthy functioning is closely dependent on the chemical and physical properties of the internal environment. In fact, the health of the human body is linked to the precision with which these physical and chemical prop-

[1] Bioclimatology is often considered as part of the study of biometeorology.

erties are regulated. Wide variation from the usual levels result in the functional deterioration of the organism. The precision of regulation is identified by the term *homeostasis*.

The properties of the internal environment and the nature of homeostatic controls—the nervous and endocrine systems—are more the concern of the medical than the meteorological researcher. It is the actual homeostatic responses to atmospheric variation that mostly concerns us here. One of most significant aspects in this respect concerns thermoregulation.

Thermoregulation

Man is a warm-blooded animal whose body temperature is often equated to the rectal temperature of 37°C (98.6°F), although this temperature is not representative of the entire body. To obtain a representative value, the following formula is sometimes used:

$$T_b = \frac{T_s}{3} + \frac{2}{3} T_r$$

where T_b is the mean body temperature, T_s is representative of the surface temperature, and T_r is the rectal temperature.

Maintenance of the body temperature calls for a balance to be attained between heat loss and heat gain. The body metabolism relates the intake of food to the production of energy (and waste) and the rates vary according to the physical status of the body. The heat produced by a resting (but awake) person approximates 50 k-cal/hr/cm² of body surface, a value designated as one metabolic unit, 1 MET. As indicated in Table 7-1, there is a considerable variation of metabolic heat productivity, the amount varying with the body activity being carried out.

Apart from heat production through basal processes, gains of body heat from the environment occur because of absorption of long-wave radiation and conduction from the surrounding air if it is above skin temperatures. Likewise, heat losses occur through radiation, conduction, and importantly, evaporation of moisture from

Table 7–1

METABOLIC HEAT PRODUCTION RELATED TO HUMAN ACTIVITIES[a]

Kind of Activity	No. of METS	Equivalent kcal/m² hr
Sleeping	0.8	40
Awake, resting	1.0	50
Standing	1.5	75
Working at desk, driving	1.6	80
Standing, light work	2.0	100
Level walking, 4 km/hr, moderate work	3.0	150
Level walking, 5.5 km/hr, moderately hard work	4.0	200
Level walking, 5.5 km/hr with 20 kg load, sustained hard work	6.0	300
Short spurts of heavy activity (e.g., climbing or sports)	10.0	500

[a] After Landsberg (1969).

the skin surface (Table 7–2). These losses and gains can be shown by

$$M \pm R \pm C - E = 0$$

where M = metabolic heat
  R = radiation
  C = conduction
  E = evaporation

If more heat is lost than is gained, then the equation is no longer balanced ($\neq 0$), implying that the body temperature must fall. Conversely, if more heat is gained than is lost, then the body temperature must rise.

Under conditions of imbalance, a number of physiologic responses occur. Cold conditions cause vascoconstriction so that blood flow to the periphery is reduced and less heat will pass out of the body. While this reduced outward flow of blood conserves heat and keeps up the core temperature of the body, the reduced blood flow to the extremities can have dire effects. Under extreme conditions, frostbite occurs and results in frozen tissue and cell destruction.

Shivering is another physiologic response to cold conditions. Its function is to increase the metabolic rate. While this certainly occurs, the

Table 7–2

HEAT BALANCE OF THE BODY[a]

Gains

1. Heat production by:
   a. Basal processes
   b. Activity
   c. Digestive, etc., processes
   d. Muscle tensing and shivering
2. Absorption of radiant energy
   a. From sun
      direct
      reflected
   b. From glowing radiators
   c. From nonglowing hot objects
3. Heat conduction toward body
   a. From air above skin temperature[b]
4. Condensation of atmospheric moisture (occasional)

Losses

5. Outward radiation
   a. to "sky"
   b. To colder surroundings
6. Heat conduction away from body
   a. To air below skin temperature[b]
   b. By contact with colder objects
7. Evaporation
   a. From respiratory tract
   b. From skin:[b] perspiration, sweat, and applied water

[a] From Lee (1958).
[b] Hastened by air movement—convection.

shivering function is not a highly efficient process. It does increase the rate by bringing more blood to the surface layers of the body, but this in turn increases heat loss by radiation and convection from the body to the surrounding environment.

The response of the body to an excessive heat load is variable. It leads, of course, to dilation of the skin blood vessels, which causes a rise in skin temperature. Sweating also aids in heat loss through higher evaporation—with the resulting cooling effect—from the surface. Excessive salt depletion can lead to cramp and, ultimately, to circulatory failure. Dehydration through excessive sweating in hot, dry conditions can result in death.

Dehydration is, of course, intimately related to the intake of water. The normal body requirement is about 3000 cc/day. This is partially derived from the intake of foodstuffs, most of which are hydrated. It is generally found that the body excretes 10% more water than it takes in, the excess being derived from oxidation of foodstuffs. This is shown in Table 7-3, where body water is shown to be lost in a number of ways. It occurs in feces, urine, perspiration, and respiration.

Water is needed then, not only as a medium in which body chemical activity can occur, but also as an excretory medium. Excretion of water from

skin, lungs, and kidneys never ceases. When this inevitable loss takes place, and water is not replenished, dehydration occurs. Slager (1962) described the effects of dehydration as producing ". . . a loose and wrinkled skin, sunken eyeballs and acidosis. The blood becomes concentrated and circulation becomes inadequate. The decreased renal circulation results in the inability of the kidneys to excrete all waste products of metabolism and urea begins to accumulate in blood. Eventually, the blood may sludge and may form clots and thrombi in the various areas of the body."

Tolerance of water shortage by the body is in the order of days. The extent to which survival occurs depends not only upon water intake, but upon prevailing temperature, clothing, and

Table 7-3[a]

a. Metabolic Requirements for Life Support of One Man per Day

	Required Supplies				
	Energy (cal)	Food (g)	Oxygen (ml)	(g)	Water (g)
Carbohydrate	1200	300	244	342	1000[c]
Protein	320	80	90	114	(food) +
Fat	1350	150	300	420	1500 (liquid)
Total	2870[b]	530[b]	634	876	2500 = 3.906 Kg

b. Waste Production

	Solids (g)	Carbon Dioxide (g)	Water (g)
Feces	—[d]	Total for diet assumed	200
Urine	68	above	1300
Respiration	—	1032	400
Perspiration	6		600
Metabolism	—		300
Total	74	1032	2800 = 3.906 Kg

[a] After Slager (1962).

[b] Calculations assume complete digestion, absorption, and oxidation of this food material. Such complete combustion is not achieved with natural foods. Hence they will weigh somewhat more than indicated here.

[c] Natural food is about 2/3 water. Thus to obtain 500 g of protein, carbohydrate and fat, 1500g of steaks, peas and bread, which contain 1000 g of water must be consumed.

[d] Since complete digestion and absorption is assumed, there is no indigestible residue, and no fecal solids appear in these calculations.

physical activity. Without water, death may result in two days with temperatures at 120°F (49°C). At 70°F (21.1°C) a person may last 10 days if motionless. By walking, even at night, the time decreases to seven days.

Numerous workers have investigated water needs under different environments. Adolph (1947) and Woodcock et al. (1952) have published comprehensive studies in this area. Woodcock's work is of interest for he made use of a "copper man" in which results could be monitored electrically. A typical experiment completed by Grande (1962) is described:

"They (the subjects) walked for two hours each day at 3.5 miles (5.6 kilometers) per hour on a 10-per-cent grade, with food intake restricted to 1000 kilocalories a day and with three levels of water intake—as much as they wanted whenever they wished it, 1800 cubic centimeters, or 900 cubic centimeters per day. . . . The increase of rectal temperature during exercise was twice as large in the low water group as in the controls, but very little greater in the medium-water group. The rate of weight loss was somewhat greater in the medium-water group in the first few days than among those who drank freely, but it was approximately the same after that. It did not diminish when they were permitted unlimited water intake. . . . Sweat loss during exercise was reduced in both low- and medium-water groups and did not return to the initial values when rehydration was permitted. It would seem . . . that readjustments can occur that permit the body to operate at somewhat lower levels of hydration than those to which it is accustomed."

A summary of the responses, consequential disturbances, and regulation failure that can result from thermal stress are shown in Table 7–4. Quantitative assessment of such factors have been extensively investigated, and numerous empiric equations devised (for example, Burton and Edholm, 1955, Winslow et al., 1949).

Biometeorological Indices

It is evident that the heat load upon a body is the function of a number of interacting variables. To estimate the effect of these a number of

Table 7–4

SUMMARY OF HUMAN RESPONSES TO THERMAL STRESS[a]

To Cold	To Heat
Thermoregulatory responses	
Constriction of skin blood vessels	Dilation of skin blood vessels
Concentration of blood	Dilution of blood
Flexion to reduced exposed body surface	Extension to increase exposed body surface
Increased muscle tone	Decreased muscle tone
Shivering	Sweating
Inclination to increased activity	Inclination to reduced activity
Consequential disturbances	
Increased urine volume	Decreased urine volume. Thirst and dehydration
Danger of inadequate blood supply to skin of fingers, toes, and exposed parts leading to frostbite	Difficulty in maintaining blood supply to brain leading to dizziness, nausea, and heat exhaustion. Difficulty in maintaining chloride balance, leading to heat cramps
Increased hunger	Decreased appetite
Failure of Regulation	
Falling body temperature	Rising body temperature
Drowsiness	Heat regulating center impaired
Cessation of heartbeat and respiration	Failure of nervous regulation terminating in cessation of breathing

[a] From Lee (1958).

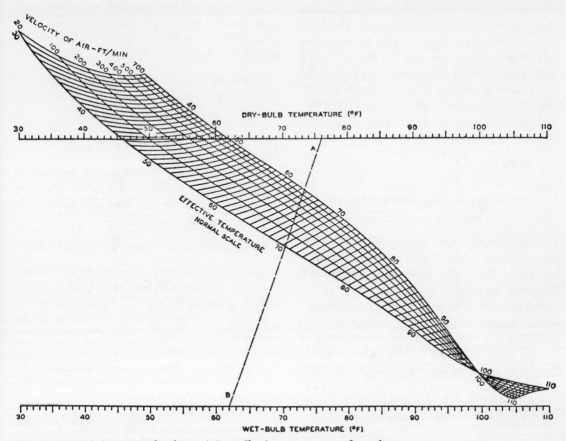

Figure 7–1. Nomograms for determining effective temperature for sedentary, normally clothed individuals from measurements of dry-bulb and
wet-bulb temperatures and from wind speed. To use the chart, draw line A-B
through measured dry- and wet-bulb temperature; read effective temperature
or wind velocity of desired intersections with line A-B. For example, given
dry-bulb temperature 76°F and wet-bulb temperature of 62°F, the *ET* at
wind velocity of 100 fpm is 69; a wind velocity of 340 fpm is required for
an *ET* of 66. (From Munn, R. F., *Biometeorological Methods*, 1970, Academic
Press. Used with permission.)

biometeorological indices have been derived.
Generally, these serve to allow prediction of
various responses to the sensation of warmth and
to further assess the physiologic strain imposed
by combined atmospheric variables.

The *effective temperature* (ET)[2] is used to establish thermal sensation by equating prevailing
conditions with a set standard. The standard is
obtained from the effect of still, saturated air at a
given temperature. Derived from dry and wet
bulb readings and wind speed the *ET* may be

derived graphically (Figure 7–1), or, as a first
approximation, from the formula

$$ET = 0.4(T_d + T_w) + 15$$

where T_d and T_w are dry and wet bulb temperatures, respectively (°F).

A similar index is the *corrected effective temperature* (*CET*) in which a black globe thermometer
is used instead of the dry bulb temperature. The
black bulb thermometer consists of a thermometer inserted in a blackened, hollow copper
sphere of variable diameter, although often in
the order of 7.5 cm. The globe temperature

[2] Not to be confused with *ET*—evapotranspiration—
described earlier.

integrates all heat exchanges, including radiant energy exchange, in the atmosphere. The black globe thermometer is also used to derive the *wet bulb globe temperature index* (*WBGT*), which is used as an index of heat stress. In practice, this uses a black globe that is kept moist to simulate evaporative processes; difficulties are experienced in obtaining this and instrument error can result. The *WBGT* index is given by

$$WBGT = 9.7\,T_w + 0.1\,T_d + 0.2\,T_g$$

where T_w = wet bulb temperature, °C
T_d = dry bulb temperatute, °C
T_g = globe temperature, °C

Tests have shown that outside activities should be suspended when $WBGT = 31$°C (87.8°F), for at and above this level, heat disorders will frequently occur.

Probably the best known of the biometeorological indices is the *temperature-humidity index* (*THI*) for it is often cited in printed or broadcast weather conditions. An empiric index, it can be derived in a number of ways using either wet or dry bulb thermometers (T_dT_w), dry bulb and dew point (*Tdew*) temperatures, or dry bulb and relative humidity (*R.H.*) values.

$$THI = 0.4(T_w + T_d) + 15$$

or

$$THI = 0.55\,T_d + 0.2\,Tdew + 17.5$$

or

$$THI = T_d - (0.55 - 0.55\,R.H.)\,(T_d - 58)$$

Statistical samples have indicated that *THI* values 60 to 65 are accepted as comfortable by most people. At a value of 75 at least one-half of the people are uncomfortable. At 80 or above almost all feel uncomfortable and the attainment of this level can provide grounds for closing (non airconditioned) businesses. The *THI* serves well to indicate the importance of humidity in the response to thermal stress. Consider a day on which T_d (dry bulb temperature) is 85°F (29.4°C) and *R.H.* (relative humidity) is 20%. Application of the formula

$$
\begin{aligned}
THI &= T_d - (0.55 - 0.55\,R.H.)\,(T_d - 58)\\
&= 85 - (0.55 - 0.55 \times 0.2)\,(85 - 58)\\
&= 85 - (0.44 \times 27)\\
&= 73 \text{ (approx.)}
\end{aligned}
$$

Most persons will feel comfortable at this level. Compare this to the index derived when, at the same temperature of 85°F (29.4°C), the relative humidity is 85%.

$$
\begin{aligned}
THI &= 85 - (0.55 - 0.55 \times .85)\,(27)\\
&= 85 - (0.11 \times 27)\\
&= 82 \text{ (approx.)}
\end{aligned}
$$

The higher humidity results in marked discomfort.

An equation used for predicting clothing requirements and the effect of wind at low temperatures is given by the *wind chill factor* (K_o). The equation expressing this is

$$K_o = (\sqrt{v \times 100} - v + 10.5)\,(33 - T_d)$$

in which T_d is the dry bulb temperature (°C) and v is wind velocity in m/sec. The equation can be solved graphically and K_o values, given in kcal/m² hr, can be verbalized in relation to the sensation that occurs.

K_o Value	Sensation (or Result)
1000	Very cold
1200	Bitterly cold
1400	Exposed flesh freezes
2000	Exposed flesh (areas of face) freeze in 1 min

In public weather forecasts the wind chill factor is often expressed in terms of equivalent temperature. For example:

Temperature (°F)	Wind Speed	Sensible Temperature
20	45 mph	−20°F (−28.9°C)

Other indices have been derived for use in industry to facilitate optimum working conditions. Examples include cooling power (CP), which is measured through the rate of heat loss from a specially constructed katathermometer (see Stone, 1943) and the predicted 4-hr sweat rate (P4SR), which is an index of heat stress expressed in terms of strain on thermoregulation.

A number of writers have used biometeorological indices to classify different climatic environments that exist over the earth. One such example is that devised by Terjung (1966), who

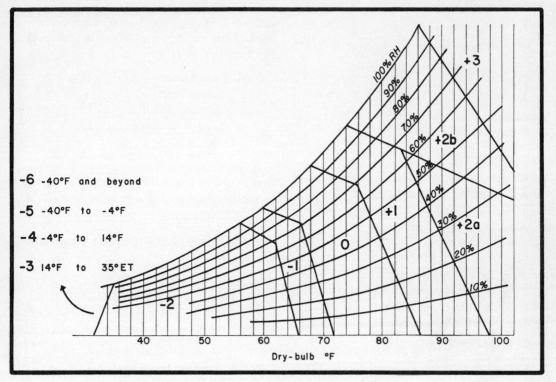

Figure 7-2. Graph for determination of the comfort index. *Key:* −6 ultra-cold; −5 extremely cold; −4 very cold; −3 cold; −2 keen; −1 cool; 0 comfortable; +1 warm; +2a hot; +2b oppressive; +3 extremely hot. (From Terjung, W., *Annals A.A.G.*, 56, 1966. Reproduced by permission from the *Annals* of the Association of American Geographers.)

has outlined a "Bioclimatic Classification Based on Man." According to him, the utility of such a classification is multifold:

1. It might be of value in medical geography by specifying best living areas for persons suffering from weather-related illnesses.

2. It could provide a realistic guide for vacationers and allow better regional evaluation by the tourist industry.

3. It provides a more realistic picture of climates for students, that is, by allowing students to equate their own experiences of certain climatic conditions to those identified elsewhere.

4. It simplifies knowledge regarding housing needs, heating requirements, and so on, for identified regions.

5. It provides a useful guide in estimating the climatic potential of a given region.

The classification is based upon the utilization

of two indices determined from prepared nomographs. The first index is derived by superimposing limits onto a psychrometric chart to give the *comfort index* (Figure 7-2). Depending upon the location of plotted temperature and humidity data on the graph, the comfort index is described according to the zone into which the data fall. For example, at a temperature of 90°F (32.2°C) and relative humidity 70%, the index +2b is derived. This is described as oppressive. At the same temperature, but with relative humidity at 20%, the derived index is +1, which is described as warm.

For any given month, both a daytime and nightime comfort index is derived. Thus for San Francisco in July[3] for the day:

Average maximum daily temperature = 64.4°F (18.0°C)

Average minimum daily R.H. = 75%

[3] All data used in this analysis are from Terjung (1968).

Comfort index (Figure 7–2) = −1 (cool) for the night:

Average minimum daily temperature = 53.4°F (11.9°C)
Average maximum daily R.H. = 92%
Comfort index (Figure 7–2) = −2 (Keen)

A summation for the month of July at San Francisco is

Comfort index = −1/−2

Interpretation of the combination is derived from a whole spectrum of combinations, shown in Table 7–5. In this case the resulting index is designated C_2, which forms part of the cool group of combinations.

A second index is derived from a *wind-chill* chart, Figure 7–3. Again, this chart is divided into sections but here the differentiated zones are designated by letters (Table 7-5I). A location experiencing a temperature of 30°F (−1.1°C) at a windspeed of 45 mph has a wind-chill factor of 1200, which is designated −g; at a temperature of 30°F (−1.1°C) and windspeed of 5 mph, the factor is 800 and the designation is −d.

The actual evaluation of the wind-chill factor and wind effect index is rather complicated for daytime effects. Again consider the example of San Francisco in July.

Data: Temperature = 64.4°F (day)
  Average wind speed = 11.3 mph
  Length of possible sunshine = 14.5 hr
  Average percentage of length of sunshine = 64%

a. From Figure 7–3 use temperature and wind data to obtain a value of −400. This is expressed in kcal/m²hour, so for 14.5 hr value is

$$−400 \times 14.5 = −5800$$

b. But actual average sunshine length is only 64% of maximum, that is, 64% of 14.5 hr = 9.4 hr.

c. Wind chill is reduced by 200 kcal/m²hr for every hour of bright sunshine, in this case, 9.4 hr

amount of reduction = 9.4 × 200 = +1880

Find the difference between sum determined in *a.*

$$(= −5800) \text{ and } +1800 = −3920$$

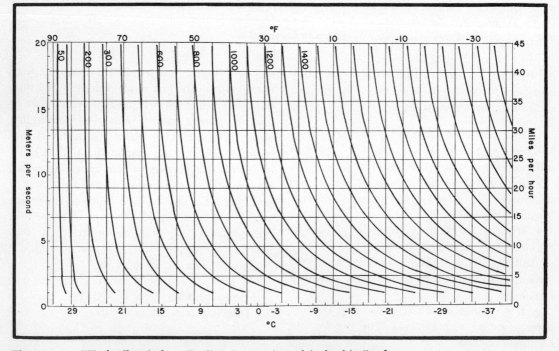

Figure 7–3. Wind effect index. Cooling is experienced in kcal/m²hr for various temperatures and wind velocities. Values based on individual in a state of inactivity with a neutral skin temperature of 91.4°F. (From Terjung, W., *Annals A.A.G.*, 56, 1966. Reproduced by permission from the *Annals* of the Association of American Geographers.)

Table 7-5
CLASSIFICATION[a]

Step I (Single Factor Maps)

Comfort Index (use Fig. 7-2)

Symbol	Sensation felt by majority
-6	Ultra cold
-5	Extremely cold
-4	Very cold
-3	Cold
-2	Keen
-1	Cool
0	Comfortable
$+1$	Warm
$+2a$	Hot
$+2b$	Oppressive
$+3$	Extremely hot

Wind Effect Index (use Fig. 7-3)

Symbol	Kcal/m²hr	Sensation felt by majority
$-h$	-1400 & beyond	Exposed flesh freezes
$-g$	-1200 to -1400	Bitterly cold wind chill
$-f$	-1000 to -1200	Very cold wind chill
$-e$	-800 to -1000	Cold wind chill
$-d$	-600 to -800	Very cool wind chill
$-c$	-300 to -600	Cool wind chill
$-b$	-200 to -300	Pleasant wind effects
$-a$	-50 to -200	Warm wind effects
n	$+80$ to -50	Neutral wind effect
a	$+160$ to $+80$*	Warming sensation to skin
b	$+160$ to $+80$**	Discomforting heat addition
c	$+160$ and above***	Very discomforting heat addition

Step II (Monthly Physiological Climates)

Comfort Index Possible Day-Night Combinations

	Group		Group
$+3/+2b$	EH_1	$0/0$*	M_1
$+3/+2a$	EH_2	$0/-1$*	M_2
$+3/+1$*	EH_3	$0/-2$*	M_3
$+3/0$*	EH_4	$0/-3$*	M_4
$+3/-1$	EH_5	etc.	etc.
etc.	etc.	$-1/-1$	C_1
$+2b/+2b$	S_1	$-1/-2$*	C_2
$+2b/+2a$	S_2	$-1/-3$*	C_3
$+2b/+1$*	S_3	etc.	etc.
$+2b/0$*	S_4	$-2/-2$*	K_1
$+2b/-1$*	S_5	$-2/-3$*	K_2
etc.	etc.	$-2/-4$*	K_3
$+2a/+2a$	H_1	etc.	etc.
$+2a/+1$*	H_2	$-3/-3$*	CD_1
$+2a/0$*	H_3	$-3/-4$*	CD_2
$+2a/-1$*	H_4	$-3/-5$*	CD_3
$+2a/-2$*	H_5	etc.	etc.
etc.	etc.	$-4/-4$*	VC_1
$+1/+1$	W_1	$-4/-5$*	VC_2
$+1/0$*	W_2	etc.	etc.
$+1/-1$*	W_3	$-5/-5$	EC_1
$+1/-2$*	W_4	etc.	etc.
etc.	etc.	$-6/-6$	UC_1

Wind Effect Index Possible Day-Night Combinations

	Group		Group
$c/-a$*	c_1	$-b/-b$	$-b_1$
$c/-b$*	c_2	$-b/-c$	$-b_2$
$c/-c$*	c_3	$-b/-d$*	$-b_3$
$b/-a$*	b_1	$-b/-e$*	$-b_4$
$b/-b$*	b_2	$-c/-c$*	$-c_1$
$b/-c$*	b_3	$-c/-d$*	$-c_2$
$b/-d$*	b_4	$-c/-e$*	$-c_3$
$a/-a$*	a_1	$-c/-f$*	$-c_4$
$a/-b$*	a_2	$-d/-d$	$-d_1$
$a/-c$*	a_3	$-d/-e$*	$-d_2$
$a/-d$*	a_4	$-d/-f$*	$-d_3$
$n/-a$	n_1	$-d/-g$*	$-d_4$
$n/-b$*	n_2	$-e/-e$*	$-e_1$
$n/-c$*	n_3	$-e/-f$*	$-e_2$
$n/-d$*	n_4	$-e/-g$*	$-e_3$
$-a/-a$	$-a_1$	$-e/-h$*	$-e_4$
$-a/-b$*	$-a_2$	$-f/-f$	$-f_1$
$-a/-c$*	$-a_3$	$-f/-g$*	$-f_2$
$-a/-d$*	$-a_4$	$-f/-h$*	$-f_3$
$-a/-e$*	$-a_5$	$-g/-g$	$-g_1$
		$-h/-h$	$-h_1$

Step III (Annual Physio-Climatic Extremes)

Possible July-January Combinations
(Assemblies)

EH/EH	etc.	M/M
EH/S	H/H	M/C*
EH/H	H/W	M/K*
EH/W	H/M	M/CD*
EH/M*	H/C*	etc.
EH/C*	H/K*	C/C
EH/K*	H/CD*	C/K*
etc.	etc.	C/CD*
S/S	W/W	etc.
S/H	W/M	K/K
S/W*	W/C*	etc.
S/M*	W/K*	CD/CD
S/C*	W/CD*	etc.
S/K*	W/VC*	VC/VC
S/CD*	etc.	etc.

[a] Key: EH, extremely hot; H, hot; W, warm; M, moderate; C, cool; K, keen; CD, cold; and VC, very cold.

* With dry-bulb 86° F. to 91° F.
** With dry-bulb 91° F. and above.
*** With dry-bulb 96° F. and above.

d. Reducing this to hourly values:

$$-3920/14.5 = -270 \text{ kcal/m}^2\text{hr}$$

For a qualitative evaluation of this result (-270), refer to the wind effect index in Table 7-5; it is given the symbol of $-b$, which is described as "pleasant wind effects."

In evaluating the nighttime wind effect index,

it is obviously not necessary to allow for the length of sunlight so that the index is read directly from the chart.

Temperature $= 53.4°\text{F} (11.9°\text{C})$
Mean hourly wind speed $= 11.3$ mph
Wind effect index $= -550 (-c$ from Table 7-5$)$

As in the case of the comfort index, the day-night

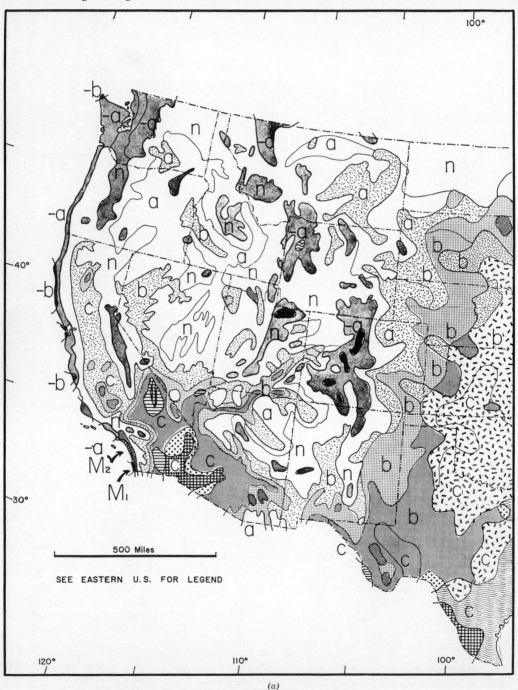

(a)

values of the wind effect index are related. As can be seen from Table 7-5, the combination $-b/-c$ falls into the $-b_2$ category, describing the cool wind effects.

The physiologic climate of San Francisco in July is thus represented by the combination $C_2/-b_2$, which is described by Terjung as "cool day with pleasant wind effects, keen night with keen wind chill." The physiological climates of the United States, evaluated in the same way, are shown in Figure 7-4. Terjung takes the evaluation one step further, such that it is possible to evaluate the annual physioclimatic extremes by assessing the assemblies of the January and July combinations. These are shown in the last column of Table 7-5.

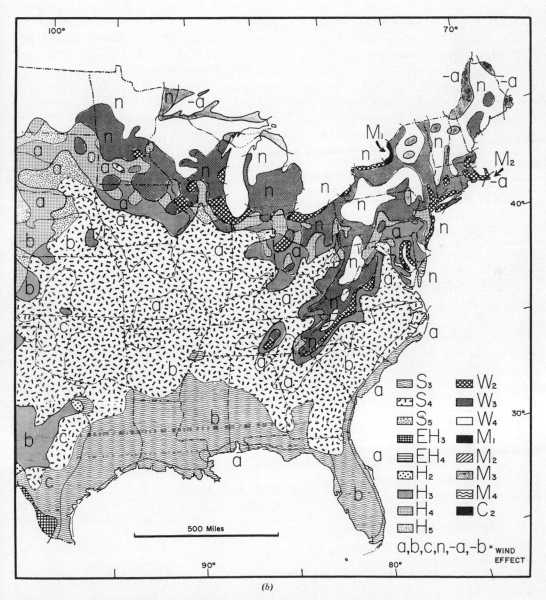

(b)

Figure 7-4. (a) Physiological climates of the western United States in July. (b) Physiological climates for the eastern United States in July. (From Terjung, W., *Annals A.A.G.*, 56, 1966. Reproduced by permission from the *Annals* of the Association of American Geographers.)

Clearly, while it is possible to criticize this system on the selection and treatment of variables, it does provide a meaningful guide in terms of the original objectives of the author. It is significant that Terjung mapped only North America, actually the United States, according to his classification. The physiological (and psychological) response used in establishing criteria for the system is based upon ". . . healthy white young men and women living under current conditions of climate, clothing and housing." Thus, in applying the criteria used to, for example, South East Asia, the resulting classification would provide a clue to the response of "healthy, white, young Americans" to the climatic conditions of that area. The process of acclimatization cannot be expressed in a system such as this and a world map showing the distribution of climatic types according to the system would not be applicable to persons living outside the temperate realm of westernized societies.

The Effects of Radiation

The screening effect of the atmosphere effectively reduces the amount of radiation reaching the earth that is harmful to man. Absorption of short-wave radiation causes man to be influenced by natural radiation at wavelengths longer than 0.3 μ. Man is thus subjected to a constant barrage of near ultraviolet, visible, and infrared radiation. Each of these directly affects the body and body functions, because they determine both the effects of thermoregulation and the photochemical responses that occur in the skin.

Since photochemical reaction can only occur when radiation is absorbed by the skin, the surface albedo is significant in determining the ultimate effects. It is estimated that the albedo of white skin is about 40%, while in heavily pigmented skin it falls to about 20%. The absorbed radiation has a number of effects. Of prime importance is the production of vitamin D, the vitamin necessary for the prevention of bone disease. As discussed on p. 216, the formation of this vitamin has given rise to a complete theory of the cause of differences in the color of skin.

A well-known effect of ultraviolet radiation on the skin is sunburn. Radiation at wavelengths around 0.32 μ produces a capillary dilating chemical (the nature of which is not known precisely at this time) which immediately induces a reddening of the skin and ultimately blistering. A "nontanned" white-skinned person will show traces of skin reddening after 3 to 20 min exposure on a clear midsummer day in middle latitudes.

Further exposure leads to a marked photochemical response of the skin. Irradiation at wavelengths between 0.3 and 0.4 μ produces melanin, which passes into the Malpighian layer. Extensive production over a long period of time leads to skin that appears to age rapidly and, possibly, to skin cancer. It has been shown that skin cancer occurs much more frequently in the southern United States, because compared to the northern states, exposure to radiant energy is greater. Similarly, skin cancer also appears more prevalent in outdoor than indoor workers. (See Auerbach, 1961).

The effects of radiant energy become much more marked with altitude. The less dense atmosphere, the lack of particulate matter, and the frequent bright skies cause greater exposure. People working or vacationing in upland areas find that tanning (or sunburn) readily occurs. In high altitudes the effects of reflected short-wave radiation are also very marked. Bright, white snow reflects large amounts of radiation and this, too, enters into photochemical reactions. Skiers often find themselves sunburned beneath the chin. Snow, too, can result in absorption of solar radiation through the eye. Radiation near 0.3 μ causes inflammation of the cornea. Glass spectacles or goggles exclude this radiation and are a necessary precaution against snow blindness.

Terjung and Louis (1971) have published a method whereby the potential solar radiation climates of man can be evaluated. They correctly point out that most work in energy-balance relations has been related to the effects upon horizontal surfaces. Man is not a horizontal surface and his ratio of height to width is significantly large.

The initial study of Terjung and Louis concerns the spatial effects of total solar radiation absorbed by man as represented in the equation

$$(Q + q)_m = (Q_m + q_s + q_g)\,(1 - a_m)$$

where $(Q + q)_m =$ total solar radiation absorbed by man

$Q_m =$ direct beam solar radiation on man

$q_s =$ diffuse sky radiation on man

$q_g =$ diffuse solar radiation reflected from the surface

$a_m =$ skin-clothing albedo of man

An estimating equation for each of the variables was obtained. In deriving Q_m, solar radiation falling on a horizontal surface was modified by introducing variables that modify it for "vertical man." Shadow area, for example, is highly variable and varies as a function of the angle of the overhead sun. At low sun angles, much more of the body is exposed to radiation; as the sun gets higher, considerable variations are found (Table 7–6). In the same way, it is necessary to take the size of the body into account.

Table 7–6
SELECTED VALUES OF THE PERCENTAGE SHADE AREA TO BODY AREA[a]

Zenith Angle °	Solar Altitude °	Percent of Shade Area to Body Area
60	30	87.5
50	40	52.4
40	50	31.4
30	60	18.8
20	70	11.3
10	80	6.7
0	90	4.0

[a] From Terjung and Louis (1971).

Equations for q_s and q_g also use modified horizontal values that take into account angle of sun and albedo. The final equation is quite complex.[4] Using derived values, the authors evaluated the effects of solar radiation on man at various latitudes and altitudes. The results of the latitudinal analysis are shown in Figure 7–5.

The conclusions, drawn by the authors from the analysis, are:

1. Each latitude has a unique daily variation of direct beam solar radiation on man. Bimodal daily curves result when man's area exposed to the sun declines more rapidly than the increasing radiation on a horizontal surface. Bimodality is more common in the lower latitudes. The opposite condition results in bell-shaped curves at higher latitudes. A transitional regime, having both types of curves, exists in the middle latitudes.

2. In lower latitudes, the daily noontime direct solar radiation on man is considerably lower for high-sun periods than for low-sun periods

3. Near the summer solstice, the maximum direct radiation on man shows little change between latitudes 0° to 80°. Latitudes 75° to 80° potentially receive the highest noontime input on earth.

4. Latitudes 0° to 30° show little seasonal variation in daily sums of direct radiation on man, but the remaining latitudes undergo great fluctuations. The largest potential energy input occurs at the poles between May and July.

5. The ratio of direct solar radiation on man to that on an equivalent horizontal interface shows rather dramatic seasonal and spatial changes. Because of increasingly lower solar altitudes, the higher latitudes and low-sun months show the highest positive departures from unity. Throughout the year, a vertical object at the equator can receive only 29 to 37% of the direct radiation incident on the horizontal surface, whereas vertical objects in the higher latitudes can receive over three times the equivalent horizontal amount. The largest possible annual input occurs between latitudes 30° and 40°, and the lowest at the equator.

6. When man's total solar heat load (the sum of direct, diffuse, and reflected solar radiation minus the effects of man's albedo) is considered during most of the year, differences in surface albedo are of great importance only in low latitudes. However, the effect of the surface albedo becomes important at all latitudes near the summer solstice. If the assumption is made that man's skin-clothing albedo is 40%, man's

[4] The equation is given as

$(Q + q)_m = \langle 1.05 \times 600 Q_h\, 7.57 \times \exp[-0.0512(90 - z) + [0.5 q_h\, 600 \times 1.86] + \{0.5[(Q_n + q_h)a]\, 600 \times 1.86\}\rangle (1 - a_m)\, K\, cal.\, hr^{-1}$

where Q_h and q_h are direct and diffuse solar radiation, respectively, on a horizontal surface and other variables are as given in text.

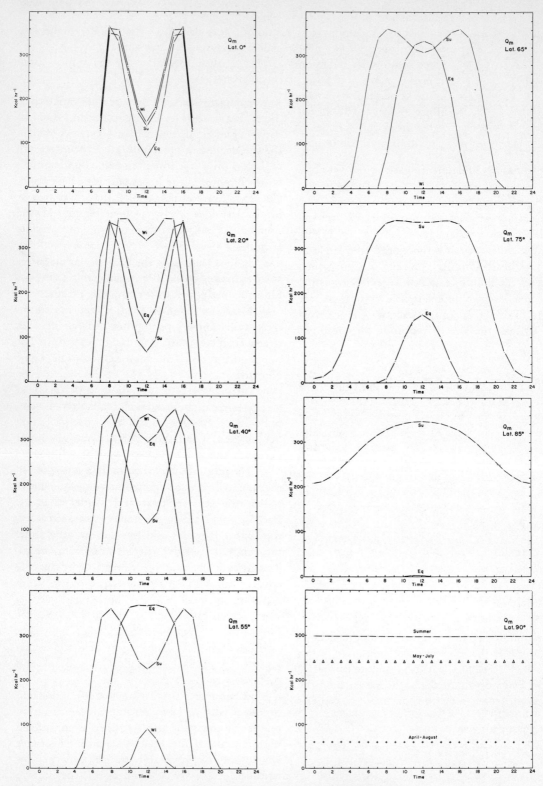

Figure 7-5. Daily direct beam solar radiation on man, Q_m, in kcal hr^{-1}.
Su = summer solstice; Wi = winter solstice; Eq = equinoxes. (From
Terjung, W. H. and S. S-F, Louie, *Annals A.A.G.*, 61, 1971. Reproduced by
permission from *Annals* of the Association of American Geographers.)

total or global solar heat load will never exceed the values received on an equivalent horizontal surface. Relatively small increases in man's albedo could change this, resulting in a total radiation load that would exceed that of the horizontal surface.

7. Except for summer, the ratio of direct (horizontal surface) solar radiation at elevations above sea level to that same radiation at sea level increases exponentially from the equator to higher latitudes and from noontime to early morning (or late afternoon) hours. During the summer months lowest ratios occur at noon in the lower latitudes, but poleward, such low ratios are increasingly moved toward the earlier (or later) part of the day. The ratio of diffuse sky radiation (horizontal) at elevations other than sea level to that at sea level shows an inverse trend.

8. During the major part of the year, the ratio of solar radiation at some height to that at sea level results in highest ratios at low latitudes and early (or later) parts of the day, but the higher latitudes show the opposite characteristic. Summer is more complex; for instance, the equator, at 10,000 ft (3048 m) elevation, receives 1.4 times the amount of horizontal radiation as sea level, but at the same elevation, the pole could receive nearly twice the amount of radiation at sea level.

The Effect of Altitude

Since the metabolism of the human organism depends, in part, upon the consumption of oxygen, a modification of the amount available to the body has marked physiologic effects. The gas mixture forming the atmosphere (78% nitrogen, 21% oxygen, 1% minor constituents) remains in the same ratio up to about 75 km; with the decrease in pressure with increasing height, the partial pressure of oxygen (pO_2) decreases. At sea level, pO_2 is about 212 mb, at a height of 10 km it is only 55 mb.

The body's need for oxygen does not alter with altitude, with the result that a rapid change from sea level to higher altitudes can cause severe body stress. Even at moderate altitudes, the effect of hypoxia (oxygen deficiency) can cause nausea. Above 6 km the lack of oxygen can seriously affect the brain.

In recent years man has learned how to cope with reduced air pressure at high altitudes—but only as a result of reproducing the environment to which he is accustomed. Pressure-controlled cabins of high-flying aircraft and the use of oxygen by mountaineers provide graphic examples. The space program has also produced a significant amount of research into the effects of reduced pressure on the human body. Chambers from which air can be evacuated to produce any degree of pressure have been extensively used in research. The vacuum chamber of Litton Industries, for example, is an 8 by 15 ft chamber in which it is possible to simulate a pressure of 10^{-4} mm of mercury, the equivalent of an altitude of 155 km (93 miles). More complex simulators, such as that at the School of Aviation Medicine, incorporate food and waste product facilities so that it is possible to control conditions over fairly long times.

Apart from artificial environments, the role of pressure has been closely studied in terms of the way in which the human body adjusts to living conditions at different altitudes. Such acclimatization results when the physiologic functions of the body become adjusted to a new set of climatic conditions, in this case, pressure. Full acclimatization is reached when the body is as efficient at high altitudes as it is at the pressure to which it is native. Investigations in the highlands of South America have shown that acclimatization to high altitudes by men native to sea level may last months, or even years. In some cases it may well be that a new steady state is never attained. The symptoms that occur during the acclimatization to high altitudes make an impressive list (Sargent and Tromp, 1964):

". . . hypotonic collapse, sleeplessness, headache, increased excitability; lower threshold of taste, pain, tendon reflexes; gastro-intestinal disturbances; loss of weight; thyroid deficiency; lung oedema; severe and fatal infections; dysmenorrhea or amenorrhea; psychological and mental disturbances."

People living in high altitude areas may suffer similar consequences when they descend to lower altitudes. The Spanish controllers of the South American Indians were aware of this in the sixteenth century. According to Monge (1948), in 1588, Philip II of Spain decreed that ". . . the Indians which we allow to be distributed shall not be from distant provinces nor from air-tempers notably contrary to the climate of the place where they are to work." While the

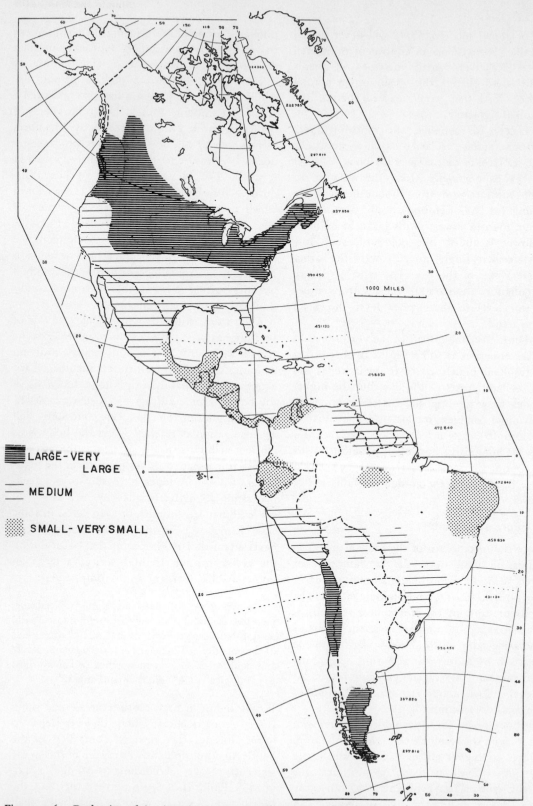

LARGE-VERY
        LARGE

MEDIUM

SMALL-VERY SMALL

Figure 7–6.   Body size of the American puma, an illustration of Bergman's
Rule. (After Newman, 1952.)

Spaniards had the wisdom to realize the effects of such change, they themselves suffered from the effects of living at unaccustomed heights. It has been noted, for example, that they lived above 10,000 ft for 53 yr before a child (who lived) was born.

Long-term acclimatization to reduced pO_2 is, according to von Hagen (1957), well demonstrated in many of the Andean Indians. He notes that they tend ". . . to be thickset, of medium height, with large hands, small wrists, disproportionately large chests, well developed legs and spreading feet." It is suggested that the squat nature of the body provides the best geometry to minimize surface area to volume ratio, thereby reducing heat loss, while the deep chest seems well developed for breathing at high altitudes. Ferris (1921) has also noted that the thoracic index of the Quechua Indians of Peru is 79 as compared to a value of 72 for Causasian children.

The acclimatization process is not restricted to altitudinal responses. Indeed, the movement from hot to cold realms, and vice versa, produces physiologic imbalances. The relationship between such responses has caused much controversy in interpretation, particularly as they relate to the so-called ecological rules.

The Ecological Rules

The relationship between the physical characteristics of warm-blooded vertebrates and their environments has been a controversial topic for many years. Disagreement generally centers on the applicability of the classical ecological rules formulated by Bergman (1847) and Allen (1877).

Bergman's rule holds that within any single, wide-ranging species of warm-blooded animals, the subspecies (or races) living in colder climates attain a greater body size than the same subspecies living in warmer climates. Basic to this rule is the fact that a greater body size increases the body mass-body surface ratio and that an increasing ratio reduces radiative body heat loss. Thus a high ratio should be found in colder climates. The classic example illustrating this rule is the American puma (Figure 7–6).

Allen's rule also relates body mass-body surface ratio to thermoregulation, but assumes that warm-blooded animals living in cold climates decrease their heat radiating body surface by decreasing the size of appendages and extremities. Example of animals that illustrate this rule are shown in Figure 7–7.

The applicability of these rules has divided anthropologists and zoologists into three camps, which Schreider suggests hold the following viewpoints:

1. Supporters of the classical rules who believe that variations in size and body-build of homeotherm species, observed in different parts of its geographic range, must be considered as evolutionary adaptations to the thermal environment.

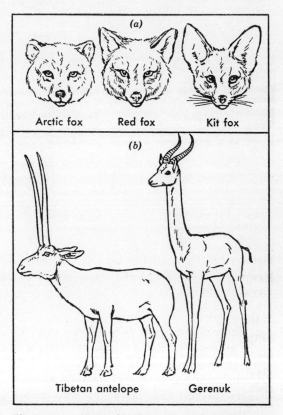

Figure 7–7. Examples illustrating Allen's Rule. (*a*) Three fox species from progressively more southern, warmer ranges; the Arctic fox (*Canis lipis*), the red fox (*C. vulpes*), and the kit fox (*C. velox*). Note the shorter nose and ears in the Arctic fox. (*b*) The gerenuk (*Litocranius*) from Abyssinia and Tanzania, and the Tibetan antelope (*Pantholops hodgsoni*). Note the shorter legs and thickset neck and face of the Tibetan species. (After Hesse et al. from Simpson G.G, C.S. Pottendrigh, and L.H. Tiffany, *Life: An Introduction to Biology*, 1957. Used with permission of Harcourt Brace and Jovanovich.)

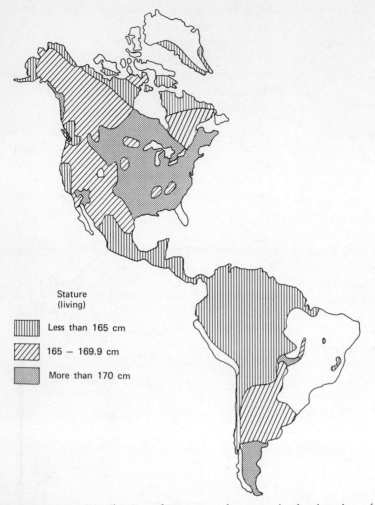

Stature
(living)

||||| Less than 165 cm

///// 165 – 169.9 cm

▦ More than 170 cm

Figure 7–8.   Distribution of average male stature in the Americas. (After Newman, 1952.)

2. Some physiologists view the very existence of the gradients as doubtful because there are so many exceptions. They further assume that the gradients could not be explained by adaptive changes because adaptation to the thermal environment is attained through physiologic instead of anatomical means.

3. Some hold that while the rules may apply to some animals, they certainly do not apply to man. They assume that man modifies his natural surroundings so efficiently that anatomical changes would not occur.

The three preceding interpretations are admirably demonstrated in a series of papers published in *Evolution* between 1955 and 1956.[5]

[5] These series and papers are included in J. Bresler (ed.), *Human Ecology*, Addison-Wesley, 1966.

Newman, a strong supporter of the ecological rules, showed that a number of clines (ecological gradients) could be found for natives in the New World. He used the distribution of male body dimensions and indices to show that the rules did, indeed, apply to man. Using stature as a measure of body size (Figure 7–8), the concentration of short people in equatorial latitudes was assumed to bear out Bergman's rule. The exception of the Eskimo, whose short stature did not fit into the scheme, was attributed to their shorter legs, for, as Figure 7–9 shows, their sitting height is large. The shorter legs (extremities) appear to conform to Allen's rule. Similar clines were found for head size, face size, and comparable patterns of facial and nasal proportions. Newman concluded that "In all likelihood these adaptive

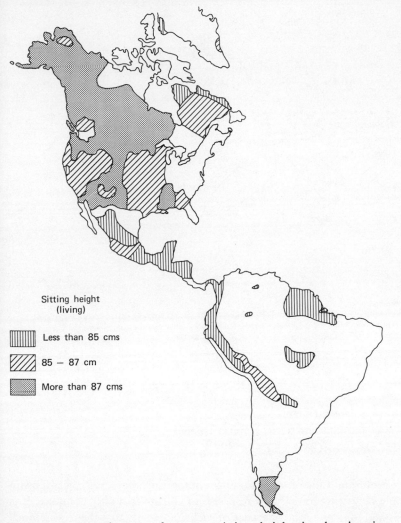

Sitting height
(living)

||||| Less than 85 cms

//// 85 – 87 cm

▓▓▓ More than 87 cms

Figure 7–9. Distribution of average sitting height in the Americas. (After Newman, 1952.)

changes took place since the New World was first peopled about 15,000 years ago."

A completely different viewpoint is expressed by Scholander (1955). He points out that a fundamental measure of the overall adaptation to cold is to determine the *critical temperature* of an animal. This is the lowest air temperature at which an animal can maintain its basal metabolic rate without losing body temperature. In tropical man, it is 25°C (77°F) to 27°C (80.6°F); for Arctic animals such as the fox or husky, it is probably −40°C (−40°F). The manner in which the critical temperature is avoided is through body insulation and adaptation of peripheral parts to remain functional at low temperatures. Figure 7–10 shows that body insulation increases

from the tropics to the Arctic, but it should be noted that the distribution of insulation over the body of the animal is highly uneven. The extremities and the face are usually poorly insulated, and it is assumed that essential heat dissipation, required with increased activity, takes place through the poorly insulated parts. In effect, the physiologic response in phylogenetic adaptation to cold need not be related in any way to size and extremities as indicated in Bergman's and Allen's rules. Methods of heat dissipation supply the only requirements for response to extreme climatic conditions.

Applying these criteria to humans, Scholander suggests that there is no basis at all for application of the rules, ". . . it would seem that tall

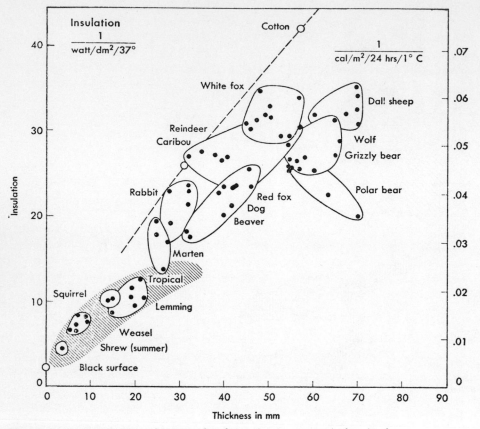

Figure 7–10. Insulation of winter fur from Arctic to tropical animals. (From Scholander, P. F., *Evolution*, 9, 1955. Used with permission.)

Eskimos could easily cheat evolution out of shortening arms and legs simply by getting long enough sleeves on their parkas and long enough legs on their pants. Extra mittens and socks could also be had if needed. Where then would be the stimulus to drive the selection?"

Both Newman (1955) and Mayr (1956) responded to this and their comments were again countered by Scholander. Newman, for example, noted that Eskimos have not yet fully attained physiologic adaptation to cold and cited evidence that the most common operation performed on Eskimos is amputation of frozen limbs, particularly frozen toes. Scholander (1956) offered another viewpoint: "The argument that Eskimos frequently suffer loss of toes from frostbite seems to be controversial; it was termed utter nonsense by a physician stationed among the Thule Eskimos. . . ."

There seems little doubt that the controversy over the application of the ecological rules will continue for some time, for man's relationship and response to climate is highly complex. It is because of such complexity that there is so much misunderstanding and misinterpretation of other man-climate relationships, particularly when dealt with in a deterministic framework.

CLIMATIC DETERMINISM

While it is evident that different people react differently, in a physiological sense, to the various climatic regimes on earth, extension of the observation to include cultural differences creates difficulties. The response to the question, "Is there causal relationship between the physical environment in which man lives and his level of civilization, his culture, his personality traits and his psychological outlook?" depends upon the degree to which one accepts the tenets of environmental determinism.

Determinism has been variously defined. Wooldridge and East (1967), for example, suggest it is "... an attempt to deduce the human resultant from the physical causes," while a dictionary definition states "The philosophical doctrine that human action is not free but necessarily determined by motives which are regarded as external forces acting upon the will." If, however, the concept is resolved to the question of the extent to which the physical environment affects man, then it is an old one. Many ancient writers—Hippocrates, Aristotle, Strabo—all concerned themselves with the relationship between types of people and the environment in which they lived. Hippocrates, whose work *On Air, Waters and Places* might be regarded as a "guidebook" for fellow physicians traveling abroad, provides observations typical of interpretations of those times. He contrasts the easy-going Asiatics with the overactive Europeans and relates the contrast to a comfortable versus a harsh environment.

The role of the environment in shaping man's destiny was not accorded much attention during medieval times for, as Tatham (1957) notes, "Firm belief in the biblical account of creation did not encourage a study of why men differed physically from each other. Differences of climate, relief and human form were all the work of God, and, as such, no fit subject for investigation." With the awakenings of the Renaissance, writers again became concerned with the problem. It was not until the rapid development of biological sciences and the advent of Darwinism in the nineteenth century, however, that environmental determinism came under the scrutinies of "scientific" investigators. In understanding and interpreting the physical world, it became possible to use natural laws; by extension, through determinism it became possible to apply similar natural laws to human culture and development. Historians and geographers used such an approach to account for the differences that occur, over both space and time, in cultural development. Eminent writers such as Haeckel, Buckle, Demolins, and Ratzel were foremost in the field of deciphering such relationships. In the United States, Ellen Semple, a student of Ratzel's, probably had the greatest impact and her forceful writings are classics of deterministic thinking.

It was not unnatural that climate, a major component of the environment, should be regarded as preeminent in such studies, and this gave rise to a school of climatic determinists. Probably the best known and most influential writer in this field was Ellsworth Huntington. In a series of works extending over many years, Huntington investigated the causal relationship between climate and all manner of human affairs. In *Climate and Civilization*, for example, he established areas of "climatic energy"—areas in which physical and mental health are conditioned by prevailing climates. From limited data input, Huntington drew broad conclusions, many of which can be shown to be statistically inaccurate. In the same book, he also constructs a map showing the world distribution of civilization. To compile such a map, Huntington "... accordingly, in the autumn of 1913, (I) asked over two hundred people in twenty seven countries to help. . . . The response was derived from 25 Americans, 7 British, 6 Germanic Europeans, 6 Latin Europeans, 1 Russian and 5 Asiatics." While some correspondents voiced concern over the proposed methodology, Huntington did construct a map. Not surprisingly, it correlated highly with his concept of climatic energy! The map represented a westernized concept of "civilization" based upon the concepts of few people.

In *Mainsprings of Civilization*, which might be considered as the culmination of his works, Huntington describes at length the role on climate in determining such diverse human characteristics as religion, the rise and fall of civilizations, and racial characteristics. Much of his theory makes fascinating reading. In relating religious beliefs to environment, for example, he suggests that the very nature of the desert environment allows comprehension of a single, omnipotent god. To forest people such a concept has no real meaning, and in the enclosed environments of forests it is much easier to conceive of many spirits, each of whom rules his own realm of the forest. It follows that animistic beliefs are more likely to be found in the religions of forest people, for the existence of one great god who rules over all is difficult to comprehend.

Huntington, like other deterministic thinkers, can be criticized in terms of sweeping generalizations based upon limited empiricism and for often ignoring evidence that was contrary to the case under analysis. Most certainly, few would be willing to accept the findings of Huntington at the present time. Unfortunately, though, Huntington is decried and dismissed by more persons who have never read his work than by those that have. Indeed, while few would accept Huntington's work today, many still draw upon the work of others who were equally deterministic. Perhaps the best example illustrating this is the fact that the historian Toynbee can be shown to present many similar ideas (Spate, 1952) although in a much more literary fashion.

The reaction against determinism has led to a whole spectrum of thinking concerning man in relation to his environment. Ranging from determinism to probablism, through possibilism to voluntarism, a whole host of interpretations can be put on man-environment relationships. Such variable interpretation is indicated by the fact that for every writer who thinks that the Middle and Near East became a desert because of agricultural mismanagement (resulting from cultural differences of successive occupants) there is another who puts the change down to physical or climatic differences. As pointed out elsewhere, "culture vs climate" as cause of regional decline is an intriguing aspect of study.

Race and Pigmentation

Another problem, one that Claiborne (1970) has called "climate's ugly legacy," relates to race as applied to physiological and cultural differences found among people of the earth. No small part of the academic interpretation of race can be traced to the early climatic determinists who equated race and civilization, where civilization is perceived in the framework of the culture of westernized European societies. Even in recent years, race, as interpreted by the color of peoples' skins, has been the center of heated controversy in scholastic literature. The ideas of Shockley, for example, who relates race to intelligence and thereafter suggests that some peoples are "evolutionary adolescent," have stirred heated debate and raised the spectre of "racism."

Despite the racist overtones that often accompany work concerning the color of peoples' skins, there is a considerable amount written about the cause for skin-color differences that is both interesting and objective.

One recent idea about the evolution of skin color has been postulated by Loomis (1967), who relates color to the human body need for vitamin D. Unlike most vitamins, vitamin D is rarely found in natural foods (many foods today are vitamin D enriched by ultraviolet irradiation). It is synthesized in the skin where pigment-producing cells are stimulated by sunlight to produce melanin. Such a process is a basic requirement of body growth and development, for calcium production and bone formation. Deficiency of vitamin D leads to rickets, while excesses can cause problems ranging from calcium deposits in the aorta to kidney stones.

Loomis suggests that early man, an inhabitant of the tropical world, originally had black skin. This was a result of the necessary protection against excess vitamin D production; as already noted, the albedo of black or dark skin is appreciably lower than that of white. Under hot tropical conditions, early man did not suffer from excess vitamin D production. As man migrated away from the tropical realm he reached regions where the black skin filtered out too much ultraviolet light, with the result that disability from lack of vitamin D occurred. Calling upon classical Darwinian natural selection theory, Loomis points out that this might well lead to the survival of the lighter-skinned people. As a result of the relative ability of the skin to transmit ultraviolet light, white people would become dominant.

To explain the "yellow" races, Loomis suggests that if an evolutionary change from black to white occurs in moving from one area to another, then a reverse effect could occur. By addition of keratin (a horny material on the outer skin layers), the effect of screening of ultraviolet light would be achieved. Drawing upon evidence of migration routes of early people, it is supposed that yellow skin is a partial response to control of vitamin D production.

Like most theories concerning man, this has deficiencies. Many biochemists would argue, for example, that additional keratin is not responsible for screening in yellow races. Nonetheless, the idea is an interesting one and points to the enormous complexity in accounting for the differences that occur in man. The simplified deterministic explanations no longer can satisfy the complex aspects that exist.

CLIMATE AND HEALTH

Escaping the weather has been a gambit employed by people—at least those who are wealthy enough—for many years. Summer palaces, Mediterranean yachts, and Florida homes, all testify to this. While such seasonal migration is, perhaps, touched with affluent societal vogues, there is little doubt that it points to the living stresses that given climates can evoke. Intense heat makes great demands upon the circulatory system and it makes good sense for a person suffering from such problems to avoid regions of stress. Diseases of the respiratory system are worsened in places with high airborne particle counts. Asthma, bronchitis, and tuberculosis cannot be adequately treated in such environments. There is, in fact, a substantial amount of literature on such facts as the climatic conditions most suitable for sanatoria specializing in the treatment of specific diseases. (Sargent and Tromp, 1964, supply both an appraisal and an extensive bibliography of the climate of health resorts.)

While many common diseases show a relationship to seasonal and other climatic variations, a direct cause-effect correlation is difficult to obtain. This results from both the problem of differentiating weather-induced illness from other illnesses and from problems of derivation and treatment of data. That is not to say, however, that the relationship cannot be shown; the following cases, drawn from a symposium held in Paris in 1963, show that some highly meaningful results can be obtained.

Some seasonal variations in respiratory diseases are exemplified by the findings shown in Figures 7–11 and 7–12. Greenberg et al., show that reported incidence of asthma increases markedly with the onset of cold weather (Figure 7–11), while Andrewes and Cross clearly indicate that respiratory diseases occur much more in the winter than in the summer, this being particularly true for cool climates. An interesting study by Goldsmith and Perkins also shows the influence of seasonality in relation to illness and, in this case, mortality. Using the ratio

$$\frac{\text{mean number of daily deaths for month}}{\text{mean number of daily deaths for entire period}}$$

they show that mortality rates for the northern and southern hemispheres are out of phase by six months. Figure 7–12 shows their results.

In the same paper, the authors also illustrate that long-term changes in cultural conditions can affect the pattern of seasonal mortality. Using data for Chicago, they show that for selected causes of death, mortality rates evolve from a summer to a winter maximum (Figure 7–13). This results from a lessening influence of diseases whose toll was in the summer months (e.g., infective and parasitic diseases) to an increase in respiratory diseases, most of which cause mortality in winter. In the same way, it can be shown that a complete shift in maximum mortality—from summer to winter—has occurred in New York City. Such findings indicate the type of problem encountered in relating climate to mortality because, as in these cases, changes in sanitation, health education, and medical research have resulted in a totally different picture of mortality in relation to weather and climate.

Of considerable concern at the present time is the deleterious effect of air pollution upon health. While in qualitative terms the effects of air pollution seem quite apparent, problems are encountered in the interpretation of data connecting long-term air pollution to illness. McCarroll (1967), for example, in his study of New York City mortality, shows that high mortality rates most certainly occur when pollution levels are high; but there are times when air quality is poor and mortality rates are low. While it is possible to obtain a correlation between air quality and death rates, it is a somewhat tenuous one.

Despite such problems, it is clear that air pollution does pose a threat to health, particularly in relation to respiratory diseases, and air pollution potential should act as an important guide in selection of living areas for persons suffering such defects. As shown in Chapter 8, site and geographic location play an important part in determining air pollution, while air stagnation in valleys is frequently the cause of low-level inversions (Figure 7–14). In autumn and winter the inversions may restrict the upward motion of air and may persist for an appreciable period. It is in such valleys, so well located for communications and settlements, that old-founded industries are located. The Meuse Valley, Trail in British Columbia, parts of London, and The Tooele

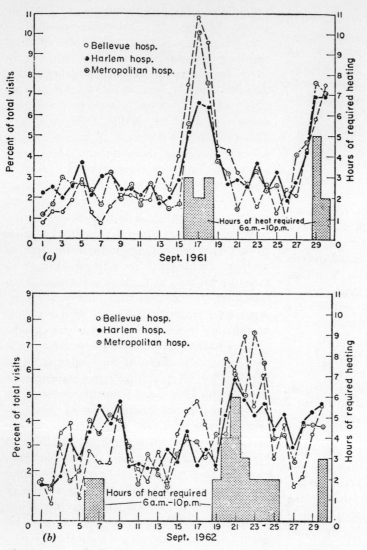

Figure 7–11.   Daily emergency clinic visits for asthma as percent of total
visits for the same period at three New York hospitals. (*a*) September 1961.
(*b*) September 1962. Note the relationship to the onset of cold spells
signified by heating requirements. (From Greenburg, L., et al., *Biometeorology*, Tromp, S. W. and W. H. Weihe, eds., 1967. Reprinted with permission of Pergamon Press.)

Valley of Utah provide apt examples. All of these
have experienced air pollution problems in rather
serious degrees. Few would question the disastrous effects of pollution episodes on health
(Table 7–7). The case of a similarly located town,
Donora, Pennsylvania, provides a fine example
of what can happen when industrial effluent is
poured into the air under inversion conditions.

Donora is located about 28 miles south of
Pittsburgh on the River Monongahela. It is a

sheltered location; bluffs up to 450 ft rise from
the river and enclose it on the north, east, and
south sides. To the west, high hills complete the
topographic encirclement. In the industrial town,
most people are employed in the steel plant and
plants that produce wire and sulfuric acid. These
factories have stacks about 100 ft high, well
below the level of the surrounding hills. The
conditions are ideal for both inversions and for
entrapment of stack output—and this is precisely

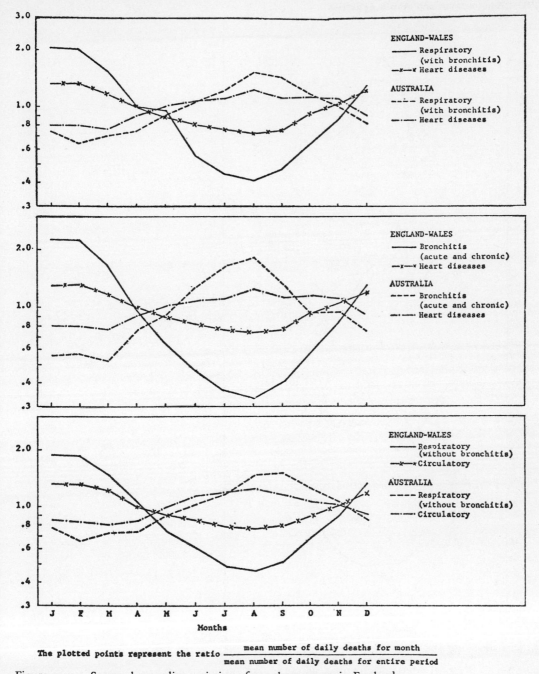

The plotted points represent the ratio $\dfrac{\text{mean number of daily deaths for month}}{\text{mean number of daily deaths for entire period}}$

Figure 7–12. Seasonal mortality variations from three causes in England, Wales, and Australia. (From Goldsmith, J. R., and N. M. Perkins, *Biometeorology*, Tromp, S. W. and W. H. Weihe, eds., 1967. Reprinted with permission of Pergamon Press.)

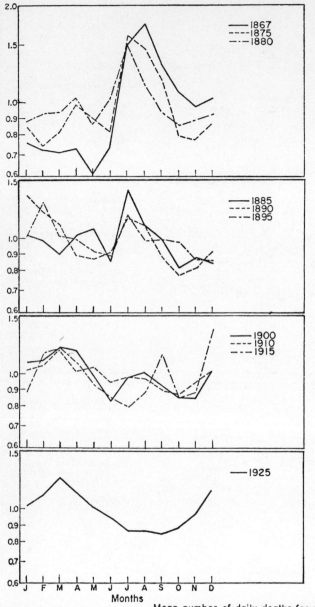

The plotted points represent the ratio $\dfrac{\text{Mean number of daily deaths for month}}{\text{Mean number of daily deaths for entire period}}$

Figure 7-13. Seasonal mortality variations, all causes, at Chicago. (From Goldsmith, J. R., and N. M. Perkins, *Biometeorology*, Tromp S. W. and W. H. Weihe, eds., 1967. Reprinted with permission of Pergamon Press.)

what occurred. The situation has been described by Roueché as follows:

"The fog closed over Donora on the morning of Tuesday, October 26th. The weather was raw, cloudy and dead calm, and it stayed that way as the fog piled up all that day and the next. By Thursday, it had stiffened adhesively into a motionless clot of smoke. That afternoon, it was just possible to see across the street, and, except for the stacks, the mills had vanished. The air began to have a sickening smell, almost a taste. It was the bittersweet reek of sulfur dioxide. Everyone who was out that day remarked on it, but no one was much concerned. The smell of sulfur dioxide, a scratchy gas given off by burning coal and melting ore, is a normal concomitant of any durable fog in Donora. This time it merely seemed more penetrating than usual."

And so, for six days, Donora endured such conditions. At the final count it was estimated that nearly 6000 had been made ill by the conditions; 20 of them, 15 men and 5 women, died.

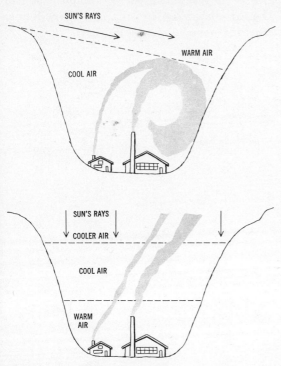

Figure 7–14. Schematic diagram illustrating the relationship between sun angle and inversion formation in valleys.

Table 7-7

MAJOR AIR POLLUTION EPISODES[a]

Date		Place	Excess Deaths
February	1880	London, England	1000
December	1930	Meuse Valley, Belgium	63
October	1948	Donora, Penn.	20
November	1950	Poca Rica, Mexico	22
December	1952	London, England	4000
November	1953	New York, N.Y.	250
January	1956	London, England	1000
December	1957	London, England	700–800
December	1962	London, England	700
January/ February 1963		New York, N.Y.	200–400
November	1966	New York, N.Y.	168

[a] From Bach (1972).

Table 7–8

AIR QUALITY STANDARDS (SULFUR OXIDES AND PARTICULATE MATTER) IN A NUMBER OF STATES[a]

State	Particulate Matter ($\mu g/m^3$)		Sulfur Oxides (ppm)	
Illinois	Annual	75	Annual	.015
	24 hr	260	24 hr	.17
			1 hr	.42
Pennsylvania	Annual	65	Annual	.02
	24 hr	195	24 hr	.10
			1 hr	.25
New Jersey	Annual	65	Annual	.02
	24 hr	195	24 hr	.10
			1 hr	.20
Delaware	Annual	70	Annual	.03
	24 hr	200	24 hr	.13
	(urban areas)		1 hr	.30
Colorado	Annual	55	Annual	.009
	24 hr	180	24 hr	.05
			1 hr	.50
Massachusetts	Annual	80	Annual	.031
	24 hr	180	24 hr	.105
			1 hr	.280
District of Columbia	Annual	65	Annual	.02
	24 hr	140	24 hr	.08
			1 hr	.25
Hawaii	Annual	55	Annual	.007
	24 hr	100	24 hr	.03
Missouri	Annual	75	Annual	.02
	24 hr	200	24 hr	.10

[a] From Bach (1972).

Disasters such as this have prompted many air quality indices to be used as a guide potential stress on the inhabitants of cities.[6] New York City, for example, has an air pollution alert where, once given values of pollutants occur, various city activities are prohibited. Beginning with a restriction on garbage disposal and ending with exclusion of all motor vehicles, it can reduce air pollution problems. Obviously though, there needs to be some level that is meaningful that is

[6] As indicated in Table 7-8, standards vary appreciably from state to state.

designated as "acceptable" or "poor" air quality. To meet this, many researchers have attempted to derive a single convenient air pollution index. This is quite difficult, because types of air pollution differ from location to location and depend both upon a high production rate and a high pollution potential.

The methods used are variable. Lowry and Reiquam (1968) devised pollution potential index that uses temperature at different atmospheric levels and pressure readings derived from radiosondes. Pasquill (1962) draws upon hourly readings of wind, sunshine, and cloudiness to classify stations into six classes of air pollution potential. These two examples are merely indicative of ongoing research to comprehend and combat climatological aspects of air pollution. (See, for example, Stern, 1962, Bach, 1972.)

CHAPTER 8

Climate, Architecture, and the City

Food and shelter are the mainstays of man's life on earth. His food requirements are ultimately met through conversion of solar radiation to usable energy, a conversion in which the role of climate is highly significant. The nature of his shelter depends largely upon the conditions of the environment in which he lives, and climate is one base on which the type of shelter needed is determined.

Primitive peoples of the world—primitive in the technological and preliterate sense—using the limited resources at hand, often developed shelters that were in perfect harmony with the climatic conditions under which they lived. A recurring theme in the literature concerning climate and architecture is that modern, westernized societies do not show the same type of response. Fitch and Branch (1960) admirably describe this when they write:

". . . with urbanization of the Western world, there is a growing tendency to minimize the importance and complexity of the natural environment. Not only is the modern architect removed from any direct experience with the climatic and geographic cause-and-effect, he is also quite persuaded that they "don't matter any more." Yet the poor performance of most modern buildings is impressive evidence to the contrary. Many recent buildings widely admired for their appearance actually function quite poorly. Many glass-walled New York skyscrapers have leaked badly during rainstorms and have had to be resealed at large cost. The fetish of glass walls has created further problems. The excessive light, heat and glare from poorly oriented glass places insuperable loads on the shading and cooling devices of the building. . . ."

It has been noted, too, that while modern architects are well aware of such fundamental factors as the type of construction required and the cost limitation, it is seldom that they are provided with the climatological data applicable to the building site, although this is an equally basic factor. Often, the appearance of the building is put ahead of its utility, and aesthetics provide the excuse for any shortcoming in design.

While it seems that modern architects might do well to study the climate-design relationships expressed in primitive dwellings, the fact should not be overstressed. Although "Primitive man often builds more wisely than we do, and follows principles of design which we ignore at great cost . . . we must not romanticize his accomplishments. With respect to many of our standards of size, amenity, safety and performance, the actual forms of many of his buildings are totally unsuitable, and it has been pointed out repeatedly how unhealthy and unhygienic such buildings might be." (Rapoport, 1970.) By the same token, it should be noted that it is very easy to become highly deterministic when dealing with primitive architecture in generalized terms. While climate certainly does impose limits upon the type of shelter used, and while the construction is limited by the materials at hand, many different groups of people living under similar climatic conditions construct buildings of quite diverse design. Often the nature of the building depends upon the culture and cultural restraints that may be intimately concerned with the use to which the building is put. While a discussion of primitive architecture tends to assume that in similar climate similar buildings occur, what is actually being stressed is that the principles of design are the same. The end products might be quite dissimilar.

CLIMATE AND PRIMITIVE CONSTRUCTIONS

It is possible to theorize on the nature of the buildings best suited to a given set of climatic conditions and, in this respect, a number of writers have suggested the optimum type of construction and design principles to be used. Drew and Fry (1956), for example, suggest that the ideal room for a hot-dry climate is a cave; the ideal for a hot-wet climate is an arbor of leaves.

223

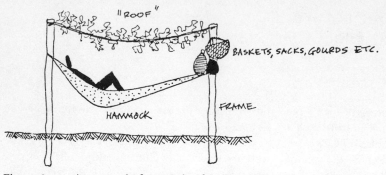

Figure 8–1.   An example from Colombia illustrating the minimum shelter required in the humid tropics. (From Rapoport, A., *House Form and Culture*, 1969. Used with permission of Prentice-Hall, Inc., Englewood Cliffs, N.J.)

This suggests that a cave possesses the quality of minimizing annual and diurnal temperature ranges and, in a desert realm, will provide an internal temperature close to the annual mean of the outside desert regime. An arbor of leaves in the hot-wet climate would reduce the radiation load and inhibit entry of rain while at the same time it provides an open framework to allow passage of cooling breezes. Both of these methods are used. The subterranean principle is found in North Africa where many groups, the Siwa for example, are troglodytes. The whimsical drawing (Figure 8–1) shows the minimum shelter used in Colombia, South America.

In a somewhat more analytical fashion, Banham (1969) suggests that the difference between shelter in hot-wet and hot-dry environments can be considered in terms of the conservative versus the selective modes. The conservative mode concerns the practice of maintaining internal conditions of uniform temperatures, be they warm or cool, in comparison to prevailing external conditions, over extended periods. In the hot-dry situation this concerns the maintenance of a nonfluctuating core interior temperature. The selective mode makes necessary the screening of undesirable elements on the one hand, while it allows entry of those that are desirable for optimum living conditions. The tree arbor is in line with this; it selectively screens rain and direct solar radiation while it allows entry of cooling breezes. A similar summary of the requirements of habitats under different climatic conditions have been given by Lee (1958), as shown in Table 8–1.

In actuality, many of the homes found in hot-wet and hot-dry climates fulfill the preceding objectives. In the hot-wet tropics, climate is characterized by small annual variation of temperature, intense solar radiation, high humidity, and heavy rainfall. Houses should thus provide maximum ventilation and shade, while the walls should be open or supplied with movable shades. Roofs, of necessity, must be waterproof. The profuse vegetation of such areas supplies ample raw material with vines, poles, and bamboo readily available. In the permanent settlements, where agriculture or fishing is the livelihood, houses are constructed on a skeletal frame, often stilted, with impervious thatched roofs that overhang to provide shade (Figure 8–2).

In the tropical deserts, climatic problems are related to the intense solar radiation and heat during the day and to the low temperatures at night. Humidity and precipitation are low and present no problem in housing comfort. Some vegetation is available as building material, but, for the most part, mud and straw are the basic materials. These are ideal, because their high heat capacity helps maintain an even temperature in the building. Figure 8–3 compares the temperature inside an adobe building with the outside and clearly indicates the significance of this. The thick walls, with minimum window space, are admirably demonstrated in many desert dwellings. Figure 8–4 shows an example of this and further indicates how grouped dwellings in dry areas are often built close together for maximum shade, a fact well exhibited in the narrow streets of many Arab communities.

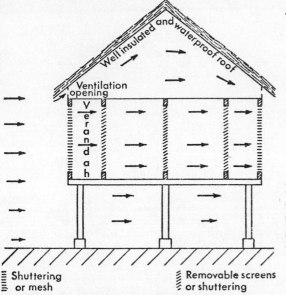

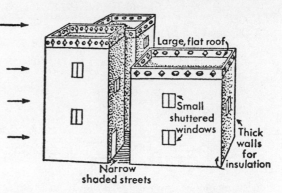

Figure 8–4. Example of buildings suited to hot, dry conditions. Note the close grouping of individual units to provide maximum shade. (From Griffiths, J.F., *Applied Climatology*, 1966, Oxford University Press. Used with permission.)

Figure 8–2. Basic house design for hot, humid climates. (From Griffiths, J.F., *Applied Climatology*, 1966, Oxford University Press. Used with permission.)

Modern, expansive boulevards are not so well suited to such conditions.

The regions that abut the wet and dry zones present special problems of house design. The inhabitants are faced with the dual problem of seasonal climates. In the Mediterranean climatic zone, for example, it is necessary to combine the requirements of buildings of the arid regions with those of places that experience a cool, wet season. In effect, to cope with the cooler wet period, the houses of the Mediterranean climate must be more substantial than those designed for desert locations only. Older dwellings in such regions, ranging back to Roman times, are often identified by a central open courtyard (Figure 8–5). These are shaded during the day, for even at the height of summer the sun is never directly overhead in these latitudes. At night they are effective as radiators. The microclimate might be further modified, for fountains and pools were often located in the courtyards. During the winter, the substantial homes could be heated to combat cooler periods.

Just as the Mediterranean regions experience a rainy season, so do the savanna areas located between the desert and the humid tropics. In such locations homes are frequently dome or cone-shaped to facilitate drainage during wet spells, and are constructed of grass, mud, and branches, although animal skins may be used.

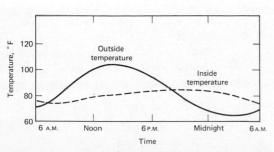

Figure 8–3. Typical temperature variations that may occur inside and outside an adobe in a hot, dry climate.

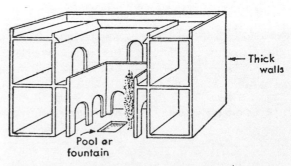

Figure 8–5. Cross section through a Mediterranean house showing the typical courtyard. (From Griffiths, J.F., *Applied Climatology*, 1966, Oxford University Press. Used with permission.)

Table 8–1[a]

PROPRIOCLIMATES OF MAN AND DOMESTIC ANIMALS

a. Thermal principles of housing

Objective	Principles	Application
In relation to a hot dry environment		
Reduce heat production	Minimize heat and vapor addition from cooking and other procedures	High operating efficiency in cooking and other heat-liberating devices
	Functional convenience	Isolation of 'wild' heat by insulation or conductive-convective removal
Reduce radiation gain and promote loss	Minimize solar projection	Labor-saving layout and facilities
	Shade	General design cubical
	High reflection of shorter but emission of longer wavelengths	Trees, bushes suitably placed
	Convection over surfaces heated by radiation	Eaves, shades on sun-exposed walls
	Insulation	White paint on sun-exposed surfaces
		Heat storage insulation in roof and sun-exposed walls
Reduce conduction gain	Insulation	Permit air-flow over sun-exposed surfaces, vents in hot-air traps
	Wind exclusion	
	Air cooling	Wall openings closable on hot days, openable on cool nights
Promote evaporation	Wetting	Evaporative cooling devices
	Maintain sufficient convection over evaporating surface	Air cooling by refrigeration or earth heat pump
In relation to a warm humid environment		
Reduce heat production	Minimize heat and vapor addition from cooking and other processes	High operating efficiency in cooking and other heat and vapor-liberating devices
	Functional convenience	Isolation of 'wild' heat and vapor by insulation or conductive-convective removal
		Labor-saving layout and devices
Reduce radiation gain	Shade	Trees, bushes suitably placed
	High reflection of shorter wavelengths	Eaves, shades, verandas on sun-exposed sides
	Convection over sun-exposed surfaces	White paint on sun-exposed surfaces
	Insulation of sun-exposed surfaces	Heat-storage or ventilated insulation of roof
		Permit air-flow over sun-exposed surfaces, vents in hot-air traps.
Promote evaporation	Air movement	Extensive wall openings unless air drying used
	Air drying	Fans or other turbulence devices
		Air drying by refrigeration or absorption in closed spaces

(Continued)

Table 8–1 (*continued*)

b. Applicability of hot dry design principles under alternative conditions

Item	Applicability to	
	Cool temperate	Warm humid
Reduced heat production by man	Not important	Still important
External shade	Not desirable	Still important
Reduced ground radiation	Not desirable	Still important
Attached shade	Minimal effect desired	Still important
Water cooling	Not desirable	Operative on roof
Minimal solar projection	Greater projection required	Still important
High reflectivity	Not desirable	Still important
High reemissivity	Not desirable	Immaterial
External convection	Not desirable	Still important
Insulation	Desirable against reverse heat flow	Still important in roof, immaterial in walls
Internal convection	Not desirable	Still desirable
Low internal emissivity	Desirable	Immaterial
Controlled ventilation	Closed	Wide open
Roof space ventilation	Closed	Open
Ground cooling	Now operates for heating	Nonoperative
Evaporative cooling	Not required	Relatively ineffective
Refrigerant cooling	Not required	For dehumidification
Reduction of heat liberation	Heat may be required	Still important

[a] From Lee (1958).

Outside the tropical realm, primitive shelters were necessary to overcome the problem of cold. The best-known construction in this respect is the igloo, a house form that is rarely found today. Using extremely limited resources, the igloo represents a building that is well adapted to hostile environmental conditions. It is hemispherical, which both minimizes the proportion of area exposed and represents a streamlined building capable of withstanding high winds, and it is constructed of dry snow blocks. These are piled one upon the other in an inward spiral. The snow blocks have a low conductivity to help conserve the interior temperatures (Figure 8–6); the interior temperature may be as high as 65°F— a marked contrast to the temperature prevailing outside. Heat is generated by both oil lamps and body temperatures and it is not unusual to find animal skins lining the interior of the dwelling.

Igloos are winter dwellings. In summer the moderate temperatures and high solar radiation (the sun is in the sky continuously) require

dwellings of different characteristics. The available materials—turf, earth, and driftwood—are used to construct sod-roofed dugouts.

Middle-latitude continental climates, with severe winters and warm moist summers, have been countered by some admirable dwellings. Nomadic people of such regions required a home suited to the extremes of climate, but also portable. Such structures are seen in the yurt

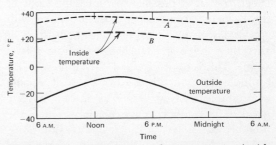

Figure 8–6. Comparison of temperatures inside and outside an igloo. *A* is air temperature, *B* is floor temperature.

(or Kazak) tent of Asia and the familiar Indian tepee. The yurt is a lightweight construction whose willow walls fold for easy dismantling. The skin or felt covering of the structure rolls up easily. Often, this covering is two layered with an air space between. The tepee is in many ways similar but uses hide covering that can be closed tightly or left as wide open as is needed.

It is impossible to describe all of the variations upon the few types of dwellings described above. Man's ingenuity and his comprehension of the environment in which he lives has given rise to dwellings of amazing diversity. Equal diversity is also found in modern buildings, but for quite different reasons.

CLIMATE AND MODERN STRUCTURES

In many westernized countries, technological development and the growth of mass communication has resulted in a partial erosion of regional differences. While, in some instances, relative remoteness or intense regional pride has negated such an effect, there is little doubt that in many countries a certain cultural homogeneity has occurred. In some ways modern architects have contributed toward this effect, because over wide areas there is a distinctive "sameness" in house design. Modern business districts, with their impressive skyscrapers, provide evidence of this, and it is not unreasonable to suggest that one downtown center looks like any other. The effect is perhaps most marked, however, in the rapid post-World War II of suburbia. As one writer has noted, one could easily move from Fullerton, California to Oshkosh, Wisconsin, and find exactly the same type of house design (Wagner, 1971). Housing vogues have had their effect, too; it is possible to find Cape Cods of the 1940s intermingled in close proximity with split-levels of the 1950s, and with bungalows of an earlier vogue and ranch houses from a later vogue. This is to be expected. Given present-day technology, a ranch house can be made equally comfortable in New Jersey as it can in New Mexico. The difference is seen, however, in the monthly fuel bills for the respective locations. A house designed for a warm, dry area cannot be expected to function at greatest efficiency in a different climate.

There is an obvious gap between house form and house function over wide areas. This is not because of a lack of awareness of the climatic environment by architects. Excellent summaries of the role of the climatic environment in architecture have been given by Aronin (1956), Olgyay (1963), and Fitch (1972). The function-form gap is probably more the result of limitations that are imposed upon the architect in the sketch-design stage. Cost is, of course, paramount and the designer must look to methods of reducing excessive expense in every possible way. This is most applicable in low- to middle-cost housing where, given a plot, a method is generally needed whereby maximum space utilization occurs.

Nonetheless, even with the limitation problems there are many ways that use can be made of the climatic advantages of a given site, and ways in which climatic problems and hazards can be mitigated. Consideration of layout, spacing, home design, and building materials in relation to the climate that occurs can provide considerable easement, both psychological and physiological, for the people who will eventually use the structures.

Layout, Spacing, and Design

Of basic importance in the layout design of a building or building cluster is consideration of the effects of solar radiation. Direct solar radiation on an exterior surface and that which is passed into the structure via openings is, for the most part, a marked asset in cold climates and should be used to full advantage; in hot regions the effects of direct radiation should be avoided as much as possible. Since the sun's path in the sky, its altitude, and resulting radiation intensity vary from location to location over the year, an understanding of earth-sun relationships is essential to formulating a rational design plan.

The position of the sun in the sky at any given latitude can be approximated using a simple representation. As shown in Figure 8–7, the apparent movement of the sun in the sky can be illustrated by considering the observer (O) located at a given latitude (in this case, 41°N, the latitude of New York City) surrounded by an horizon (H). The large, surrounding circle (S) represented the celestial sphere across which the sun can be assumed to pass. At the equinoxes

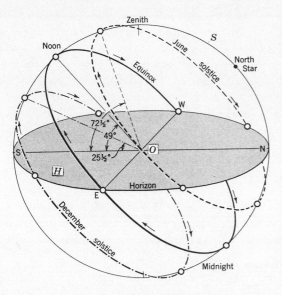

41° N New York City

Figure 8-7. Apparent path of the sun in the sky at New York City. To the earth-bound observer, the earth's surface is a flat horizontal disc. The sun, moon, and stars appear to travel on the inner surface of a hemispheric dome .(From Strahler, A.N., *Physical Geography*, Copyright © 1969 by John Wiley and Sons. Used with permission.)

the sun will rise in the due east, attain its highest elevation at noon, and set in the due west. The angle of the noon sun is given by

$$\text{Altitude} = 90° - \text{latitude}$$
$$(\text{in this case, } 90° - 41° = 49°)$$

At the winter solstice when the sun is overhead 23½°S, the altitude of the noon sun is given by

$$\text{Altitude} = 90° - (\text{latitude} + 23\tfrac{1}{2}°)$$
$$(\text{for } 41°N, 90° - (41° + 23\tfrac{1}{2}°) = 25\tfrac{1}{2}°)$$

and when the sun is overhead Cancer, 23½°N, in summer

$$\text{Altitude} = 90° - (\text{latitude} - 23\tfrac{1}{2}°)$$
$$(\text{at New York, } 90° - (41° - 23\tfrac{1}{2}°) = 72\tfrac{1}{2}°)$$

The diagram also shows that the sun is in the sky (i.e., above the horizon) for different lengths of time at these three intervals. At the equinoxes, the plane of the horizon bisects the plane representing the solar path. There are, excluding twilight, 12 hr day and 12 hr night. At the winter solstices, the part of the arc representing the time

when the sun is below the horizon is appreciably greater than that above the horizon; at the summer solstice the reverse holds true.

Architects have at their disposal several devices that allow them to calculate the path of the sun in the sky at different locations. Using the sun-path diagram, for example, and a shadow angle protractor,[1] not only can the relative angles be calculated but also the requirements for effective shading from direct solar radiation. A simple example of this is illustrated in Figure 8-8, where the correct angle and location of a roof overhang can exclude direct solar radiation in periods when it is undesirable but allows penetration through openings when needed. Clearly, in order to facilitate such treatment it is necessary to have the structures oriented so that benefits of solar radiation can be maximized or the detrimental effects minimized. Using the data derived from the sun-path diagram, it is possible to obtain an ideal orientation and the optimum can be worked out for any given location or site. Olgyay (1963) provides precise instructions on a methodology that could be used.

There are a number of ways that the architect can approach the problem of too much or too little light. Fitch (1972) has summarized these in terms of orientation and structural approaches (Table 8-2). Although most buildings are static, designs and techniques are available whereby the structure itself or moving screens or louvers can be used. Figure 8-9 shows a design in which either a whole building or an opaque screen can be used to manipulate light. Such designs are not feasible in low-cost housing tracts, but Fitch shows that here, too, there are methods by which layout can be used to full advantage. Houses on small lots can be manipulated so that the units get maximum exposure to sunlight all year, and, at the same time, maximize the effects of cooling summer winds while minimizing those of winter.

The thermal environment within a building is obviously closely related to the solar radiation factor. The object of a building is to establish and maintain an internal thermal environment conducive to habitation in terms of the prevailing external conditions. The modern approach to

[1] See Appendix 2.

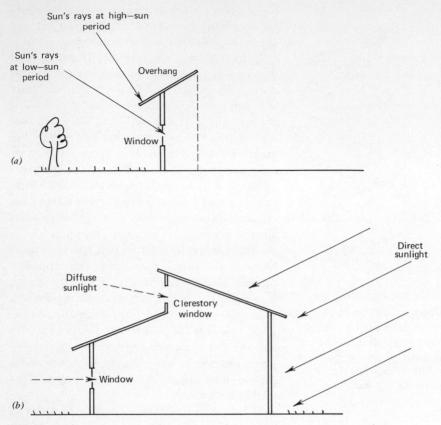

Figure 8–8   Correct use of overhangs (*a*) can allow penetration of sunlight when needed, yet exclude it at other periods and (*b*) shows the principle of the clerestory. Scattered and reflected light is admitted through the clerestory and wall windows, while direct sunlight is excluded by opaque walls on the sunlit side. (After Critchfield, 1966.)

maintaining this is to rely, for the most part, upon air-conditioning and heating systems. Obviously, such systems are needed at times, but over-reliance on them has often resulted in unbearable strains on power resources. As considered in the following material, the most important aspect in considering the thermal environment is the membranes that separate the internal and external environments.

Much thought has been given to the use of solar energy as a source for heating and cooling in buildings. This has actually been put into effect in, for example, Albuquerque, New Mexico, where a solar-powered office building has been constructed. The south facing wall has been designed so that it utilizes the high quantity of solar radiation typical of the southwest; the wall, sloping at an angle of 30° from the vertical, is constructed of glass-faced boxes containing

coils; absorbed energy is transported through water as a medium.

Layout design also has a marked effect upon air motion. Wind in hot, moist climates is an asset and should be exploited as natural air conditioning, for it tends to reduce temperatures and excessive humidity. In colder regimes, the main effect of the wind is to induce convective cooling and so lower the sensation of temperature.

Many of the climatic effects on buildings require both macro- and microanalysis. This is particularly true of the role of wind, for in many situations local topographic effects give rise to highly specialized conditions. High winds frequently occur, for example, on the windward sides of slopes in higher latitudes. If such situations are used for building, it then becomes necessary to design the structure in relation to wind breaks or natural topographic barriers.

Table 8-2 [a]

BUILDING STRATEGIES RELATED TO THE LUMINOUS ENVIRONMENT

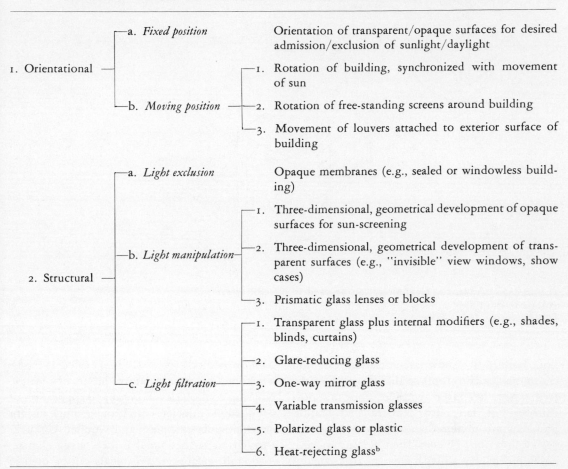

1. Orientational	a. *Fixed position*	Orientation of transparent/opaque surfaces for desired admission/exclusion of sunlight/daylight
	b. *Moving position*	1. Rotation of building, synchronized with movement of sun
		2. Rotation of free-standing screens around building
		3. Movement of louvers attached to exterior surface of building
2. Structural	a. *Light exclusion*	Opaque membranes (e.g., sealed or windowless building)
	b. *Light manipulation*	1. Three-dimensional, geometrical development of opaque surfaces for sun-screening
		2. Three-dimensional, geometrical development of transparent surfaces (e.g., "invisible" view windows, show cases)
		3. Prismatic glass lenses or blocks
	c. *Light filtration*	1. Transparent glass plus internal modifiers (e.g., shades, blinds, curtains)
		2. Glare-reducing glass
		3. One-way mirror glass
		4. Variable transmission glasses
		5. Polarized glass or plastic
		6. Heat-rejecting glass [b]

[a] From Fitch (1972).
[b] Although used for thermal reasons (i.e., to reduce transmission of infrared radiation), such glasses also affect optical behavior of the wall.

Note: Energy input-output relationship between building and its luminous environment is a function of: (1) location and shape of building with reference to path of sun; (2) area, geometry, and exposure of external surfaces of building; (3) physical characteristics of these surfaces—opaque, transparent, or selective filtration.

The channeling of wind through valleys also requires special design. The local building response to the cold Mistral winds of the Rhone Valley provides a good example. Small windows, or even no windows at all, are characteristic of the northward-facing sides of buildings where wind breaks are commonly found.

A knowledge of prevailing wind direction is also of value in evaluating potential windborne pollution. Odor as well as particulate matter is a common hazard, for the odors generated by some industrial centers can prove quite overwhelming. Many home owners, originally unaware of the problem, have been forced to relocate as a result of such pollution. A similar effect is seen in nonurban areas, but here lessons learned over time are more evident. It would be quite unusual, for example, for a farmer to locate his farmhouse on the downwind side of animal stalls and pigpens.

Precipitation regimes also play an important role in design and layout of housing. Precipita-

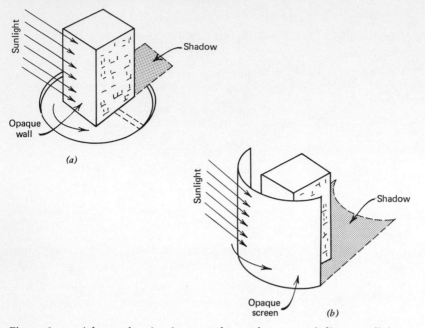

Figure 8–9. Advanced technology can be used to control direct sunlight falling on a building surface. (*a*) Rotating buildings and (*b*) rotating screens have been suggested to mitigate effects. (After Fitch, 1972.)

tion leading to snow accumulation can add enormous loads to roofs so that the pitch of the roof must be given consideration in snow climates; in hot, dry climates precipitation problems are minimal, although the danger of flash floods must be considered. Recent housing development on alluvial fans in arid-semiarid regions are faced with a problem almost equal to that of buildings located on flood plains: in such locations storms often give rise to disastrous mud-flows.

Building Materials

The amount of energy needed to maintain a required internal temperature in a dwelling depends upon both the external climatic conditions and the fabric of which the walls are made. As already noted, both the adobe and the igloo seem well suited to the environments in which they occur.

These two examples clearly indicate the importance of construction materials under different climatic conditions. In these days of prefabricated houses or uniform building materials the natural thermal properties of the material are sometimes overlooked. This is in accord with the philosophy that with artificial heating and cooling methods, most building problems can be overcome by manipulating input of energy. Nothing could be less correct, and application of a few simple principles can provide much insight into optimum thickness and required conductance characteristics suitable for a given climate. Although prefabricated materials can be used in such diverse climatic regions as New England and the American southwest, the manner in which they are arranged needs to be quite different.

To understand how a building reacts to a given external climate, it is necessary to estimate a number of thermal characteristics. Basically, the main concern is the flow of heat through the wall or roof in terms of both quantity and time. For a single material the flow is given by

$$Q = \frac{k}{x} A \Delta t$$

where k is the conductivity of the material, x is the thickness, A is the surface area, and Δt is the difference in temperature on either side of the wall. If the wall consists of several layers of material, it becomes necessary to use

$$Q = UA\Delta t$$

where U is the reciprocal of the properties of individual layers and air spaces in the wall or roof.[2] When materials are considered in layers and the U value required, it is often convenient to express thermal properties as the R value, the heat resistance; this is merely the reciprocal of U. It is found that the R value varies considerably. For example,

Material/Layers	R Value (Heat Resistance)
Galvanized metal over open frame	0.6
Galvanized metal over 1 in. insulation board	3.6
Galvanized metal over 3⅝ in. wood shavings with galvanized metal inside	10.5

While a simple galvanized metal shed may appear to be a potential oven in a hot climate, simple modification of the structure and rational use of building materials can mitigate the conditions.

The flow of heat calculated using the preceding method, that is, the internal/external temperature difference, does not take into account the effects of direct solar radiation falling on the structure. This is included in the determination of a sol-air temperature, which is defined as ". . . the temperature of the outside air which would give the same rate of heat transfer and the same distribution of temperature through a construction as the combined effects of solar radiation and air temperature." (United Nations, 1972.) On sunlit surfaces the sol-air temperature will be higher than the air temperature and, where these are concerned, the following formula can be used:

$$\theta sa = \frac{aI}{f_o} + \theta_o$$

where θsa = the sol-air temperature, °C
a = the absorptivity of a surface to solar radiation—given as a fraction
I = intensity of solar radiation

[2] $U = 1/f_o + L_1/k_1 + L_2/k_2 \ldots L_n/k_n + 1/C + 1/f_1$

f_o = conductance of outside surface (W/m °C)
f_1 = conductance of inside surface (W/m °C)
$L_1, L_2, \ldots L_n$ = thickness of individual layers (m)
$k_1, k_2, \ldots k_n$ = conductivities of layers (W/m °C)
C = conductance of air spaces or cavities

f_o = outside surface conductance
θ_o = outside air temperature

and from this the rate of heat transfer due to radiation can be calculated and a solar heat factor (q/I) derived.

$$q = U(\theta sa - \theta_o)$$

so that $q = \dfrac{UIa}{f_o}$ (substituting from $\theta sa = \dfrac{aI}{f_o} + \theta_o$)

and the solar heat factor

$$\frac{q}{I} = \frac{Ua}{f_o}$$

It is evident that the higher the q/I value, the greater the addition to the heat load by solar radiation.

Once the relationship between building material and heat flow is derived, another important factor must be considered. The transfer of heat through a structure will not occur instantaneously and there will be a time lag between the maximum outside and maximum inside temperatures. The time lag can vary from about 3 hr for a light wall made of two layers of sheet material separated by an air cavity, to at least 8 hr for a compacted earth wall about 300 mm thick. Not only will time lag occur, but there will also be an interior dampening of the extremes found in the exterior climate. As Figure 8-10 shows, the differences can be visualized as two cycles; the time lag causes them to be out of phase while the dampening (or decrement) causes the interior cycle to have less amplitude than that outside. Use of the correct wall materials and thicknesses can result in an ideal situation, where the phase and amplitude are designed to meet the needs of the climatic realm in question.

In damp climates, building materials also need to act as vapor barriers. Lack of proper use of vapor barriers is seen in peeling surface paint, stains on inside walls of heated buildings, and damaged wood or metal structures. The resistance of material to moisture penetration is given by its *vapor resistance*, the reciprocal of its permeability (P_1 measured in hr/ft²/psi).

The Mahoney Tables

From the preceding brief outline, it is evident that climate has a great deal of significance in

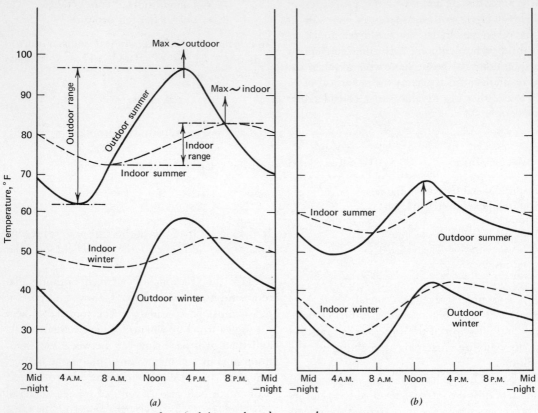

Figure 8-10. Comparison of typical internal and external temperatures in (*a*) an adobe in hot, dry climates and (*b*) a wood structure in middle latitudes.

architectural design. It is also evident, however, that in-depth analysis is highly time consuming and requires much climatic information. This is not always available and, considering all of the other factors that need to be considered, climatic analysis may be underemphasized. To alleviate this problem and to take some of the guesswork out of the interpretation, Carl Mahoney developed a methodology that can be used to analyze the basic climatic variables. These are highly interesting both from a climatic and architectural standpoint, and are given in detail in Appendix 2.

URBAN CLIMATES

If the object of building construction is to create an artificial climatic environment in which the optimum physiologic needs of man are met, then it has been highly successful. Given present-day technology and sufficient funds, it is possible to produce an interior environment in which light, heat, and humidity conditions can be regulated to meet any contingency. The construction of such an internal environment, however, cannot be achieved without modification of the preexisting external conditions. As a by-product of meeting human living requirements, a new set of climatic conditions are created. Ryd (1970) has used the term "climatic sheath" to describe the modified climatic space that surrounds a building. Within this sheath many modifications, ranging from temperature and humidity anomalies to a rain shadow effect, will be found. When, as in cities, buildings occur in large concentrations, the modification results in a "climatological dome." This dome provides the milieu in which urban climate changes occur.

While the climatic environment is often considered as a variable in siting new buildings and building clusters (Landsberg, 1970), little attention has been paid to the modified climate that will result. Perhaps the reason for this relates to the lack of predictive models allowing evaluation of such changes. It has become a prime task of climatologists to examine the changes that occur and to attempt to assess quantitatively the nature of the variables involved.

The study of urban climatology has, of course, ramifications that extend far beyond the confines of the city. The changes that occur form an integral part of the whole spectrum of ecologic changes resulting from the growth of urban industrial societies. In the city, however, the results of polluted air and massive surface modification provide a working ground in which man-induced environmental changes can be studied in detail. Changes in the city climate resulting, for example, from increased turbidity[3] provide insight into what might occur on a worldwide basis. It has been shown by many workers that even relatively remote areas experience an increase; the amount of dust in glaciers in the Caucasus showed little variation up to 1930, but in the period 1930 to 1963, it increased almost twentyfold (Davitaya, 1969, Figure 8–11); similarly, McCormick and Ludwig (1967) found marked turbidity increase in Davos, Switzerland, while Bryson and Kutzbach (1968) found an increase of 30% in the turbidity at Mauna Loa, Hawaii, between 1957 and 1967. The long-term effects of such increases can be evaluated using the city as a laboratory.

The Problem of Data

The problem of evaluation of climatic changes in urban centers is compounded, because meaningful data have only recently become available. Much of the current research is centered around gathering as well as interpretation of data. Admittedly, most urban centers have had weather stations for a number of years, but such stations are usually located so that recording equipment is as removed as possible from modifying influences such as buildings. Typical locations are parks, tops of high buildings, and airports. This is to be expected, since the purpose of most weather data collection is to facilitate synoptic meteorology. The data should be representative of wide areas rather than

[3] Atmospheric turbidity is an index of the transparency of the atmosphere to solar radiation.

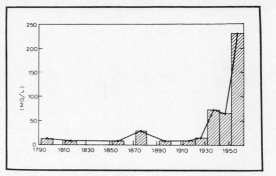

Figure 8–11.   Dust fall trapped in the firn of glaciers of the high Caucasus according to Davitaya. He suggests that the rapid increase in recent decades is due to human activities. (From Bryson, R., *Weatherwise*, 21: 56–61, 1968. Used with permission.)

indicative of a microclimate. Even so, such data are used to obtain a basic evaluation of the trend of city climates. Figure 8–12, for example, shows city-airport comparative data in terms of the number of days above or below a given threshold limit. While useful, such data must be treated with care, for an airport environment is just as artificial as the city.

In collecting data, some ingenious methods have been devised. In evaluating wind direction, Okita (1960, 1965) studied the distribution of rime ice that formed on the windward side of trees; he has also used observed smoke direction at various locations to gain insight into complex air motions. Moving vehicles of all types have been employed to obtain horizontal and vertical gradations. Helicopters and cars are most common, but even bicycles (Tamiya, 1969) have been used to good effect. Clearly, such methods cannot be used where point data are concerned; this is a major problem of evaluation of some climatic variables. In the case of precipitation, it is quite questionable how far a single rain gauge is representative of an urban area and, unfortunately, emplacement of rain gauges at varied locations in a city provides a splendid outlet for vandalism. Apart from such problems there is also some question of the inhomogeneity of the data that do exist. As an example, after study of the precipitation records of Central Park, New York, Spar and Ronberg (1969) demonstrated that an apparently significant trend in precipitation is anomalous.

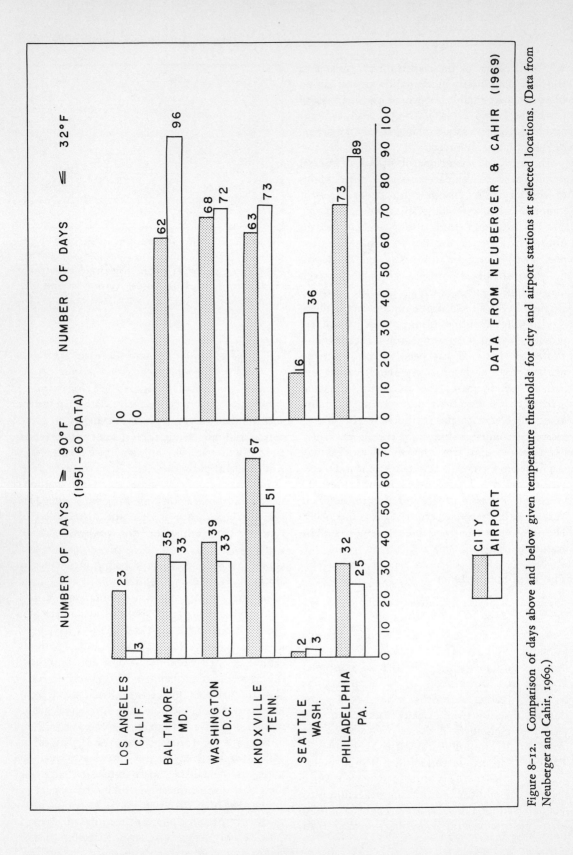

Figure 8-12. Comparison of days above and below given temperature thresholds for city and airport stations at selected locations. (Data from Neuberger and Cahir, 1969.)

Partly because of the need for data, most investigations of urban climates have concerned a single climatic variable. Using such studies, it is possible to generalize on the nature and amount of climatic change induced by cities (Landsberg, 1956, 1970), but it must be recognized that each study actually relates, in quantitative terms, only to the area that was observed. To utilize such findings for prediction and correlation, sophisticated models are required. While important work in the construction of models has been completed (Neiberger, 1970, Terjung, 1971, Myrup, 1969) much remains to be completed.

A CONCEPTUAL MODEL OF URBAN CLIMATE

In order to assess the complexities of the urban climate, any conceptual model must point toward the interaction and interpenetration of the variables involved. To stress this, it proves convenient to explore the nature of the city climate using the Venn diagrams. These represent a simple way of illustrating relationships between sets representing members of the urban climate.

Figure 8–13 shows the conceptual model defined at three levels. Level 1 indicates the processes responsible for the climate that occurs at any point on the earth's surface. It is through the modification of energy and mass budgets that the climate of the city differs from its neighboring nonurbanized area. As indicated, the alteration can occur through man-induced changes of both atmosphere and surface. The amount of change of the variables in Level 1 depends upon the character of the city, its geographic location, its size (in terms of both population and areal extent), and its basic functions; these are considered in Level 2 of the model. The use of the term function raises the specter of city classification through industrial specialization; this is not intended here. The term is used as a surrogate for the gross energy that is imported into the city to maintain it as a functioning entity. Level 3 of the model concerns the way in which selected climatic elements in the city differ from those of nonurban areas. The nature of these elements depends, of course, upon the modified mass and energy budgets in Level 1 of the model. It is useful to consider them separately, however, for in many ways they represent

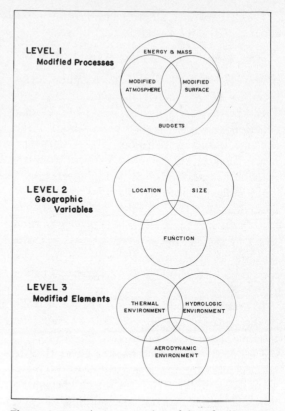

Figure 8–13.   A conceptual model to facilitate the study of urban climates.

a characterization that allows the sensible climate of cities to be viewed. The nature of the elements are considered in terms of the thermal, hydrologic, and aerodynamic environments.

While any division of a holistic unit is quite artificial and most arbitrary, the various levels of the conceptual model attempt to retain the unity of the city climate. The schematic representation using Venn diagrams emphasizes the point.

Level 1. The Modified Processes

The individuality of any climate relates to the flux of energy that arrives at a surface and the manner of its subsequent distribution. The amount of energy recieved at a surface depends upon earth-sun relationships together with the way in which the solar beam is modified as it passes through the atmosphere or is reflected from the surface. Man can do little to modify earth's astronomical location, but can do a great deal to modify the atmosphere and surface; it is in this way that the energy balance can be altered.

The receipts and ultimate disposition of net radiation (R) can be expressed as follows:

$$(Q + q)(1 - a) - I = R = LE + H + G$$

where the left-hand side of the equation represents net all-wave radiation input. In this, Q is direct beam solar radiation, q is diffuse solar radiation, a is the surface albedo, and I is the effective outgoing radiation ($I\uparrow - I\downarrow$, i.e., terrestrial radiation minus atmospheric counter-radiation). The disposition of R is given by the "heat budget" shown on the right-hand side of the equation. The variables include LE, the latent heat flux (L is latent heat of vaporization and E is the amount of evaporation), H is the sensible heat flux and G is the flux of heat into the ground—the storage of energy (see Chapter 1).

In the urban environment, each of the variables in the equation can be modified. Consider, for example, the effect of man-induced aerosols on the input ($Q + q$). One might deduce that increased turbidity over a city decreases the short-wave radiation input and that the city becomes a gloomier place than the surrounding countryside. This is true, and several studies (e.g., Chandler, 1965, McCormick and Kurfis, 1966) bear out Landsberg's (1956) conclusion that annual total solar radiation over cities is lower than nonurban areas. Legislative action against pollution has permitted "before and after" effects of particulate matter on solar radiation. Subsequent to the Clean Air Act of 1956 in the United Kingdom, Monteith (1966) found that total solar radiation in London increased significantly. Similarly, Jenkins (1969), in evaluating the average number of hours of bright sunshine in London, found an increase of 50% from 1958 to 1967 as compared to 1931 to 1960.[4]

The role of clouds in reducing the amount of energy arriving at the surface is well known. For a number of reasons, including the contribution of aerosols as hygroscopic nuclei in cloud formation, cities are generally cloudier places than neighboring nonurbanized areas. The increased cloud cover, as much as 10% higher in winter,

[4] Note that some cities not affected by the Clean Air Act have experienced a decline in particulate matter and an increase in solar radiation. Thus, the Act itself may not be the major contributing factor toward the changes cited.

further modifies the energy input, at the same time causing a lowering of effective outgoing radiation (I).

The role of particulate matter in modification of solar radiation is not, however, a simple case of reflection of the solar beam. Mitchell (1971) shows that the assumption that atmospheric aerosols are most significant in scattering and attenuating short-wave radiation is not entirely correct and that absorption may be more significant. The significance of this is that if absorption is greater than backscatter, then aerosols in the troposphere would act to increase rather than cool the surface layers. The distinction between the relative warming-cooling effect can be approximated by comparing the observed aerosol absorption-to-backscatter ratio (a/b) with that of a critical absorption-to-backscatter ratio $(a/b)_o$. Mitchell notes that in urban areas the relationship is marginal and depends upon the properties of the aerosols that prevail.

Modification of a surface can also modify the net radiation by altering the albedo (a). Different surfaces have different reflective characteristics and the city comprises a multitude of surface varieties. Not only do the characteristics vary, but it is also noteworthy that the vertical extent of city buildings will further cause it to vary markedly from nonurban areas (Figure 8-14). Airborne surveys (Kung, Bryson, and Lenschow, 1964) have shown that the urban albedo is generally about 10% lower than that for rural surfaces.

It is evident that the modified atmosphere and surface in the city grossly modify the radiation budget. The same is true of the eventual way in which radiation receipts are budgeted. One important aspect of this is well illustrated by the ratio of energy between the sensible heat flux and that of the latent heat flux, described by the Bowen ratio (H/LE). Table 8-3 gives the Bowen ratios for various surfaces. It can be seen that a high value occurs in urban areas, indicating that more energy is utilized in the sensible rather than latent heat flux. The reason for this partly relates to the modified hydrological cycle in urban area.

Figure 8-15 gives a schematic representation of the disposal of precipitation falling onto an urban and nonurban area. Except in limited green areas and parks, the pathways taken by

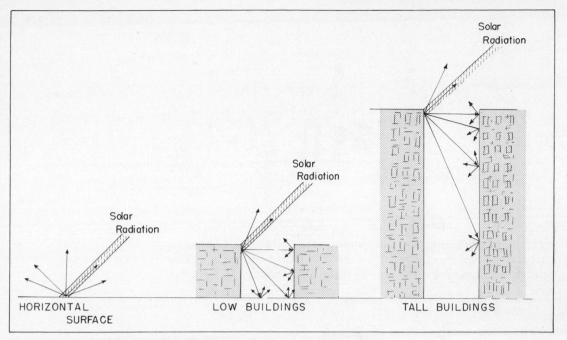

Figure 8–14. The effects of a horizontal surface and vertical surfaces of different heights on incoming solar radiation.

Table 8–3

REPRESENTATIVE VALUES FOR THE BOWEN RATIO AND SENSIBLE HEATING INDEX FOR VARIOUS SURFACES[a]

Surface	Bowen Ratio (B)	Sensible Heating Index (C)[b]
Deserts	20.0	0.95
Urban areas	4.0	0.80
Grasslands/farmlands (warm season)	0.67	0.40
Coniferous forests	0.50	0.33
Deciduous forests	0.33	0.25
Stable snowfields	0.10	0.09

[a] After J. Murray Mitchell, Jr., (1971).
[b] The ratio of sensible heating to total (sensible + latent); derived from the Bowen Ratio (B) and given by $C = B/(B + 1)$.

moisture is quite limited in the city. Transpiration, infiltration (to both ground and soil water), and evaporation are radically reduced. In fact, in most urban areas precipitation is channeled into storm sewers as runoff. The amount of surface standing water is reduced to a minimum with the result that energy available for sensible heat is increased significantly. Of course, it must be noted that the production of water vapor and steam as a by-product of various combustion processes does modify the diagrammatic representation shown in Figure 8–15, a factor that is intimately concerned with the energy production within the city itself. So significant is this contribution that the heat budget need be rewritten:

$$R + Rm = H + LE + G$$

where Rm is man-produced energy.

The amount of Rm, as described in the following material, varies appreciably, but a gross estimation of its effect has been given by Harte and Socolow (1971) who write "To get an idea of the magnitude of urban energy production, we can compare the energy output in New York City with the solar flux directly on the city. Approximately 7.8 million people live within an area of 365 square miles in New York City. If we assume that New Yorkers consume energy at the national per capita rate of 7.5×10^{10} calories per year, then every year 5.8×10^{17} calories of energy are consumed within 365 square miles. In other words, energy production in New York City is roughly 1.6×10^{15} calories per square mile per

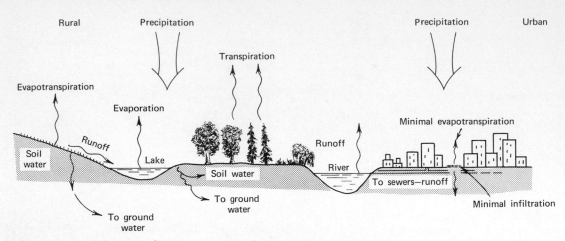

Figure 8–15. Schematic diagram showing disposal of precipitation on rural and urban surfaces. Note that through contribution of water vapor by man, the evaporation rates may be higher than indicated in the urban area.

year. . . . The solar flux affecting the earth's climate and biosphere, averaged out over the surface of the earth, is, in these units, 4×10^{15} calories per square mile per year. This is only $2\frac{1}{2}$ times larger than our estimate for the energy consumption by man in New York City. Modern cities would experience catastrophic climatological repercussions from the high density of waste heat production, except for the fact that they are still small enough for winds to dissipate their heat adequately." Given the present 4 to 5% annual increase in man's energy production, some writers feel that our average global temperature rise of more than 1°F will occur within the next 135 yr (Holdren, 1971). Derivation of G, the ground heat flux is a difficult task in the urban setting. Terjung (1970) found this to be the case in his study of the energy balance climatology in the Los Angeles basin when he used slabs of macadam to derive values. The great variability of the surface poses many problems in its evaluation.

Level 2. The Geographic Variables

A brief survey of the bibliography concerning urban climates will indicate that almost all studies on city climates have been completed for middle-latitude locations. By contrast, with some notable exceptions (e.g., Nakamura, 1967), low- and high-latitude stations have not been accorded the same coverage. Despite this inequality, it is evident that geographic location does play an important role in determining the nature of a city climate. The role that gross location plays relates to the frequency of air mass dominance and rate of air renewal in the city. It has been shown, for example, that the urban "heat island" cannot come into existence above a critical wind speed (Table 8–4). The obvious corollary to this is that areas experiencing constant high winds or rapid air mass motion will not exhibit a climate appreciably different from that of a neighboring nonurban area, while cities that experience stagnant air for extended periods should show a well-defined difference.

This latter point is further stressed by the role of pollution in modifying the climate of cities. As indicated, the contamination of the atmosphere plays a multipurpose role, and the amount and frequency of pollution is a significant factor in the nature of the urban climate. The most common meteorological condition contributing to high pollution levels is the presence of an inversion. This occurs when warm air overlies cooler air and limits vertical motion. As Hosler (1961) notes "Since the general spatial problems of slow atmospheric dispersion which affect large areas unquestionably arise at times when a stable stratified layer of air exists near the surface, a knowledge of the frequency of low-level stability for geographical and/or climatological

Table 8–4

CRITICAL WIND SPEEDS REQUIRED FOR THE ELIMINATION OF THE HEAT ISLAND
EFFECT IN VARIOUS CITIES[a]

City	Year of Survey	Population	Critical Wind Speed (m/sec^{-1})
London, England	1959–1961	8,500,000	12
Montreal, Canada	1967–1968	2,000,000	11
Bremen, Germany	1933	400,000	8
Hamilton, Canada	1965–1966	300,000	6–8
Reading, England	1951–1952	120,000	4–7
Kumagaya, Japan	1956–1957	50,000	5
Palo Alto, California	1951–1952	33,000	3–5

[a] From Peterson (1969).

areas will be helpful for assessing the potential for air pollution inherent to a given region." Figure 8-16a shows the marked differences that occur in the frequency of low-level inversions over the continguous United States. Following Hosler, the area is divided into gross regions each characterized by the prevalence of inversions. More recently, and as shown in Figure 8-16b, maps have been prepared in terms of air-pollution potential. Clearly, geographic location plays a significant role in the frequency of such meteorological events.

Location, as it relates to topographic situation, is also significant in relation to air motion and the cleansing of polluted air. Many old founded industrial centers are located for economic reasons; their sites are best suited for factors such as transport or raw material availability. Unfortunately, many such centers lie in the heart of valleys, and such locations are highly susceptible to locally induced inversions, and pollutant-laden stagnant air. The well-known Donora, Pennsylvania, disaster provides graphic evidence of this (see pp. 218–221).

Figure 8-16a shows that west coast stations, such as Los Angeles, have a high incidence of inversions, with low-level characteristics and subsidence contributing to the total. Yet when one consults the list of the most polluted cities in terms of particulate matter, Los Angeles is low on the list. In explaining this apparent anomoly, Chandler (1970), describes the situation succinctly: "The pollution which typifies most industrial cities of Europe and eastern Northern America, what we might almost call 'traditional'

air pollution, is a mixture of mainly carbon and sulphur compounds. But there are more modern brands of pollution of increasing importance in most modern cities, these are the gaseous brews of the internal combustion engine and what London is, or rather was, to smoke, Los Angeles is or was to hydro-carbons, nitrogen dioxides, carbon monoxide and such other products. . . . In a sense, therefore, every city is unique in the chemistry of its urban atmosphere. . . ." Function of the city thus becomes an important variable from the standpoint of the mode of modification of the urban atmosphere.

As outlined here, in some cities, excess heat of combustion contributes significantly toward the modification of the thermal environment. This factor is closely related to the size and form of the city. In modern middle-latitude cities, enormous amounts of fuel are imported for heat and power; the amount used is largely a function of the density of population of the city. Various estimates of the amount of imported energy are available. Kratzer (1956) has suggested that the amount of energy produced in Berlin and Vienna are one-third and one-sixth to a quarter, respectively, of that received from solar radiation. Garnett and Bach (1965) show that in Sheffield, England, the amount produced is one-third of all wave radiation. Working on the night temperature distribution of a new dormitory town (population 10,000) near Tokyo, Tamiya (1969) estimated the heat produced by human activities at 6.4×10^9 cal/hr, an amount evaluated almost equal to the heat needed to raise the air temperature by 1°C.

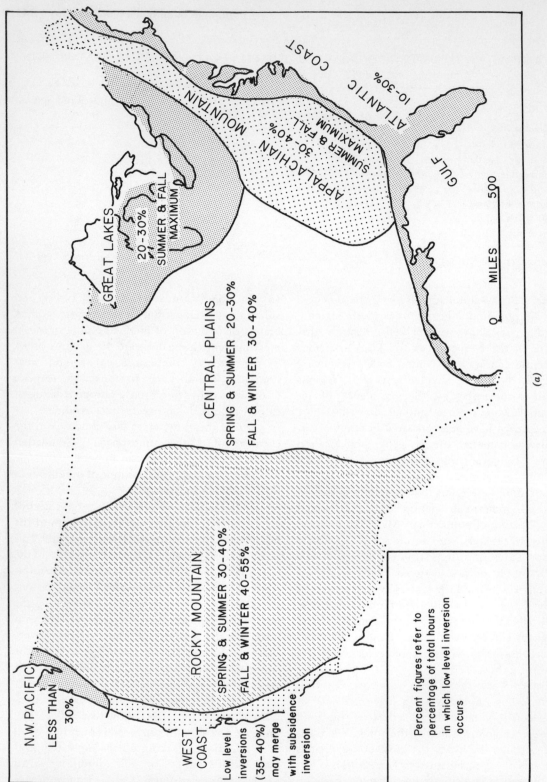

Figure 8–16. (a) Regional distribution of low-level inversions (by percent) in the United States. (After Hosler, 1961).

N.W. PACIFIC
LESS THAN
30%

WEST
COAST

Low level
inversions
(35-40%)
may merge
with subsidence
inversion

ROCKY MOUNTAIN

SPRING & SUMMER 30-40%
FALL & WINTER 40-55%

CENTRAL PLAINS

SPRING & SUMMER 20-30%
FALL & WINTER 30-40%

GREAT LAKES
20-30%
SUMMER & FALL
MAXIMUM

APPALACHIAN MOUNTAIN
30-40%
SUMMER & FALL
MAXIMUM

ATLANTIC COAST
10-30%

GULF

Percent figures refer to
percentage of total hours
in which low level inversion
occurs

0   500
MILES

(a)

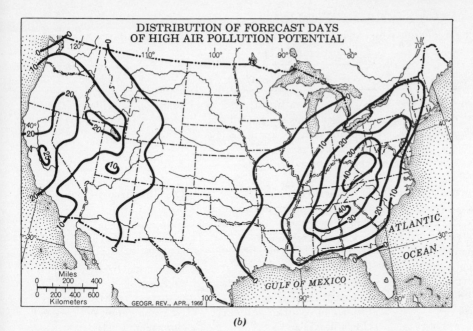

(b)

Figure 8-16 *(continued)*. (*b*) The air pollution potential advisory forecasts of the division of Air Pollution, U.S. Public Health Service, began August 1960 for the eastern United States and October 1963 for the western United States. The numbers shown on the contours indicate the number of forecast days from the initiation date in each case through December 1964. (From Leighton, P.A., *Geographical Review*, 57, 1966. Copyright by the American Geographical Society of New York.)

Apart from the import of energy, the areal size of the city also determines the extent of surficial covering of man-made material. Different reflective and thermal characteristics of a surface play a very important role in the first level of the model—the modification process. By the same token, the vertical extent of buildings will not only alter the disposition of solar radiation, but will also determine the modified nature of wind flow over the city.

Level 3. The Modified Elements

A third level that needs to be considered in the study of urban climate is the amount of change that occurs in the individual climatic elements. While it is of prime importance to consider the climate of a city holistically, it also remains true that some problems posed by the modified climate can best be dealt with in terms of the single element. Thus, while a researcher in plant growth in cities is concerned with the energy budget, he might also focus his attention upon elemental controls, such as the length of frost-

free season. Similarly a worker concerned with fluid effluent from a city needs a good assessment of precipitation characteristics—particularly in relation to maximum precipitation rates.

As shown in Figure 8-13, it is convenient to treat such "sensible" effects in terms of the thermal, hydrologic, and aerodynamic environments.

The thermal environment of many cities has been studied. Table 8-5 indicates some of the studies that have been completed. From such studies a generalized picture of the thermal environment can be deduced and a number of authors, notably Lowry (1969), have synthesized such findings to gain an overall view of what takes place. One of the most clearly defined aspects to emerge is the fact that cities create an "urban heat island."

This effect is most readily distinguished at night when a marked temperature gradient exists between the city and its surrounding countryside. While in some instances the gradient exists during the daytime hours, it is less easily detected. The result of the intensity of the heat island is to

Table 8–5

LIST OF SELECTED CITIES FOR WHICH IN-DEPTH TEMPERATURE STUDIES HAVE BEEN COMPLETED

City	Nature of Study	Authors
1. Cincinnati, Ohio	Nocturnal boundary layer	Clark and McElroy (1971)
2. Corvallis, Oregon	Heat island effect in a small city	Hutcheon et al. (1967).
3. Columbus, Maryland	Location of study in microclimate of buildings	Landsberg (1970)
4. Dallas, Texas (with Fort Worth)	Urban temperature fields	Ludwig (1970).
5. Hamilton, Ontario, Canada	Urban heat island	Oke and Hannel (1970)
6. Hibariaga-Oka., Japan	Night temperature distribution in a "new town"	Tamiya (1969)
7. Leicester, England	Nocturnal temperatures	Chandler (1967)
8. London, England	Heat island	Chandler (1956)
9. Louisville, Kentucky	Lapse rate studies	DeMarrais (1961)
10. Montreal, Quebec, Canada	Size and gradient of heat island	Oke (1968)
11. Manchester, England	City versus airport temperatures	Lawrence (1968)
12. Nairobi, Kenya	Estimation of urban heat effect	Nakamura (1967)
13. New York, New York	Urban heat island	Bornstein (1968)
14. San Francisco, California (with San Jose and Palo Alto)	Temperature gradients	Berg (1954)
15. Sheffield, England	Artificial heat generation	Garnett and Bach (1965)
16. Tokyo, Ogaki, and others	Comparison of temperature patterns	Kawamura (1966)
17. Toronto, Canada	Local effects on heat island	Munn (1969)
18. Vienna, Austria	Microclimatic study	Mahringer (1967)
19. Various	Urban heat island related to size of city	Mitchell (1961)
20. Washington, D.C.	Nature of temperature patterns	Woollum and Canfield (1968)

produce an average annual warming effect (Table 8–6), but differences are usually more marked on a seasonal basis. Wollum (1964) substantiates Landsberg's (1956) observation that in many centers the greatest differences may occur in winter, but some British authors have found that the greatest differences may be found in the summer or early autumn (in London, Chandler, 1965; in Reading, Parry, 1966).

Table 8–6

ANNUAL MEAN URBAN-RURAL TEMPERATURE DIFFERENCES FOR SELECTED CITIES, (C°)[a]

Chicago	0.6	Washington	0.6
Los Angeles	0.7	Philadelphia	0.8
New York	1.1	Paris	0.7
Moscow	0.7	Berlin	1.0

[a] From Peterson (1969).

In the investigation of the vertical temperature profile over cities, some interesting observations have been made. On clear nights, with light winds, it is commonly found that ground radiation chills the air immediately above the surface to form a low-level temperature inversion. Over cities, the heat island often negates this effect, and lapse rate conditions occur. That is, temperature decreases with increasing altitude (Figure 8–17). With slight air motion, in areas downwind of the urban-center, inversions are again encountered; but above the low inversion a plume of warm air exists and lapse rate conditions might be found. This effect designated "the urban heat plume" has been described in some detail by Clarke (1969).

The hydrologic environment. It would seem reasonable to assume, as did Landsberg (1956), that precipitation over cities should be increased.

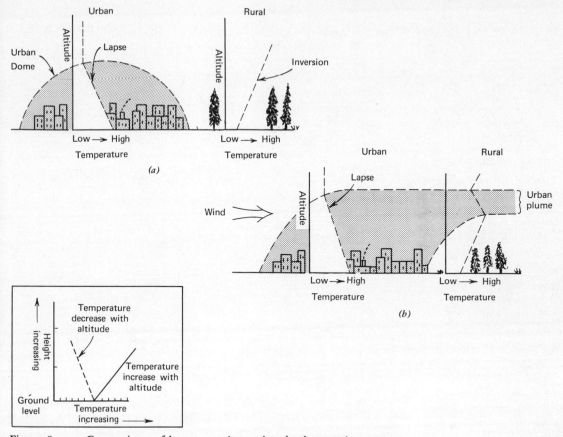

Figure 8–17.   Comparison of lapse rates in rural and urban environments under different atmospheric conditions. (*a*) On a still clear night, lapse rate conditions occur in the city and inversions in rural areas. (*b*) When a regional wind is blowing the rural lapse rate is modified by the urban plume. Inset shows characteristics of temperature-height graphs.

While some climatologists question this (Weisse and Kresge, 1962, Spar and Ronberg, 1968, Sekiguti and Tamiya, 1970), intensive studies by Changnon, 1970, tend to bear out the assumption. The controversy is, in many ways, related to the problems of precipitation data collection in cities. Development of airborne nuclei measuring devices and the introduction of in-depth precipitation studies such as METROMEX (Changnon et al., 1971) will certainly help clarify the situation.

Table 8–7 summarizes some earlier findings of Changnon for four American cities. Clearly, one of the most impressive findings is the existence of the LaPorte anomaly. This effect, widely covered in popular press, indicated that the town of LaPorte, Indiana, located some 30 miles downwind of the Chicago-Gary industrial center,

experienced a high incidence of precipitation and thunderstorms. The timing of these events correlated strongly with factory output and pollution levels in the Chicago area. The findings of this study have, however, been challenged. Holzman (1971) has pointed out that "The rainfall anomaly began in 1927 when a new cooperative climatic observer was appointed and ended in 1964 when an automatic rain gauge at LaPorte replaced him." Recent precipitation patterns have not been entirely consistent with the pre-1960 distributions, but it is evident that the controversy will continue. In the meantime, other workers are approaching the problem using alternate methods that overcome the relative paucity of data in the area. Harmon and Elton (1971) use tree-ring analysis in an attempt to clarify the situation. Their analysis shows that

Table 8-7

INCREASES IN PRECIPITATION AND RELATED CONDITIONS FOR SELECTED URBAN AREAS
IN THE UNITED STATES[a]

	Urban-Rural Differences (Increase) Expressed as Percentage of Rural Value			
	Chicago	LaPorte	St. Louis	Champaign-Urbana
Annual precipitation	5	31	7	5
Warmer half-year precipitation	4	30	—	4
Colder half-year precipitation	6	33	—	8
Raindays 0.01 or 0.1 in.				
Annual	6	0	—	7
Warmer half-year	8	0	—	3
Colder half-year	4	0	—	10
Raindays 0.25 or 0.5 in.				
Annual	5	34	—	5
Warmer half-year	7	54	—	9
Colder half-year	0	5	—	0
Number of thunderstorm days				
Annual	6	38	11	7
Summer	13	63	21	17

[a] From Changnon (1970).

tree-ring variation does indicate a pattern that corresponds, in some ways, to the anomaly but that can be explained in terms other than industrial output.

A number of studies about how precipitation is related to the urban dome have been completed. Results again give divergent viewpoints. In a study of three Japanese cities (Tokyo, Osaka, and Nagoya), Sekiguti and Tamiya (1970) have noted that ". . . it often happens that no rain has been observed in the big cities, but in their outskirts, fairly good amounts of rain (have been) measured." It has been suggested that because of the high concentration of condensation nuclei over cities, water droplets never become large enough to attain their terminal velocities. So, while cloud amounts would be increased, total measured precipitation might decrease. This, of course, contradicts the idea of induced precipitation because of atmospheric contamination.

A quite opposite effect to reduced precipitation was found by Atkinson (1971). In studying a storm that formed about 20 miles west of London,

he found that as the storm passed over the city there was rapid cloud growth and increased precipitation. The suggested cause of this was that warm, moist air that lay over London was drawn into the storm. The same author (Atkinson, 1970) has also indicated that in calm conditions, with prevailing high temperatures, spontaneous convection can initiate a storm over urban regions. Perhaps the cautions of drawing too many conclusions from such results need be stressed and Atkinson writes ". . . the urban effect is real, but is reflected in daily precipitation patterns only when all conditions are 'just right'!"

The humidity characteristics of cities have not been widely investigated (Peterson, 1969), although qualitative assessments can be found. Generally, while fog occurs more frequently, it is agreed that because of rapid runoff with resulting low evapotranspiration, the humidity in the city is lower than that of a nonurban area. Note that evaluation of humidity through use of relative humidity is open to question. Relative humidity is a function of both moisture content and temperature, and as noted, temperature is higher in cities. Investigations into a more comparable measure—absolute humidity (e.g., Chandler, 1965), show that variations are often a function of extreme localized conditions and, at times, city humidity can be higher than in neighboring rural area.

An appraisal of the water budget of urbanized areas has been completed by Leopold (1968), whose study clearly shows the amount of increase of runoff resulting from the urban process. Muller (1966), using the Thornthwaite approach, has also shown how the water budget of urban areas is grossly modified.

The Aerodynamic Environment. Munn (1970) has provided a useful evaluation of airflow in urban areas. He shows that while increased surface roughness reduces speed of city winds, it also induces greater turbulence. Table 8–8 shows that the once-in-50-years wind speeds in selected U.S. cities is appreciably lower than that in neighboring open areas. Note, though, that while the average wind speed is decreased, local effects, such as chaneling of winds through "city canyons," often gives the opposite impression. Indeed, the microeffects of the wind in cities is extremely complex.

Table 8–8

COMPARISON OF ONCE-IN-50-YEARS WIND SPEEDS AT CITY OFFICE AND AIRPORT FOR SELECTED UNITED STATES CITIES[a]

	City Station		Airport	
	Height of Anemom- eter (m)	Wind Speed (m/sec^{-1})	Height of Anemom- eter (m)	Wind Speed (m/sec^{-1})
Boston	57.3	32.2	19.2	46.0
Chicago	—	25.1	11.6	31.3
Omaha	36.9	29.1	20.7	40.7
Knoxville	33.8	25.5	21.6	39.8
Spokane	33.5	22.8	8.8	34.9

[a] From Munn (1970), after Davenport (1968).

This complexity is further abetted by the fact that when regional winds are light, the city creates its own wind field. Evidence for this has been accumulated by Okita (1960, 1965), and Pooler (1963). Not only does the wind respond to the heat island, but it is also influenced by surface anomalies occurring in the city, and on still nights numerous variations occur. Munn (1970) has aptly described this effect. "It is plausible to believe that if the surface air motions in a city should be recorded from above with time-lapse photography during light geostrophic winds, the patterns would resemble the motions of an amorphous, slowly-pulsating jelly-fish."

In more windy conditions the concentration of pollutants is decreased and the wind will transport material away from the city. An immediate effect of this is the already mentioned urban heat plume. This influences areas in the downwind vicinity of the city. This obviously is not the end of the city output, and satellite photographs have shown a drift of haze from eastern United States to the central Atlantic. The nature of the urban climate is intimately related to the impact of man and his activities on a global level.

The Significance of Urban Climatology

In a world in which urbanization is rapidly expanding, it is necessary to examine all aspects of "cities." At present, and quite correctly so, the social ills are obtaining priority. At the same time the environmental hazards of city living are being intensively studied and appropriate legislation is slowly being effected (Bach, 1972). But what of the urban climate, and how should it be perceived in terms of man's occupancy of the earth?

At first glance it might seem that the best way to rationalize about the urban climate is to actually draw upon its benefits. One might cite the case of the urban heat island. Its effects are already—although unwittingly—put to use. The modified thermal characteristic reduce snow cover and thereby aid transportation; it also serves to decrease the number of heating degree days and so helps in fuel conservation. The higher temperature might also be expected to extend the growing season of plants—presupposing they can exist in a polluted environment.

Such effects, however, are insignificant compared to other aspects; perhaps it is best to view the urban climate from two standpoints. First, and as has been stressed in this review, the urban climate provides a laboratory in which the effects of man's concentrated activities can be monitored and studied. To study the urban climate for its own sake is very interesting, but not really meaningful; it is in relation to the potential for modification that could occur on a global scale that is of concern. To the human ecologist, as to everyone interested in man-environment relationships, the city climate is an important measure of man's ultimate impact on the ecosphere.

Second, the nature of the urban climate can be used as an input into the plans concerned with mitigation of effects caused through urban industrialization. It is in planning that the climatic aspect becomes highly significant. Ecologic planning, so forcibly set forward by McHarg (1969) will become a necessity in the future. Many attractive elements of such planning—green belts, parks, estuarine conservation—are sometimes difficult to substantiate in terms of statistical-economic parameters, and unfortunately aesthetics are difficult to quantify.

By using quantitative effects of climatic modification both on a local and global level, strong cases can be made for rational planning. As an example, the cumulative effect of microclimatic modification of poorly designed buildings (from an environmental standpoint) can point toward the utilization of materials and shapes that mitigate extreme changes. In effect, anyone investigating the designed habitat of man needs to be aware of its potential for climatic modification. This is well summed up by Landsberg (1970), who writes, "If we plan carefully, urban climates need not differ radically from rural conditions, if they offer more healthful and pleasant conditions. Nor need costs to accomplish this be prohibitive. But we will achieve neither economy nor climatic amelioration without careful assessment of all relevant macro-, meso-, and micro-meteorological facts."

CHAPTER 9

Climate
and Agriculture

Early man functioned as part of his environment, probably having no greater impact than any other animal species. The economy of his society was based upon hunting and gathering, a mode of life still practiced by some primitive people. It was after man developed tool making, and learned to control fire and construct primitive shelters that the revolutionary change to tillage and agriculture occurred. The circumstances of this change are not well known and controversy still exists over both the nature and the site of the beginnings of agriculture. Much evidence, however, points to its origin in the "Fertile Crescent" of western Asia, a viewpoint admirably expressed by Braidwood (1960).

From such remote origins, agriculture has evolved into a highly technical field. At first, cultivated plants were similar to those found in the wild but through man-induced changes they became highly specialized; today, it is often difficult to trace the original strains from which some widely grown plants are derived. While, through selective breeding and trial and error techniques, specialization of cultivated crops took place over the ages, most of the dramatic changes in plant productivity have occurred in the twentieth century. Discoveries of genetics have opened new fields for agriculture. Plants have been made more responsive to fertilizer, more resistant to cold and drought and more productive in terms of protein content. Dramatic results are seen in the development of new high-yielding varieties of wheat and rice for tropical and subtropical regions. The early maturing rice IR8[1] (120 days versus 150–180 days) allows multiple cropping to occur with resulting higher annual yields.

Although genetics has played such an important role, the vagaries of climate still give rise to massive agricultural problems. Climatic limits,

periodicities, and disasters still face agriculture because, no matter how much expertise exists, the agricultural system is still an ecosystem (albeit a man-made one) that depends on climate to function.

An agricultural system is, in fact, an ecosystem that operates through energy flow as does a natural system. Figure 9-1 shows the components of a "universal" model of ecological energy flow, and it clearly applies equally to an agricultural as to any other system. However, as noted below, the agricultural system lacks the diversity of natural systems and depends upon inputs of energy, pesticides, and fertilizer to maintain its stability. In a natural ecosystem, autotrophs fix solar energy, some of which is stored and accumulates as new tissues through the growth of organic matter; the remainder is used in respiration, supporting the existing plant tissues. For a plant the "budget" of fixed energy might be represented as

$$NP = GP - Rsa$$

in which NP is net production—the plant growth that can be measured by its dry weight, GP is gross production—the total amount of energy fixed, and Rsa is the energy used in plant respiration.

In an intact "natural" ecosystem, heterotrophs, organisms that derive their energy via the autotrophic plants, also contribute toward total respiration. Net ecosystem production (NEP) might be represented as

$$NEP = GP - (Rsa + Rsh)$$

where Rsh is respiration of the heterotrophs. As Woodwell (1970) notes, solution of the components of this equation ". . . establishes the important distinction between a 'successional' or developmental ecosystem and a 'climax' or mature one. In the successional system the total respiration is less than the gross production, leaving energy that is built into the structure

[1] Subsequent testing has shown IR8 to be susceptible to pests and disease and more resistant strains have been developed.

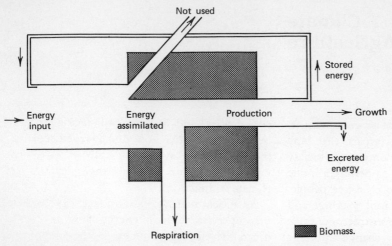

Figure 9–1.   Schematic energy-flow chart for a "universal" ecological system. The components of the model apply equally to both agricultural and natural systems. (After Odum, 1968.)

and adds to the resources of the site . . . . In a climax system . . . all the energy is used in the combined respiration of plants and the heterotrophs. *NEP* goes to zero."

In an agricultural ecosystem, the energy "diverted" by the heterotrophs is undesirable, for the aim is the attainment of the highest net production. The exclusion from the system of heterotrophs, be they carnivores or herbivores, is achieved through inputs of pesticides while herbicides decrease loss of energy through respiration of other plants. This suggests that in an agricultural system, the aim is

$$NAP = GP - Rsa$$

The net agricultural production (NAP) is the gross production less resiration of plants under cultivation. This is, of course, a gross simplification of what actually occurs. Figure 9–2, for example, shows the aboveground production of dry matter in a rice crop. Note how, with the passage of time, grain production decreases while gross productivity continues to increase. While biologic productivity might be large, yield of crops for use by man is not necessarily so. The representation does, however, indicate the basic fact that while natural systems maximize gross production, agricultural systems attempt to maximize net production. But even under favorable growing conditions, the percent

of net to gross production in intensive agriculture is quite variable (Table 9–1).

The high annual yields that are obtained in some modern agricultural systems are actually maintained through man's addition of energy. As Odum (1971) notes "Fuel used to power farm machinery is just as much an energy input as is sunlight." Any form of energy that is added and eventually converted to increase production—and there are many such additions in agricultural systems—is termed an energy subsidy. The significance of energy subsidies is clearly seen in Figure 9–3, which shows estimated net primary production for different crops grown with and without fuel subsidies. The differences are enormous.

The agricultural system is in marked contrast to the diversity exhibited by a self-regulating climax ecosystem. To maintain such a system requires many inputs by man; even this is at times insufficient to maintain the system. Monoculture, compared to diversity, often results in plant losses on a grand scale through a host of natural disasters. Most of the widely grown crops have, at one time or another, been subjected to destruction. Wheat growers have had to combat black rust and root diseases such as "eye-spot"; cocoa production has been decreased by swollen root; cotton has suffered from the boll weevil. The list of agricultural disasters is a

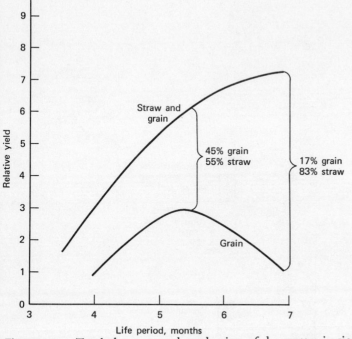

Figure 9–2. Total above-ground production of dry matter in rice shows that yield of the grain content decreases with duration of vegetative period. (After Odum, 1971.)

lengthy one, testifying to the proneness of the simplified ecosystem to all manner of problems.

Since man's very existence depends upon agriculture, it is not surprising that agroclimatology is the focus of much research. Obviously, the factors applicable to plant growth and development outlined in Chapter 5 are of considerable importance for light, warmth, and water are necessary in any plant complex, be it wild vegetation or a carefully controlled irrigation system. To view the many aspects of climatology as it relates to plant production, it proves expedient to consider the following:

1. The relationship between climate and crops in terms of the climatic conditions that contribute toward optimum crop yields.

Table 9–1

RELATIONSHIPS BETWEEN SOLAR RADIATION AND GROSS AND NET PRODUCTION IN CROPS UNDER INTENSIVE CULTIVATION[a]

	kcal/M²/day					
	Solar Radiation	Gross Production	Net Production	% Gross/Solar	% Net/Solar	% Net/Gross
Sugar cane, Hawaii	4000	306	190	7.6	4.8	62
Irrigated maize, Israel	6000	405	190	6.8	3.2	47
Sugar beets, England	2650	202	144	7.7	5.4	72

[a] From Odum (1971), after Monteith (1965). Expressed on a daily basis during favorable growing season conditions.

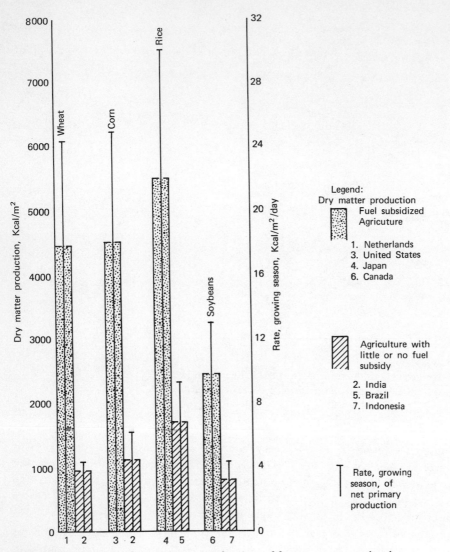

Figure 9–3. Estimated net primary production of four crops at two levels.
At Level 1 agriculture is heavily fuel subsidized, and at Level 2, little or no
fuel subsidy occurs.

2. The distribution of crops in relation to gross climatic limits.

3. The methods that man has used to overcome climatic limitations placed upon crop production.

CLIMATE AND CROP YIELDS

The purpose of agricultural production is to produce as much foodstuff from a given area as possible. In carrying this out, however, the farmer is required to maintain production such that his efforts become an economic success; the line between success and failure in farming often depends upon the yield per acre in relation to the costs of maintaining that yield. Much effort has thus gone into the problem of increasing yields so that farming can be economically viable.

The climatic aspects of input-yield relationship essentially concerns the determination of the optimum conditions that produce the maximum yields. To this end, work in both the laboratory and field has been completed.

The Light and Heat Factors

Laboratory investigation has been responsible for many of the important findings concerning the heat and light requirements of cultivated plants. Some of the most significant results have been attained through simulation of thermal/light conditions in giant "greenhouses" called phytotrons. These are large installations in which plants can be subjected to variable climatic conditions that can be controlled with precision. Pioneered by Went in Pasadena, phytotrons are now found in countries ranging from Australia to Holland. The utility of such large installations has been put to good use in other, related studies. At Washington State University an air pollution phytotron is used; at the University of Wisconsin the concept of the phytotron has been expanded into a biotron in which the relationship between climate, animals, and man as well as between climate and plants can be examined. The significance of the findings of work in such facilities can be shown by citing a few of the results derived by Went (1957).

The yield of tomatoes varies enormously from place to place in the United States. In an evaluation of why this should occur, Went first subjected the warmth-loving tomato plant to a constant temperature of 79°F (26.1°C). The plants did poorly and little difference was found when the humidity was varied. Better results were obtained when the temperature was reduced to 64°F (17.8°C); in view of the higher temperatures at which tomatoes usually grow, this was somewhat surprising. Continued experimentation ultimately showed the 64°F temperature was, in fact, related to night temperatures of tomatoes grown outdoors. Went concluded that the optimum yield of tomatoes depends upon a daily cycle in which best results were obtained if the night temperature was about 64°F. The potato was found to have a similar response. Best results were obtained when a marked daily cycle exists in which night temperatures range from 50°F (10.0°C) to 57°F (13.9°C). In some ways, this explains the high yield of potato crops grown in Idaho, Maine, and northern Europe.

As has been noted, the light and temperature requirements of plants vary as they pass through their developmental and growth periods. This poses problems in relation to some plants in which different processes, at different times, are needed to provide the plant that best fits the needs of the consumer. The sugar beet provides one such example because growth and sugar production are greatest under different temperature conditions. Went found that plant growth is greatest when night temperatures are above 68°F (20.0°C); sugar production in the beet occurs most efficiently when they are below this value. The ideal conditions for beet growth is therefore found in regions where warm summers, promoting the growth of the plant, are followed by cool autumns that allow development of high sugar content.

Laboratory research has provided innumerable other factors significant in plant growth and development. For example, the German plant physiologist Gassner developed the concept of vernalization. He found that the flowering of cereals could be influenced by controlling the temperature at which the seeds germinate. Winter rye, for example, when planted in the autumn, germinates during winter and flowers the following summer. If it is planted in spring, there is insufficient time to flower in the remaining growing season. Gassner found, however, that by keeping the seeds near freezing during germination, winter rye would have sufficient time to flower when planted in spring. This discovery allowed the method to be adopted for transformation of winter cereals into spring cereals. Similar experimentation in the germination, flowering process, and ripening of fruits are well described by Janick et al. (1970) and further indicates the great strides made in this area.

The significance of light-plant relationships has already been stressed. Such an important aspect has, of course, been the subject for extensive research. The basis of much of the laboratory research in this area centers around problems concerning the fixation of energy during photosynthesis. It has been found that photosynthesis is most effective at selected wave lengths.

Reifsnyder and Lull (1965) provide a summary of the effects of different wavelengths upon plants that is based upon these defined by the Dutch Committee on Plant Irradiation. In this summary, eight wavelength bands are distinguished:

Band 1
more than 1 μ   No specified effects upon plants

Band 2
0.7 to 1 μ   Specific elongation effect in plants

Band 3
0.61 to 0.7 μ   Strongest absorption by chlorophyll—strongest photosynthetic activity

Band 4
0.51 to 0.61 μ   Low photosynthesis

Band 5
0.4 to 0.51 μ   A second peak of photosynthetic activity with chlorophyll formation greatest at 0.455 μ

Band 6
0.315 to 0.4 μ   Strong photographic and fluorescence

Band 7
0.28 to 0.315 μ   Strong photographic and germicidal action

Band 9
less than 0.28 μ Below limit of atmospheric transmission

Identification of the varying effects of different wavelengths has proved valuable in determining the nature of supplemental illumination of plants for artificial light sources vary in their spectral distribution. Tungsten lamps, for example, whose filaments are heated to several thousand Kelvins, produce a red-blue continuous spectrum. Unfortunately, tungsten lamps are inefficient because of the infrared radiation lost as heat. Ordinary fluorescent lamps do not provide radiation in the red and far-red

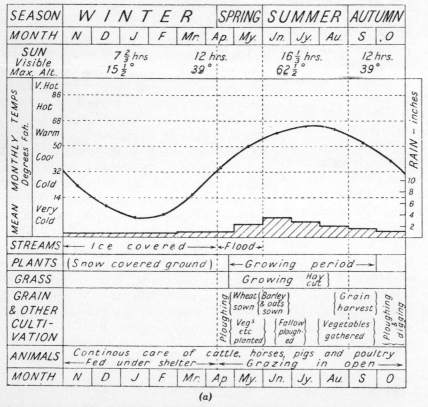

Figure 9-4. Agricultural rhythms distinguished by Unstead. (a) Rhythms in a middle-latitude grain and grazing land; growers of hardy crops in Canada.

parts of the spectrum so if they are used, for example, in experiments in photoperiodism, they must be supplemented by bulbs rich in red. In phytotrons a combination of light sources might be used to produce a spectrum similar to that of natural sunlight.

It is not laboratory research alone that provides information concerning the role of climate in optimizing crops yields. While experimental agricultural stations are actively engaged in field tests, much has been learned over the years by experience and farmers have learned to adapt their activities and mode of cropping accordingly. In optimizing light availability to field crops, farmers have found the requisite plant density providing highest yields. This is no easy task; initially a high population of plants is desirable so that maximum use of light is available for early stages of development. As competition for

light increases, a lower plant density may provide higher yields. In considering these the agriculturalist must balance all factors one against the other and ultimately decide upon the optimum plant density for his crops.

Not only has the farmer been concerned with individual climatic elements in relation to his crops, he also adapts his activities to a yearly cycle of climatic events. Figure 9-4 shows, for example, two farmers' years. Representing seasonal farming activities in middle-latitude and tropical wet-dry environments, they indicate how the activities are geared to seasonal climatic events. Obviously, such representations are somewhat simplistic and merely represent a summary of what might occur under a given set of conditions. The important point that emerges from such diagrams is that farmers have long been aware of the relationship between

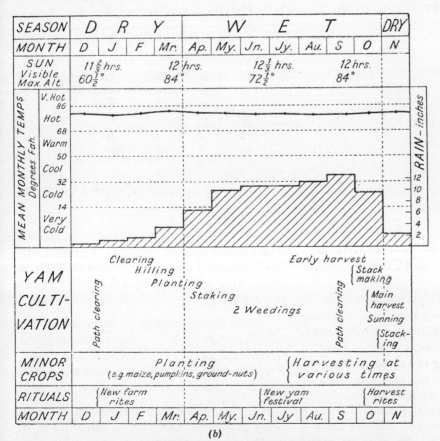

(b)

Figure 9-4 (continued). (b) Rhythms in a tropical crop-land; yam growers in Southern Nigeria. (From Unstead, J.F., A World Survey from the Human Aspect, 5th ed., 1957, University of London Press. Used with permission.)

Table 9–2a
DATES OF CHERRY BLOSSOMINGS (A) AND FULL BLOOMS (B) (1916 TO 1965)[a]

I[b]						II					
Year	A	B	Year	A	B	Year	A	B	Year	A	B
1916	4– 1	4–15	41	14	—	1916	—	—	41	1	7
17	9	11	42	4	11	17	—	—	42	3–25	4
18	6	11	43	19	24	18	—	—	43	5	14
19	3–28	5	44	22	—	19	—	—	44	9	15
20	3	10	1945	—	—	20	—	—	1945	—	
21	7	15	46	4	16	21	—	—	46	3	10
22	3	10	47	12	22	22	—	—	47	12	18
23	1	10	48	4	11	23	—	—	48	1	10
24	15	22	49	—	—	24	—	—	49	8	19
1925	16	20	50	14	18	1925	—	—	50	2	7
26	11	17	51	14	21	26	—	—	51	3–31	8
27	11	17	52	18	29	27	—	—	52	7	11
28	13	16				28	—	—	53	5	11
29	3	16				29	—	—	54	3–29	5
30	8	13				30	—	—	1955	3–28	7
31	5	15				31	—	—	56	4– 1	4–10
32	9	20				32	—	—	57	7	12
33	14	18				33	—	—	58	2	8
34	17	22				34	10	12	59	3–23	2
1935	10	15				1935	4	11	60	3–29	5
36	18	22				36	15	19	61	3–31	6
37	10	15				37	3–28	11	62	2	7
38	5	7				38	3–29	3–31	63	4	9
39	6	14				39	4	7	64	5	6
40	14	19				40	8	12	1965	9	17

[a] From Sekiguti (1970).
[b] I and II indicate different varieties of cherry trees. II is the most common one, SOMEI-YOSHINO (Prunus yedoenis) in Japan.

climate and the periodic phenomena in organisms; that is, they are well aware of the basic concepts of phenology.

The importance of phenological data in relation to agriculture is evident from the many aspects that relate to the dates at which planting, germination and mergence of seeds occur or dates at which flowering and ripening take place. Much work of interest has been completed by Japanese researchers in interpreting phenological dates of cherry blossoms. Records for recent years are relatively complete (Table 9–2a) and these can be related to the dates of cherry festivals going back to the eighth century (Table 9–2b). Drawing upon such information

it is possible to infer variations in climate that have influenced the dates of the events. Sekiguti (1970) using the data given in the tables suggests the following:

9th and 10th centuries	warm and dry
11th century	cool and dry
12th century	warm and wet
13th to 16th centuries	cooler and wetter
17th to 20th centuries	warmer and drier

Phenological data are often shown on maps using isophenes, lines that connect places at which a similar phenological event takes place at the same time. Some examples of significant agricultural dates are illustrated in Figure 9–5.

Table 9-2b

HISTORICAL DATES OF CHERRY-FESTIVAL AT KYÔTO IN GREGORIAN CALENDAR[a]

A.D.

A.D.		A.D.		A.D.		A.D.		A.D.	
705	4-24	1006	4-10	1322	16	1478	6- 1	1544	16
		16	20	23	5- 4	81	16		
744	7	18	23			84	6	1560	9
747	18			26	17	85	4	63	8
748	26	29	26	31	22	86	9	65	24
750	14	41	13	32	24	87	3	66	7
				44	11	95	4		
755	10			46	17	98	7	79	12
		1105	24	47	19	99	21	80	6
812	1	24	5			1500	7	85	15
815	15			57	17			88	15
		27	9	66	22			94	19
831	6			69	4	1501	13	97	24
		54	16			02	5-12		
851	18	56	3-30			05	16		
853	14			1406	15	06	16	1603	13
864	9	79	13	07	19	07	18	04	9
866	16	84	15	22	19	08	20	05	15
				24	8	09	12	06	16
912	4					10	20	09	18
917	5	1212	3-27	26	9	11	21	12	3-27
		25	1	30	18	12	18	13	16
926	8			32	13	13	20	22	16
941	19	30	12	43	17	14	15		
949	17	33	8	50	10	17	16	33	8
		46	3-29			18	25		
957	22	47	16	51	13	19	21	51	17
958	19			52	1	20	14		
961	3-28	63	30	56	15	21	17		
963	4			57	15	22	13		
965	14	78	23	58	22	24	18	1846	7
966	8	85	15	59	23	25	14	47	17
967	13	86	22	60	21				
974	18	95	7	62	16	26	5- 8	51	12
975	17			63	1	27	15	53	12
				64	13	28	15	64	14
977	22	1302	6	65	8	29	17		
		07	14	66	15	30	17		
		16	12	68	10	31	13		
		17	12			32	17		

[a] From Sekiguti (1970).

In middle-latitude areas it was observed that perennial plants begin their growth-development activities as soon as a temperature threshold value occurs. Similarly, it was seen that a late or early spring had an appreciable effect upon the relative rates of growth of domesticated plants. Such observations caused extensive inquiry into the relationship between temperature and phenological events. The correlation was dubious and it was found that many other factors need be considered. Thus, while many phenological events seem entirely dependent upon prevailing temperature, such factors as photoperiodism, antecedent conditions of both the preceding summer and winter, and precipitation regimes also need to be recognized.

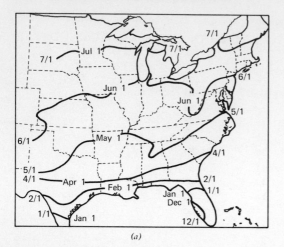

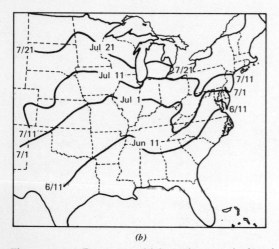

Figure 9–5. Dates at which various agricultural activities take place in the United States. (*a*) Date at which strawberry picking usually takes place. (*b*) Date when winter wheat harvest usually occurs. (After Visher, 1954.)

The Water Factor

Crop yield is affected markedly by water deficiencies. In dire circumstances, if soil moisture is below wilting point for prolonged periods, plant life cannot be sustained. Even when mosture is available, but in limited quantities, plant growth is modified. Under such conditions, the ratio of root to stem is increased while the leaf area of the plant may be reduced, with the leaf sometimes becoming appreciably thicker. As a result, the agricultural yield will suffer both in quantity and quality.

Water deficiency not only modifies the patterns of plant growth, it also affects the gross pro-

duction of the plant. The lowered rates of photosynthesis results from a lower capacity in dehydrated protoplasm while loss of plant turgidity limits the intake of carbon dioxide. Ashton (1956) has demonstrated this effect by measuring the rates of photosynthesis of sugar cane under different soil moisture conditions. As shown in Figure 9–6, the rate of synthesis dropped markedly when soil moisture reached a value of less than 40% of its capacity.

Water is, of course, required for plant transpiration and it has been shown that decreased transpiration in plants has adverse effects. While the role of transpiration in plant function is not totally understood, it is known that it prevents excessively high temperatures in the leaf through its cooling effect. It has also been shown, that under field conditions there appears to be a linear relationship between crop yield and transpiration rates (Figure 9–7). The results of such findings, in relation to the consequent action to be taken, poses the problem of how much water does a plant need to produce the highest yield. In attainment of an answer, models based upon the plant water budget need to consider:

1. The increment to soil moisture through precipitation; this, in turn, requires data on precipitation intensity and infiltration rates.

2. The amount of moisture returned to the atmosphere through the combined effects of transpiration and evaporation.

The first set of factors, the introduction and disposition of water, depend upon both the local precipitation regime and the texture and structure of soils. While constructed models allow assessment of these two factors under given sets of conditions, they will obviously vary considerably from location to location. In the same way, the amount of evaporation that occurs will be spatially variable; but the fact that evapotranspiration ultimately depends upon input of solar radiation makes models and empiric formulas used in its determination much more meaningful in agricultural terms.

Problems involved in evaluating evapotranspiration have already been discussed (Chapter 2) and a number of empiric solutions outlined. Other estimates, in which actual plant types are considered, are also available. The method

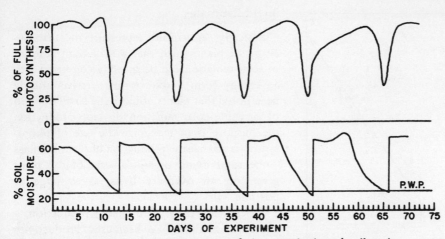

Figure 9–6. Relationship between rates of photosynthesis and soil moisture content determined in Ashton's study of sugar cane. (After Ashton, 1956, from Chang, Jen-hu, *Climate and Agriculture*, Aldine Publishing Co., 1968. Copyright © 1968 by Jen-hu Chang. Reprinted by permission of the author and Aldine-Atherton, Inc.)

proposed by Blaney and Criddle (1950) uses the equation

$$U = KF = kf$$

where U = the consumptive use (evapotranspiration) in inches

F = the sum of the monthly consumptive use factors (the sum of f, the product of mean monthly temperature and monthly percentage of daytime hours)

K = the empiric coefficient for the plant in question (k being the monthly coefficient)

Estimates of the plant coefficient (K) are derived from observed data for crops in arid and semiarid regions. Typical values are alfalfa 0.85, corn 0.80, citrus trees 0.60, and rice 1.20. When used in humid regions the coefficient values need be decreased by 10%. Use of the method is discussed later in this chapter when it is applied to irrigation needs.

Papadakis (1966) has used evapotranspiration under different climatic regimes to formulate a "climate classification based upon agricultural potential." Although the classification he proposes is somewhat complex, it contains much that is interesting regarding climate and water needs. For example, in evaluating evapotranspiration he uses saturation vapor pressure corresponding to monthly temperatures. This is given by

$$E = 0.5625 \, (e_{ma} - e_d)$$

where E is the monthly potential evapotranspiration in centimeters, e_{ma} is the saturation vapor pressure corresponding to the average daily maximum temperature, and e_d is vapor pressure corresponding to the mean monthly temperature.

To facilitate use of the formula a prepared table is used, from which values are derived from a slightly modified formula. In this table, the expression $e_{m1} - 2$ is substituted for e_d. The new expression corresponds to the vapor pressure equivalent to the average daily minimum temperature minus 2; Papadakis bases this upon the observation that the "normal" difference between the average daily minimum temperature and the dew point is 2°C (3.6°F). As an example of the use of this method, Papadakis gives the case of Salvador, Brazil. In July the average maximum temperature is 29.9°C (approx. 86°F), the average minimum 23.2°C (73.7°F) with the corresponding vapor pressures (multiplied by 0.5625) 238 and 142 mm, respectively. This gives a monthly potential evapotranspiration of 96 mm.

In tests (as indicated in Table 9–3) of various estimates of potential evapotranspiration, the Papadakis formula appears to hold up well.

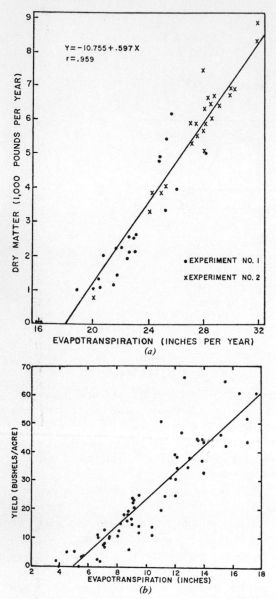

$Y = -10.755 + .597 X$
$r = .959$

• EXPERIMENT NO. 1
× EXPERIMENT NO. 2

DRY MATTER (1,000 POUNDS PER YEAR)

EVAPOTRANSPIRATION (INCHES PER YEAR)

(a)

YIELD (BUSHELS/ACRE)

EVAPOTRANSPIRATION (INCHES)

(b)

Figure 9-7. Relationships between crop yield and evapotranspiration. (*a*) Results obtained from a number of crops grown in a lysimeter in South Carolina. Data show an almost perfect linear correlation. Note that the first 18 in. of evapotranspiration were enough only to sustain plant survival. (*b*) Relationship between wheat yields and evapotranspiration in tanks. The linear relationship indicated became curvilinear when observed in open field study. (From Chang, Jen-hu, *Climate and Agriculture*, Aldine Publishing Co., 1968. Copyright © 1968 by Jen-hu Chang. Reprinted with permission of the author and Aldine-Atherton, Inc.)

Yield Estimates

Basically, agriculture represents the manipulation of plants such that a maximum amount of solar radiation can be speedily converted into an energy form usable by man. It has already been noted that plants utilize but a small fraction of available solar radiation; in crop production the purpose is to maximize the rate of photosynthesis so that the production of plant biomass —the total living organic matter of the plant given on an oven-dry basis—is as large as possible.

To estimate crop (or biomass) production, a number of methods have been used. Productivity can be experimentally determined either by measuring the carbon dioxide used or oxygen released in photosynthesis. Estimation of chlorophyll in a given amount of vegetation can also provide a productivity value. More recently, work has centered upon energy availability and energy incorporated into plant cover.

Some of the investigations in this energy approach use quantum theory, which states that light is composed of tiny particles without mass, called quanta or photons. They are considered "particles of energy" that travel from a light emitting source until they interact with matter. This concept is of value in plant studies because it helps to clarify the way in which chlorophyll absorbs the energy and converts it into chemical form. Baker and Allen (1965) describe the process as follows: "The chlorophyll absorbs the photons of light. These photons react with chlorophyll, transferring their energy in definite little packages. The chlorophyll then converts this energy into a form which can be used by the plant. Within certain limits, the more photons which are delivered to a leaf, the more energy the cholorophyll can absorb. By knowing the chemical characteristics of chlorophyll, and the wavelength of the light to which it is exposed, it is often possible to predict how much sugar will be produced within given periods of time." This is the methodology used by Loomis and Williams (1963) in their investigation of the theoretical maximum yield that could be obtained in the United States.

The energy content of one mole of quanta (6.02×10^{23} quanta), designated one einstein, is equivalent to 2.854×10^7 g-cal divided by

Table 9–3

MONTHLY VALUES OF MEAN DAILY POTENTIAL EVAPOTRANSPIRATION COMPARED
WITH ESTIMATIONS BY DIFFERENT FORMULAE. RATIO OF ESTIMATIONS TO VALUES
DEDUCED FROM MEASUREMENTS ARE GIVEN IN BRACKETS[a]

Month	Potential Evapotran-spiration[b] (mm)	Penman (mm)	Papadakis (mm)	Thornthwaite (mm)	Hamon[c] (mm)
January	2.2	1.7 (77)	2.8 (127)	0.7 (32)	1.2 (55)
February	2.9	2.6 (90)	2.8 (97)	0.8 (28)	1.4 (48)
March	3.7	4.1 (111)	3.7 (100)	1.5 (41)	2.0 (54)
April	5.3	5.6 (106)	5.3 (100)	2.7 (51)	2.9 (55)
May	6.1	6.8 (111)	6.4 (105)	4.0 (66)	3.9 (64)
June	6.9	7.4 (107)	7.2 (104)	5.4 (78)	4.9 (71)
July	6.5	7.0 (108)	6.5 (100)	5.5 (85)	4.9 (75)
August	5.9	6.4 (108)	6.1 (103)	5.3 (90)	4.5 (76)
September	5.5	5.2 (95)	5.4 (98)	4.1 (75)	3.5 (64)
October	4.4	3.9 (89)	4.8 (109)	3.1 (70)	2.6 (59)
November	3.2	2.5 (78)	3.4 (106)	1.8 (51)	1.8 (56)
December	2.3	1.6 (70)	2.7 (117)	0.9 (39)	1.3 (57)
For year	54.9	53.8 (98)	57.1 (104)	35.8 (65)	34.9 (64)
For summer (May-August)	25.4	27.6 (109)	26.2 (103)	20.2 (80)	18.2 (72)
For spring and autumn (March, April, September, October)	18.9	18.8 (100)	19.2 (102)	11.4 (60)	11.0 (58)
For winter (November-February)	10.6	8.4 (79)	11.7 (110)	4.2 (40)	5.7 (54)

[a] From Omar (1968).

[b] Based upon observations at Giza, U.A.R.

[c] Method using $E = 0.0055D^2P_t$, where E = average potential evapotranspiration (in/day); D = day length in units of 12 hr; P_t = saturation absolute humidity (g/m^3).

the wavelength of the photon in millimicrons. Given this and assuming that in summer most parts of the United States receive 500 ly/day—222 of which are in the visible portion of the spectrum, Loomis and Williams calculated net production as 71 grams/m^2 per day. Their evaluation (shown in Table 9–4) assumes a number of constants and the final estimate is, of course, very high because not all radiation falling on a surface is used efficiently. In fact, the highest yield of corn ever obtained in the United States was in California where the yield was 52 grams/m^2/day. Chang (1970) provides a useful summary of the energy approach while he also outlines a method for establishing potential photosynthesis over the globe.

In his evaluation, Chang notes that not all of the gross photosynthetic product is retained in the plant; large amounts are lost in respiration. Through application of established empiric formula (e.g., Thomas and Hill, 1949) rates of respiration can be estimated. It is noteworthy too that a one-to-one correlation between agricultural yield and potential photosynthesis does not always exist, since local factors concerning the production of specific plant needs must be considered. However, with these points in mind, Chang uses temperature and solar radiation values for 386 world stations to compute world potential photosynthesis over different time periods.

Figure 9–8 shows the potential photosynthesis for four-month, eight-month, and annual periods. Interesting results are seen in each of these. In the four-month map, lowest values occur in the tropics (except for polar realms) where, on an average, the values are 25% lower than the temperate realms. Of considerable surprise is the fact that the highest values are found in eastern Alaska, the upper MacKenzie River,

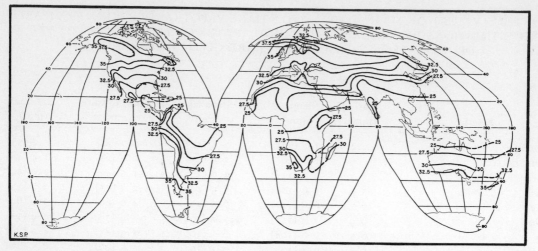

(a)

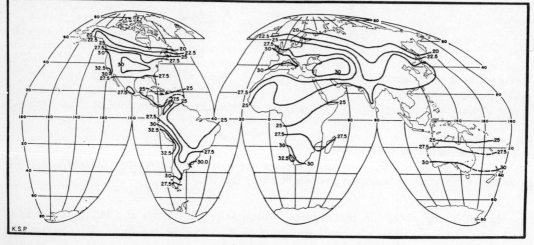

(b)

Figure 9–8.   (a) Potential photosynthesis (g/m²/day) for the four months from May through August in the northern hemisphere and November through February in the southern hemisphere. (b) Potential photosynthesis (g/m²/day) for the eight months from March through October in the northern hemisphere and from September through April in the southern hemisphere.

southern Scandinavia, and Iceland. This distribution is, however, probably less meaningful than the eight-month period. In this, much of the area representative of the Mediterranean (Köppen Cs climate) has the highest potential photosynthesis. Mean daily values are in excess of 30 grams/m². A similar distribution is seen in the annual pattern; again this shows the Mediterranean climates as one of the most productive converters of solar radiation.

Such findings prompted Chang to relate agricultural production of a single crop to the potential photosynthesis rates. As Figure 9–9 indicates, there is a high correlation between productivity and the derived rates. Similar correlations were found for most crops grown in tropical and temperate realms. Such results might lead one to suggest that the low productivity of "underdeveloped" lands of the tropics results from low photosynthesis instead

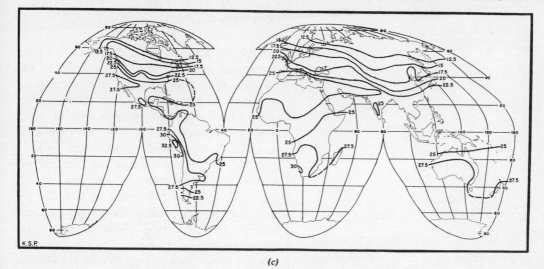

(c)

Figure 9–8 (*continued*). (*c*) Mean annual potential photosynthesis (g/m²/ day). (From Chang, Jen-hu, *Annals A.A.G.*, 60, 1970. Reproduced by permission from the *Annals* of the Association of American Geographers.)

of the numerous other factors often cited.

At a more local level, there have been many studies of climate and crop yields. Figure 9–10 shows the way in which Visher (1944) determined the optimum climatic conditions for high corn yields in Indiana. Studies that attempt to relate climatic conditions to production over long periods must take into account the input of technology, because this often causes an upswing in production despite adverse climatic conditions that sometimes exist.

This input is well illustrated in a number of studies. Chagnon and Neil (1968) have used regression analysis to show how technology has caused an upward trend in corn yields between 1955 and 1963. Indeed, they found that during this period (a time at which there was no marked shortage of water), high plant yields correlated to nitrogen application and plant population rather than any specific climatic variables. A similar finding was made by Arakawa (1957) working on rice production in the Tohuko

Table 9–4

EVALUATION OF THE POTENTIAL DAILY PRODUCTIVITY BY A CROP SURFACE RECEIVING 500 CAL/CM²/DAY

1. Total solar radiation per day	500 cal/cm²
2. Resulting visible radiation (400–700 mμ)	222 cal/cm²
Quanta equivalent of 2.	4320 μ einsteins/cm²
Albedo loss 360	
Inactive absorption loss (10%) 432	-792
Total quanta (400–700 mμ) available for photosynthesis	3528 μ einsteins/cm²
Carbohydrate produced (assume a quantum requirement of 10)	353 μ moles/cm²
Respiration loss	-116 μ moles/cm²
Net production of carbohydrate	237 μ moles/cm²
Net production assuming 30g/mole carbohydrate	$237 \times 30 = 71 g/m^2$

[a] After Loomis and Williams (1963).

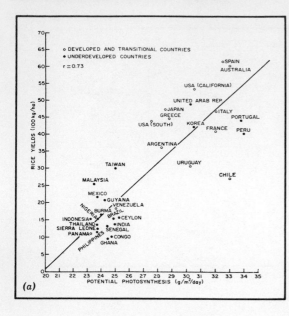

(a)

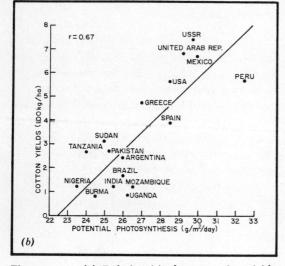

(b)

Figure 9-9. (a) Relationship between rice yields and estimated potential photosynthesis for the four summer months. (b) Relationship between cotton yields and estimated potential photosynthesis for the eight-month period. (From Chang, Jen-hu, *Annals*, *A.A.G.*, 60, 1970. Reproduced by permission from the *Annals* of the Association of American Geographers.)

district of Japan. While, as indicated in Figure 9-11, crop failures did occur as a result of abnormalities of the July and August temperatures, there is a marked upward trend of yield over the time studies. This is attributed to improved technology.

CLIMATE AND THE DISTRIBUTION OF CROPS

A glance at most world maps, showing the distribution of crops (e.g., those given in Van Royen, 1954), indicates that the distribution of crops have marked spatial limits. These reflect the demarcation between areas in which yields are high enough to allow the crop to be economically produced, since crop yield is probably a decisive factor in determining whether or not a crop will be commercially produced. It is not infrequent to find that the limit corresponds closely to a climatic limit because, as noted, crop yields depend highly upon climate. It is found that some crops simply will not grow in areas outside the climatic limits, while others will grow but will not flower. Deciduous fruit trees, for example, grow in the tropics but do not bear fruit. In agricultural terms, they have a negligible yield.

The study of a few widely grown crops provides much insight into the problems and potentials of climate in relation to agriculture. Consider the example of wheat.

Because there are many cultivated varieties, wheat is widely distributed and has wide climatic limits. Although grown far south in India, wheat has an equatorial limit for constant high temperatures, and humidity makes it highly susceptible to disease. Toward the polar limits, wheat is replaced as a grain crop by oats, rye, and barley, all of which give a higher yield under cooler conditions. Moisture requirements, like those of temperature, are also variable. Wheat can be produced in quite dry climates through dry-farming methods; the optimum amount, however, appears to be in the order of 30 in./yr, because high amounts of rainfall—particularly in cool conditions—inhibit its growth.

The wide distribution of wheat means that yields vary enormously. Some insight into the most productive areas can be obtained by assessing the quality of the wheat, rather than its quantity, because high-quality wheat—those high in protein content—bring the best price to the farmer. Figure 9-12 shows that the protein content of wheat is highest in the wheat belts of North America and the black-steppe area of the U.S.S.R. This might be accounted for in a number of ways. Such areas are, of course,

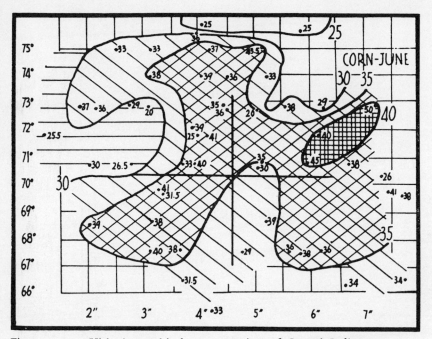

Figure 9–10. Visher's graphical representation of Central Indiana corn yields (1887-1931) as related to June rainfall and temperature conditions. The figures within the climograph are average yields in bushels per acre. Shaded areas indicate yields of more than 40, 35 to 40, 30 to 35, and less than 30 bushels/acre. The heavy crossed lines indicate normal temperature and precipitation. (From Visher, S.S., *Climate of Indiana*, Copyright © 1944, Indiana University Press, Bloomington. Reprinted with permission of the publisher.)

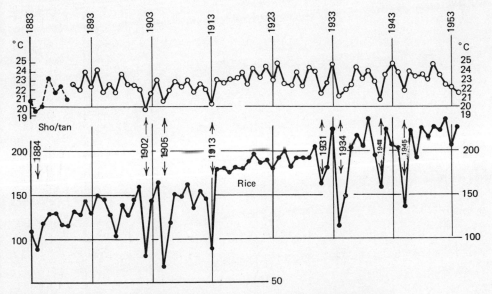

Figure 9–11. The relationship between rice yield in the Tohoku district of Japan and the July-August temperatures for the period 1883-1954. Rice yield is shown in sho/tan, Where 1 sho/tan = 0.2024 bushels/acre. (After Arakawa, 1957, from Munn, R.E., *Biometeorological Methods*, Academic Press, 1970. Used with permission.)

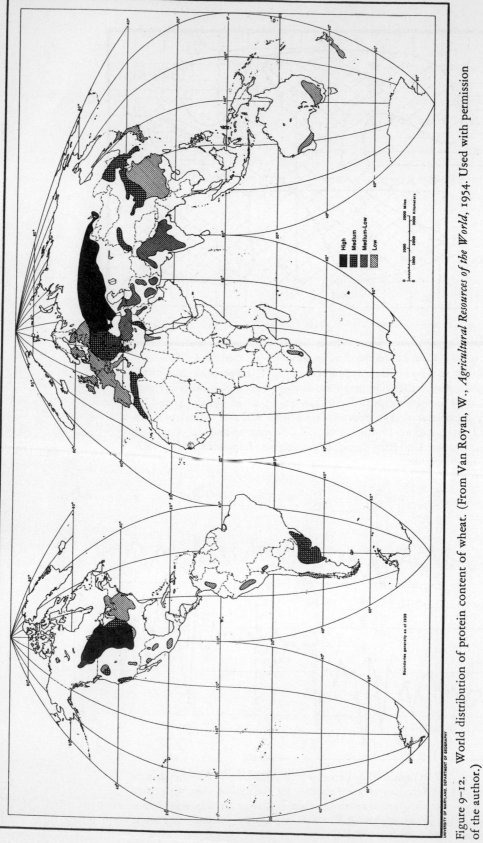

Figure 9-12. World distribution of protein content of wheat. (From Van Royan, W., *Agricultural Resources of the World*, 1954. Used with permission of the author.)

technologically advanced, and utilize the latest advances in agricultural techniques. They also correspond to the middle-latitude grassland biomes of the world; highest productivity is associated with a biome in which grasses, to which wheat is related, comprise the natural vegetation.

Despite the apparently ideal conditions for growing wheat in the grassland biomes, such areas are not without climatic problems. The continental locations make such areas highly prone to summer thunderstorms with their associated hail. Hailstones can achieve considerable size and, when falling onto standing wheat fields, can cause enormous damage. It is estimated that more than $50 million (80% of which results from damage to crops) is lost annually through hail damage in the United States (much of this occurring in the wheat-growing areas). One storm in Nebraska destroyed three million bushels of standing wheat.

To overcome such losses, there has been much research into efforts to decrease the incidence of hail. The Soviets have been particularly active in this field and, during a single trial year, their efforts resulted in losses of only 3.1% of the wheat crop as opposed to 19% loss in unprotected areas. They base their hail modification upon cloud seeding, using refined techniques in identifying potential hail clouds and distributing silver iodide through the cloud. Radar is used to identify potential hail storms, and for quick delivery of silver iodide into the critical area of the cloud they use nonsplintering artillery shells that are effective up to a distance of 22 miles.

The wide-open plains are also adversely affected by strong winds, and shelter belts have become common features of the scenery. These are mostly associated with farmsteads and watering ponds and, as Figure 9-13 indicates, marked decreases in wind speeds occur behind the shelter belts. It is possible that they may become more prevalent over wider areas, since Pelton (1967) has shown that wind breaks can increase the yield of wheat. Working in Saskatchewan between 1960 and 1964, he shows how wind breaks (erected after the snow period because of the differences in soil moisture that would result from differential snow melt) alter the wind speeds markedly and this, in turn, affects the yield. Increases occurred on both sides of the windbreak reflecting the fact that the wind was not always from one direction. Figure 9-14 outlines his findings.

Many agricultural plants are now produced in areas where they do not occur naturally. The banana was native to Southeast Asia but the main production area is now tropical America; the potato, with its origins in America, is now widely produced in Europe; sugar cane from Southeast Asia is widely produced in tropical America. The list could be extended considerably. In many cases the relocation of agricultural crops often results in better yields. The reason for this is twofold; first, in the area in which they were native, plants probably formed an integral part of the environment in which they were

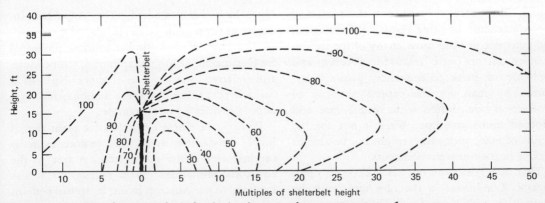

Figure 9-13. The decrease of wind velocity (expressed as a percentage of that in the surrounding open country) using a half-solid shelter belt. Density of the shelter belt modifies total effect on wind velocity.

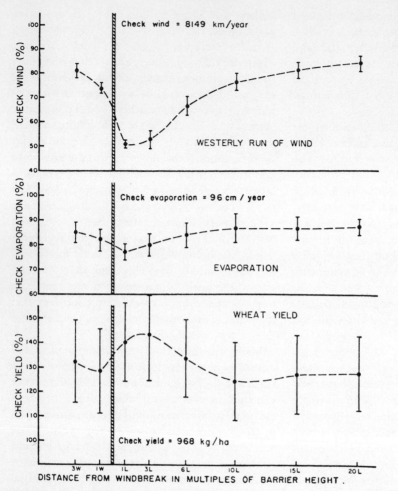

Figure 9–14.   Effect of a wind break on run-of-the-wind, evaporation, and wheat yield. Study area is open prairie near Swift Current, Saskatchewan. (After Pelton, 1967, from Munn, R.E., *Biometeorological Methods*, Academic Press, 1970. Used with permission.)

found. They were susceptible to the pests and the competition of other plants of that region. On transport to other areas, many of the natural limiting factors were removed. Second, and probably of more consequence, plants transported by man for the express purpose of production, are treated with special care and afforded much attention. Were it not for the inputs of man, such plants might not be able to survive in their new environments.

That such relocation can occur depends upon climate. Conditions in the new area must be compatible with those found in the region to which the crop is native. To find such areas, climatic anologs—specifically agroclimatic ana-

logs—have been studied. The American Institute of Crop Ecology has produced many such studies, most of them similar to those pioneered by Nuttonson. Using precipitation, temperature, and relative humidity as indices Nuttonson formulated analogs of many regions; his 1947 paper provides a good example.

Perhaps the finest example of a plant that is highly productive in an area remote from its original area is the rubber tree. A tree of the equatorial rain forest (*Hevea braziliensis*), it was confined to the Amazon basin; its spread perhaps inhibited by the dissimilar climates that occur on either side of the biome. Problems of collecting latex from trees scattered throughout the extensive forest led, through somewhat underhanded means, to its introduction as a plantation crop in Southeast Asia. The intriguing story of how rubber caused a boom in South America, and how it was smuggled from there to Kew Gardens, London and, ultimately, to Asia makes fascinating reading.

Obviously, there is a similarity between the climate of the original rubber area and Southeast Asia. For optimum growth, the rubber tree requires a temperature regime with a mean minimum of above 75°F (23.9°C). At high temperatures, especially with low humidity, latex ceases to flow. Rainfall in abundance is needed with at least 70 ins./yr; long dry spells (unless remedied by man) inhibit growth.

Climate places other restrictions upon crop growth in that plant pests and plant diseases seem to thrive under some conditions and not under others. It has already been noted that wheat is limited in hot, moist realms because of disease. Cotton is also highly susceptible to disease, and more than 500 pests are known to attack it. The boll weevil is probably the best known pest; although originally confined to Mexico, it is now widespread. This highly destructive pest is killed off by low winter temperatures. Unfortunately, cotton requires an extensive frost-free season—from 180 to 200 days—so that it is widely grown in areas in which boll weevils can survive.

The relationship between climate and pests is well demonstrated in the case of locusts. In passing through their life cycles, locusts need a variety of surface vegetation forms to survive. These range from bare desert conditions to well-vegetated surfaces. They are frequently found, therefore, in a contact or transition zone between two types of biomes. The female lays eggs in sand or loose soil that must contain moisture because, in their development, the eggs must absorb their own weight in water. After hatching, the locusts do not fly but pass through a hopper stage that lasts from five to six weeks. In this stage, green plants are necessary to sustain them. Thereafter, once they swarm they can reduce a cultivated crop to bare stubble in a very short time.

The rather specialized environmental conditions needed by locusts would appear to make their control easy. The requirement of a loose surface that has been moistened would indicate that regions of the desert, in which a periodic shower has occurred, would be the ideal location for the female to lay her eggs. This is frequently the case. Unfortunately, potential conditions of this type occur over enormous areas. To ascertain and pinpoint such locations is extremely difficult with the result that preventive control at this stage is minimised.

An alternate method of control might be to find the factors that cause locusts to swarm. Rainey (1958) offered one explanation (Figure 9-15) that relates swarming to migration of the ITC. In May 1950, swarms moved from the Red Sea coast toward what was French Equatorial Africa—following the direction of the northeast monsoon. Rainey suggested that the movement toward the ITC was a response of the swarms to the conditions of convergence experienced in this zone. Despite the evidence that he cites in this case, many disagree with the relationship between swarming and prevailing pressure conditions (e.g., Kraus, 1958).

Clearly climate plays a most significant role in determining what crops are grown in any region. Some of the climatic limitations placed upon agriculture are summarized in Table 9-5. While such limitations exist, many ways to overcome climatic problems have been devised and some crops are now grown in areas because of manipulation of that environment by man.

EXTENDING THE CLIMATIC LIMITS

If a crop is susceptible to frost, then the obvious place for it to be grown is where frost does not occur. It is sometime found, though, that some plants grow exceptionally well in areas where only an occasional frost occurs. The farmer is thus faced with a problem; should he take the risk and grow the crop aware of the fact that high returns are equally balanced against total loss if a frost should occur? In modern agriculture, large investments often require that the end product is guaranteed, and chances should not be taken. To meet this, methods to overcome climatic limitations on crop growth have been implemented. They not only concern the danger of frost, but also the problem of an inadequate water supply.

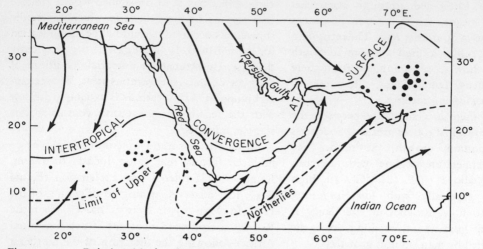

Figure 9–15.   Relationship between locust swarming and the location of convergence areas. Dot size is approximately proportional to the number of swarms reported. (After Rainey and Bodenheimer, from Critchfield, H.J., *General Climatology*, 1966. Used with permission of Prentice-Hall Inc., Englewood Cliffs, N.J.)

Frost Protection

For most commercially important crops, active growth ceases when the temperature approaches freezing. While many cold-region varieties merely cease to actively function, many others are actually killed by frost. Indeed, as indicated in Table 9–6, critical temperatures are often above freezing and vary in the different stages of plant development.

To overcome the problem of frost destruction, it is obviously necessary to be aware of the way in which frost can occur. Two main types can be identified. Radiation frost results from rapid cooling of the air layer above the ground when heat loss through ground radiation causes the ground to be cooler than the air above. This will occur mostly on cool, clear nights when there is little air turbulence. A second type of frost occurs through advection. This occurs through horizontal motion when cold air is introduced by winds into an area. This is most frequently found in colder regions where frost-prone crops are not grown outdoors. Thus, radiation frost is the major type to be combated in frost protection of crops.

When the layer of air next to the ground is chilled, an inversion exists. If the temperature at the ground surface is low enough, frost will develop. The association between frost and inversions has been apparent to farmers for many years. Citrus fruit growers, for example, have learned from experience that it is unwise to plant their trees in the bottom of a valley. The cold air next to the ground is denser than the air above. In situations like that shown in Figure 9–16, the cold air will flow down the valley side to form a pool at the bottom of the valley. This makes trees in the valley bottom more likely to be affected by frost.

The purpose of frost protection is essentially to mix the air near the ground so that an inversion is weakened. This might be achieved through a number of means.

One form of modification is the use of heaters. This is probably the most effective method of modifying radiation frosts by supplying heat and creating movement in the still air (Figure 9–17). It has been found that it is more practical to use a number of heaters strategically located throughout the field than it is to have one or two large ones. As shown in Figure 9–18, heaters can modify the temperature appreciably. For the most part, oil burners are used and, although expensive to maintain, have been found much more efficient than use of other fuels. The effectiveness of heaters in frost protection means that the farmer must have access to detailed

Table 9-5
TEMPERATURE AND PRECIPITATION REQUIREMENTS FOR SELECTED COMMERCIAL CROPS

Crop	Temperature	Precipitation	Notes
Cocoa	Since temperatures of between 50 and 60°F may be harmful, the crop cannot be profitably grown in regions where mean maxima of coldest month falls to 57°F or where absolute minima of less than 50°F occurs	Tree not resistant to dry weather so generally restricted to areas where dry season does not exceed 4 months	A tree of the humid tropical lowlands —mostly grown within 20° of equator below 1500 ft
Citrus Fruits	Little or no growth where temperature below 60°F. Dormancy in cooler months of subtropical climates. Temperatures slightly below freezing are highly damaging	Requires high soil moisture content. Orchards often irrigated even in fairly moist areas	Can be grown on a variety of soils with high humus content
Coffee	A tropical crop whose temperature requirements varies with species. Generally, optimum temperatures are between 60 and 78°F	Depending upon temperature, optimum amounts vary from 50 to 90 in./yr. The distribution is important with an ideal minimum in the flowering season. Too much water can promote tree disease	Generally does best on a well-drained loam soil. Thus some species are highland variety. Others do well in lowlands
Cotton	Needs a frost-free growing season of from 180 to 200 days. Does not grow below 60°F and optimum temperatures are from 70 to 72°F during the growing season. Four to five months of uniformly high temperatures are beneficial	Can tolerate a wide range in annual precipitation, the distribution during the growing season is of critical importance. Frequent, but light, showers immediately following planting an attribute	Needs sunshine. Lack will prevent ripening of the boll in full maturity
Rubber	For optimum growth and yield a mean maximum of over 75°F, the maxima should not, however, exceed 95°F	Evenly distributed rainfall of more than 70 in./yr. Lengthy dry periods inhibit growth. Good drainage essential	See text p. 268
Sugar cane	Susceptible to low temperatures. Little or no growth below 50°F while optimum is appreciably higher. Frost very dangerous to young cane	During vegetative growth requires a considerable amount of moisture and is sensitive to drought. In ripening period should be relatively dry to maintain high sucrose level	Often grown in cleared areas formerly occupied by tropical forests
Tea	Optimumly, temperatures should not fall below 55°F nor exceed 90°F	Can tolerate high amounts of rainfall (100 to 150) inches per year, if rain fairly evenly distributed throughout year	Often grown as an upland crop in tropical areas

Table 9–6

RESISTANCE OF SELECTED CROPS TO FROST IN DIFFERENT DEVELOPMENTAL STAGES[a]

	Temperature (°C) Harmful to Plant in Phases of		
	Germi-nation	Flower-ing	Fruiting
Highest resistance to frost			
Spring wheat	−9, −10	−1, −2	−2, −4
Oats	−8, −9	−1, −2	−2, −4
Barley	−7, −8	−1, −2	−2, −4
Peas	−7, −8	−2, −3	−3, −4
Resistance to frost			
Beans	−5, −6	−2, −3	3, −4
Sunflower	−5, −6	−2, −3	−2, −4
Flax	−5, −7	−2, −3	−2, −4
Sugar beet	−6, −7	−2, −3	—
Medium resistance to frost			
Cabbage	−5, −7	−2, −3	−6, −9
Soy beans	−3, −4	−2, −3	−2, −3
Italian millet	−3, −4	−1, −2	−2, −3
Low resistance to frost			
Corn	−2, −3	−1, −2	−2, −3
Millet	−2, −3	−1, −2	−2, −3
Sorghum	−2, −3	−1, −2	−2, −3
Potatoes	−2, −3	−1, −2	−1, −2
No resistance to frost			
Buckwheat	−1, −2	−1, −2	−0.5, −2
Cotton	−1, −2	−1, −2	−2, −3
Rice	−0.5, −1	−0.5, −1	−0.5, −1
Peanuts	−0.5, −1	—	—
Tobacco	0, −1	0, −1	0, −1

[a] After Chang (1968).

weather forecasting data, without which the method would prove impractical.

Temperature inversions can also be modified by wind machines of which, according to Angus (1958) there are three main types:

1. A type that consists of two airscrews mounted on a turn table atop a tower, usually at a height of about 40 ft. Ideally, the air-screws should have a diameter of about 21 ft and be tilted at an angle of between 70 and 80°. The efficiency of these screws depends upon the horsepower of the motors used, and it has been estimated that at least 10 hp is needed for each acre.

2. A type that comprises a fan similar to a helicopter rotor. In this case, the rotor is located horizontally above the ground so that it exerts a direct "suction" on the air and so induces mixing. Large diameter, slow speed rotors are needed, and it is estimated that at least 20 hp/acre are required.

3. Ducted fans that consist of a sweeping metallic duct about 3 ft in diameter through which air is forced. These are not widely used.

Typical results of the modification of nocturnal inversions using wind machines are illustrated in Figure 9–19.

Sprinkling and flooding are also used to combat frost. The object in this method is to reduce excessive cooling and increase thermal conductivity of the ground. Because of latent heat, when water freezes the temperature of the plants will not fall below freezing as long as the change of state occurs. The method does have limitations; even through it retards loss of heat during the night, it also limits heat gain during the day, and successive applications of water become less effective. Of the two approaches, sprinkling is probably better although problems do occur in determining the amount and frequency of water to be added. It has been found that areas not receiving water from the sprinkler are damaged, while loss also occurs in those areas receiving limited added water.

Other methods to overcome the problem of frost range from "brushing"—the addition of a protective covering of kraft paper over plants to reduce nightime radiation loss—to the use of wind breaks.

Irrigation

Water is essential for plant growth. As such, it would seem that agriculture must be restricted to areas in which water is available at the various stages of plant development. This is obviously not the case because man has long since learned that by transporting water to his crops agri-

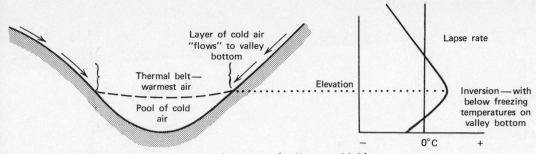

Figure 9-16. On cold, still nights, the bottoms of valleys are highly susceptible to frost formation.

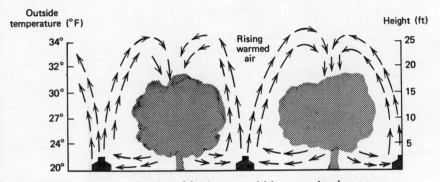

Figure 9-17. Currents developed by heaters within an orchard to protect it from frost. (From Longley, R.W., *Elements of Meteorology*, Copyright © 1970 by John Wiley and Sons. Reproduced by permission.)

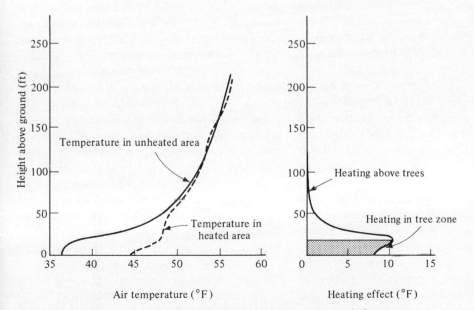

Figure 9-18. The effect of orchard heating. Results obtained from 33 gal of fuel burned in a 15-acre orchard with 45 heaters/acre. (After Kepner, from Miller, A. and J.C. Thompson, *Elements of Meteorology*, 1970, Charles E. Merrill Publishing Co. Used with permission.)

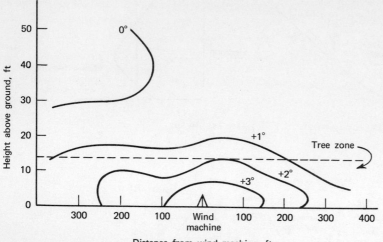

Figure 9-19.   Temperature change (°F) 35 min after start of a wind machine. A temperature inversion of 6°F extended up to 50 ft prior to modification. (After Crawford and Leonard, from Miller, A. and J.C. Thompson, *Elements of Meteorology*, Charles E. Merrill Publishing Co. Used with permission.)

culture can be successfully carried out. Indeed, irrigation is an integral part of the story of man's development of agriculture and the great aqueducts and complex water diversion schemes of classical antiquity are well known. Some writers have suggested that the basic cultural aspects of large societies is partly a result of man's control of water (Wittfogel, 1956).

It is possible to identify three types of regions in which precipitation must be augmented by man. There are the great deserts of the world in which a perpetual drought exists. There are regions in which the distribution of rainfull causes marked deficits of water to occur seasonally. Then, there are the usually well-watered areas where, because of periodic droughts, water must be added to maintain the normal production. It is possible then to recognize permanent, seasonal, and periodic drought.

Recognition of these different types of water shortages immediately points to the problem in using the term "drought." Perception of drought in the desert differs from that elsewhere; in fact, quite diverse definitions of the term exist, often reflecting the area in which it occurs. In the United Kingdom, for example, drought has been defined as time when no measurable precipitation has fallen for 15 days. Such a definition

would be ludicrous if applied, for example, to the grasslands of Africa.

The problem is further compounded in that drought has different meanings in different disciplines. It is possible to identify agricultural drought in which the state of drought is determined by moisture content in the soil. This would differ from a meteorological drought that might be determined through deviations from the mean. The hydrologist concerned with runoff might not find either of these definitions practical for his purposes. In outlining such problems, Palmer (1965) provides an absorbing account of how drought might be perceived. In the same article, he also proposes a method that brings in many of the variables needed to express drought. Although the method is rather long and complex, it certainly provides a rational approach to the problem.

Regardless of the problem of the definition of drought, the major problem that faces the farmer using irrigation is how much water to add and when to add it. To attain an answer to this question, it is necessary to evaluate the amount of water required by a plant to function at its maximum capacity. It is generally assumed that this amount is equivalent to the evapotranspiration that occurs. As already noted, there is no single method capable of deriving evapo-transpiration with exactitude. Nonetheless, application of empirically derived formula is probably the most effective way of determining irrigation needs.

The Blaney-Criddle formula (see p. 259) provides an example of how water additions to crops are derived. Table 9–7 shows the method used to compute the seasonal amount of irrigation water to be added on a farm in Colorado. For each crop listed the consumptive use (U) is derived using F, the consumptive use factor, and the appropriate crop coefficient K. The rainfall that occurs over the area is subtracted

Table 9–7[a]

a. Computations of Rates of Consumptive Use for Crops in the Montrose Area, Colorado

Culture	Growing Season	Consumptive Use Factor (F)	Consumptive Use Coefficient (K)	Consumptive Use (U) Inches	U Minus R Inches	Feet
Alfalfa	5/6–10/6	31.12	0.85	26.45	21.73	1.81
Grass hay	5/6–10/6	31.12	.75	23.34	18.62	1.55
Corn	5/6– 9/6	26.21	.75	19.66	16.02	1.33
Small grain	5/6– 8/6	19.83	.75	14.87	12.51	1.04
Orchards	5/6–10/6	31.12	.65	20.23	15.51	1.29
Seeped land	5/6–10/6	31.12	.80	24.90	20.18	1.68
Natural vegetation	5/6–10/6	31.12	1.20	37.34	32.62	2.72

$U = KF$ = Consumptive use for growing or irrigation season.
K = Empirical consumptive-use coefficient.
F = Sum of monthly consumptive-use factors (f) for the growing season.
R = Sum of monthly rainfall for growing season.

b. Illustration of the Method Used to Compute the Normal Seasonal Amount of Irrigation Water Required at Headgate at a Typical Farm, Montrose, Colo.

Classification and Crop	Area, Acres	Consumptive Use Minus Rainfall (U − R) Acre-Feet per Acre	Farm Irrigation Efficiency, Percent	Water Required at Farm Headgate Unit Acre-Feet per Acre	Total Acre-Feet
Irrigated:					
Alfalfa	35	1.81	60	3.02	105.7
Grass hay	20	1.55	50	3.10	62.0
Corn	10	1.33	55	2.42	24.2
Orchard	10	1.29	60	2.15	21.5
Miscellaneous:					
Roads	3	0	—	—	0
Natural vegetation	1	2.72	—	2.72	2.7
Seeped lands	1	1.68	—	1.68	1.7
Total water delivery required at farm headgate for normal season					217.8

[a] From Blaney (1955).

from the derived consumptive use $(U - R)$ and this gives a value used to determine the amount of water added. The value cannot be used directly for different crops (with different coefficients) occupy different areal extents and nonirrigated vegetation in the area must be taken into account. Further, irrigation systems are not 100% efficient, so that the relative efficiencies of distribution systems must be accounted for. As Table 9–7 shows, the final computation shows the amount of water required at the farm headgate for the normal growing season.

Such careful computation is not always completed and it often occurs that either too much or too little water is added. In either case, the results can be disastrous, leading to decreased rather than increased productivity.

Application of too much water can cause a rise in the water table. This interrupts plant growth in a number of ways. The waterlogging inhibits the development of roots while the evaporation of water from the surface can result in layers of salt being deposited at the surface. This has occurred widely in West Pakistan. Indeed, in the 1960s it was estimated that in the region 60,000 acres of fertile cropland was being lost each year because of waterlogging and salinization. A preventive measure was instituted after study by a U.S. team. They suggested that tube wells be used to lower the water table. The water could then be discharged on to the fields to flush downward the saline deposits. The scheme appears to have met with success.

Irrigation agriculture has been the subject of many research efforts and a great deal of literature exists on the topic. These cover such aspects as the methods of irrigation water application, the economics of irrigation, the possibilities of using seawater or desalinized water, disease vectors arising from irrigation schemes; the list is a lengthy one and contains much material beyond the scope of climatology.

In terms meaningful from a climatological aspect, however, there are many studies concerning the way in which the creation of irrigation systems modify the local climates. It will be appreciated that large irrigation schemes mostly occur in dry climates, for it is here that

irrigation is needed. In many of the schemes large dams have been built and enormous amounts of water impounded behind them. Man, in effect, creates oases in the middle of dry regions.

The effect of such large man-made bodies of water on climate has been interpreted in a number of ways. Thornthwaite (1956) reports that in the area of the Rubinsky Dam in Russia changes in temperature along the shores resulting from the created expanse of water are minimal. The wind speed over the water is, on an average, doubled, but its effects are felt only along the shoreline and even there only as a local modifying factor. In the arid southwest of the United States, the creation of the 300 mile² Salton Sea and the 175 mile² Lake Mead has little effect upon climate modification. In the Salton Sea area, for example, the moisture content of the air merely 2000 ft from the sea is not different from that existing elsewhere in the dry region.

Another viewpoint is expressed in the SMIC study (1971), which suggests that irrigation changes can induce climatic modifications on a long-term global basis. It is suggested that "All of the heat used in evaporation is returned to the atmosphere where condensation occurs. Apart from local effects, *which are large*, (author's italics), the global climatological impact will depend upon such indirect influences as changes in cloud cover and resulting effects on the energy budget. The estimation of such effects is an extremely complex problem, and, . . . we do not yet have any reliable method for making quantitative calculations of them." The study goes on to note that an irrigated area usually has a lower albedo than the ground cover it replaces. Thus, while there is a tendency for a lowering of local temperatures because of increased evaporation, the changes induce a rise in global temperatures because of decreased reflection of incoming solar radiation. (That the heat budget is modified when land is used for agricultural purposes is shown by data in Table 9–8.) One area of agreement is reached, however, in the observation that large bodies of water in arid regions will not increase precipitation amounts. Thus many of the grandiose schemes for making the desert flower through construction of large inland "seas" are exercises in futility.

Table 9-8

CHANGES OF HEAT BUDGET AFTER CONVERSION FROM FOREST TO AGRICULTURAL USE[a]

	Albedo Assumed as Representative	Bowen Ratio	Q_S (W/m²)	F_A (W/m²)	F_L (W/m²)	ET [b] (mm/month)
Coniferous forest	0.12	0.50	60	20	40	41
Deciduous forest	0.18	0.33	53	13	39	40
Arable land, wet	0.20	0.19	50	8	42	43
Arable land, dry	0.20	0.41	50	15	35	36
Grassland	0.20	0.67	50	20	30	31

[a] From SMIC (1971), after Mitchell, 1970; Flohn, 1971.
[b] Q_S is net radiation; F_A is convective heat transfer; F_L is latent heat transfer; ET is evapotranspiration.

CLIMATE AND ANIMAL HUSBANDRY

The emphasis of modern animal husbandry is to produce protein, of animal extraction, to supply as many people as possible. As in the case of crop production, yield again assumes importance.

High-yield cattle have been reared in western Europe for hundreds of years. Within the region, slight differences of environmental variables led to breeding animals of quite distinctive characteristics. In Britain, for example, French (1966) lists no less than 24 different breeds of cattle. With the opening of "new lands," explorers, settlers, and entrepreneurs of European extraction were often dismayed by the low-yield local stock that they found. In their attempt to remedy the situation, breeds of European cattle were introduced into many areas of the world. In some instances, the introduced cattle did exceedingly well and Jerseys, Holsteins, Herefords, and the like form the base for many dairying and ranching operations in quite diverse parts of the world.

The results of such introduction were not always successful and some, particularly in hot, moist climates, were a total failure. At first the failure was related to lack of quality feed; but eventually the inability of the animal to adapt physiologically became apparent. Indeed, until quite recently it was assumed that high-yield European cattle could not do well because of "tropical degeneration." It has been demonstrated, however, that proper attention to a multitude of factors, from adequate housing to parasite control, could result in successful introduction. The high cost of such requirements often make this uneconomic and ". . . it has been found more economical to introduce animals of tropically adapted breeds and to improve the environment to the level necessary for lower productive cattle to achieve optimal rates." (FAO, 1966.)

Even to attain such conditions, if they were economically viable, it is necessary to evaluate the physiologic response of animals to climate; animals, like man, exist in a proprioclimate that provides an external stimulus to which they respond. In fact, as Lee (1958) points out the climatic environment provides a stress that may cause the animal functions to be displaced from the state of normal equilibrium and a subsequent strain, which is the actual displacement that occurs.

Climatic stress can result from a thermal environment that is too hot or too cold. Most research has been centered upon the reaction to hot conditions, with necessary variations that reflect humid or dry conditions. The stress of cold climates has been less intensively investigated and it has been suggested that there is a need for research in managing farm livestock in cold climates. Perhaps the lack of emphasis in this respect is due to the fact that beef cattle, for example, can survive in temperatures as low as $-40°C$ if they are provided with a shelter of the most primitive type. Further, animal husbandry has been carried on in cool climates for many years and farmers have learned by experience the optimum methods of livestock farming. Obviously, until quite recently with the advent of air conditioning, it has proved much easier to warm a building than to cool it.

Mahavedan (1968) suggests that the success of an animal experiencing high temperatures depends upon:

1. Its ability to promote heat loss by such means as increased evaporative cooling of the body surface.

2. Its ability to reduce heat production by lowering its metabolic rate and effecting a more efficient means of energy utilization.

3. Its ability to cope with a rise in body temperature or with the consequences of compensatory reactions.

The varying ability of different animals with respect to these functions has resulted in many studies.

Animals in an open area are subjected to a proprioclimate that depends mostly upon the effects of radiation on their flanks, and the different stresses and strains that result are partically conditioned by the coat or skin of the animal. MacFarlane (1958), working in Australia, found that the heat absorption and transmission properties of the coats of cattle depends upon the arrangement, length, and color of the coat. Data from his study (Figure 9–20) clearly shows that smooth, light-colored coats are most suited

for cattle in open fields under strong sunlight. Reflection from the black coat of the Angus was only a quarter of that from the Zebu; the smooth and shiny-coated Shorthorn was found to experience far less stress than the wooly Shorthorn. In relating the effects of coat color to physiologic responses it was found that respiration rate—an index of heat balance—is closely correlated to texture and color of coat. Attempts at modifying the coat artificially, by wetting, for example, show that body temperature can be lowered significantly but that differences in the coat are still important (Figure 9–21).

The effects of high temperature on different breeds of cattle show that there are marked differences between breeds. Of note in the results of experiments is the varying intake of water under different thermal conditions. With temperatures between 70°F (21.1°C) and 90°F (32.2°C), the water consumption of each animal increases 100%. Over a longer period, however, studies have shown that water intake increases enormously in the early period of heat stress but then falls off (Figure 9–22). At no time, however, is it lower than twice the amount needed at lower temperatures. Such findings indicate the problem

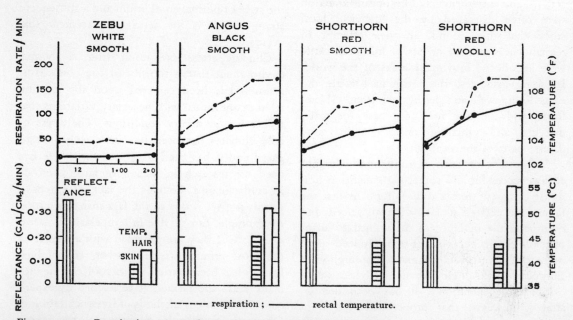

Figure 9–20. Respiration rate, reflectance, rectal, skin and hair temperatures of cattle exposed to bright sunlight at a wet bulb temperature of 91°F. (From Macfarlane, W.V., *Arid Zone Research*, XI, UNESCO, 1958.)

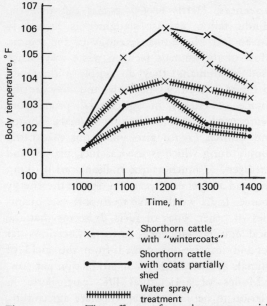

Figure 9–21. The effects of wetting coats with spray water on body temperature regulation for shorthorn cattle with different coats. (After Mc-Dowell and Weldy, 1968.)

of providing adequate water in hot arid climates, where short supply can have marked adverse effects.

Like water, food intake varies with temperature and this, of course, influences body weight and ultimate yield. This applies not only to the larger ruminants but also to other farm animals.

While much has been learned about the physiology of animals and their response to

different climatic conditions, innovations in animal husbandry do not alone depend upon the availability of specific scientific knowledge. Any marked change in mode of life of people over the earth is not completed merely by introduction of new techniques, for cultural variables of necessity need to be considered.

THE CULTURAL VARIABLES

"First, the weather; secondly, the fixed equipment of dams, windbreaks, bunds and sluices for dealing with it; thirdly, the resources of techniques, materials and equipment available for agriculture; fourthly men with a certain outlook, a certain set of values, a certain set of economic and social organization—they are the four variables which determine the relative efficiency of an agricultural system . . . . New tools, new techniques, new breeds, can in most situations be easily discovered and prescribed. The change of their application, however, depend upon the fourth factor—man, his economy and society."

In a succinct paragraph, Dore (1968) outlines one of the major variables that limits the application of agroclimatic advance in many areas of the world. No matter what breakthroughs occur in research and design for agricultural systems, unless the methods can be applied they are of no value. As a generalization, it is found that such advances are most readily implemented in technologically advanced countries, areas where people are conditioned to changes in

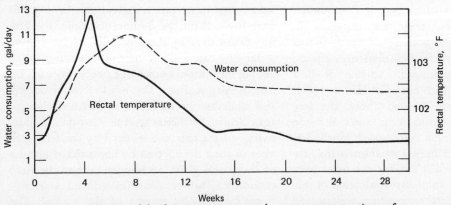

Figure 9–22. Means of body temperature and water consumption of shorthorn cattle exposed to a 90°F temperature (vapor pressure, 22 mm of mercury) for 30 weeks. (After Dowling, 1968.)

strategy, for the implementation of modified agricultural techniques often means a whole new approach to cultivation methods. In some "underdeveloped" countries, the entire cultural patterns of life are geared to agriculture, and changes may require adaptation to a completely new mode of life.

We find many of the agricultural patterns in western Europe and the United States, for example, have changed considerably in the last century. In the United States, the growth of "agribusiness" has caused farming to become a large-scale business enterprise, and the farmer with a few acres and a few cows is now the exception rather than the rule. Time-honored traditions have changed, too. In mountainous countries of western Europe, the practice of transhumance, the seasonal movement of farm animals to different grazing areas, has been modified over time. For centuries, the annual migration saw farm animals taken from their winter quarters in lowland areas to the high summer pastures of the mountains. In Switzerland mountain chalets became the summer quarters of herdsmen, while in Scandinavia it was the saeters. In recent years the streamlining of modern agriculture has seen a diminishing number of animals moved. In Switzerland, Alpine pastures are now mostly used for young animals, others are permanently housed and stall-fed in lowland areas. Such sights as the Draille du Languedoc, the movement of sheep from lowland to highland pastures in southern France, still takes place but the numbers of sheep involved are decreasing every year.

While the implementation of modern agroclimatic methods is often beneficial in underdeveloped countries, there are instances in which lack of understanding of local factors might cause them to lead to disastrous effects. The oft-cited "ground-nut" scheme of East Africa provides one such example (Hance, 1967). Much too has been written about the low productivity of cattle in some areas and the "sacred cow" of India has caused much discussion. The large numbers of animals that are allowed to roam at will are, for the most part, low producers. The immediate evaluation of this situation by agricultural experts from westernized societies was to slaughter many of them and improve the breed. Subsequent evaluation suggests that this might have far-reaching consequences. Harris (1965) points out that the Hindu taboo against slaughter is intimately woven into the human ecology of the region. Most of the animals feed on waste that is not used by humans; they do supply milk, albeit in limited quantities per cow, and they are the source of energy in agricultural production, being used widely as draught animals. Ehrlich and Ehrlich (1970) also point out that cattle supply dung which is used as fuel, plaster, and fertilizer—as much as 300 million tons of dung are used as fuel each year. To replace the energy source, India would have to import vast quantities of other types of fuels. It seems that the local cultural influences have ramifications far beyond impressions gained from the yield of animals as seen in charts of production per cow.

Modes of agricultural life, particularly in relation to herding versus sedentary agriculture, have provided the focus for some highly interesting arguments concerning the role of man in climatic change. There is one school of thought that feels the decline of agriculture of a region is a result of climatic change; another suggests that it is related to cultural change. As an example, the lands of North Africa which, in Roman times, were a granary, are now largely areas of limited agricultural production. Some writers, with Huntington well to the fore, suggest that the decline is due to progressive desiccation of the climate. Others, notably Murphey (1951), feel that it results from changing land-use patterns of people of different cultural backgrounds. Murphey argues that during Roman times, the sedentary agricultural people were well versed in irrigation agriculture. Crop yields were high. With the decline of Rome, the area was taken over by Arab peoples whose culture is largely based upon nomadic herding; such people had little knowledge, or little use, for irrigation agriculture. This resulted in the neglect and underuse of irrigation and the decline of agriculture. Furthermore, the "natural productivity" of the area was reduced by the felling of trees or their destruction by unrestricted grazing of sheep and goats, an integral part of the Arab economy. This resulted in reduced soil productivity through erosion of the topsoil. The argument for and against human versus climatic change as a factor in the agricultural decline of regions is an intriguing one that is perhaps best considered by full examination of both climate and culture evolution over time. This is well illustrated by considering what has happened to the grazing areas of the United States over the past 100 years.

Dasmann (1968) outlines a Forest Service survey completed during the 1930s that evaluated the effects of range livestock in the western United States. As Hollywood has often reported, the period of expansion of the west was a colorful part of U.S. history, and cattle played an important part in this. A phenomenal growth in the number of cattle occurred between 1870 and 1890, when the number went from between 4 and 5 million to 26½ million in the 17 western states. This number far exceeded the carrying capacity of the land. It was estimated that the original capacity dependent upon the native vegetation was 22½ million animal units— where one unit equals one cow or horse, five sheep or five goats. Overstocking decreased the capacity considerably. By 1930, it was estimated that the capacity had fallen to 10.8

million animal units, but even then there was a cattle population in excess of 17 million. The land was seriously eroded in places and elsewhere erosion problems were evident.

The misuse of the land has been put down to overgrazing through ignorance (many farmers from the east attempted to apply humid farming methods to the semiarid west), or to pure exploitation for quick profit. Dasmann points out, however, "Climate has been, and remains, a major cause of range damage, although it is often blamed for man's mistakes. Droughts are normal on the western ranges and grazing capacity fluctuates with wet years and dry. A range properly stocked for a high rainfall year, may be dangerously overstocked if drought follows." At the same time, he notes "Economic factors remain a major cause of range damage. High prices for beef, mutton or wool encourage heavy stocking; falling prices make it difficult to dispose of animals without great financial loss." Again, the interplay between climate, culture, and economic conditions ultimately result in destruction of natural resources.

CHAPTER 10

Climate, Industry, and Transport

"Climate is one of the most important factors in industrial development . . . . In some localities it has been treated—and rightly so—as a natural resource." (Landsberg, 1961.) Evidence supporting this statement is quite easily derived. One need think only of climate in relation to the tourist industry, or the costs of heating or cooling industrial facilities, to appreciate climate as a resource. But beyond these and other "outdoor" activities, the role of climate in the location and functioning of industry is a more subtle one. Often, in analyses concerning industrial location, the word climate never appears and it is only by closer inspection that climate, as an industrial resource, becomes apparent.

Industrial activity can be considered in terms of primary, secondary, and tertiary industries. Primary industries are those concerned with the exploitation of raw materials and foodstuffs directly from the physical environment. Agriculture, forestry, and mining fall into this category. The secondary industries utilize the resources gained from primary activities for further processing, and activities range from iron and steel works to foodstuff processing. Tertiary industries are the service industries; they supply the services required by people who might well be employed in any level of industrial activities; professional services, trading, and tourism provide examples. Climate plays a role in activities carried out at each of these levels.

INDUSTRY

Primary Industries

The significance of climate in one of the major primary industries, agriculture, has already been discussed, and it is clear that climate is of outstanding importance in the nature and distribution of agricultural activity. Its role in another primary activity, the extractive industry, is not so clear, because earth resources are not prone to differences in climate over the earth; a mineral or fossil fuel deposit may be the result of palaeoclimatic events, but the prevailing climatic conditions have no effect upon their distribution.[1] While, at the location level, climate is not a significant factor, its impact is certainly felt at the operational level.

Once a mineral deposit is discovered, its utilization as an ore depends upon both economic and technologic factors. If it is determined that exploitation of the ore is economically viable, then mining or quarrying will occur irrespective of the climatic region in which the deposit occurs. The role of climate here is to either increase or decrease the cost of working the deposit and transporting the ore to its market. Many examples of the development of mineral resources in areas of severe climates are found. The massive iron ore deposits of northern Scandinavia provide an example of mining under subarctic conditions. In the Kiruna area of Sweden, iron ore mining is carried on throughout the year despite the severe winter cold and the long hours of darkness. Additional production costs are incurred through the necessity of long periods of floodlighting and by payment of relatively high wages for such working conditions.

Until quite recently the exploitation of oil went on regardless of where the wells were located. More recently, though, a new problem faces the oilmen, a problem that is a direct result of the new wide awareness of man-environment problems. The rich oil fields of the North Slope of Alaska might, a few years ago, have been rapidly developed. The area occurs in the fragile tundra biome, however, and the threat that this ecological realm might suffer irreversible consequences if developed has caused conservation-

[1] There are exceptions to this; in some arid regions of the world, salts are being actively deposited while guano phosphate deposits occur in localized conditions partially depending upon present-day climatic belts.

ists to block immediate action. Whether sufficient care and design in the extraction and transport of the oil can satisfy the environmentalists, or whether the economic demands for oil will be the overriding factor has yet to be determined. One thing is clear, however; disregard for the environment in which resources occur is no longer acceptable, and the location of a given resource, in terms of the climatic region/world biome in which it occurs will be of prime importance in determining the nature of exploitation in many areas of the world.

Such a case is already evident in another area of oil production. The development of offshore wells with their attendant platforms is a case in point. Oil escape from submarine wells can lead to extensive marine pollution, and the developers need to take massive precautions against this. In certain areas of the world the offshore rigs are subjected to seasons of extremely stormy weather that can result in loss of platforms and lives. The experiences in the North Sea and the Gulf of Mexico have shown the dangers that occur. Note, too, that while the climate and weather play a role in the safety of the platforms, the costs involved in constructing them are directly related to the marine conditions that exist.

Secondary Industries

In their examination of the location of industry, geographers and economists have, in recent years, turned to highly sophisticated methods of locational analysis. There is little doubt that such methods are highly meaningful, but the role of environment is not always well defined in the resulting mathematical models. Indeed, many of the classical theories of location (e.g., Weber, 1929) are defined in terms of a uniform climate. As such, an assessment of the role of climate in industrial location may be best attained by turning to the descriptive methodology used. A number of writers have outlined factors contributing toward the location of given industrial activity; Pounds (1961) provides one such example. He suggests that the basic factors contributing toward a given industrial location are the historical influence, the provision of raw materials, the availability of fuel and power resources, the supply of labor and market considerations, and

the transport facilities. While the word climate does not appear in any of these categories, a brief examination of each brings out the role that is played by climate.

Historical influence. The present-day distribution of world industry is highly variable, with some countries having achieved a high degree of industrialization while others are still in their early stages. Such differences of industrialization are intimately concerned with the designation of developed and underdeveloped areas; perhaps this in itself can be attributed to climatic influences, for it has been suggested that the climatic environment plays a role in relative rates of development. But such an approach is highly deterministic and little is achieved through its discussion. Indeed, the whole sequence of the interpretation of industrial development over time, in terms of manufacturing or in terms of agricultural industry, is fraught with problems. The location of an industry cannot always be adequately explained in terms of present-day factors; at the same time, no old-founded industry is entirely removed from the historic factors that contributed toward its development. Industrial inertia, human skills, and individual choice all play a role that causes precise analysis of changes over time to be highly difficult.

The role of climate in the historical aspects of industrial location is, itself, problematic. An explanation of the location of the old-founded textile industry of England provides a good example. After the industrial revolution, the textile industry in Britain was dominated by the industries in the counties of Yorkshire and Lancashire. These two counties, like many other regions, developed textile works in medieval times, the industry being based upon local availability of wool and suitable supplies of water for water wheels and dyeing. With the development of steam power, both areas, located on major coal fields, became predominant. Yorkshire continued to thrive on woolens but Lancashire became the cotton manufacturing center of the world. It has been suggested many times that one factor that caused this was climate. In spinning cotton thread, there is a tendency for it to snap under dry conditions. A moist environment is necessary, and Lancashire, located on the western side of England, possessed such a climate. Commenting upon the possible relationship, Estall and Buchanan (1961) write, ". . . that the climate of Lancashire played the fundamental part in the location there of the cotton industry (has) to be dismissed for the oversimplification that (it is)."[2] While the "fundamental" role is questionable, there is little doubt that climate is important, for the same writers later remark, "Climate may also affect the industrial process itself . . . . Where natural conditions are not as good as would ideally be required the requisite conditions can often be provided artificially. Thus artificial humidifiers are used, especially for fine textile work . . . ."

Other often-cited examples of the role of climate are also difficult to substantiate. The growth of the precision instrument industry in Switzerland has been related to the detailed work completed by Swiss artisans who were confined to their homes during long, cold winters. The development of craft industries most certainly did occur. Whether such an inheritance has resulted in a highly skilled labor market for work in small components is, of course, another question.

Although the role of climate in the development of some industries is difficult to assess, there are examples in which climate can be shown to play a major part. The rapid growth of industry in southern California illustrates the point for the development of the aircraft industry was initially directly related to climate. Early aircraft manufacturers were attracted to the location because "Its mild winters and light winds (summed up by the term "flying weather") found favor with early flying enthusiasts, who were also early plane manufacturers." (Nelson, 1959.) A further factor of importance was that the large hangers required in aircraft manufacture did not require expensive heating. This latter point is well demonstrated when the number of heating degree days for California are compared with those of other areas of the United States. The degree day is used by heating engineers to estimate the amount of fuel likely to be required

in given locations. When the average temperature of a day falls below 65°F (18.3°C), its value is subtracted from that base and the difference gives the number of heating degree days. Thus if the temperature of a day is averaged at 45°F (7.2°C), there is a total of 20 degree days. The degree days for the entire heating period are added to give a cumulative value, for fuel consumption is taken as proportional to the accumulated heating degree days. As shown in Figure 10-1, the cumulative number of heating degree days is appreciably less in California than in many other parts of the country.

Another interesting example of the development of an industry because of the climatic conditions in southern California is the movie industry. Established at a time when camera equipment was not highly sophisticated, the long hours of sunshine and light proved ideal for outdoor shooting of films. The location was further aided by the myriad of diverse climates and related vegetation associations that occur within a fairly small area around southern California. It was not necessary to travel far to film desert conditions, while mountains provided a backdrop for stories associated with everything from Yukon miners to avid Alpinists. As with other industries, this historic inducement was negated with the development of sophisticated equipment. Lightweight cameras and improved transport meant that the situation film no longer be simulated; the area about which the film was made could actually be visited. Technology and the development of other film-making areas led to the demise of Hollywood as the film capital of the world. Interestingly, and perhaps because of the concentration of expertise and facilities, television production films remain installed in Hollywood.

Historic development in industrial location has played a detrimental role in air pollution problems. The conditions that contribute toward high-level pollution have already been outlined; it merely serves to point out here that many old-founded industries are poorly located in relation to pollution potential. The site location of an industry usually reflected optimization of transport facilities, the situation in relation to access to raw materials, and the proximity of a labor or consumer market. Frequently, sheltered valleys were considered ideal locations. They

---

[2] The climate of Lancashire is not, in fact, more humid than any other western part of Britain. Periodic spells of dry continental easterlies occur there as elsewhere. The argument was hence demonstrably inadequate from the first.

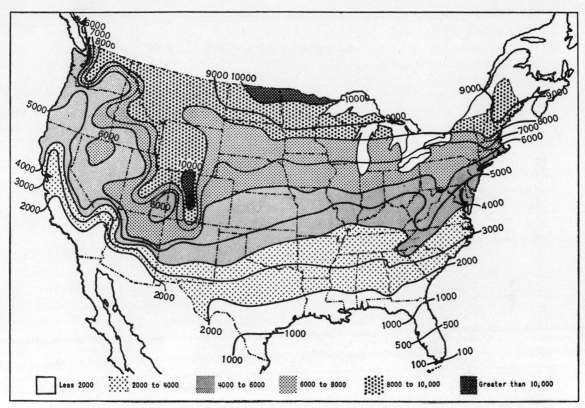

Figure 10-1. Average annual heating degree days (base 65°F) in the United States. (Courtesy of U.S. Weather Bureau.)

have not proved ideal from the standpoint of air pollution, and some sites proved ideal for the accumulation rather than the dispersal of airborne material. The problem is not a new one. Pollution in London was described in the diaries of John Evelyn as early as 1640. The smog potential of the Los Angeles Basin was apparent to early explorers. They noted that smoke from the fires of settlements hung as a pall in the sky. The shallow layer of marine air responsible for inversions in the area existed, and even in those early days the name "Bay of Smokes" was most applicable. Today, these meteorological conditions still exist, but now they are combined with all of the outputs associated with the population growth and related urban-industrial development, so that the problems are heightened. As shown in Figure 10-2, effects range from health problems to vegetation deterioration over wide areas.

Raw materials. The most obvious influence of the role of climate upon industrial location as

conditioned by raw materials concerns the industries that are based upon agricultural and forest products. In many cases, large industrial processing plants appear where raw materials are found. The meat-packing and grain-processing industries of the American midwest and the location of pulp and paper mills in the forest belt of Canada provide apt examples. Such location is to be expected because the value per unit volume of raw materials is less than that of finished products. As with many primary processing activities, transportation costs are the key to industrial location. Of course, there are variations from this pattern, many of them of historic origin. The location of the early cotton industries in Lancashire and New England, for example, would not follow such a pattern, nor would the processing of much tropically grown agricultural produce. There are many reasons for the differences, but, climate is sometimes the key. This is true particularly for some of the tropical products.

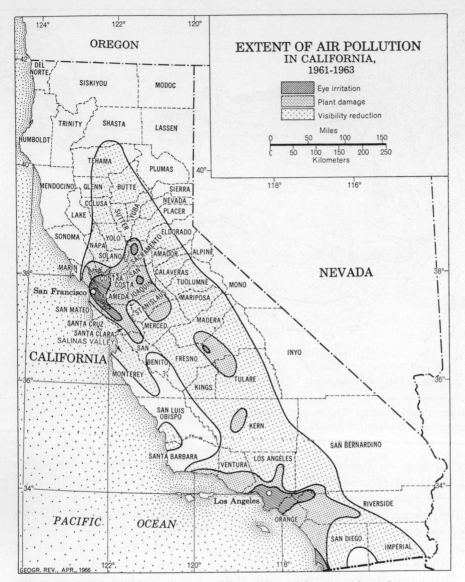

Figure 10-2. Extent of general air pollution in California, 1961–1963. The plant damage areas are specific, but the eye irritation and visibility reductions may be due in part to forms of general pollution other than photochemical. Sources: For plant damage, J. T. Middleton, California Department of Public Health, 1961; for eye irritation and visibility reduction, local reports and observations up to December 1963. (From Leighton, P. A., *Geog. Rev.*, 56, 1966. Reprinted from the *Geographical Review*, copyrighted by the American Geographical Society, New York.)

While much has been made of the exploitation of natural resources of tropical lands by European powers, there is good reason for the treatment of some products away from their areas of production. An example is shown by the fact that while West Africa is the major producer of cocoa, western European countries are among the main producers of chocolate and chocolate products. It does not take an expert economist to work out why this is so. The hot, humid conditions of the production areas would necessitate high-cost facilities to stop chocolate products from melting; air-conditioned storage and ships would be required to stop its spoiling. Obviously, in this case it is more expedient to ship the raw rather than the finished product, particularly since the market for the finished product is far away from the cocoa-producing area. Until fairly recently, the same was true of wood products supplied by the tropics. Plywood, for example, could not be easily made, because most glues would not hold; after a short period, the wood disintegrated. Research has provided a more suitable glue product and since the climatic problem has been negated, the situation is changing. It is clear that climatic factors need to be considered in evaluating the state in which such raw materials are shipped.

Reduction of bulk prior to shipment of mineral ores is also a common practice. In some cases, this merely involves beneficiation of the ore; in other cases, it might be totally refined. The latter is often completed if refining the ore requires large inputs of energy. Production of alumina from bauxite, for example, requires enormous inputs of energy so that refineries are often associated with location where ample hydroelectric power can be generated. This location is thus climatically oriented, because hydropower production requires a sustained flow throughout the year for maximum benefits. This is not the only way that water availability influences industrial location, since water itself might be considered as one of the basic "raw materials" in industry.

Table 10-1 shows the amounts of water required for selected industrial activities. Where water is not available in sufficient amounts for industrial use, special plans are required to fulfill the need. The iron and steel plants of California, for example, need to recycle water; where local supplies are not up to the amount needed, it is sometimes necessary for the industry to create its own reservoir. This occurs at the enormous Llanwern steel works near Newport in South Wales.

The need for such emphasis on water relates not only to the amounts used in manufacturing but also to the problems of the industrial effluent. While such problems are often concerned with treatment of effluent prior to discharge, the climatic effect of seasonality also plays a role. In the discussion of river discharge (Chapter

Table 10-1

INDUSTRIAL WATER REQUIREMENTS[a]

Item	Unit	Water Required per Unit (gal)
Aluminum	lb	160
Brewing	gal	470
Canning—Corn	100 cases	2,500
—Tomato products	No. 2 cans 100 cases No. 2 cans	7,000
Cement	Ton	750
Electric power	kW	80
Iron ore (brown ore)	Ton	1,000
Meat—slaughter house	100 hogs killed	550
Oil refining	100 bbl	77,000
Paper—paper mill	1 ton	39,000
Paper—pulp, sulfate	1 ton	64,000
Steam power	ton of coal	60,000– 120,000
Rayon manufacture	1000 lb produced	135,000– 160,000
Woolens	1000 lb finished	70,000

[a] After Kazmann (1965).

2) the seasonal patterns were shown to be highly variable in different parts of the world, and clearly the amount of water in a river will, in part, help determine the concentration of pollutants. The adage "dilution is the solution to pollution" is true, and effluents that remain constant throughout the year are appreciably more damaging in the low river stage than in the high.

Fuel and Power. It has already been noted that present-day climate has little bearing upon the location of fossil fuels, while for hydroelectric power it is of major importance. It is also of extreme importance in the potential development of another source, solar energy.

Direct use of the sun's energy by man is not a new idea. Most schoolboys know that paper can be set on fire by use of a magnifying glass; it is even said that Archimedes temporarily saved Syracuse from the Romans by using large mirrors to set the invading fleet aflame. The sun's energy has been used to evaporate seawater to obtain salt, and as a means of producing dried foods for hundreds of years in some places. Such benefits are, for the most part, small scale, and only recently has technology become available to use solar energy as a power source.

French scientists have perhaps been the most active in this area, and the work of chemist Felix Trombe is most outstanding. His greatest feat has been to engineer the construction of a solar furnace near Odeillo in the Pyrenees. Here, more than 200 days of sunshine can be expected each year. The constructed system consists of a 140-ft fixed parabolic reflector containing 9000 individual reflecting mirrors. This parabola, to which laboratories and offices are attached, is much too large to follow the sun. The sun's rays are reflected onto it by means of 63 smaller mirrors—heliostats—arranged in tiers on a terraced slope (Figure 10–3). Each of these heliostats is controlled by its own photoelectric cell to retain a "fix" on the position of the sun. They follow the sun throughout the day and merely have to be reset to face the sunrise at the end of each day.

The rays of the sun are focused onto a furnace where total rays are concentrated in a circle only 12 in. in diameter. Material melted in this furnace is extraordinarily free of contaminants. For example, the solar furnace is free of contami-

nating carbon associated with high-intensity electric arc furnaces. In a test using the furnace, bauxite and ceramic produced high-voltage insulators of highest purity.

Use of solar energy as a power source is also being vigorously investigated. In describing work at the University of Arizona, Hammond (1971) writes, "The proposed system would capture the sun's energy extremely efficiently by means of specially coated collecting surfaces, which would be heated by the resulting super greenhouse effect to temperatures as high as 540°C (1004°F); the heat energy would be collected and stored in a thermal reservoir, to which conventional steam boilers, turbines, and electrical generating equipment would be attached." The key to the method is the production of highly selective surfaces that act as black bodies to visible light but that are poor emitters of infrared light. They must be able to absorb all incident sunlight but give off reduced amounts of radiation at ordinary temperatures. Research has shown that such surfaces can be produced. For instance, layers of a metal and a completely transparent material such as quartz can be alternated with thicknesses adjusted to wavelength so that it absorbs visible light but is a mirror in infrared or ultraviolet light. Such surfaces would be enclosed in a vacuum to eliminate cooling through convection. While

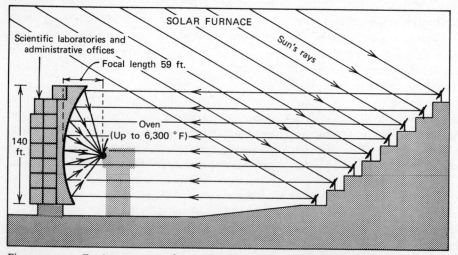

Figure 10–3. Design aspects of the solar furnace near Odeillo. Sixty-three mirrors in eight tiers reflect sun's rays to a parabolic reflector and hence to solar oven.

still in the experimental design stage, it is evident that such energy-producing systems would be highly beneficial. For example:

1. Although requiring large capital investment, operating costs are low.

2. The system would be relatively pollution free. The main problem would be thermal pollution of water used in the process.

3. Although the generating system would be restricted to areas with high potential sunlight, recent developments in superconducting power transmission lines would appreciably widen the area served.

Other direct uses of the atmosphere are available. Paradoxically, one source—that of using the wind—has declined markedly over the past few hundred years. The windmill is now a curiosity rather than the integral part of many landscapes that it once was. Obviously, wind-driven pumps are still used and the wind as a potential power source for generating electricity is currently being investigated.

Besides its importance for providing a basic source of power, climate also directly influences the amount of energy required to maintain plants at their optimum working temperatures; furthermore, it gives rise to many problems in the transmission of power from generating stations to factories (and homes).

Table 10-2 shows the suggested optimum indoor operating conditions of selected industries. It shows quite clearly that in many of the highly industrialized areas of the world it is necessary to either heat or cool the factory area and artificially modify the humidity conditions.

Not only does climate influence the production of energy, it also plays a role in energy distribution. The problems facing such a supply have been aptly summed up by Critchfield (1966), who writes:

"Construction of power lines for the distribution of electricity, whether that electricity is developed by steam or water power, must take into account a great number of climatic effects on equipment. Wind causes the greatest damage to power poles and lines; strong, gusty winds blow over poles and snap lines, or blow trees and other debris into the lines. Hurricanes are by far the worst for overhead power lines. Moisture conditions affect electrical

Table 10-2

OPTIMAL INDOOR OPERATING CLIMATIC RANGE FOR SELECTED INDUSTRIES[a]

Industry	Temperature (°F)	R.H (%)
Textile industry		
Cotton	68–77	60
Wool	68–77	70
Silk	71–77	75
Nylon	85	60
Orlon	70	55–60
Food industry		
Milling	65–68	60–80
Flour storage	60	50–60
Bakery	77–81	60–75
Candy	65–68	40–50
Process cheese production	60	90
Miscellaneous industries		
Paper manufacturing	68–75	65
Paper storage	60–70	40–50
Printing	68	50
Drug manufacturing	68–75	60–75
Rubber production	71–76	50–70
Cosmetics manufacturing	68	55–60
Photographic film manufacturing	68	60
Cosmetics storage	50–60	50
Electric equipment manufacturing	70	60–65

[a] From Maunder (1970), after Landsberg, and Grundke.

transmission and the functioning of insulators. Wet snow, sleet, or freezing rain add weight to lines and poles and make them more vulnerable to wind. To reduce ice from lines, power companies employ various methods of temporarily disconnecting a line from the main current and heating it with an artificially high power load. . . . Temperature fluctuations influence the operations of switches, transformers and other equipment. In hot weather, lines expand and sag and are more easily damaged by high winds. Thunderstorms bring not only the dangers of wind and precipitation but also of lightning, which causes at least a temporary power failure if it strikes a line."

Note, too, that extreme climatic conditions can cause such a demand for power that "brownouts" and "blackouts" are becoming the rule rather than the exception during such periods. Clearly, the industries that depend upon an input of electricity for their function must consider climate carefully in relation to their manufacturing needs.

Labor Markets. In the past, little attention was given to the choice of the individual in terms of the climatic regime under which he found employment. With the better-educated, less-fixed labor market that now exists in many industrialized countries, a new trend is apparent in respect to industrial location. Ullman (1954) pointed out that, beginning in the United States, the affluence of sectors of the society has allowed families to migrate, not because of economic opportunity, but for the prime purpose of finding a "better" place to live. It appears that the amenity most sought is that in which the winter season is essentially snow-free and where outdoor activities are carried on throughout much of the year. Thus there is a large labor market that is readily attracted, or already available, in such climatic areas as the desert southwest, and the Pacific and Gulf coasts of the United States. Such a trend is not only applicable to the United States. In Great Britain, too, there has been a distinct migration of the labor force away from the north toward the south, which experiences warmer, sunnier summers. Of course, climate is not the sole energizer of such a trend; the relative unattractiveness and diminishing importance of the older industrial centers also contribute toward such migration.

Just as climate can act as an asset in attracting a labor market, it can also act as a deterrent. Alaska and areas of the eastern U.S.S.R., although rich in resources, have minimal concentrations of population. In order to attract a working populace to such locations, special considerations are needed and methods of inducements required. Obviously, the labor market is only one part of the need of industry, it is equally significant that a market for the finished products is readily accessible; again, climate is one factor that acts against their growth. The importance of *transport* and the climatic influences upon it are of the utmost significance and are dealt with in more detail in a later section of this chapter.

The Tertiary Industries

Perhaps the most obvious instance in which climate can be considered as a natural resource in relation to the service activities is tourism. While, initially, such an industry might seem quite secondary to other sectors of industry, it is extremely important in terms of both the number of people it serves, the number of people it employs, and the earnings it generates.

Climate is not, of course, the sole energizer of the tourist industry. Some vacation centers, often urban locations, owe their significance to purely cultural factors. It would be difficult to establish, for example, that New York City is a "summer festival" on account of its summer climate. The hot, moist, summer climate is not really conducive to touring the city. But some urban centers, while serving cultural or recreational needs, also depend upon climate to attract their visitors. Such centers as Miami, Las Vegas, Monte Carlo, and Honolulu, while offering many facilities, also benefit from an attractive climate.

As a gross generalization, climate appears to offer two forms of attraction; first, the need for sun and warmth for those living where winters are cold, a factor well demonstrated by the southern migration of flocks of visitors in the northern hemisphere. Florida and the Caribbean are enjoyed by Americans living in cooler latitudes; the Mediterranean countries are invaded by thousands of vacationers from northern Europe; the Black Sea resorts represent a central focus for many tourists from Europe and particularly from Russia. A second attraction offered by the climate is that special conditions allow specialized activities. The most obvious are related to winter sports, for, although artificial snow can be generated, the success of winter sports resorts depend upon the climate.

The fact that the tourist industry is a seasonal one points to the role played by climate. The concept of climate as a resource is subject to human perception and cultural trends. Bermuda is a good example of the influence of such changing social variables on the tourist industry. Before it became fashionable to swim and use beaches for sunbathing, the island was considered a winter resort. With changing social values, the beach activities became the vogue; since Bermuda winters are not really hot enough for this, the island became a summer resort.

The assessment of climate in relation to recreation is not an easy task. In his account of the influence of weather on outdoor recreation, Clawson (1966) notes that there is a woefully inadequate body of empirical findings, although the situation is improving. Thus, while it is possible to generalize about the significance of climate, and while many resorts emphasize the nature of their climates, it is rather difficult to separate the importance of climate from other variables, particularly when attempting to establish why one person selects one resort over another.

Just as a holiday resort can lose large amounts of money if the climate differs from that expected, so can many other outdoor industries. Such losses are most frequently put down to weather patterns rather than long-term climatic conditions, a situation well demonstrated by data pertaining to the construction industry (Table 10-3). Extending these results to a long-term pattern, it is clear that the limitations imposed are much more significant in areas where critical limits occur more frequently. The total dollar loss to the construction industry has been estimated at between $3 and 10 billion annually because of imposed weather restraints. Costs are considerably higher in some areas where, in order to facilitate outdoor work, heated shelters are needed.

A comprehensive account of the importance of weather in relation to consumer sales, and to the value of weather decision making in tertiary industries has been given by Maunder (1970), who provides many examples of the impact. The stress throughout the studies cited is on short-term weather events rather than long-term effects. This is to be expected, because day-to-day weather certainly is important in the variations that exist. Less attention has been given to the climatic aspects, although, of course, no large retailer is unaware of the seasonal demands that are made. A summer visit to a department store in the northeast United States shows a marked emphasis on outdoor wear for pleasure and accoutrement for outdoor sports. In winter, the sleds, skis, and ski materials

are evident. Designers, too, follow the seasons; a fall display of new dresses is obviously quite different from that of winter. Such instances are well known, but, unfortunately, few studies have been completed to provide empiric data and in-depth analysis. Much interesting and significant work can be completed in this area. An interesting study could be made, for example, of the year-to-year variations in snowfall and how it affects the skiing industry; in the winter of 1971 to 1972, for example, snowfall on the ski slope areas of the northeast was far below normal, and the skiing season was appreciably shortened, leading to losses of income to those in the industry. Knowledge of the frequency of such variations from the norm would prove extremely valuable and could lead to a more rational approach to the establishment of ski resorts.

One tertiary activity intimately concerned with weather and climate is the insurance industry. Hendrick and Friedman (1966) show how weather-damage insurance is an adaptation by man to an element over which he has little control. The losses due to storms (Table 10-4) are fortunately so distributed in time and space that insurance coverage is a practical means of spreading the risk (and cost) over a large number of property owners. This is not true of secondary weather effects, and hazards such as floods and tidal inundation are not usually covered by insurance. The propensity for such hazards to be restricted to river valleys and costal strips means that risk spreading is not applicable. The importance of weather and climate to the insurance business is testified to by the research and resulting publications sponsored by the insurance industry in these fields (e.g., Collins and Howe, 1964, Russo et al., 1965).

Climatic Analysis for Industrial Operation

From the foregoing descriptive analysis it is clear that climate plays a significant role in industrial location. Of equal significance is the role played by climate in industrial operation on a year-to-year, season-to-season basis. Unlike the climate-industrial location discussion, it is possible to assess this relationship in more quantitative terms. Benefits arising from a

Table 10–3

CRITICAL LIMITS OF WEATHER ELEMENTS HAVING SIGNIFICANT INFLUENCE ON CONSTRUCTION OPERATIONS[a]

Low temperature/high wind equivalent to Chill Factor = 1000 — 1200 (Very Cold — Bitter Cold)

Temperatures above 90°F

High temperature/high humidity equivalent to temperature—humidity index (THI) = 77

Operation	Rain	Snow and Sleet	Freezing Rain	Low Temperatures (°F)	Low Temperature and High Wind (Chill Factor)	High Temperature	High Temperature and High Humidity (Temperature–Humidity Index)	High Wind (mph)	Dense Fog	Ground Freeze	Drying Conditions	Temperature Inversion	Flooding	Abnormal
Surveying	L[b]	L	L	0 — -10				25	x[c]	—	—	—	—	—
Demolition and clearing	M	M	L	0 — -10				15—35	x	x	—	x	—	—
Temporary site work	M	M	L	0 — -10				20	x	x	—	—	—	—
Delivery of materials	M	M	L	0 — -10				25	x	—	—	—	—	—
Material stockpiling	L	L	L	0 — -10				15	x	—	—	—	—	—
Site grading	M	M	L	20 — 32				15—25	x	x	x	—	—	—
Excavation	M	M	L	20 — 32				35	x	x	x	x	—	—
Pile driving	M	M	L	0 — -10				20	x	x	—	—	x	—
Dredging	M	M	L	0 — -10				20	x	x[d]	—	—	x	x
Erection of coffer dams	M	L	L	32				25	x	—	x	—	x	x
Forming	M	M	L	0 — -10				25	—	x	—	—	—	—
Emplacing reinforcing steel	M	M	L	0 — -10				20	—	x	—	—	—	—
Quarrying	M	M	L	32				25—35	x	x	x	x	—	—
Delivery of premixed concrete	M	L	L	32				35	x	—	x	—	—	—
Pouring concrete	M	L	L	32				35	x	x	x	—	—	—
Stripping and curing concrete	M	M	L	32				25	—	x	x	—	—	—
Installing underground plumbing	M	M	L	32				25	—	x	x	—	—	—
Waterproofing	M	M	L	32				25	x	x	—	—	—	—
Backfilling	M	M	L	20 — 32				35	x	x	x	x	—	—
Erecting structural steel	L	L	L	10				10—15	x	—	—	—	—	—

Operation[a]	[b]	[b]	Low temperature/high wind equivalent to Chill Factor = 1000 – 1200 (Very Cold – Bitter Cold)	Temperatures above 90°F	High temperature/high humidity equivalent to temperature–humidity index (THI) = 77				
Exterior carpentry	L	L	0— -10	15	—	—	—	—	—
Exterior masonry	L	L	32	20	x	x	x	—	—
External cladding	L	L	0— -10	15	—	—	—	—	—
Installing metal siding	L	L	0— -10	15	—	—	—	—	—
Fireproofing	L	L	0— -10	35	—	—	—	—	—
Roofing	M	L	45	10—20	—	x	x	—	—
Cutting concrete pavement	M	M	0— -10	35	x	x	x	—	—
Trenching, installing pipe	M	M	20— -32	25	x	x	x	—	—
Bituminous concrete pouring	L	L	45	35	x	x	x	—	—
Installing windows and doors, glazing	L	L	0— -10	10—20	—	—	—	—	—
Exterior painting	L	L	45— 50	15	x	x	x	—	—
Installation of culverts and incidental drainage	M	L	32	25	—	x	x	—	x
Landscaping	M	L	20— 32	15	x	x	x	—	—
Traffic protections	M	M	0— -10	15—20	x	x	—	—	—
Paving	L	L	32— 45	35	x	x	x	—	—
Fencing, installing lights, signs, etc.	M	M	0— -10	20	x	x	—	—	—

[a] From U.S. Dept. of Commerce (1966). The survey data reported in this table is furnished for purposes of illustration only and does not constitute a representation applicable to particular construction or projects.

[b] L indicates light; M indicates moderate.

[c] Indicates operation affected by this condition but critical limit is undeterminable.

[d] Indicates water freeze.

Table 10–4

ESTIMATED RANGE OF ANNUAL INSURED PROPERTY LOSSES BY STORM TYPE IN THE UNITED STATES[a]

Storm Type	Range of Annual Insured Property Losses (Millions of Dollars)
Hurricanes	250– 500
Tornadoes	100– 200
Hail/Wind thunderstorms	125– 250
Extratropical wind storms	25– 50
Total	500–1000

[a] From Hendrick and Friedman (1966).

rational treatment of climate as a resource base can be measured in monetary terms.

Landsberg and Jacobs (1951) in their review of applied climatology stress the practical use of climatic data in the operation of manufacturing industries. They note that in order to arrive at meaningful decisions regarding the impact of climate, it is necessary to establish the answers to two basic questions. First, it must be determined what climatic elements most effect the industry in question; second, analysis must be completed so that the available climatic information can best serve the industry in question.

To establish meaningful answers to such questions they provide two questionnaires. The first, Table 10–5a, analyzes the needs of the industry, and the second, Table 10–5b, provides a checklist for required data needed in the analysis. In-depth study of the responses to the questionnaires allows the climatic input to form an integral part of the whole planning and operational purposes.

Although the paper just cited was published some time ago, the logic of the analysis and the benefits derived by following results of simple climatic analysis have been borne out by many subsequent studies. For example, an analysis by Bickert and Browne (1966) of five manufacturing firms in Colorado show that quite diverse industries are highly sensitive to climatic variations. (Note that their study was carried out to determine "weather" sensitivity, but the results show the importance of long-term observations.)

The initial response of company executives pertaining to the impact of weather on the various industries was that ". . . there was a general initial lack of awareness of the magnitude or the existence of weather variables which affected each company." Generally, the respondents' first reaction was to state that weather had a negligible effect upon operations. Upon further exploration, the researchers found precisely the reverse to be true. In relation to costs, for example, it was found that a brick and ceramics producer had, for some time, over-scheduled production in winter while under-scheduling it in summer. This led to overtime costs and high inventories. The seasonal nature of the industry that the company served had not been taken into account. It was found, too, that a beer-producing company incurred costs that resulted directly from weather-related conditions. Insulated railroad cars had to be used in winter and, because of drought potential of the area, much money was spent searching for ground water supplies. The role of prevailing atmospheric conditions also made its mark in a precision instruments company. Thousands of dollars had to be invested to overcome excessive high humidity to prevent rust and corrosion. Such examples are indicative of the long- and short-term effects of atmospheric conditions on industrial operation.

The acquisition of climatic data and use of modern statistical techniques now allows many weather-related, decision-making processes to be completed quantitatively. Miller and Thompson (1970) provide a good outline of examples of this and the following is drawn from their presentation.

"Consider the case of a prospective user of a weather forecast faced with the problem of deciding whether or not to take protective measures against a certain adverse weather element. The user may be, for example, a farmer, a business man, or an airline operator, and the weather element may be snow, rain, low ceiling, or poor visibility, or the like. In general, the user should take protective measures against the occurrence of adverse weather if, in the long run, an economic or other gain will be realized; otherwise, no protective measures should

Table 10–5*a*

ANALYSIS OF OPERATION

(*Climatic Anamnesis*)

1. Class of enterprise (business, industry, agriculture, etc.)?
2. What operational subdivisions exist?
3. Class of problem?
   a. To lead to a design.
   b. To affect a procedure.
   c. To make an operational decision as to "where" or "when."
4. How does the prospective user of the weather information define his *weather* problem?
5. How does he define his *operational* problem?
6. Will the climatic or weather information the user believes he needs furnish a solution to the operational problem?
7. Have solutions previously been obtained for analogous operational problems?
8. Can the operational problem be redefined in more realistic climatological terms?
   a. What periodic fluctuations (diurnal, seasonal, etc.) exist?
   b. What are the critical climatic and weather limitations on the operation?
   c. Is it of importance to stay below critical limits or are there known or desirable optimum conditions?
9. Are operational figures (production, yield, etc.) available for correlation purposes? For how many years (cases)?

Table 10–5*b*

ANALYSIS OF CLIMATIC DATA

(*Climatic Diagnosis*)

1. Class of climatological technique required?
2. What types of data are desirable?
   a. Elements (pressure, temperature, winds, etc.)?
   b. Are combinations of elements required? If so, what combinations? Are their relationships known?
   c. What length of record is needed for an adequate solution?
3. What climatic data are available?
   a. At the spot; in the area; over the region?
   b. In what form are the data? Are existing summarizations suitable?
   c. What length of record is available?

Table 10–5*b* (*continued*)

d. Are climatic data available for the same periods covered by the operational information? (See Table 10–5*a*, Item 9.)

4. What are the local peculiarities in climate that may affect the usefulness of existing climatic data?
5. What are the physical or statistical limitations imposed on the data and their interpretation?
6. Does theoretical information (or a suitable technique) exist that can serve to supplement, evaluate, or be substituted for, the inadequate observational materials?
7. Are solutions available for analogous climatological problems?
8. Is there a need to elaborate a new theory of interrelations?
9. What form of presentation of the conclusions is desirable?
   a. Graphs.
   b. Tables.
   c. Formulas.
   d. Alignment diagrams.

be taken. In order to derive a criterion for making this decision, the following terms are defined:

f_w = Frequency (number of cases) of adverse weather.

f_{nw} = Frequency (number of cases) of favorable (non-adverse) weather.

N = $f_w + f_{nw}$ = total frequency (adverse + non-adverse weather cases).

G_p = Total expected gain for N occasions if protective measures are taken.

G_{np} = Total expected gain for N occasions if protective measures are not taken.

C = Cost of protection on each occasion that protective measures are taken.

L = Loss suffered on each occasion that adverse weather occurs and no protective measures have been taken.

T = Average net profit for each occasion, exclusive of the cost of protective measures, or the loss which may have been suffered.

If, now, protective measures are taken on N occasions, the total gain will be the profit minus the cost of protection, both multiplied by the number of occasions. Or, in symbolic form,

$$G_p = (T - C)N \qquad (1)$$

If protective measures are not taken, the gain can be divided into two parts—the reduced profit during cases when adverse weather occurred and the profit for cases of no adverse weather; in symbolic form,

$$G_{np} = (T - L)f_w + Tf_{nw} \tag{2}$$

Since the user should take protective measures only if the gain by so doing will be greater than the gain by taking no protective measures, protection should be provided when $G_p > G_{np}$ or, from Equations (1) and (2).

$$(T - C)N > (T - L)f_w + Tf_{nw} \tag{3}$$

which reduces to simply

$$f_w/N > C/L \tag{4}$$

The left-hand side of this expression defines "P," the probability that adverse weather will occur. Note that P is the "relative frequency" of adverse weather, or the number of occurrences of adverse weather divided by the total number of cases. In a similar manner, it may be shown that protective measures should not be taken if $P < C/L$, and either course may be followed if $P = C/L$. Thus, a criterion for making a decision to protect, or not protect, against adverse weather may be expressed.

$$P \left\{ \begin{matrix} > \\ = \\ < \end{matrix} \right\} C/L \quad \begin{matrix} \text{Protect} \\ \text{Either course} \\ \text{Do not protect} \end{matrix} \tag{5}$$

The value $P = C/L$ thus represents a critical ratio, above which protection should be provided, and below which it should not. It is interesting to observe that for C, L, and T, as defined here, the profit drops out and need not be considered in the decision. Alternative, but generally more complex expressions may be derived by defining these terms in a different manner.

Now it is clear that the user of the weather information must provide the numerical values of C and L. They are the costs and losses associated with the weather-dependent portions of the operation. The value of P, the probability of adverse weather, must be provided by the meteorologist. This may be done by an analysis of past experience. For example, with a particular set of initial surface- and upper-air conditions, the meteorologist knows, or can estimate quite well, the relative frequency with which adverse weather has occurred under similar conditions in the past. If a given weather map has resulted in only a relatively few occurrences of adverse weather in the past, the probability of adverse weather will be low. If adverse weather has occurred frequently under these conditions in the past, the probability of adverse weather will be high."

TRANSPORT

Air Transport

All phases of air travel are influenced by atmospheric conditions. The effects are felt at all levels of operation, from the construction of a landing strip to the trip en route and the landing and take-off conditions. The significance of atmospheric studies in relation to air travel is illustrated by the simultaneous development of the two. As air transport improved, it became necessary to understand flying weather; as planes flew higher, the altitudinal variations of the atmosphere became a necessary area of study. Development in one led to research in the another.

Many of the problems of air transportation are concerned with conditions that will be encountered at a given time and, as such, aviation weather is of more immediate significance than aviation climatology. But the two are obviously interrelated, and long-term weather conditions, or expected frequencies of events, need to be considered. This is well illustrated in the problems of airport runway construction. Riehl (1965) provides a number of interesting examples in this respect.

At Benghazi, Libya, the single airport runway proved highly satisfactory during the summer, but difficulties in landing and take-off were experienced in winter. In reviewing the climate of the location, it was found that winter winds of more than 10 mph occurred from 30 to 50% of the time, blowing at an angle across the runway. From the frequency distribution of the winds, a second runway was designed so that during the winter, planes did not have to land in strong crosswinds. Thus, a seasonal climatic effect was very important in the design of the enlarged airport. Similar climatic analyses have proven valuable in Arctic regions where snow on the runway is a problem. Analysis of prevailing winds in relation to snowfall allows the proper location of snow fences at right angles to the snow-bearing winds. Runways can be kept remarkably clear using such techniques.

The design of the runway is not the only aspect that needs to be considered in the light of climatic analysis. One of the major problems concerning take-off and landing facilities is visibility. Although the development of electronic landing aids has reduced concern about visibility problems, most pilots still like to see during the last few seconds before touchdown. Thus the incidence of fog or low ceiling at a location is of prime concern in consideration of airport locations.

While such factors are variable on a day-to-day basis, some locations are much more prone to incidence of fog and low clouds than others. This can be the result of localized climatic conditions or as a result of the long-term weather patterns that exist. In either case, a fog-bound airport is either closed or the number of flights is reduced markedly. Commercial flight regulations require that an aircraft carry enough spare fuel to reach an alternate airfield in case the original destination is closed. Obviously, this creates enormous problems when fog is widespread and necessitates closing of runways over wide areas. This leads to enormous logistical problems and to considerable loss of income to the airlines, as well as inconvenience to the passenger. A natural outcome of this is that many fog-dispersal methods have been tried.

As early as World War II, the British used oil burners to burn off fog from landing strips. This was a case of dire necessity, for the aircraft were often returning from missions over Europe and if not damaged, were often very low on fuel. The program initiated, called FIDO (Fog Investigation Disposal Operations) essentially concerned generating sufficient heat to raise the temperature of the air above its dew point and so promote its dissipation. The method was obviously expensive and dirty. Its principles have, however, been used more recently. At Orly Airport in Paris, some success has been obtained using turbo engines placed at strategically located positions. Halacy (1968) shows that their use has enabled the airport to remain open when, without artificial means, it would have been closed. The fogs that provide the problem in these locations are "warm" fogs; that is, they are comprised of water droplets rather than ice crystals and supercooled moisture. These types of fog pose the most problems,

because the cold fogs can be dispersed through "seeding," using the same principles as applied to cloud seeding. To clear the warm fogs, other ideas have been generated.

One experiment, carried out by the Cornell Aeronautics Laboratory, had 700 lb of salt spread over a fog at Elmira, New York. A hole, ¼ by ½ mile, was opened in the fog for 15 min. Salt, of course, is not the ideal medium to disperse over airports, because it is corrosive to metals. Alternate chemicals are now being tested. Another approach has been to "blow away" the fog. A helicopter with large rotors causes breaks in the fog cover when it hovers over a fog-bound field. At Nantucket, where fogs are a frequent event, a fog-suction machine has been tried. These methods obviously require a considerable amount of research to become practical and are, at best, temporary measures. Other lines of research into the problems of clearing fogs use sound waves and ultrasonic vibrations. Again, these are in an early stage of development. The various programs in fog dispersion, well described by Beckworth (1966), will probably continue until some adequate solution is derived. That this is so, is well shown by the cost of flight cancellations due to weather conditions. As shown in Table 10–6, the amount is considerable.

En route, aircraft face many problems. Many of these are concerned with meterological conditions and include such factors as icing, in-cloud turbulence, and the selection of optimum cruise altitude in relation to aboveground winds. The development of high-altitude aircraft, flying above the troposphere, initially led to hopes that many of the problems associated with turbulence would be negated. This was an erroneous assumption, and the problem of clear-air turbulence now is experienced.

Clear-air turbulence (CAT), which is not normally visible to pilots, poses a severe safety hazard because of the violent motions it induces; furthermore, aircraft that experience CAT are usually flying at high speeeds that will increase the potential for damage or injury. The probability of encountering CAT increases with increasing altitude, being most typical at 9 to 13 km. It is most often associated with the presence of strong vertical wind shears, that is, locations where large vertical gradients of wind speed or

Table 10–6

ESTIMATED COST OF FLIGHT CANCELLATIONS IN THE UNITED STATES DUE TO WEATHER: 1959 AND 1962[a]

Costs (Dollars)	1959	1962
Passenger revenue loss due to flight cancellations[b]	14,989,000	22,730,000
Expense of operating nonrevenue ferry mileage[c]	605,000	628,000
Passenger service-interrupted trip expense[d]	2,443,000	3,116,000
Duplicate reservations, ticketing, and accounting expense[e]	652,000	978,000
Gross costs	18,689,000	27,452,000
Less: savings in aircraft operating expense	12,165,000	18,941,000
Net costs	6,524,000	8,511,000
Number of flight cancellations	54,309	61,101
Cost per flight cancellation	120	139

[a] From Maunder (1970), after United Research, Inc. (1961).
[b] Represents the portion (47%) of total passenger revenue booked on cancelled flights that is lost to other modes of transportation and to discontinuance of travel.
[c] Estimated as 5% of nonrevenue mileage; nonrevenue mileage was estimated as 1.6% of revenue mileage projected for U.S. domestic carriers in National Requirements for Aviation Facilities, 1956 to 1975.
[d] Based upon 1956 to 1959 cost per flight cancellation.
[e] Estimated as 2% of passenger revenue lost to airlines as a result of flight cancellations.

direction occur. Experiments have also shown that where such conditions occur there may be a significant temperature difference between the two air masses on either side of the shear. In an effort to allow detection of CAT, researchers have tried microwave radar, optical (laser) radar, and infrared sensors. Of these, the infrared sensor (despite the fact that it is ineffective when flying directly into the sun or may record temperature effects caused by hot jets of other aircraft), appears to be a most promising method of detection.

Water Transport

The significance of climate to the development of sea transport is aptly demonstrated in the terminology used to describe the major wind and pressure belts over the earth's surface. Figure 10–4 shows the names that were until quite recently used to describe them. The calms associated with the major pressure belts indicate the problems that they posed to sailing vessels. "To be in the doldrums" became an expression indicative of state of mind as well as global location. It is said that the "horse latitudes" derived their name from the fact that when becalmed, ships often dispensed with any

horses on board in an effort to both lighten the load and save on feed.

Like the pressure systems, the winds, too, indicate their effects upon past mariners. The trade winds aptly describe their utility, with no embellishment required. Any reader of sea stories knows of the fickleness of the westerly variables; expressions such as the roaring forties give clear indication of the climatological problems posed to ships in the days of sail.

Today there are few sailing ships, and sailing routes are not so dependent upon the nature and direction of the wind. Nonetheless, the climatic characteristics of the oceans still provide major problems in oceanic transport. Floating ice, fogs, and violent storms have all taken their toll of modern ships, while least-time tracks are still designed with climatic principles in mind.

The shortest distances between any two points on the surface of the earth is described by the arc of a great circle, where a great circle is a circumglobal path that divides the earth into two hemispheres. The North Atlantic great circle route between North American and European industrial areas is perhaps one of the busiest routes in the world. The route passes over the northern waters of the Atlantic Ocean (Figure 10–5); it is therefore subject to periodic

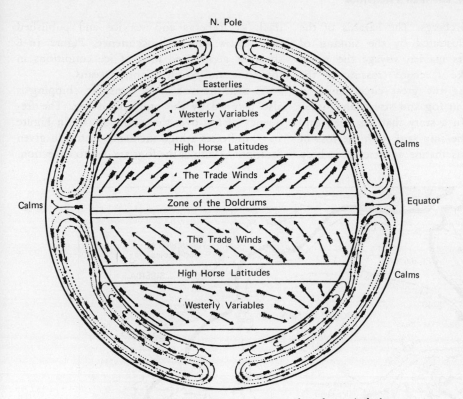

Figure 10-4. Although the general characteristics of surface winds have been known for many years, interpretation of upper air circulation has been considerably revised. Pattern shown here is based on an 1857 depiction by Thompson.

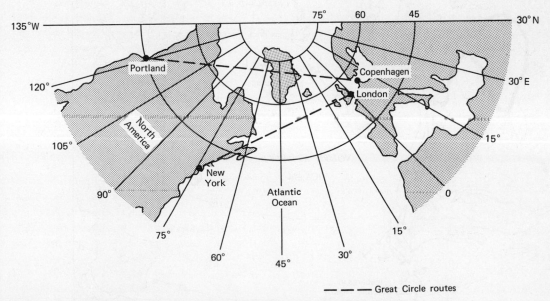

Figure 10-5. Examples of Great Circle Routes. The routes are shown as a straight line on the Gnomonic projection that is used. This does not provide the true bearing of the route.

floating ice and icebergs. The hazards of the route are well illustrated by the sinking of the *Titanic*. On its maiden voyage the vessel attempted to make a record crossing of the Atlantic. Following the great circle route, it became shrouded in fog and struck an "unexpected" iceberg. In a story that has been told countless times, the ship sank with the loss of many lives. Various marine agencies now keep track of icebergs and sea ice and published maps show expected occurrence. Figure 10–6 shows a generalized map of ice conditions in the Arctic and North Atlantic Oceans.

As with the case of air transport, shipping is also plagued by the problem of fog. The frequency of fog over the world, shown in Figure 10–7, indicates a high incidence of fog in given locations. Many of the fogs are due to advection.

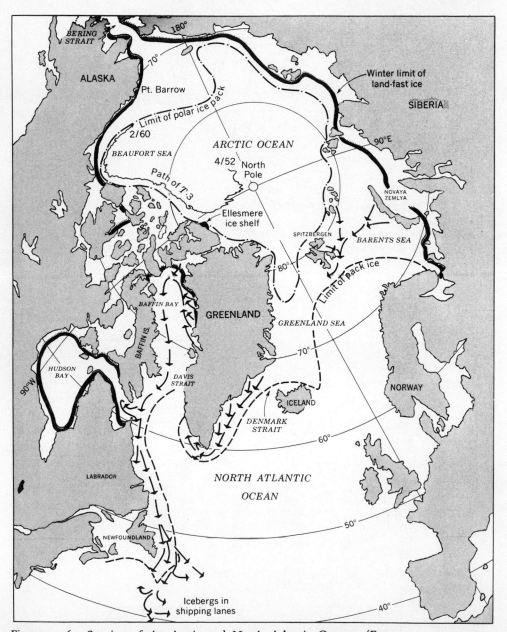

Figure 10–6. Sea ice of the Arctic and North Atlantic Oceans. (From Strahler, A. N., *Physical Geography*, copyright © 1969 by John Wiley and Sons. Reproduced by permission.)

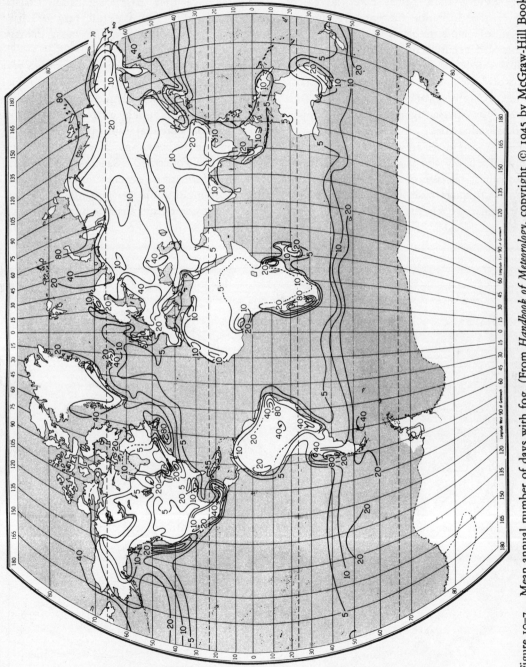

Figure 10-7. Mean annual number of days with fog. (From *Handbook of Meteorology*, copyright © 1945 by McGraw-Hill Book Company. Used with permission.)

The explanation of the infamous fogs off the Newfoundland Banks, for instance, is related to the movement of warm air off the warm waters of the North Atlantic Drift and its sudden chilling as it passes over the colder waters of the Labrador Current. Despite the adoption of modern navigation techniques, accidents still occur in fog. The distribution of fogs also clearly shows that the coastal deserts of the world are among the foggiest that occur anywhere. It is little wonder that they have been designated by special symbols in the Köppen climatic classification.

Violent storms at sea are another hazard facing shipping. Probably the most dangerous storms are hurricanes, because the effects are felt over thousands of square miles of the oceans. Hurricanes originate over the warm seas in low latitudes. Thereafter, in the northern hemisphere, they take a westward and then northeastward path (Figure 10–8). Deriving their energy from the oceans over which they originate, they soon dissipate once they pass over land. Coastal regions, however, feel the full fury of these storms, and it is here that the most damage and loss of life occurs.

The number of hurricanes that occur each year is highly variable and the path that any individual storm may take is erratic (Figure 10–8). Through careful surveillance, both by earth-based monitors and satellites, potential hurricanes are spotted and appropriate warnings given to shipping in the area concerned. Success in utilizing satellite information depends upon adequate surface communications and organized "escape" action in the coastal area that the hurricane might

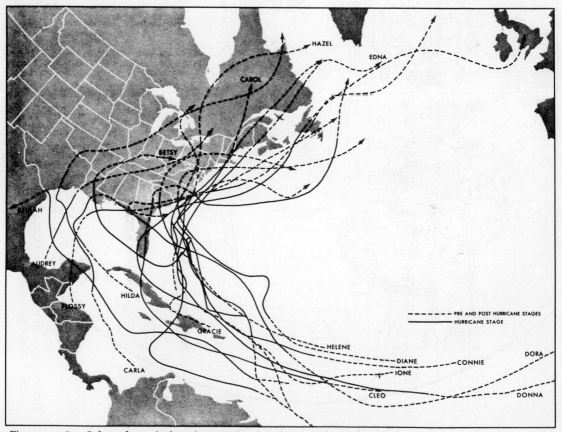

Figure 10–8.   Selected tropical cyclone paths showing erratic tracks, with general recurvature occurring toward the north and east. (After U.S. Weather Bureau, from Cole, F. W., *Introduction to Meteorology*, copyright © 1970 by John Wiley and Sons. Reproduced by permission.)

hit. This is not always available. This was the case in the disastrous storm that struck the coastal lowlands of West Bengal in 1970. The flood wave associated with high seas generated by the storm caused the death of more than 150,000 persons.

While careful monitoring and forecasting have lessened the loss of life due to hurricanes, they still cause millions of dollars worth of damage every year. In an attempt to comprehend the incredible complexity of the interaction between the sea and the air, which is inextricably linked to the energy of a hurricane, ESSA has established a research program. This sea-air interaction laboratory (SAIL) uses satellites, research ships and aircraft, rockets and balloons, as well as remote islands and buoyed stations to obtain the necessary data for analysis.

Another approach to the hurricane problem is the hope that the power of the storm can be decreased. Project Stormfury, which has been in existence for some years, has conducted a number of cloud-seeding experiments on both hurricanes and tropical cumulus clouds. The theory behind the seeding of hurricanes is "based on a chain theory which predicts what should happen to the storm as a consequence of the seeding. Theoretically, seeding alters the cloud structure in such a way that the balance of wind-controlling forces near the eye is upset. Injection of silver iodide particles into the eye wall upstream of the hot towers—the thermal chimneys of the hurricane—should transform supercooled water droplets to ice crystals, releasing heat energy into the storm system near the warm core. The effect of this additional heat should reduce atmospheric pressure adjacent to the low-pressure center of the storm, and reduce differences in pressure across the eye-wall—the steep pressure gradient which produces winds of hurricane force. The smoothing of the pressure gradient should cause hurricane winds near the center to diminish. The storm should expand its spiral bands and lose some portion of its fury." (ESSA, 1969.)

A number of hurricanes have been seeded as part of project Stormfury (Esther, 1961, Beulah, 1963, Debbie, 1969, Ginger, 1971), and while results have not been spectacular, they at least seem promising. Problems of evaluating the results of seeding are due to differentiating man-induced and natural changes. The small number of hurricanes seeded probably results from the fact that it has been necessary to demarcate an area of the Atlantic from which no hurricane on record has ever struck a highly populated coast within 36 hr. While ostensibly such a restriction is to allow time for data collection, it is obvious that damage caused by a seeded hurricane could lead to extensive litigation.

With the problem of fog, ice, and storms, the selection of a least-time track requires climate and weather information at a number of levels. The route that a ship selects is based upon the great circle route, but with allowances for adverse weather conditions to be experienced when following that route. In bad weather, it is often advantageous and less time consuming to deviate from a stormy great circle route. To enable the best route to be determined across, for example, the Pacific Ocean, seasonal changes are first considered. The great circle route between San Francisco and Japan passes as far north as 50°N. In the winter, storminess and high seas are characteristic of this latitude. A more southerly route is thus preferable. While this adds many miles to the route, the time it takes is usually shorter—and much more comfortable. Apart from the knowledge of seasonal differences, both long-range (at least five days) and short-range (up to 48 hr) forecasts are used in determining a least-time track. The information might be used to determine how closely the great circle route might be followed; the latter gives information concerning immediate measures to be taken. The significance of least-time tracks is very well described by Alvin Moscow in his book, *Collision Course.*[3] This provides an in-depth coverage of the story behind the sinking of the Italian luxury liner S.S. *Andrea Doria*. This vessel collided with the M.V. *Stockholm* in July 1956 close to the Nantucket Lightship, despite the fact that both ships were equipped with radar and other modern navigational techniques. Moscow writes:

"Captain Calamai (of the *Andrea Doria*) had taken the shortest route across the North Atlantic, the Great Circle route, passing through the Azores Islands and heading almost due west for the Nantucket Lightship, which served as a substitute

[3] A. Moscow, *Collision Course*, Putnam, New York, 1959.

landfall for the United States. During the winter months, the *Andrea Doria*, like other Italian Line ships traveled a longer, more southerly route in an effort to follow the sun across the Atlantic. But in the summer, the Great Circle route offered sunshine as well as economy of fuel consumption. . . . The *Andrea Doria*, for instance, burned ten to eleven tons of fuel oil every hour underway, the equivalent of what the average homeowner used to heat his home for two years."

Climate not only influences shipping routes, it also needs to be considered in relation to cargo. One typical problem that has resulted in enormous losses is "sweat." This term is used by mariners to describe condensation that occurs in the holds of ships. "Ship sweat" occurs where moisture condenses on the inner structure of the ship and drips onto the cargo; "cargo sweat" is moisture that condenses directly onto the cargo itself.

The sweat problem results from the fact that air is used to ventilate holds. A ship traveling from Belem to New York in winter experiences a number of temperature and humidity regimes. Initially the air entering the hold is moist, damp, maritime tropical air. As the ship moves northward the air is chilled and considerable condensation—sweat—takes place. Damp cargoes often result in considerable losses.

Efforts to overcome this have largely been mechanical and the meteorological aspects were given little attention. For example, to overcome the corrosion problems new automobiles being shipped in the late 1940s were completely wrapped in a plastic film. For further protection, a dehydrating agent, silica gel, was enclosed with each car within the plastic. All of this was completely effective, but the cost far outran the savings and the idea was dropped. Air conditioning has been used to overcome the problem in recent years, but this itself is not entirely effective.

Just as cargo on board is subjected to climatic stress, so are the ports that handle the cargo Climate affects port facilities in a number of ways. The most obvious are the climatic extremes that result in dock work—loading and unloading. While automation is widely used in many modern ports, the facility with which the ships can be handled is certainly influenced by prevailing conditions. Excessive cold, rain and snow, and hot, humid conditions obviously

create problems. But this is a relatively minor aspect of the impact of climate. It may happen that the entire port facility may be closed for long periods because of ice or other hazards.

The role of climate in this respect is well shown by considering North Atlantic ports. Figure 10-9 shows January temperatures over the North Atlantic. The 32°F (0°C) isotherm is located almost at 40°N over the northeast United States. In Scandinavia it is located north of 60°N. Most of the ports located north of the isotherm thus experience subfreezing temperatures for much of the winter and many will be ice-bound. The location of the freezing isotherm is, of course, a partial result of the influence of heat transport by ocean currents. Norway is affected by the warm North Atlantic Drift; waters off Newfoundland are associated with the cold Labrador current.

The economic benefits of ice-free ports are illustrated by the utilization of Norwegian ports, such as Narvik, for export from Sweden. Swedish ports on the Baltic Sea, an inland sea, are frozen in winter. Iron ore from the northern ore fields in Sweden passes through the ice-free ports of Norway in winter.

Wind-driven currents are also partially responsible for problems in other areas of the world. Until 1950, the Ivory Coast, located on the Gulf of Guinea in Africa, did not have a modern port. Following the prevailing wave direction, longshore drift carried sand along the coast, a coast lacking marked indentations. Offshore bars separated the mainland port from the ocean. From 1904 to 1907, a canal was constructed through the bar; it silted up in a few months. In 1933 a ditch, 1 yd wide, was cut along the former canal to drain flood waters from the lagoon. Within a week the ditch had widened to 375 yd. Six months later, it had practically disappeared.

The most recent attempt to overcome the problem of drifting sand was the construction of the Vridi Canal. The western dyke of the canal was extended further to the sea than the eastern dyke. This created a current that directed sand to a deep offshore area, the *trou sans fond*. The canal has remained open since (Figure 10-10).

Inland waterways suffer many similar climatic problems as those of the ocean. In discussing the problems of ice formation in the Great

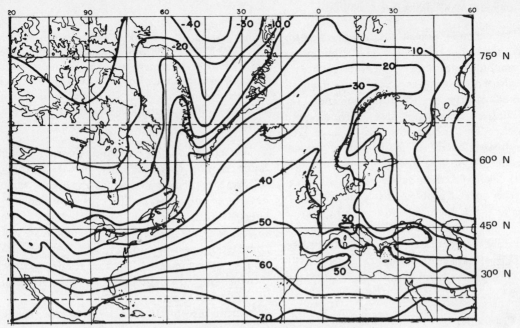

Figure 10-9. Mean January temperatures (°F) over the North Atlantic.

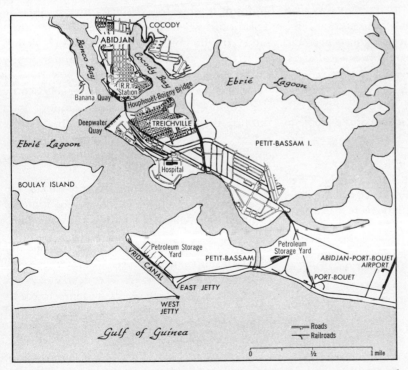

Figure 10-10. The port of Abidjan, Ivory Coast. Note the extension of the Vridi Canal's western dike. (From Hance, W. A., *Geography of Modern Africa*, 1964. Used with permission of Columbia University Press.)

Lakes, for example, Seawall (1971) notes that the ice-free period on Lake Superior is slightly over eight months although its entire surface is rarely frozen in winter. Ship movement is prevented mostly by ice formation near the shore and at constricted areas such as Sault Ste. Marie. This is not the case in Lake Erie, which freezes coast to coast; such differences reflect the role of heat storage in the lakes. Erie is much shallower than Superior.

A rather special problem that concerns waterway traffic is the fluctuation in levels of the route because of changing river levels. If the water is very low, cargo vessels cannot be loaded to their normal capacity. Floods or exceptionally high water poses obvious problems for shipping on rivers.

Land Transport

Unlike air and sea transportation, land transport relies upon fixed routeways. The significance of climate thus concerns not only the hazards met en route, but also the construction and maintenance of the route itself. It follows that engineers responsible for the selection and construction of a route must be highly conversant with the climate and climatic extremes of areas through which the route passes.

Problems associated with one climatic regime, the tundra, have already been discussed. It serves here to reiterate the nature of the problem by citing observations by Haynes (1972). He writes,

"During the winter of 1968 a road was built between Livengood, north of Fairbanks, and Prudhoe Bay (Alaska). In reality, it is little more than a path across the tundra. Ice was used to build bridges along the route. Wet planks were laid across areas to be spanned, and with water added to form layer upon layer of ice, ice bridges resulted. The bridges thaw in summer and have to be rebuilt if used, each winter. . . . It has been pointed out that the so-called Hickel Highway[4] has extensive areas where trucks have damaged the tundra insulation. These places have eroded deep ruts and ravines throughout the summer; as a result, winter highway crews were required to cut new roadbeds around the eroded sections, perhaps to perpetuate the problem. . . . After the spring thaw the land be-

comes saturated and lakes and streams appear everywhere. . . . The tundra softens, and surface transportation becomes impossible unless over specially constructed roads of gravel. . . . Tracks can become ditches that will be visible for years. In one summer an 18 inch caterpiller track can erode to a 50 foot gorge."

The effect of freezing and thawing upon roads is not restricted only to far northern realms. Most temperate regions experience some degree of frost, and this causes detrimental effects. Frost action on pavements can cause frost heave and settlement that results in loss of compaction, rough surfaces, and deterioration of the surfacing material. The resulting thaw provides both drainage problems and restriction of subsurface drainage.

In frost-susceptible soils, ice may grow in the form of lenses or veins. As the ice crystallizes, water from below is attracted to the freezing mass and increases the volume of ice. The resulting surface raising is termed frost heave. The heave is most problematic when it is not uniform, leading to extremely uneven surfaces. This occurs when subgrade material varies, from sand to silt for example, or where drains, culverts, or utility lines break the uniformity of the subgrade material.

During the thawing period, water released by the topmost melting portion cannot drain through the still-frozen lower soils. It therefore tends to move upward (Figure 10–11) to emerge through cracks at the surface and to further weaken the roadbed.

To obtain an estimation of the effects and extent of subfreezing temperatures, the highway engineer uses a number of indices. These include:

Degree-Days[5] The degree-days for any one day equal the difference between the average daily air temperature and 32°F. The degree-days are minus when the average daily temperature is below 32°F (freezing degree-days) and plus when above (thawing degree-days).

Freezing Index. The number of degree-days between the highest and the lowest points on a curve of cumulative degree-days versus time

---

[4] So named for former U.S. Department of Interior Secretary, 1969 to 1970, Walter J. Hickel of Alaska.

[5] Note the comparison with degree days of the heating engineer who uses 65°F (18.3°C) as a base (p. 284). Three types are thus distinguished: heating degree-days, freezing degree-days, and thawing-degree days.

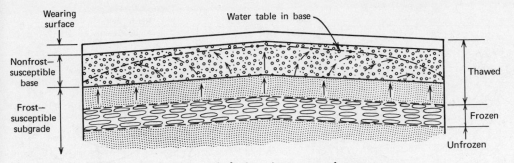

Figure 10–11. During the thawing period, there is an upward movement of moisture into the road bed. (After *Encyclopedia of Highway Design*, McGraw-Hill, 1962.)

for one freezing season. It is used as a measure of the combined duration and magnitude of below freezing temperatures occurring during any given freezing season. The index determined from air temperatures is called the *air freezing index*; that determined for temperatures immediately below a surface is the *surface freezing index*.

Mean Freezing Index. The freezing index determined on the basis of mean temperatures.

Thawing Index. The number of degree-days between the lowest and highest points on the cumulative degree-days-time curve for one thawing season. It is used as a measure of the combined duration and magnitude of above-freezing temperatures occurring during any given thawing season. Both an *air thawing* and a *surface thawing* index are used.

Mean Thawing Index. The thawing index determined on the basis of mean temperatures (Woods, 1960).

These concepts are illustrated in the graph shown in Figure 10–12, while isolines showing the distribution of the mean air thawing and freezing indices are shown in Figure 10–13. Comparison of the two provides insight into the intensity of freezing that occurs; this is obviously related to the depth of freezing when

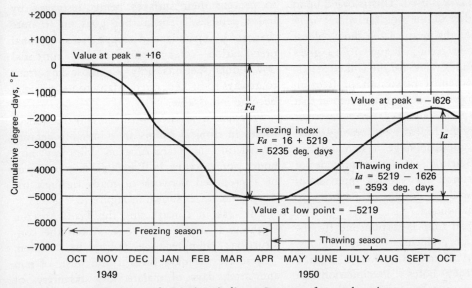

Figure 10–12. Freezing and thawing indices. See text for explanation. (After *Encyclopedia of Highway Design*, McGraw-Hill, 1962.)

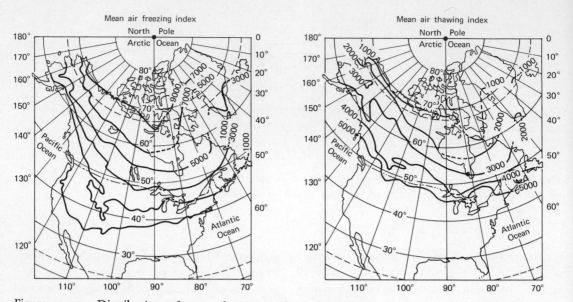

Figure 10–13. Distribution of mean freezing and thawing indices in North America. (After *Encyclopedia of Highway Design*, McGraw-Hill, 1962.)

a point is reached where the depth of summer thaw is less than that of winter freezing, and permafrost will occur.

In the design and construction of highways, emphasis must be given to soils on which the road is to be built, the temperature, particularly as it relates to depth of frost penetration, and the role of water, both in a frozen and nonfrozen state. As a guide to potential frost hazards a *Design Freezing Index* is used. This is based upon the freezing index of the coldest winter in a 10-yr period or as the mean of the three coldest winters in a 30-yr record. Figure 10–14 gives design freezing index values for the United States.

Use of the potential maximum of freezing in the design stage is typical of the method that must be used in designing drainage systems for highways. As the *Highway Engineering Handbook* notes, "A good drainage design involves an accurate prediction of the magnitude of peak rates of runoff for various intervals of expectancy as well as design of facilities to accommodate that runoff."

Rain rather than frost is responsible for the many land transport problems in tropical regions. In an evaluation of transportation in tropical Africa, Hance (1967) notes, "Precipitation is a climatic element of even greater importance, both directly and indirectly, than temperature.

In the rainforest belt, heavy year-round precipitation makes construction of roads and railways very costly. I have seen new roadbeds seriously eroded before it was possible to surface them. Laterite roads in rainforest areas fortunately tend to harden on exposure, but they cannot stand up to heavy truck traffic under the climatic conditions that prevail. In many places roads are legally closed whenever it rains to prevent their surfaces being destroyed by heavy vehicles." The author goes on to note that in the savanna and steppe areas, seasonal precipitation—often comprised of torrential downpours—permits only the most important routeways to be kept open. Many routes actually become impassable.

Problems associated with the construction of routes in tropical realms are admirably demonstrated in the Trans-Amazon Highway now under construction in Brazil. This route, part of a proposed network of roads through the Amazon basin, is shown in Figure 10–15.

The task of constructing the Trans-Amazon Highway is not an easy one. About the construction, Hummerstone (1972) has written, "These are the days of heat and dust, of rain and mud, days of malaria and dysentery, of stinging ants, mosquitoes and flies in one of the earth's most hostile regions. Infrequently

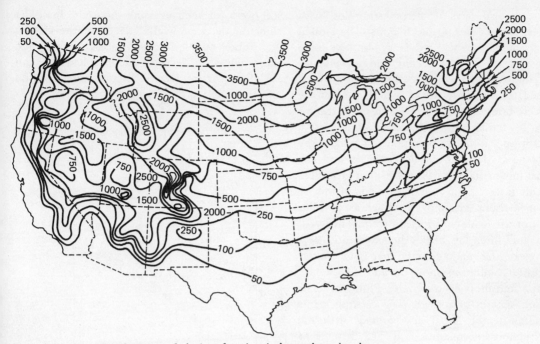

Figure 10–14. Distribution of design-freezing-index values in the continental United States. *Note:* Design-freezing-index values are cumulative degree days of air temperature below 32°F for the coldest year in a 10-yr cycle. Marked local variations, particularly in mountainous areas, may occur. (After *Encyclopedia of Highway Design*, McGraw-Hill, 1962.)

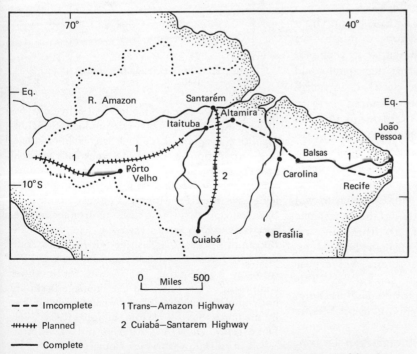

Figure 10–15. Sketch map showing highways, in various states of development, in the Amazon Basin.

there are also days of fiercer struggles, with snakes, jaguars and hostile Indians." By United States standards the road is a primitive one, consisting of a two-lane dirt and gravel construction. Countless bodies of water need to be crossed, and initially, these will be crossed by wooden bridges, or, where more than 100 yd wide, by ferries. Although the route is not a difficult one, from the standpoint of terrain, the remoteness of the construction sites and the almost impenetrable forest make it an incredibly difficult task. Initial clearance is completed by four-men teams who move as much as 50 miles ahead of the main construction team. Their task is to cut, often by hand, the enormous trees and clear the undergrowth that may occur. Thereafter bulldozers move in and remove the stumps and heavier materials. Problems exist in both spheres of activity. The clearing teams are often remote from any contact with the construction headquarters and face dangers of all types. Hummerstone reports that contact with Indians can be harrowing; he also reports that jaguars have been known to kill at least one man. The bulldozers are extremely difficult to operate in the wet periods because they are easily bogged down in the thick red mud that results from clearing of the forests.

The construction of the Trans-Amazon Highway has been criticized from a number of viewpoints. Some contend that it is economically foolish, pointing out that the capital could be used to more advantage elsewhere in Brazil. Others feel that insufficient planning, both in terms of the road itself and settlements that will result, has occurred. Ecologists, too, have expressed concern. It has been suggested, for example, that cutting back of enormous areas of the Amazon forests, which may supply as much as one-fifth of the world's oxygen, will seriously alter the composition of the atmosphere and result in extended effects. This, however, has not been substantiated. There is little doubt, though, that clearing the forests will expose large areas of lateritic soil to sun and rain. This might result in marked erosion and leaching, together with other problems of using lateritic soils (see Chapter 4).

Accepting the fact that routes can be constructed and stand up to the climatic extremes that exist in a given region, climatic hazards

still occur for vehicles using the route. In colder climates, deep snowfalls can cause the closing of a route, and it is often necessary to design snow fences to decrease the problem. Ice can also be a major hazard. Many municipalities

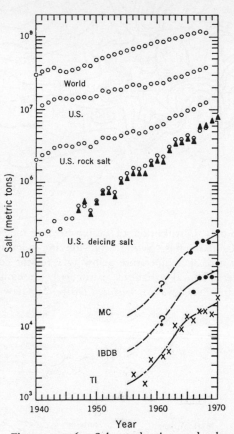

Figure 10–16. Salt production and salt usage for deicing. *Key.* World: Total salt production including, brine, evaporated and rock salt; U.S.: total national production as above; U.S. rock salt: national production of rock salt; U.S. deicing salt: national use of salt for deicing; M.C., IBDB, and TI, deicing salt used in Monroe County, Irondequoit Bay drainage basin and town of Irondequoit respectively. Open circles represent data from U.S. Bureau of Mines; triangles, the Salt Institute; solid dots, the International Salt Company; crosses, the town of Irondequoit. The records for localities comprising the county and the drainage basin are not complete, but the data shown are probably low by no more than 5%. (From Bubeck, R.C., W.H. Diment, B.L. Deck, A.L. Baldwin, and S.D. Lipton, *Science*, 172, pp. 1128-1132. Copyrighted 1971 by American Association for the Advancement of Science.)

now use salt for deicing roads, a practice that has increased tremendously over the last few years (Figure 10–16). This has, however, marked disadvantages. Brine solution is highly corrosive and wear on automobiles is increased. The salt solution, when it runs off, might cause some salinization of water supply. The extent of the problem has been investigated by Bubeck et al. (1971), who have shown that chloride concentration in a bay near Rochester, New York, has increased fivefold in the last 20 years. They attribute this to the vast amount of salt used for deicing streets in Rochester.

While in most of the foregoing account, emphasis has been on road transport, many of the problems apply to railways. Obviously, severe storms damage tracks and bridges; icing causes problems with signals; heavy snow and avalanches may close tracks. Indeed, in mountainous areas it is often necessary to use tunnels (initially to overcome steep grades) to avoid such problems. Climate must also be considered in freight handling. Perishable materials, livestock, the necessity for stockpiling, and seasonal demands are all factors influencing rail transport.

PART III

Climates of the Past and Climatic Change

CHAPTER 11

The Background
to Climatic Change

In viewing the voluminous literature concerning both climates of the past and climatic change, it is difficult to realize that only 75 years ago the very existence of great changes over geologic time was a point of disagreement. Interpretation of past climates has a history that is intimately concerned with the "Apes and Angels" controversy of the nineteenth century for, to recognize such changes, it was necessary to assume a long period of earth history. The theological arguments against evolution of life were equally effective against the assumption of evolution of the earth and its atmosphere.

The first real impetus for assuming that climates have not always been as they are today occurred in Switzerland. Here, the debris, striated rocks, and erosion features caused by ice were evident in locations well below the limit of present glaciers. In 1821, the Swiss engineer Ignatz Venitz used such evidence to suggest that Alpine glaciers were remnants of much larger ice covers; shortly after, in 1824, Jens Esmark suggested that Norwegian fjords were the results of ice gouged valleys that had been flooded. In 1832, the German scientist Reinhard Bernhardi suggested that in earlier times much of the entire continent of Europe had been covered by great layers of ice. Perhaps the most well-known advocate of the existence of earlier ice ages was Louis Agassiz. After extensive field work in Europe, he became convinced that extensive ice covers had occurred in earlier times; he introduced these ideas into the United States when he went to Harvard in 1846. Under his leadership, the concept of changes in climate over time was investigated vigorously and much evidence, particularly of ice age climates, was gathered. Not all were convinced, and as recently as 1899 there were a considerable number of scientists who felt that the evidence could be explained in terms other than climatic change.

The development of the ideas of changing climates over time has passed through a number of stages. Initially the evidence of recent glaciation of the northern hemisphere was combined with study of other geologic deposits (coral reefs, coal deposits, and evaporite beds—interpreted as the result of warm climates) to lead to the assumption that throughout most of geologic time, climate had been much warmer over the earth's surface and that the Ice Age was an isolated incident of world cooling. This idea had to be revised when evidence of older ice ages was discovered. A tillite, an indicator of glaciation, was found in Norway by Reusch in 1891. This was found above striated rocks, dated as early Cambrian, a further indication of ancient glacial activity. This Eo-Cambrian glaciation was verified when investigations in the southern hemisphere showed that glacial activity had been a recurring event throughout geologic history. The idea of a long, uninterrupted warm history was revised to allow for periodic ice ages.

Another great revision in interpreting past climates was necessary when the idea of continental drift was introduced. By accepting the theory, the study of palaeoclimates needed to be approached in terms free from prior assumptions about relative distribution of land and sea over the globe.

The theory of drifting continents was first proposed by nineteenth century geologists; Eduard Suess in particular had suggested that that the continents had, at one time, formed a single land mass, a mass he termed Gondwanaland (so named from a geologic province in India). It remained, however, for the German scientist Alfred Wegener to formulate the first comprehensive theory of continental drift. About the idea he wrote:

"The first notion of the displacement of continents came to me . . . when on studying the map of the world, I was impressed by the congruency of both sides of the Atlantic coasts, but I disregarded it at the time because I did not consider it probable. In the autumn of 1911, I became acquainted . . .

with the palaeontological evidence of the former land connection between Brazil and Africa, of which I had not previously known. This induced me to undertake a hasty analysis of the results of research in this direction in the sphere of geology and palaeontology, whereby such important con-

firmations were yielded that I was convinced of the fundamental correctness of my idea."

The inquiry led Wegener to reconstruct the continents as shown in Figure 11–1. During the late Carboniferous, land masses were attached

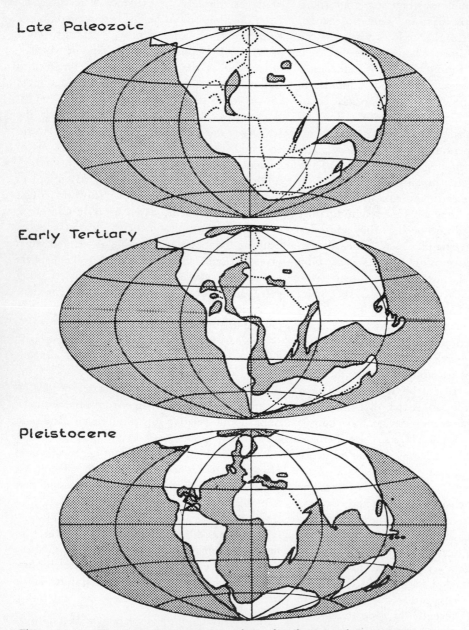

Figure 11–1. Wegener's recontruction of the distribution of the continents at three periods to illustrate his hypothesis of continental drift. Africa is shown in its present-day position merely as a center of reference. (From Stokes, W.L. and S. Judson, *Introduction to Geology: Physical and Historical*, 1968. Used with permission of Prentice-Hall, Inc., Englewood Cliffs, N.J.)

one to another in a great continent that he termed Pangaea. During the Mesozoic, the continents started drifting apart and, by early Quaternary, they resembled the pattern shown today.

Wegener's hypothesis was backed by numerous physical examples, fossil distribution, and matching mountain and structural systems on either side of the Atlantic; many scholars were convinced of its basic corrections. Critics of the concept were aided by the problem arising from explanation of the mechanism responsible for the drift. Wegener called upon the roles of gravity and centrifugal force of rotation to explain it. If a continental mass is assumed to "float" on a lower, denser material, then the continental areas will have both a center of gravity and a center of buoyancy. If these do not coincide then a remnant force will be available to provide the mechanism for movement. This mechanism is directional and the resulting force will act toward the equator. This was termed the *polfluchtkraft*—the pole fleeing force. While such a force certainly exists, various workers showed that it would never be strong enough to cause continental motion.

Continental drift became a topic of controversy between scientists. The idea was fascinating and the theory so attractive that it might be considered "geopoetry." This "geopoetry" could not become "geofact" until extensive geophysical exploration was completed and theory accordingly formulated. The development of the understanding of earth's radiogenic heat, subcrustal convection, and gravity anomalies has strongly reinforced the idea of continental drift. Indeed, modern theory of plate tectonics depends upon the idea of a motion in the lithosphere. Exploration also turned up fossil evidence indicative of areas that were once joined. From an area near the Ross Ice Shelf in Antarctica a fossil of the hippopotamuslike reptile *Lystrosaurus* was obtained. It had been thought that this animal had lived only in South Africa and Asia; its discovery in Antarctica reinforced the concept of the existence of a super continent.

Another pointer to geofact of continental drift is the findings of remanent magnetism. If the continents move, then it follows that orientation of the continents in relation to the magnetic poles must have changed over time.

Fossil magnetism shows that this is the case. Like the theory of continental drift, the idea that ferromagnetic materials in rocks would provide insight into geologic history originated in the nineteenth century. It was not until the 1940s that the idea was really thoroughly investigated.

Some materials in rocks, such as iron and nickel, become strongly magnetized when subjected to a magnetic field. In the earth, materials subjected to the earth's magnetic field should be aligned according to the location of the magnetic poles. If the magnetism remains, it will indicate the position of the poles during the time of formation; it will further indicate how the rocks have moved in relation to the former poles, or how the pole itself has drifted. Remanent magnetism can occur through a number of processes, but that resulting from the cooling of volcanic rocks provides an appropriate example.

Ferromagnetism comes into being when the exchange interaction between atoms is greater than their thermal energy. Thermal energy varies with temperature and above a critical value (the Curie point—named for Pierre Curie) thermal energy becomes greater and the material ceases to be ferromagnetic. In the formation of volcanic rocks, cooling of the molten material causes the Curie point to be reached and the material, influenced by the earth's magnetic field, becomes ferromagnetic. Even with decreasing temperatures the magnetism is strong, the reasons for this being well explained by Yakeuchi et al. (1967) who write:

"Consider a volcanic rock which has just erupted as lava. Its temperature is, of course, above the Curie point, and therefore, there is no ferromagnetic order among the magnetic moments of its atoms; the magnetic moments are oriented quite at random . . . although exposed to a geomagnetic field, the lava acquires no strong magnetism . . . (below the Curie) point, ferromagnetic control overcomes the thermal energy of the atoms; the magnetic moments of atoms begin to align parallel to one another in the direction of the geomagnetic field. It is as if a large crowd of people, hitherto uncontrolled, suddenly lined up facing one direction. Meanwhile the temperature falls lower and lower. Each atom is now, so to speak, arm-in-arm with its neighbors, facing one direction. The atoms lack the thermal energy to reverse their direction or to move at all . . . the remanent magnetism of the cooled lava is very stable. It can now be demagnetized only by raising the temperature to the Curie point—or by applying to it a particularly strong magnetic field."

Using remanent magnetism the location of the poles during geologic time can be determined.[1] Interestingly the location of the pole from North America and Europe for similar geologic formations do not coincide. This lack of coincidence can only be explained by assuming the movement of the continents.

Clearly, the location of the poles is also indicative of the location of the equator. It becomes possible to reconstruct the climatic variations that are found in given periods by assuming the relative positions of the poles and equator. Findings of remanent magnetism often point to the basic correctness of reconstructed climates derived from other forms of evidence. The location of the Carboniferous equator (Figure 11-2) provides one such example.

With the theory of continental drift now fairly well established, the interpretation of palaeoclimates can take into account different global locations as well as varying latitudinal climates. Interpretation is thus intimately related to geophysical as well as climatological findings.

THE PROBLEM OF TIME

To complete a chronology of climatic change over time, it is necessary to establish two criteria. First, the nature of the climate that existed must be determined. As shown in Chapter 12, this is completed through inference from other forms of natural phenomena. It is only very recently in, geologic terms, that man has been on earth, so that evidence from archaeology and recorded history covers only a small part of the area of concern. Once the climate is inferred, it is then necessary to locate it in the sequence of events over geologic history. Thus, understanding of the methods of historical geology is of prime concern to the palaeoclimatologist.

The development of the geologic timetable was a masterly piece of deduction. Originally divisions were based upon qualitative obser-

vations and these related to the principle of uniformitarianism. As Berry (1968) notes, "All phenomena that are related to the past history of the earth are dependent upon the principle of uniformity in nature's processes through time for their interpretation. Everything from interpreting shells preserved in rocks as remains of once-living organisms to ascertaining the passage of time by using decay rates of unstable isotopes . . . depends upon this principle." The fact that "the present is the key to the past" applies equally to climatic change as it does to geologic processes.

As Berry indicates in the preceding quotation, there are two ways of interpreting the time of the past. Time can be considered in relative or absolute terms. That a shell is found in a rock up to a given geologic period—and thereafter is no longer found—allows us to date a rock relative to the time of extinction of the shell species. Some animals had only a limited existence on earth and provide excellent indicators of a given rock series (Figure 11-3); others range over much of the geologic timetable and are of limited value in this respect. While faunal and floral records allow a relative time scale to be established, the actual dating of the absolute time needs to use specific dating methods.

The great advance in dating rocks in absolute terms came with the discovery of radioactivity and the identification of various isotopes of given elements. An atom comprising an element has a nucleus surrounded by a "cloud" of electrons. The mass of the atom is in the nucleus, which may comprise 99.9% of the total mass of the atom. The nucleus contains two types of particles, protons and neutrons. For a given element the number of protons remain the same and are expressed as the atomic weight of the element. When the mass of the neutrons are added to the protons the atomic mass of the element is determined. While the atomic number remains fixed, the atomic mass can vary; Uranium 238, for example, has 146 neutrons and 92 protons. Its atomic number is 92, its atomic mass is 238. Uranium-235 also has an atomic number of 92, but with these 92 protons are 143 neutrons, giving an atomic mass number of 235. Different varieties of the same element, such as U^{235} and U^{238}, are called isotopes.

[1] Investigation of remanent magnetism shows that the earth's magnetic field has reversed a number of times during geologic time.

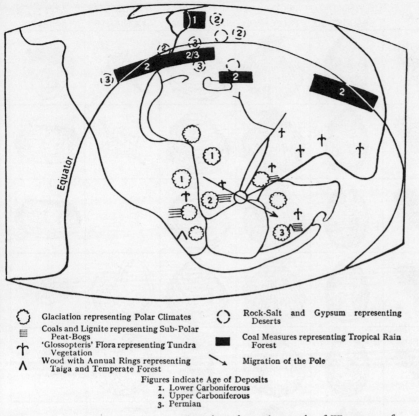

Figure 11-2. Miller's reconstruction, based on the work of Wegener, of climatic zones in Permo-Carboniferous times. (From Miller, A.A., *Climatology*, 1953. Used with permission of Methuen, London.)

Some isotopes are unstable and a small part of the nucleus may fly off, ultimately leading to the formation of a new element. Emission of alpha, beta, and gamma rays convert mass to energy in this breakdown. As the element decays, new products, called daughter products, result. The daughter product may itself be unstable and again decay. Ultimately a stable daughter product is produced. In the decay of the already mentioned U^{238}, the stable daughter product is lead-206.

Table 11-1 shows that there are a number of radioactive materials that can be used to determine the age of rocks. The age can be determined because if the rate of decay of the element is known, then the amount of new, stable daughter product in relation to the parent element gives the amount of decay. The decay is measured in half-life. For example, if we begin with 1 gram of U^{238}, then after one half-life 0.5 grams of

the material will remain. After two half-lives there are 0.5/2 or 0.25 grams remaining. Thus the decay represents the exponential decrease of the number of atoms of the parent material that remain. Table 11-1 shows that some of the half-life periods are extremely long, that of rubidium-87, for example, being longer than the history of the earth itself. Graphs indicating the rates of decay are shown in Figure 11-4. Note that in the example given, potassium-40, breaks down to calcium-40 and argon-40. As the K^{40} decreases, the calcium and argon increase.

Considering the long half-life period of most naturally occurring radioactive minerals, it is fortunate for the determination of ages of more recent materials that in 1947 Libby discovered the Carbon-14 method of dating. Above 16 km (about 10 miles) in the atmosphere, atoms of ordinary nitrogen N^{14}) are bombarded by neutrons created by cosmic rays from space.

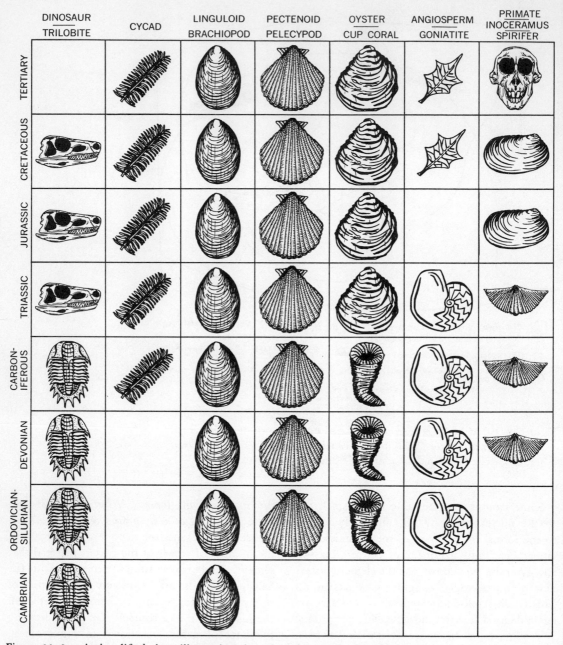

Figure 11–3. A simplified chart illustrating the principles involved in determining the correct succession of life forms on earth and how fossils are used to date rocks or periods of time. Although the drawing is diagrammatic, the individual fossil groups represented are placed according to their existence in time. For example, dinosaurs are known only from the Triassic, Jurassic, and Cretaceous, and have never been found in association with primates. (From Stokes, W.L. and S. Judson, *Introduction to Geology: Physical and Historical*, 1968. Used with permission of Prentice-Hall, Inc., Englewood Cliffs, N.J.)

Table 11-1

PRINCIPAL METHODS OF ISOTOPIC AGE DETERMINATION[a]

Isotopes	Half-Life of Radioactive Isotope, yr	Effective Dating Range, yr	Some Materials That Can Be Dated
Uranium-238/Lead-206	4.50×10^9	10^7 to T_0 [b]	Zircon Uraninite Pitchblende
Uranium-235/Lead-207	0.71×10^9	10^7 to T_0 [b]	Zircon Uraninite Pitchblende
Potassium-40/Argon-40	1.30×10^9	10^4 to T_0 [b]	Muscovite Biotite Hornblende Whole volcanic rock Arkose[c] Sandstone[c] Siltstone[c]
Rubidium-87/Strontium-87	4.7×10^{10}	10^7 to T_0	Muscovite Biotite Microcline Whole metamorphic rock
Carbon-14	5730 ± 30	0 to 50,000	Wood Charcoal Peat Grain Tissue Charred bone Cloth Shells Tufa Ground water Ocean water

[a] From Longwell et al. (1969).
[b] T_0 = age of the earth, about 4.6×10^9 years.
[c] For paleogeographic studies.
Source: After Isotopes, Inc.

The N^{14} atoms absorb a neutron and emit a proton with the resulting production of Carbon 14, an isotope of carbon. The Carbon-14 is taken up in the form of carbon dioxide by plants and animals on the earth's surface, in a fixed proportion to that of normal carbon. Upon the death of the organism, the C^{14} is no longer replenished and the C^{14} reverts to N^{14}. The proportion to C^{14} to N^{14} in the remains allows the date at which the animal died to be estimated.

As with most dating methods, there are problems in the C^{14} method. It is necessary to assume, of course, a constant rate of C^{14} production over time. If the amount does vary, then the ratios on which the dating method is based will be in error. Some discrepancies have been found in data derived from tree ring analysis and it may be that slight modifications in earth-sun relationships could cause a difference. It has also been suggested that the

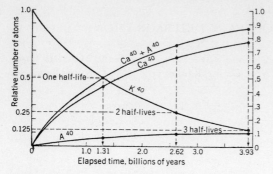

Figure 11–4. Exponential decay and growth curves for potassium-40 (K^{40}), calcium-40 (Ca^{40}), and argon-40 (A^{40}). (After Hurley, from Strahler, A. N., *The Earth Sciences*, 2nd ed., © 1971 by A. N. Strahler. Reprinted by permission of Harper and Row Publishers, Inc.)

strength of the earth's magnetic field might lead to an error of 5% in the determined results. Despite these problems, the method has proved extremely valuable in the determination of dates applicable to man's existence on earth.

THE TERMINOLOGY

Like all areas of endeavor that encompass a number of disciplines, the terms that are used in the study of past climates have become somewhat confused and in many cases, ambiguous. In an attempt to standardize terms and concepts, the World Meteorological Organization (WMO, 1966) published a work bringing together the various terms and suggested a standardization that might help overcome the problems of usage.

Figure 11–5 shows the logical relationships between recommended terms. Below each term, a symbolic graph indicates the nature of the change involved. Temperature, on the ordinate, is plotted against time on the abscissa to show how each term is used.

The diagram shows that *climatic change* is an all-embracing, general term that implies all categories of change regardless of the physical nature or time period involved. A *climatic fluctuation* (or variation) comprises any form of systematic change other than a *climatic trend*—a change characterized by smooth monotonic increase or decrease of values over a given period—and *climatic discontinuity*—a change that is rather abrupt or permanent during the period of record.

Study of climatic fluctuation is treated either in terms of *climatic oscillation* or *climatic vacillation*. An oscillation comprises a fluctuation in which the variable tends to move gradually and

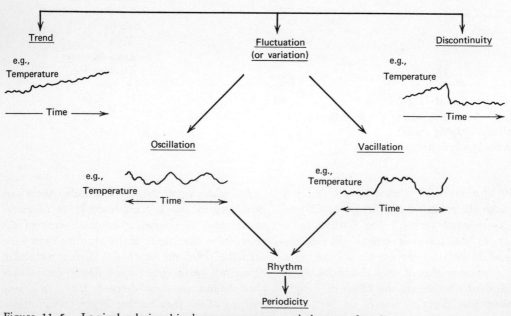

Figure 11–5. Logical relationship between recommended terms for climatic change. See text for explanation. (After UNESCO, 1966.)

smoothly between maxima and minima. This contrasts to a vacillation where the variable tends to dwell alternately around two (or more) average values and to drift from one to the other at either regular or irregular intervals. A *climatic rhythm* can be applied to vacillation or oscillation as long as the successive maxima and minima occur at about equal time periods. A *periodicity* describes the rhythm that has a constant—or nearly constant—time interval between maxima and minima.

Confusion of terminology also is found in relation to the terms used to describe the time period that is considered. This might be expected when considering the variable approaches available in climatic change. One has only to think of the term "recent" to appreciate this. To the geologist and the historian it means quite different things. It is necessary to be aware of the time scale of climatic change and to recognize, in as precise terms as possible, the period that is being considered.

Figure 11–6 shows the geologic timetable drawn on a logarithmic scale. There is little

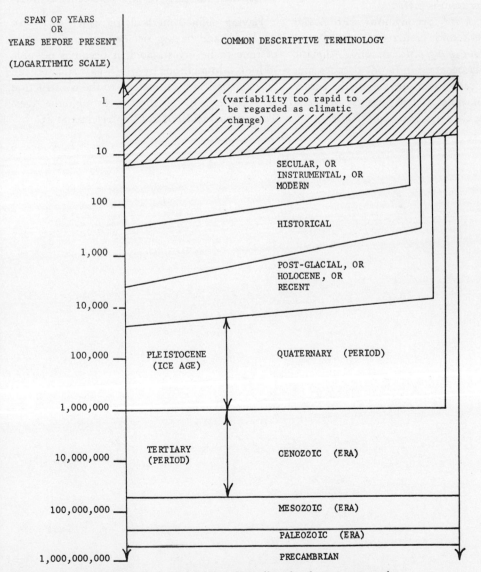

Figure 11–6. Geological timetable showing climatic change nomenclature. Note the logarithmic scale. (After UNESCO, 1966.)

problem in identifying time periods involved in old rocks and the terms Precambrian, Palaeozoic, Cenozoic, and Quaternary are quite specific. Problems arise in the terms used for describing climatic events after the Pleistocene. The WMO offers the following recommended glossary:

Historical climatic change; Historical period. Climatic change that has occurred in recent centuries (and period of time related thereto), concerning which some direct or indirect information is available from human recordings, documents, chronicles, diaries, or the like.

Holocene period. Synonymous with *Post-glacial period* (the latter term is preferred).

Instrumental period. Synonymous with *Secular period* (the latter term is preferred).

Modern climatic change; Modern period. Climatic change that has occurred in recent decades, but beginning not earlier than the twentieth century (and period of time related thereto). (Usually more restrictive than *secular climatic change.*

Post-glacial climatic change; Post-glacial period. Climatic change that has occurred in recent millenia (and period of time related thereto), following the melting of the most recent ice sheets in the Pleistocene.

Secular climatic change; Secular period. Climatic change that has occurred in recent decades or centuries (and period of time related thereto) concerning which information is available from quantitative meteorological observations at a number of geographically well-distributed observing stations. In most parts of the world the secular period is considered to have begun not earlier than the early or mid-nineteenth century.

Having defined methods by which rocks can be dated, and having decided upon the period of time to be investigated, it becomes possible to look into methods by which past climates can be estimated and, ultimately, to the construction of a chronology of climate through the ages. These are considered in Chapters 12 and 13.

CHAPTER 12

Methods
of Interpretation

To interpret climatic changes over time, a great variety of methods are available. This is to be expected, because past climates are reconstructed through interpretation of evidence derived from environmental factors other than climate itself. Insight into climates of the past can be attained only through an understanding and interpretation of the effects of climate upon given environmental systems, with the result that the study of past climates represents an area of endeavor that relies almost totally upon applied climatic principles.

The methods of interpretation often depend upon the time interval involved, and, for any given time of earth history, a variety of interpretive methods are utilized. Some methods, such as the physiologic adaptation of early plants or animals, are used to interpret ancient climates. Still other methods such as the geomorphic or geologic evidence of the work of ice can be used throughout much of the earth's history. In any analysis of past climates, whatever the time period involved, numerous methods are available, and the correlated results might allow for a fairly detailed history. While this is true, it is useful to consider the approaches systematically and such an outline is followed here.

THE FLORAL AND FAUNAL RECORDS

The study of past life on earth provides much information about the conditions that once existed. Fossils, both plant and animal, have been widely used in the reconstruction of past ecologic conditions, and a great deal can be learned about the climates under which they lived (e.g., Imbrie and Newell, 1964). Such interpretation, however, is not without problems. Nairn (1961) writes, "When the habitats of plants and animals are known it is natural to infer similar conditions for their fossilized remains. Where specific identification with living species is possible the inference seems reasonable, and

it probably also holds for closely related species in similar ecological conditions. Interest obviously centers around the plants and animals which live in a restricted environment where climate is the dominant control. Tropical plants and the coral reefs of warm seas are good examples. Many creatures are however tolerant to wide ranges of temperatures and from these and migratory animals not a great deal is to be learned. Dangerous ground is reached when assumptions are made about groups without living representatives. The rugose corals are commonly considered to have required tropical waters because of the requirements of living hexacoralids. Whether or not this assumption is correct, it is an assumption and as it is by such means that palaeontologists extend the time range over which they can contribute toward palaeoclimatology, it is essential to realize this limitation."

The Faunal Evidence

Invertebrate fossils are widely used in establishing the geological sequence of rocks (Figure 12–1), and it is natural that they prove valuable for reconstruction of past climates. While this is so, the preceding points cited must be kept in mind, because it is possible to draw biased conclusions from incomplete evidence. This may apply particularly to the invertebrates, because many fossil species lived in fresh or salt water; they were not directly affected by the atmospheric climate for the sea, lake, or river in which they lived acted as a buffer to direct exposure. Note that transport of heat in, for example, the sea, can provide relatively warm oceanic waters off a cold mainland, and this most certainly would be of consequence to sea life in that area.

It is possible, too, that the fossil record is not totally representative of life that existed under inferred conditions, because there is an inherent bias in the fossil records. Fossilization is much

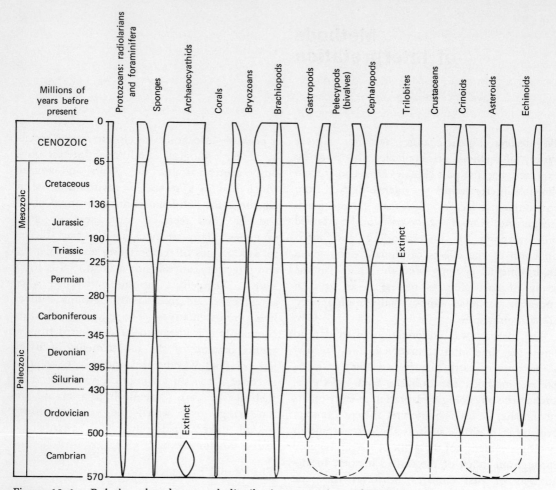

Figure 12–1.   Relative abundance and distribution over time of major groups of invertebrate marine animals. Width of band gives relative abundance of group at that time. (After McAlester, from Strahler, A.N., *The Earth Sciences*, 2nd ed., © 1971 by A.N. Strahler. Reprinted by permission of Harper and Row Publishers, Inc.)

more frequent for organisms with hard parts, and soft-bodied creatures are preserved less frequently. Since, as Newell (1959) notes, that of some 10,000 species living in a riverbank environment, only 10 or 15 are likely to be preserved as fossils in the river alluvium, the fossil record is not totally comprehensive. Note that environment also plays a role in fossilization. Animals living within a continental shelf area, for example, will produce many more fossils than those in an area (such as an upland zone grassland) where erosional rather than depositional processes are dominant.

While the physiology of fossil animals in relation to modern species are used to determine palaeoclimates, much information has recently been gained from the chemistry of invertebrate organisms. Of particular note in this respect is the work of H. C. Urey on isotopic dating.

Oxygen has three isotopes, O^{16}, O^{17}, and O^{18}, with oxygen-16, the lightest, comprising 99.7% of all oxygen molecules. In 1947, Urey presented a paper concerning the rates at which the different isotopes go off when water evaporates. Discussion following the paper led Urey, and later Emiliani, to use the observation for determining water temperatures in which fossil organisms lived. The method depends basically upon the relative abundance of O^{18} in the organism, for the amount will depend upon the ratio of O^{18} to O^{16} of the water in which the animal lived. This is a function of the temperature of the water.

Using a modern shell that had grown under known temperature conditions, isotope ratios could be established and standards acquired. The initial problems of putting the theory into practice were overcome with the development of precision instruments. At first, those available could only establish a 0.2% difference in the O^{16}/O^{18} ratio: This accounted for a 10 C (18 F) range. For the method to be practical, a value of 1°C (1.8°F) was needed to show temperature variation; this precision was ultimately achieved. The method involves heating the shell to a high temperature and then treating it with phosphonic acid. Carbon dioxide emitted is analyzed by mass spectrometer, and the ratio of oxygen isotopes is established. An early experiment using a belemnite, a cigar-shaped fossil with concentric growth rings resembling a tree, gave significant results (Figure 12–2).

Although further work has borne out the initial promise of the oxygen isotope method, there are problems that enter into its use. For example, local evaporation rates or the influx of fresh water into a depositional area will also affect the ratio. Furthermore, the shell that is used for temperature determination might well have experienced solution when, after death, it sank to the bottom of the sea; while on the ocean bed, it might also be modified by water percolating through it.

A significant study in isotopic temperature determination used cores taken from the bottom of the ocean. Study of deep-sea cores has intensified in recent years, and one of the major areas of research is analysis of the sedimentary layers within them. Such cores offer an undisturbed record of deposits from the ocean above, and, in the case in question, foraminifera within the sediments are used as a guide to prevailing climatic conditions under which they lived. Emiliani (1955), with cores from the Pacific,

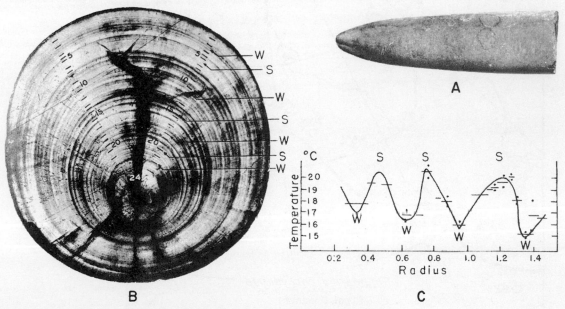

Figure 12–2. The first measurements of ancient temperatures using the oxygen isotope method was made from a belemnite. (A) Lateral view of a belemnite shell; (B) Enlarged cross section of an Upper Jurassic belemnite showing concentric growth layers; (C) Temperature curve recorded by the O^{16}/O^{18} ratios in successive layers of the shell. Evidently the creature was born in late summer and died in spring four years later (W = winter, S = summer). (After various sources, from Dunbar, C.O., *Historical Geology*, 2nd ed., Copyright © 1960 by John Wiley and Sons. Used with permission.)

Caribbean, and Atlantic, used the oxygen isotope method to establish the temperature changes of the waters. Oscillations in temperature regimes of the oceans were found; in the Caribbean, for example, the palaeo-temperatures varied by as much as 6°C (10.8°F). Using the data, a sequence of temperature change over the Pleistocene was established (see Chapter 13). The results obtained, however, were questioned by workers using similar cores, but an entirely different methodology.

Ericson (1961) and Ericson and Wollin (1964) found that temperature changes in surface water are recorded in the succession of layers of

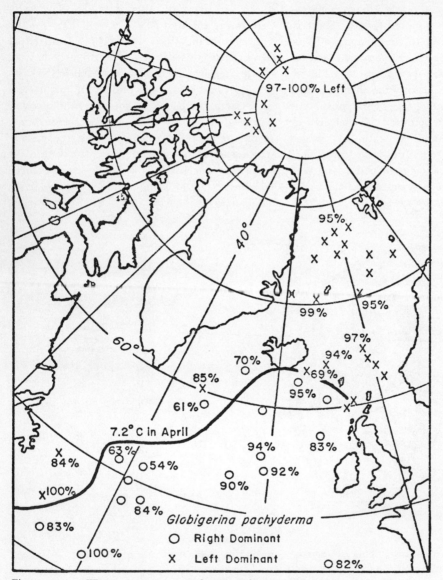

Figure 12-3. Water temperature and coiling direction of *Globigerina pachyderma*. Coiling to the right dominates in lower latitudes; coiling to left dominates in high latitudes and polar waters. The dividing line approximately coincides with the boundary between water temperatures above and below the 7.2°C (45°F) April isotherm. (From Ericson, D.B. and G. Wollin, *The Deep and the Past*, 1964. Reprinted with permission of Alfred A. Knopf, Inc.)

bottom sediments as variations in relative frequencies of planktonic foraminifera that are most sensitive to temperature. The sensitivity is gauged by the frequency of types in the topmost layers of the core. The method does have an inherent weakness in that the deeper the water, the more solution of the calcareous shells; heavy-shelled foraminifera might be found dominant, not because of relative abundance, but because lighter-shelled varieties have been dissolved. This objection is partly overcome because it is possible to compare the foraminiferal distribution of various zones in numerous cores. The relative abundance can then be used to compare the varieties within the cores, because the top and bottom sediments would have probably been subjected to the same depositional processes and the amount dissolved should be the same throughout.

In the examination of frequency, a rather important discovery was made that allowed some temperature swings to be identified with precision. Some shells of the conically spiraled *Globigerina* do not coil in the same direction. Some coil to the left and some to the right. In examining the distribution of various species they discovered that *G.pachyderma*, the only widely distributed species that lives and thrives in Arctic waters, coiled to the left in the polar realm but further south coiled to the right (Figure 12-3). The boundary where the change occurs corresponds to the 7.2°C (approximately 45°F) April isotherm.

Specimens taken from cores to the north of this isotherm coil mostly to the left. To the south, shells in the topmost part of the cores coil to the right, but between 10 and 20 cm below the tops of the cores, the coiling is to the left (Figure 12-4). This must indicate a refrigeration of waters during the ice ages and it is possible to correlate changes that occur in different places. Unfortunately, the cores containing *G.pachyderma* are not complete, so that an entire sequence cannot be obtained. Chapter 13 shows that the chronology devised using this method differs appreciably from the isotopic temperature method.

Vertebrate animals provide important clues to past climates. Much can be interpreted from their fossil distribution and their physiology as related to environment. The great changes in

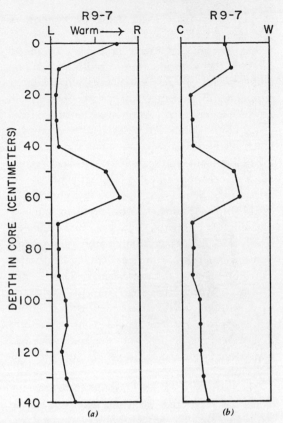

Figure 12-4. (*a*) Correlation of climate curves based on changes in percentage of left-coiling and right-coiling shells *of Globigerina pachyderma* (L = left-coiling, R = right-coiling). (*b*) The relative number of cold-water and warm-water species of planktonic foraminifera other than *G. pachyderma* in samples taken at 10 cm intervals in a deep-sea core. (C = cold climate, W = warm climate). (From Ericson, D.B. and G. Wollin, *The Deep and the Past*, 1964. Reprinted with permission of Alfred A. Knopf, Inc.)

vertebrate life over geologic time results in quite different interpretive methods.

The Devonian of the Lower Palaeozoic is sometimes referred to as the Age of Fishes, because they were one of the dominant life forms of that time. Many of the fish were air-breathing, similar to the modern lungfish. Since the conditions under which modern lungfish live are known, by analogy it might tentatively be suggested that the ancient lungfish lived under similar conditions—namely, a warm, wet-dry seasonal climate. Amphibian development, with fossil remains over widely spaced geographic areas, would tend to confirm a widespread warm climate.

Such warm conditions probably continued through the Carboniferous. Cold-blooded amphibians and reptiles, unless specially adapted, require warm air temperatures to maintain their body temperatures. Again the cosmopolitan distribution points to widespread warmth for much of the period.

The Permian, representing the rapid emergence of reptiles, with dinosaurs preeminent, is an interesting period in use of vertebrates to establish past climates. The reptilian method of reproduction—the amniote egg—allowed reptiles to live away from water sources and become much more independent of land-water distribution than the amphibians. That is not to say, however, that such animals could not exist without water. The wide distribution of reptiles in the Permian would seem to indicate that water was available on the land, and that conditions were similar over broad areas of the globe.

The extinction of the dinosaurs in Cretaceous times has promoted much discussion; a widely held view is that progressive dryness and increasing world aridity might be the cause of their extinction. This version was admirably displayed in the Disney production of *Fantasia* where, to appropriate music, dinosaurs are seen marching to their thirsty end. There are many reasons to question such an interpretation, not the least of which is that while other animals— for example, belemnites and ammonites—became extinct, related reptiles, notably crocodilians, did not. Note, too, that many plants, which are usually more responsive to changes of climatic regime, did not pass from the face of the earth.

There are many other theories to explain the demise of the dinosaurs. One, for example, invokes temperature rather than moisture change. It has been suggested that changing temperatures might influence sperm production in the dinosaurs and, because they could not reproduce, they became extinct. Other explanations call upon the effects of the Alpine orogeny. The great earth movements associated with this mountain-building period would so modify the environment of the reptiles that they could not adapt to new conditions. In investigating the extinction of the dinosaurs, Emiliani used isotopic temperatures. As shown in Figure 12–5, he found a decrease in temperature during the critical period. Unfortunately, lack of suitable fossils limited the investigation and no precise picture could be obtained.

The use of mammal fossils in determining past climates poses many problems, for homeotherms are less responsive than reptiles to air temperature. Colbert (1953) describes the prob-

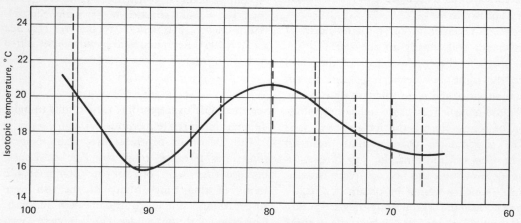

Figure 12–5. Temperature fluctuations, derived from isotopic temperature analysis at times indicated, that occurred toward time of dinosaur extinction. Maximum temperatures were reached about 80 million years ago. It has been suggested that the subsequent temperature decline contributed toward extinction of dinosaurs. (After Emiliani, 1958.)

lem as follows: "The distribution of extinct faunas containing large reptiles is an indication of the possible distribution of warm climates in past time . . . but the mammals and birds, being endothermic, are not such effective indicators of the conditions under which they live now . . . . Modern elephants are tropical and subtropical animals, but their close relatives, the extinct mammoths, were temperate and arctic animals. The same is true of modern rhinoceroses and some of the extinct rhinoceroses. Modern musk oxen are arctic mammals, but it may well be that some of the extinct musk oxen lived in temperate regions. The picture is complicated still further by the fact that many mammals and birds have very wide ranges. For instance, the modern puma or cougar of the Western Hemisphere ranges from the snows of Canada to the tip of South America, while his cousin, the Old World leopard, extends from northern China to the southern part of Africa."

While the physiology of vertebrates is probably the most widely used method for interpretation of the ecologic conditions under which they lived, important evidence is also offered by the way in which they are fossilized. Some are found, for example, "right side up," in a standing position. To explain their death, it might be assumed that they became bogged down in a swamp environment; by relating such evidence to surrounding deposits and other indicator fossils, it becomes possible to reconstruct the environments in which they lived. The fact that many fossil remains are found close together indicates death of animals through a catastrophe. Obviously, the catastrophe can take various forms—freezing or drought, for example—but by correlation to other past climatic indicators, its nature might be deduced.

The Floral Evidence

Plant distribution provides an important guide to the distribution of climate at the present time; the same is true of palaeoclimates. Identification of vegetation patterns and their changes over time are widely used to interpret past climates.

The physiology of plants, like that of animals, again provides much information. The development of drip leaves in plants is indicative of its occurrence under very moist conditions; the fossil remains of plants with thick, fleshy leaves is probably indicative of arid or semiarid climates.

The interpretation of the climate of the Carboniferous, a time of prolific vegetation that gave rise to great thicknesses of coal measures, provides examples of this use of plant physiology. Many of the fossil plants in the Carboniferous appear to be related to the horsetail (Equisetales) and the club mosses (Lycopodiales), both indicative of a marsh or swamp environment. Such an interpretation is endorsed by fossils indicating plants with layered roots such as those found in modern bog plants, and by many minor structures that suggest that some of the plants actually floated on water. Trees lacked a development of marked growth rings, indicative of a climate without marked seasonal differences, while the dominance of trees over herbaceous plants would further indicate a swamp environment. In all, the representative vegetation suggests a warm, moist climate, that favored a luxurious, if wet, plant cover.

Similar evidence has been used to help reconstruct the climates of late Palaeozoic and Mesozoic rocks. More recent deposits can now, however, be interpreted using pollen analysis and, for much more recent plant distribution, tree ring analysis is used.

The study of pollen grains or spores is termed palynology. Its success depends upon the fact that many plants produce pollen grains in great numbers (for example, a single green sorrel may produce 393 million grains and a single plant of rye, 21 million grains), and that pollen is widely distributed in the area in which plants are found. Most important, the outer wall or exine of the pollen grain is one of the most durable organic substances known. Even when heated to high temperatures or treated with acid it is not visibly changed. This is important, for pollen possess morphological characteristics that allow identification of taxonomic groups above the species level.

For pollen to be of value in interpreting the past distribution of vegetation and hence inferring the climate that occurred, it is necessary to obtain a layered sequence of the pollen. As shown in Figure 12–6, this often occurs in ancient lakes or peat bogs where seasonal pollen deposits would be covered by sediments. Subsequent cores taken would show a sequence pattern.

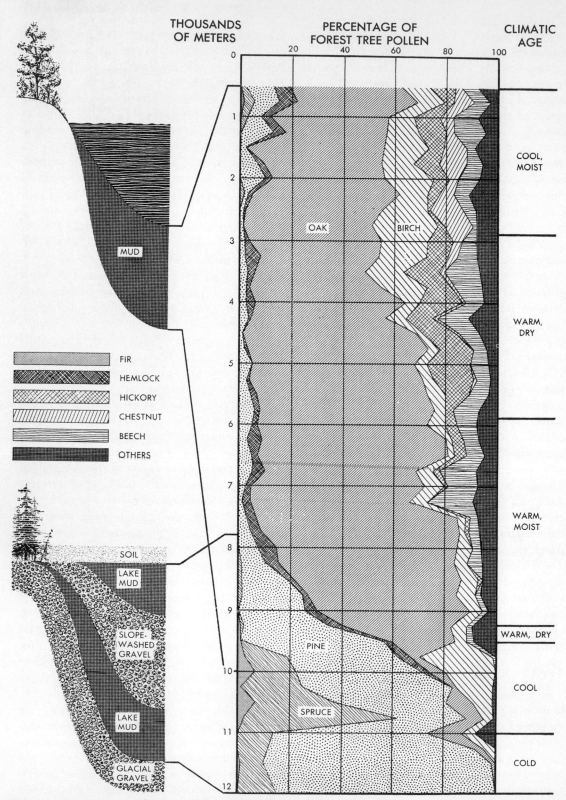

Figure 12–6. Example of interpretation of changes in climate using pollen analysis. (From Deevey, E.S., *Scientific American*, February 1952. Copyright © 1952 by Scientific American, Inc. All rights reserved.)

Pollen from the cores are identified and a frequency distribution of plant types derived. Thus, a high proportion of spruce pollen in the lower core might give way to oak pollen at higher levels. This would indicate that a vegetation change had occurred over time and that the difference could be related to a passage from cool to warmer climatic conditions. Even a relatively crude classification of pollen type, for example, those from trees compared with those from other plants, could provide a rough guide to changing climatic conditions. The change from pollen associated with the nontree climates of the cold tundra to tree climates might indicate an amelioration of climatic conditions. In-depth statistical counts obviously provide much more detail. Much work in this has been completed in Scandinavia, where the first palynological stratigraphy was devised.

Despite the important progress using this method, it does have shortcomings. A vegetation cover only attains its climax after a fairly lengthy period of time, and it is quite feasible that the vegetation established through pollen analysis represents a successional stage that is not totally indicative of the prevailing climate. In some areas the vegetation cover is mixed, and it becomes difficult to establish any dominant type that can be related to climate. It has also been pointed out that, from Neolithic times, man has interfered extensively with the forest cover, and man-induced changes might well give misleading results.

Tree ring study was pioneered by A. E. Douglass and his colleagues at the University of Arizona. Initial studies were used in an attempt to relate seasonal growth of trees to sun spot cycles, and a great deal of significant work was completed (Douglass, 1936). In the quest for ancient living trees, the *Pinus aristata* was found to be 4000 years old. Analysis of such ancient trees permitted reconstruction of climates of the American southwest during various settlement periods.

Tree ring analysis depends upon the fact that growth rings record significant events that happened during the life history of the tree. Growth rings are formed in the wood of xylem of the trees; early in the season, the xylem cells that are produced are large and light in color. Late in the growing season, the cells are smaller and darker. The abrupt change from light- to dark-colored rings delineates the annual increments of growth. Study of these rings, their size and variations, provides information about the varying environmental conditions to which the tree was subjected and provides a field of investigation known as dendrochronology. The method is, of course, most valuable in determining conditions that existed in a relatively recent part of geologic history and is widely used in archaelogical research.

While precipitation and temperature do explain much of the variation in tree rings (Figure 12–7), Fritts (1963) has pointed out that many other variables enter into the anatomical response. A large ring, for example, might result from any one, or any combination of the following:

a. Adequate soil moisture to offset transpiration losses.

b. Adequate sunlight for photosynthetic production of basic food materials.

c. Sufficient leaf area so that photosynthesis can support its food-using tissues.

d. Sufficiently high temperatures for rapid respiration rates.

e. Adequate minerals for assimilation needs.

If one or more of these factors are limited for a prolonged period during the growing season, a small ring will result. Thus the interpretation of large and small rings is not a simple one-to-one relationship between moisture deficiency or an exceptionally cold growing season; instead it is the response to a whole set of environmental conditions. Consider, for example, a tree in a semiarid environment. The size of the tree ring depends upon available soil moisture; it is quite possible for a year with normal precipitation to include a long dry spell during the growing season. This will result in a lack of soil moisture and the ring growth will be modified. Despite an annual precipitation near normal, the distribution might cause the tree to lack an entire ring, or be interpreted as a "dry" year.

In tree ring studies, emphasis must be given to statistical models for actual interpretation. A number of models for analysis of tree rings have been suggested (e.g., Matalas, 1962) and a methodology has been outlined so that the ring sequence can be used for correlation with other

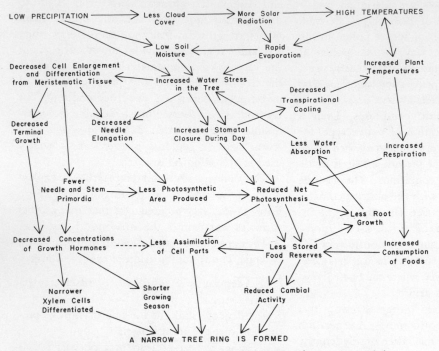

Figure 12–7. Schematic chart showing relationships between precipitation and temperature that results in narrow tree-rings in a warm, semiarid climate. (From Fritts, H.C., in *Encyclopedia of Atmospheric Sciences and Astrogeology*, R.W. Fairbridge, ed. © 1967 by Litton Publishing, Inc., reprinted by permission of Van Nostrand Reinhold Co.)

specimens. This is explained by Fritts (1967), who, in the same paper presents a series of maps showing how tree ring analysis can be used to indicate climatic change over time. Figure 12–8 shows examples of these; the maps are part of a sequence extending from 1501 to 1940 showing the regional variation in climate based upon 10-year relative departures in tree ring indices.

Although, as suggested, tree ring analysis is most often used in interpreting climates in archaeological terms, it is also a useful tool for the study of very recent climatic changes. This factor is well demonstrated in the Laporte precipitation anomaly discussed earlier. Evidence from tree rings did show marked variations in precipitation and partially helped to establish that an anomalous precipitation distribution occurred.

THE SEDIMENTARY RECORD

Sediments deposited during geologic time offer evidence of the climatic environment in which they were formed. This fact has already been stressed, because floral and faunal evidence used to interpret climates are only available when conditions are right for their fossilization, and this in part depends upon the sedimentation that is going on. Apart from fossil evidence, sedimentary beds themselves provide information. The interpretation of such evidence relies upon the doctrine of uniformitarianism, which assumes that present processes provide a guide to those active over geologic time. It follows that if a sedimentary rock is identified as having the same physical and chemical characteristics as one now being formed under, for example, arid conditions, then the sedimentary rock can be assumed to have formed in a similar way. Proof of this is offered by evaporite deposits.

Salt deposits can be formed when, on a long-term basis, evaporation exceeds precipitation but where water is available from other sources. Water will evaporate to leave salts, formerly in solution, as deposited sedimentary rocks.

In interpreting the climate responsible for the great Permian evaporite beds of Texas, Germany, and the U.S.S.R., it is fortunate that similar deposits are currently being formed.

Gypsum and halite are being actively deposited in the Sechura Desert of northwest Peru, where a 20-km marine inlet leads to rapid evaporation and eventual deposition of salts. A similar example is found in the Gulf of Akaba, on the eastern side of the Sinai Peninsula, where periodic flooding by the sea results in the deposition of gypsum and halite. The Gulf of Karabugas, a basin on the eastern side of the Caspian Sea, again has deposits of gypsum and halite, although in winter magnesium sulfate is deposited. These present-day examples provide evidence about the environment in which great salt deposits formed in geologic times. It is evident that arid, hot conditions are necessary for their formation. Thus the distribution of evaporite beds allows the reconstruction of the extent of ancient deserts. Figure 12–9 shows the distribution of Permian evaporites and the way in which these can be located in relation to the Permian equator. After examining the extent of evaporites and mapping the distribution of ancient deserts, Green (1961) notes, "It appears that the mean position of the warm arid climatic belt has moved equatorward at a mean rate of 5° of latitude each 100 million years, and that the warm arid climatic belt has always lain grossly parallel to the earth's present equator."

It is unfortunate that, unlike the evaporites, there are no well-documented, present-day equivalents of the red beds. Throughout many geologic periods, extensive beds of sedimentary rocks have been deposited whose most marked surficial feature is a red coloration. Such red beds were among the first geologic deposits to be used as climatic indices. The interpretation of the climates under which they were deposited is, however, quite variable. Early European writers, strongly impressed by the association between red beds and evaporite deposits of the Permo-Trias, assumed that the red coloration was indicative of a desert environment. In North America, the bias was different. The red soils of warm, humid climates caused their interpretation to be in terms of lagoonal or deltaic deposits, under warm conditions suitable for oxidation. Over time, other interpretations

Table 12–1

PAPERS PUBLISHED CONCERNING THE ENVIRONMENT AT PLACE OF DEPOSITION OF RED BEDS (1831 TO 1960)[a]

	Number of Papers Describing Environment as				
Decade	Marine-Estua-rine-Lagoonal	Lacus-trine	Desert	Glacial Border Con-glomer-ate	Fluvia-tile-Delataic
1831–1840	3	1	—	—	—
1841–1850	4	—	—	2	—
1851–1860	1	3	—	1	—
1861–1870	2	—	—	—	1
1871–1880	2	4	—	3	1
1881–1890	5	4	1	1	1
1891–1900	2	2	2	2	4
1901–1910	7	2	7	—	7
1911–1920	4		7	1	13
1921–1930	5	2	7	2	12
1931–1940	4	3	5	3	5
1941–1950	1	3	4	—	9
1951–1960	7	1	4	—	13

[a] Derived from data published by Nairn (1961).

were added until, as indicated in Table 12–1, the numerous papers published between 1830 and 1960 can be classed in five broad categories.

Obviously, there is no single interpretation of red beds available. Their color alone does not allow them to be categorized into any single mode of origin. The deposits in question must be analyzed for chemical and physical properties and then related to surrounding beds—both laterally and vertically—to obtain a guide to their modes of origin. Some of the climatic inferences that can be drawn, given the analytic conditions stated, include:

a. Red beds associated with coal measures probably accumulated in warm, humid climates where oxidizing conditions prevailed to preserve the red detritus.

b. Red beds deposited in piedmont valleys and rift valleys probably accumulated in tropical to subtropical environments. This is partially borne out by related flora and associated evaporite formations indicative of a seasonally dry climate.

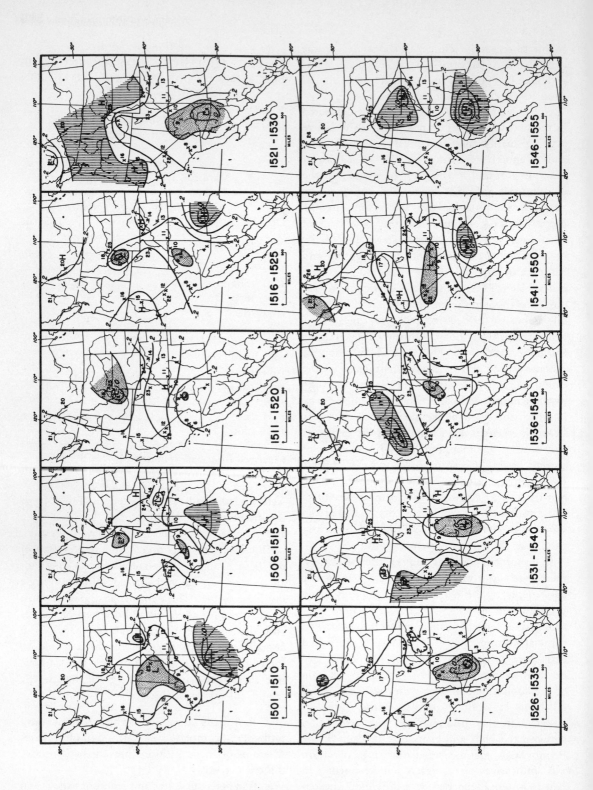

c. Red beds that accumulated over extensive areas in association with coastal plains or tidal flat areas probably developed under dry conditions. This is partially indicated by widespread evaporites and wind-deposited sediments.

The latter point, the association between red beds and evaporites, might seem highly indicative of markedly arid climates, because, as already noted, extensive evaporite deposits are often indicative of desert environment. Unfortunately, this is not always true. Van Houten (1961) writes "Lateral gradation of red bed-evaporite sequences and aeolian deposits and the repeated replacement of red bed environments by windswept deserts, as depicted during the late Triassic and Jurassic time in the western United States . . . resulted from successive shifting of environments rather than from alternating climates . . . . The rapidity with which a desert environment can overrun a vast alluvial plain is illustrated by the transformation of the western part of the well-watered and wooded Indo-Gangetic plain into the Rajputana (Great Indian) Desert within the past several thousand years."

While this might be true, the nature of wind-deposited rocks offers good evidence of ancient desert conditions; aeolian sands, in fact, provide one of the best indicators of the existence and distribution of ancient deserts. Of course, dune sands can be deposited in environments other than the desert (e.g., coastal dunes) but it is possible to differentiate the types. In desert sands, wind ventifacts, erosion features, and evaporite deposits offer confirming evidence; furthermore, because wind is a highly selective sorting agent, the size of particles can be distinguished from river-deposited sand. While both beach sand and desert dunes consist of similar grain sizes, the different environments under which they formed can be determined from the nature of their stratification. Desert dunes often dip steeply (33°), while beach sand dunes rarely exceed 12° (Nairn, 1961).

Analysis of modern dune movement and location shows that large dune formation and migration is directly related to the global winds and can thus be related to major circulation patterns. It might be assumed that analysis of ancient dunes can provide information about ancient wind systems. As shown in Figure 12-10, it is possible to reconstruct ancient winds and their prevailing direction. Such reconstruction is aided by evidence including:

a. The form of the dune itself; depending upon the type of dune (e.g., transverse, barchan, longitudinal), it is possible to determine the windward and leeward sides.

b. Sandy areas need a source for their supply, and it is possible to relate the major wind direction by ascertaining the source of the material. Although not always well defined, the source may also be verified by the fact that grain size of eroded material decreases away from the source area.

c. The polished surfaces resulting from wind abrasion are most marked on the windward side of abraided faces.

Desert dunes are not the only features of wind deposition that allow reconstruction of past wind systems. Much work has been completed on loess deposits, particularly those derived from glacial material. Kukla (1970), working on the European loess deposits, has devised an analysis that helps to elucidate the chronology of the Pleistocene in that area. Tentative correlations have also been established for American loess (Wright and Frey, 1965).

Another form of sedimentation that has proved useful in establishing the sequence of past climates are varves, deposits that reflect a

Figure 12–8. Extract from a series of maps (from 1501 to 1940) showing regional variations in climate based upon 10-year relative departures in tree-rings in the western United States. Positive departures indicate cool, moist areas; negative departures indicate warm, dry areas. *H* (shaded) and *L* (stippled) show areas of high and low growth, respectively, where departures exceed 0.6. (From Fritts, H.C., in *Encyclopedia of Atmospheric Sciences and Astrogeology*, R.W. Fairbridge, ed. © 1967 by Litton Publishing, Inc., reprinted by permission of Van Nostrand Reinhold Co.)

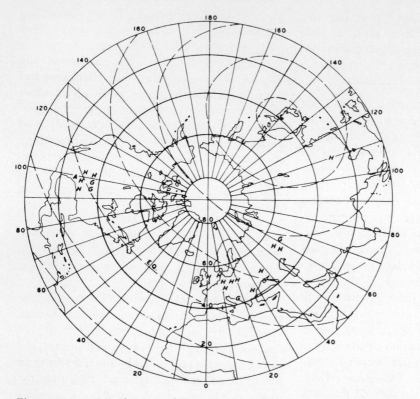

Figure 12–9.   Distribution of Permian evaporites in the northern hemisphere. Dashed lines are lines of latitude (20° interval) relative to the palaeomagnetic pole. Key: *A* = anhydrite; *H* = halite; *G* = gypsum. (From Green, R., *Descriptive Palaeoclimatology*, A.E.M. Nairn, ed., Copyright © 1961 by John Wiley and Sons. Used with permission.)

yearly cycle of events. Most work on varves has been completed on the light- and dark-colored sediments deposited on the floors of marginal glacial lakes. The alternating coloration and texture provide a history of seasonal events in the life of the lake, for each varve represents the deposit of one year. The light-colored layer of silt settles from suspension during a warm (summer) period when surplus water from the melting of ice and snow result in a relatively thick layer of sediments. In winter, when the lake is frozen over, the water is still and a darker layer of finer material, often organic debris, is laid down.

Records of hundreds of years may accumulate. The most complete sequences are found in lakes formed as a continental glacier retreat. Very old varves are found at the maximum extent on the ice, the most recent at the edge of the present-day glacier. In Scandinavia a

connection between recent and older varves has allowed a sequence going back 15,000 years to be established.

THE GEOMORPHIC EVIDENCE

Geomorphic interpretation of past climates, while of value in assessing ancient rocks, is probably most applicable to Cenozoic events. Landforms undergo constant modification and it is only where relic features are preserved that geomorphic interpretation is of vital evidence. Such relic preservation is beautifully demonstrated by the discovery of glacial grooves in the Ahaggar Region of the Sahara. Formed during the Ordovician period, such evidence points to past location of the ancient South Pole over parts of the Sahara (Fairbridge, 1971).

Geomorphic evidence was of consequence in initial deductions concerning the very existence

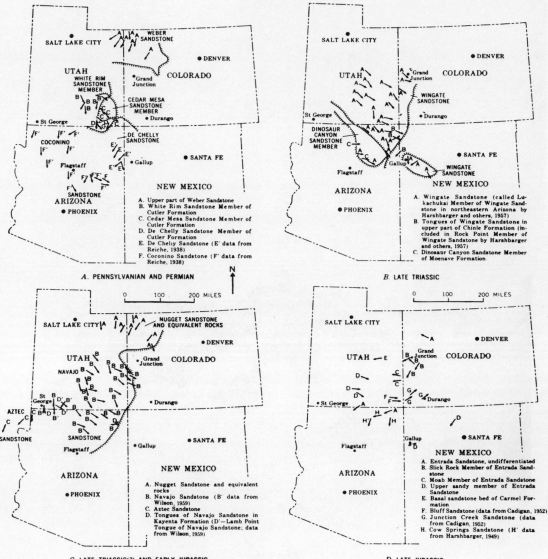

A. PENNSYLVANIAN AND PERMIAN

A. Upper part of Weber Sandstone
B. White Rim Sandstone Member of Cutler Formation
C. Cedar Mesa Sandstone Member of Cutler Formation
D. De Chelly Sandstone Member of Cutler Formation
E. De Chelly Sandstone (E' data from Reiche, 1938)
F. Coconino Sandstone (F' data from Reiche, 1938)

B. LATE TRIASSIC

A. Wingate Sandstone (called Lukachukai Member of Wingate Sandstone in northeastern Arizona by Harshbarger and others, 1957)
B. Tongues of Wingate Sandstone in upper part of Chinle Formation (included in Rock Point Member of Wingate Sandstone by Harshbarger and others, 1957)
C. Dinosaur Canyon Sandstone Member of Moenave Formation

C. LATE TRIASSIC(?) AND EARLY JURASSIC

A. Nugget Sandstone and equivalent rocks
B. Navajo Sandstone (B' data from Wilson, 1959)
C. Aztec Sandstone
D. Tongues of Navajo Sandstone in Kayenta Formation (D'—Lamb Point Tongue of Navajo Sandstone; data from Wilson, 1959)

D. LATE JURASSIC

A. Entrada Sandstone, undifferentiated
B. Slick Rock Member of Entrada Sandstone
C. Moab Member of Entrada Sandstone
D. Upper sandy member of Entrada Sandstone
E. Basal sandstone bed of Carmel Formation
F. Bluff Sandstone (data from Cadigan, 1952)
G. Junction Creek Sandstone (data from Cadigan, 1952)
H. Cow Springs Sandstone (H' data from Harshbarger, 1949)

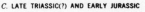

Figure 12–10.   Inferred wind directions during late Palaeozoic to Mesozoic times on the Colorado Plateau. Arrows indicate general direction of wind; hachured lines show approximate limits of deposition with barbs on side of nondeposition or erosion. (From Poole, F.C., *Problems in Palaeoclimatology*, A.E.M. Nairn, ed., Copyright © 1964, John Wiley and Sons. Used with permission.)

of glaciation in past times. The results of ice erosion and deposition are marked characteristics of large parts of the northern hemisphere. Figure 12–11 shows a schematic representation of the evolution of an area that has experienced mountain (Alpine) glaciation. The resulting features—the U-shaped valleys, the hanging troughs, the aretes, and the tarns—are well known and clearly indicative of ice activity.

As shown in Figure 12–12, glacial features associated with continental glaciation differ markedly from these of the Alpine. Depositional features such as the position of terminal moraines and the erratic boulders (Figure 12–13) are

(a)

Horn

Arête

Col

Cirque

Névé

Truncated
spur

Lake

Lateral
moraine

Medial
moraine

(b)

Tarn

Hanging
trough

Glacial
trough

Hanging
valley

Alluvial
cone

(c)

340

studied in detail and allow reconstruction of the extent and movement of the ice. The erosional features, glacial striations, ice-gouged lakes, and so forth, provide similar evidence.

While the results of the work of ice is an important interpretive device, glaciers themselves provide a measure of predicting temperature and precipitation conditions. The advance or retreat of glaciers has been used in interpreting climatic change over historic periods. About 3000 B.C. there is evidence that glaciers had shrunk and that Alpine snow level was at least 1000 ft higher than today. By 500 B.C. there was a marked readvance followed by a recession. Between the seventeenth and nineteenth centuries, a general resurgence was observed in the Alps and Scandinavia. The twentieth century has tended to be a time of glacier retreat, although, as Field (1955) notes, "At present we seem to be in an in-between stage . . . some glaciers are growing; others are disappearing."

It is not, however, features associated with the ice itself that supply all geomorphic evidence regarding ice ages. Areas not directly affected by the great ice sheets experienced a markedly different climate from today. Some areas, now quite dry, might have experienced much wetter climates as a result of modified circulation patterns. Many inland basins, for example, have been occupied by large lakes as a result of the higher precipitation. Such pluvial lakes, so named because they resulted from increased precipitation in earlier times, have been widely identified.

The western part of the United States, particularly in the Great Basin area, shows fine examples of the extent of such lakes. In this area it is estimated that as many as 120 lakes may have formed during the Pleistocene; two of them, Glacial Lakes Bonneville and Lahontan were enormous. At its maximum, Lake Bonneville occupied 19,940 square miles, an area approximately the size of Lake Michigan. Evidence of the extent of this great lake is found

in the present Bonneville salt flats and in the strandlines, the shore areas indicative of the former level of the lake.

Such lakes, of necessity, would modify the drainage systems and the formation of a lake, and its eventual overflow might lead to a totally new drainage direction. The diagrammatic sketch in Figure 12–14a shows how this might occur. Actual examples of such diversions are found in many places. In England, for example, large areas have a drainage that results from ice interference (Figure 12–14b).

Evidence of the height of former water levels also occurs along coastlines and interpreting the changing levels of the sea is an important part of establishing the extent of former warm and cool times. Obviously the changing level of the sea requires more than geomorphic investigation to find out what happened, and other interpretive methods are used.

Changes in the level of the oceans can occur either when the volume of water or the volume of the ocean basins decrease or increase. Many factors can cause either of these to occur, but most of them, such as the accumulation of sediments on the ocean floor or the extrusion of igneous rocks into the oceans, take an exceedingly long period of time. Rapid changes result mostly from the alternating accumulation and melting of ice. It has been suggested that if the present Antarctic ice sheet were to melt there would be a worldwide increase in water sufficient to cause a sea level rise of 60 m (200 ft). The melting of ice from a large area would, however, be accompanied by an isostatic readjustment of the land, while downwarping would occur in the oceans because of the weight of water added. Even allowing for this, melting of the Antarctic ice would still cause a rise of 40 m (135 ft), sufficient to flood most of the major ports of the world.

The rise and fall of sea level during the Pleistocene is an important guide to the periodic glacial and interglacial events. Marine terraces,

Figure 12–11.   Evolution of an area experiencing Alpine galciation. (a) Preglacial landscape is typical of that resulting from fluvial erosion under humid conditions. (b) Growth of mountain glaciers give rise to a new set of landforms. (c) After ice recedes, typical scenery associated with Alpine glaciation results. (From Strahler, A.N., *Physical Geography*, Copyright © 1969 by John Wiley and Sons. Reproduced by permission.)

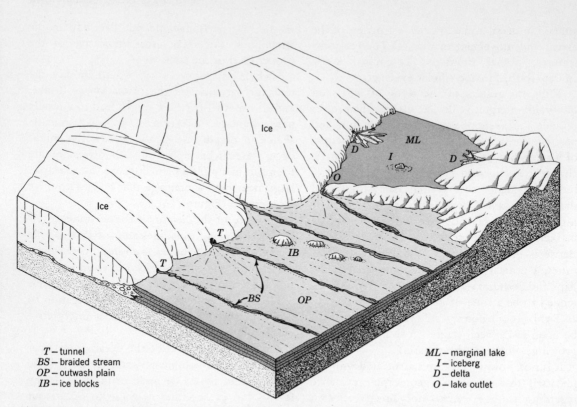

T — tunnel	ML — marginal lake
BS — braided stream	I — iceberg
OP — outwash plain	D — delta
IB — ice blocks	O — lake outlet

A. With the ice front stabilized and the ice in a wasting, stagnant condition, various depositional features are built by meltwaters.

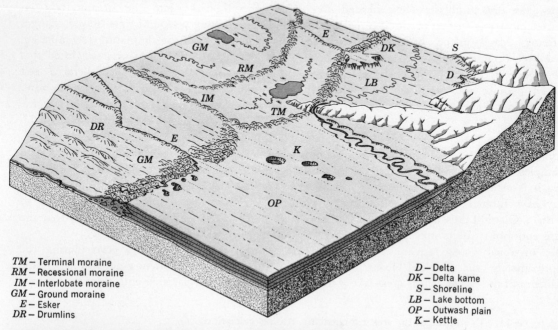

TM — Terminal moraine	D — Delta
RM — Recessional moraine	DK — Delta kame
IM — Interlobate moraine	S — Shoreline
GM — Ground moraine	LB — Lake bottom
E — Esker	OP — Outwash plain
DR — Drumlins	K — Kettle

B. After the ice has wasted completely away, a variety of new landforms made under the ice is exposed to view.

Figure 12–12.   Marginal landforms associated with continental glaciation.
(From Strahler, A.N., *Physical Geography*, copyright © 1969 by John
Wiley and Sons. Reproduced by permission.)

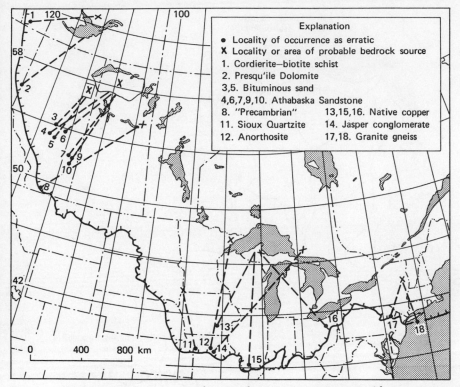

Figure 12–13. Paths of erratics can be traced to source areas to provide a guide to ice movement. (From Flint, R.F., *Glacial and Pleistocene Geology*, Copyright © 1957 by John Wiley and Sons. Used with permission.)

submergence and emergence of coastal areas all point to the amount of water tied up as ice; using appropriate dating methods, such evidence provides a guide to glacial advance and retreat. An excellent account of the significance of the changing levels of the sea has been given by Fairbridge (1963); the importance of such changes is discussed in Chapter 13.

Geomorphologists have also looked into relic landforms as guides to interpretation of past climates. One method of approach looks to the end product of erosion under different climatic conditions. Bloom (1969) writes, "In an attempt to describe the form of landscapes that have undergone long-continued weathering and erosion in humid regions, W. M. Davis in 1889 introduced the elegant work *peneplain*. He used as a root the word plain, in the geographical context of a regional surface of very low relief, near sea level. Realizing that the base level is the limit of subaerial erosion, which like a mathematical limit is approached but not achieved,

he prefixed the word plain by the Latin derivation 'pene' meaning almost." So defined, a peneplain is the end product of erosion in a humid climate. But, by analogy, a peneplain can also be attained in climatic conditions other than humid. As shown in Figure 12–15, it becomes possible to identify the "end" product of erosion under different climatic conditions:

a. The humid peneplain consists of low, convex hills with broad concave valleys.

b. In semiarid regions a peneplain might be referred to as a pediplain, indicating the formation of wide pediments ending in outcrops of resistant rock.

c. Arid climates might provide the "P'iang Kiang" landscape, derived from the Mongolian desert, which consists of large deflation hollows surrounded by alluvial fans, cliffs, and talus slopes. Such scenery is not determined by a base, such as sea level, but instead depends upon ground water levels.

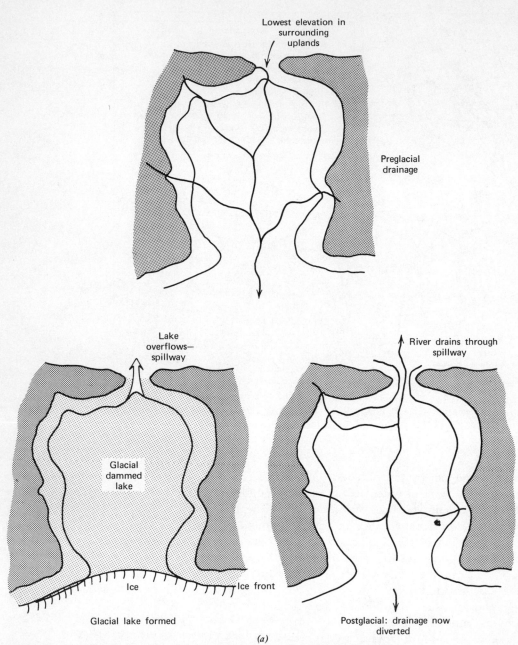

Figure 12–14. (a) Schematic representation showing diversion of drainage resulting from ice interference with preglacial drainage conditions.

d. Tropical regions, especially those of the tropical wet-dry climates, produce sugar-loaf hills or inselbergs, the equivalent of the residual hills of the peneplain.

e. Permanently frozen areas, outside ice caps, are subjected to solifluction and mass wasting. The initial irregular topography becomes filled with thick debris above which rise ridges strewn with frost-wedge fragments.

While theoretically valid, such interpretation is not totally practical, because climate change itself often prevents peneplanation to occur under a single climatic regime.

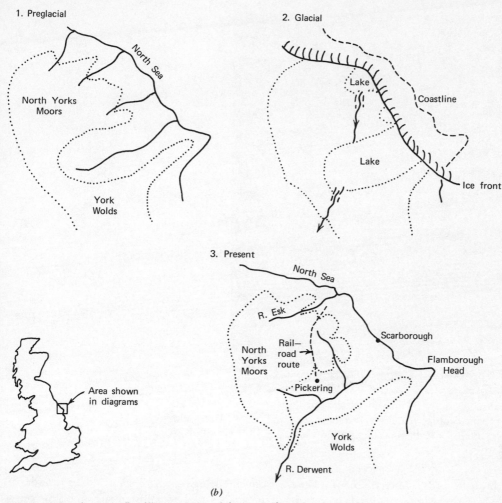

1. Preglacial

North Sea

North Yorks Moors

York Wolds

2. Glacial

Lake

Coastline

Lake

Ice front

3. Present

North Sea

R. Esk

North Yorks Moors

Rail— road route

Scarborough

Flamborough Head

Pickering

York Wolds

R. Derwent

Area shown in diagrams

(b)

Figure 12–14 (*continued*). (*b*) Example of drainage diversion in northeast England.

The nature of weathering under a given climatic regime might also be assessed if fossil soils are present. It has already been noted that red beds are used to interpret climates; the same is true of fossil red soils, the laterites. A well-known example of lateritic weathering in an area where it now does not occur is in the Giant's Causeway of northern Ireland. Here a thick laterite occurs between tertiary basalt flows. In Bermuda, eolian sands alternate with palaeosols, providing evidence of alternating climates during the Pleistocene. Controversy does exist concerning the mode of formation of these, but the marked variation of deposition, contrasted to soil formation, does allow a Pleistocene sequence to be established.

The evaluation of palaeosols depends upon the pedogenic regimes outlined in Chapter 4. Evidence of cold climates is indicated by the influence of frost and extent of permafrost; evidence of dry climates is indicated by fossil evidence of salinization. While much has been written on fossil soils, much remains to be clarified, particularly in pre-Pleistocene deposits (Kubiena, 1953).

EVIDENCE FROM THE HISTORICAL PERIOD

Many workers have used historical records to establish climatic changes that have occurred during man's brief existence on earth (Lamb,

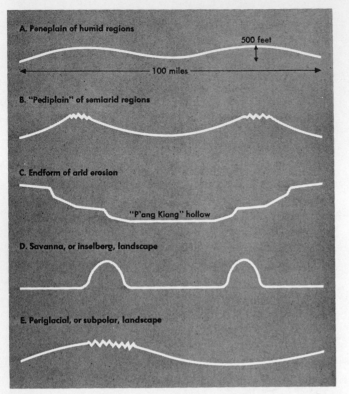

Figure 12-15. The peneplain and its climatic variants. See text for ex - planation. (From Bloom, A.L., *The Surface of the Earth*, 1969. Used with permission of Prentice-Hall, Inc., Englewood Cliffs, N.J.)

1969, Fairbridge, 1967, Butzer, 1964). Their findings have allowed fairly detailed reconstruction of climates over the past 6000 years. The type of evidence used is again highly variable and the following merely provide some guidelines to the methods.

a. Records of floods. In commenting upon the significance of floods, Fairbridge writes "A deluge such as that described in the Book of Genesis occurs in the legends and folklore of almost every ancient people. The Greeks told how Deucalion, son of Prometheus, was forewarned by his father and thus survived the flood wrought by Zeus to destroy mankind. Babylonian scripture related how the city was founded on a site where the god Ea conquered the floodwaters gushing from the mouth of the nether world, which he sealed with a giant stone. The magnificent bas-reliefs of the Cambodian temple at Angkor Wat illustrate the Hindu legend that relates how 'Manu . . . was saved

from the deluge . . .'." Such evidence cannot, of course, be used to substantiate flooding on a world scale, but many writers have related the common theme to great changes in sea levels that occurred in former times. More scientific evidence of climates can be derived from Egypt where careful records of the Nile floods have been kept for many thousands of years. By relating the occurrence and height of floods it is possible to reconstruct the pluvial conditions in the source areas of the Nile. In more recent times, evidence is available from western Europe, particularly Holland, where the level of the sea and the amount of flooding is of great concern.

b. Records of droughts. The periodic effects of great periods of drought have frequently been used in reconstruction of past climates. As an example, the growth and abandonment of large dwellings in the southwest United States have been attributed to drought. The abandoned dwellings of the Chaco Canyon and the Mesa

Verde country both show that they supported quite dense, highly active communities. By 1300 A.D., large settlements had become deserted. Reasons other than drought, have been suggested to account for the abandonment, but tree ring analysis tends to confirm that progressive aridity was a basic cause. The analysis shows that between 1276 and 1299 A.D., practically no rain fell in the areas. Droughts such as this have been used to explain mass migrations to less arid areas.

c. *Large-scale out-migration.* The mass migration of peoples from steppe to moister areas has certainly occurred over history. The interpretation of the causes for this is problematic. As pointed out earlier, both Huntington and Toynbee attribute the cause to climate, specifically progressive aridity in the outmigration region. Huntington claimed to have found a 600-year cycle of nomadic migrations that could be correlated to periods of desiccation. In relation to this, Toynbee suggested that the outpourings of Arabs under Mohammed was climatically induced, as was that of the Mongols under Genghis Khan. The determinists' view has been countered by many writers (e.g., Lattimore, 1938) although recently new interpretation has been put on Huntington's "Pulse" theory (Chappell, 1970).

d. *Contemporary literature.* Medieval chronicles contain many references to prevailing weather conditions. Unfortunately, although as might be expected, these pertain to exceptional weather events rather than day-to-day conditions. These include such unusual phenomena as the freezing of the Tiber in the ninth century or the formation of ice on the River Nile. While not supplying a continuous record, it is possible to obtain an overall view of the usual conditions by assessing the number of times given events are recorded. The freezing of the River Thames provides one such example. Between 800 and 1500 A.D., only one or two freezings per century are recorded. In the sixteenth century, four freezings are known, in the seventeenth there were eight, while six are known to have occurred in the eighteenth century. One can only suppose that a progressive cooling caused the frequency to increase.

Such chronicles also provide insight into climate through inference of other factors. Silverberg (1970) notes that William of Malmes-bury, writing about the Gloucester region of England in 1125 stated ". . . (the area) exhibits a greater number of vineyards that any other county in England, yielding abundant crops of superior quality . . . they may also bear comparison with the growths of France." Since the fifteenth century there has been no wine industry in England and it might be supposed that this is related to a worsening of the climate since those times. As noted earlier (p. 256), phenological events such as cherry blossoms in Japan, also provide an important guide.

e. *Evidence of agriculture.* Apart from documented evidence such as that described above there is considerable physical evidence of agricultural activity, of a given type, in areas where it can no longer be practiced. Perhaps the best-known type of evidence in this respect is the existence of irrigation systems that are no longer in use; as noted earlier, however, it has been pointed out that cultural rather than physical differences could account for their falling into disrepair.

Evidence of agriculture, together with other indicative pointers, has been used to reconstruct the climates that existed at the time Greenland was settled by the Vikings. Under the leadership of Eric the Red, the Vikings passed from Iceland, which they had settled in the ninth century, to Greenland. While an icy land, there was sufficient vegetation (dwarf willow, birch, bush berries, pasture land) for settlement. Two colonies were established, and farming of sorts was permitted. The outposts thrived and regular communications were established with Iceland. By 1250, Greenland was practically isolated from outside contact, with extensive drift ice preventing the passage of ships. By 1516 the settlements had practically been forgotten, and in 1540 a voyager reported seeing signs of the settlements, but no signs of habitation. The settlers had perished. Whether this was due to the deteriorating climate or to invasions of other peoples is not completely settled, although a Danish archaeological expedition to the sites in 1921 found evidence that deteriorating climate must have played a role in the demise of the people. Graves were found in permafrost that had formed since the time of burial. Tree roots entangled in the coffins indicate that the graves were originally in unfrozen ground and the permafrost had moved progressively higher. That food supplies had been insufficient was shown in examination of the skeletons found; most were deformed or dwarfed and evidence of rickets and malnutrition was clear. All the evidence points to a climate that became progressively cooler leading eventually to the isolation and extinction of the settlers.

f. The Instrumental Period. Accurate measurement of climatic variables could not occur until instruments were devised to measure them. Although some indication of prevailing weather conditions were accumulated by Greeks and Egyptians (e.g., Hippocrates "Airs, Waters and Places"—400 B.C., Aristotle "Meteorologica"—350 B.C., Ptolemy's diaries of local weather—

first century A.D.) precise observation really began with the invention of the thermometer by Galileo in 1593 and the barometer by Torricelli in 1643. Some weather data, particularly in Europe, were collected as early as 1649 and by 1686 Edmund Halley had published the first climatological map. Even by 1800, however, there were few weather stations, with twelve in Europe and five in the United States. The great impetus for collection and use of weather data came with the invention of the telegraph in the 1830s; since that time more and more data have become available. In Appendix 4, the nature and sources of current data availability are outlined in detail.

CHAPTER 13

Climate
through the Ages

In-depth study, correlation, and interpretation of the type of evidence outlined in Chapter 12 allows past climates to be reconstructed with varying degrees of exactitude. This chapter provides an overview of how climate has changed over time; perhaps it is fitting that it should be titled "Climate through the Ages," because this is the title of a classic work by Brooks (1949) who contributed much to the understanding of former climates. Indeed, Figure 13-1 is from Brooks and this, together with Table 13-1, supplies a thumbnail sketch of what has occurred. It is evident that throughout most of the earth's history, temperatures have been appreciably higher than they are today. We live in a cool period of earth history, a time forming part of one of the periodic cooler epochs that have occurred.

Reconstruction of past climatic conditions vary from highly tenuous to relatively exact. Generally the older the time period concerned, the less information is available; thus the reconstructed climates of the Palaeozoic and Mesozoic are much more general than those of the Cenozoic. For the earlier geologic periods the prevalence of a given type of climate—either warm or cold by today's standards—must be drawn from fragmentary evidence such as the discovery of ancient traces of glaciation and the cosmopolitan distribution of fossils that might point to the lack of climatic barriers.

Assessment of climate through the ages essentially resolves itself into the identification of the periodic interruptions of dominant warmer climates by glacial periods, together with the relative extents of humid or dry climates. While mild to warm climates have most certainly been the mode over much of the time, it must not be assumed that continental areas resembled those of warm lands today. For example, until late Silurian time, the lands were totally devoid of plant and animal life, and landscape everywhere would have presented the barren appearance similar, in some ways, to the driest deserts of the present. This picture did not change until vascular plants invaded the land (Figure 13-2) and even then many millions of years passed before any resemblance to present-day terrestrial ecosystems would occur.

PALAEOZOIC AND MESOZOIC CLIMATES

Not a great deal is known about the Pre-Cambrian climates. A bench mark in climatic interpretation occurs, however, at the Precambrian-Cambrian junction, where evidence of a glaciation exists. This Eo-Cambrian glaciation is indicated by tillites and striated pavements overlain by fossiliferous Cambrian rocks in Scandinavia, the British Isles, Spitzbergen, and eastern Greenland. Fragmentary evidence has also been discovered in North America. Subsequent to this glacial period, the Lower Palaeozoic rocks of the Cambrian, Ordovician, Silurian, and Devonian all show evidence of a widespread warm, sometimes dry, climate extending well into high latitudes.

Typical of the conditions that existed in the Lower Palaeozoic times are those of the Devonian (Figure 13-3). Such conditions probably continued into the Upper Palaeozoic. The northern hemisphere most certainly had widespread limestone deposits, with interbedded coral reefs extending from Europe through Eastern Greenland to Alaska. During this time, the great Carboniferous coal deposits were laid down. Note that coal deposits themselves do not necessarily point to the wide extent of hot climates. In explaining this Schwarzbach (1963) writes ". . . one can quote the modern spread of peat in the predominantly cool, middle-latitudes just as easily as the 'tropical' habitat of coal flora, and accordingly infer a cool or warm climate. Yet one thing must not be overlooked; peat accumulates if the decomposition of

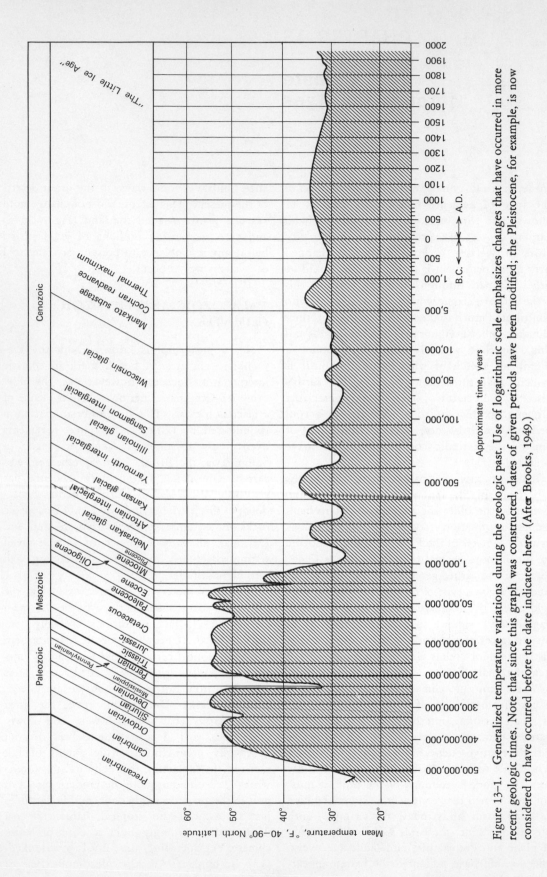

Figure 13–1. Generalized temperature variations during the geologic past. Use of logarithmic scale emphasizes changes that have occurred in more recent geologic times. Note that since this graph was constructed, dates of given periods have been modified; the Pleistocene, for example, is now considered to have occurred before the date indicated here. (After Brooks, 1949.)

Table 13-1

GEOLOGIC ERAS, PERIODS, EPOCHS, AMD REGIONS OF KNOWN MAJOR GLACIATIONS[a]

Era	Period	Epoch	Beginning of Interval (Millions of Years Ago)	Major Galaciations							
				Antarctica	South America	Africa	Australia	Asia	Europe	North America	Arctic
Prearcheozoic			4600								
Archeozoic	Keewatin		2800							x	
	Timiskaming									x	
Proterozoic	Huronian		1000		?	x	x			x	
	Algonkian										
	Precambrian					x	x	x	x	?	x
Paleozoic	Cambrian		600								
	Ordovician		500								
	Silurian		430								
	Devonian		400		x	x					
	Carboniferous	Mississippian	350								
		Pennsylvanian	330	?	x	x	x	x			
	Permian		275								
Mesozoic	Triassic		225								
	Jurrassic		180								
	Cretaceous		135								
Cenozoic	Tertiary	Paleocene	66								
		Eocene	59								
		Oligocene	38								
		Miocene	25								
		Pliocene	12								
	Quaternary	Pleistocene	0.6	x	x		x	x	x	x	x
		Holocene	0.01					x	x		x

[a] From Sellers (1965).

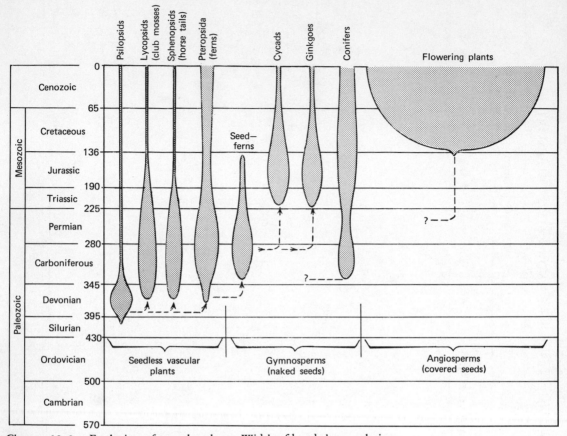

Chapter 13–2. Evolution of vascular plants. Width of band shows relative abundance of groups. (After McAlester, from Strahler, A.N., *The Earth Sciences*, 2nd. ed., © 1971 by A.N. Strahler. Reprinted with permission of Harper and Row Publishers, Inc.)

vegetable matter is slow. This happens today, predominantly in cool climate, but it is very probable that the same effect would be obtained in a warm climate if sedimentation occurred." Thus while coal deposits certainly require a moist climate, they can be indicative of both cool or warm conditions. The probable predominance of warmth is derived from other evidence, such as the enormous size of the insects that inhabited the swamps, the nature of the flora, and the fireclays—probably derived from laterites—that are found in association with coal.

The southern hemisphere experienced a glaciation during the late Carboniferous and as Figure 13–4 shows, the ice movement can be interpreted in terms of the extent of the former land mass (Gondwanaland).

It is probable that the Hercynian Orogeny grossly modified the land-sea distribution that existed over much of the Palaeozoic Era. This helps explain the notable difference between climates deduced from Permian deposits and those of earlier times. The Permian is marked by extensive development of rock salt, potash, and aeolian deposits that point to the widespread aridity that might have occurred. Indeed, the climate that existed during Permian time is more aligned to that of the Mesozoic than the Palaeozoic Era.

The oldest beds of the Mesozoic, the Trias, are so closely related to those of the Permian that it is difficult, in places, to separate them stratigraphically. Hence, the term Permo-Triassic is often used. Like the Permian, the Trias shows

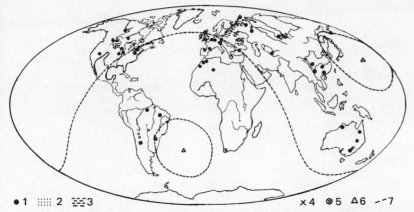

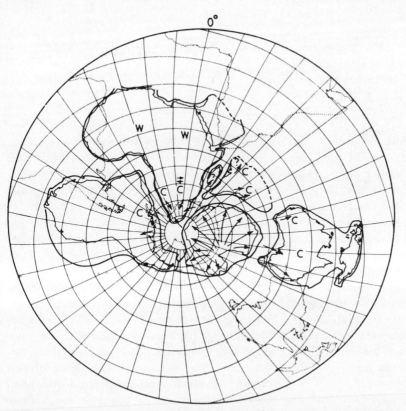

Figure 13–3. Climatic map of the Devonian. Key: 1 - Coral Limestones; 2 - Old Red Facies; 3 - Evaporites; 4 - Iapo Formation and Table Mountain Tillite; 5 - Upper Devonian Tillite; 6 - Palaeomagnetic South Pole and North Pole from which the equator and polar circles have been constructed (7); B. Bauxite. (From Schwartzbach, M., *Descriptive Palaeoclimatology*, A.E.M. Nairn, ed., copyright © 1961 by John Wiley and Sons. Used with permission.)

Figure 13–4. The Late Carboniferous: C = cold; W = warm temperate; arrows indicate movements of ice sheets, with modern directions shown for Antarctica. Compare this depiction with that given in Fig. 11–2. (From King, L.C., *Descriptive Palaeoclimatology*, A.E.M. Nairn, ed., copyright © 1961 by John Wiley and Sons. Used with permission.)

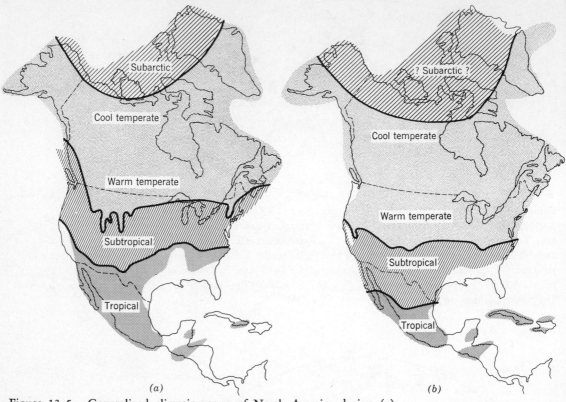

(a)   (b)

Figure 13–5.   Generalized climatic zones of North America during (*a*)
late Eocene to early Oligocene; (*b*) late Oligocene to early Miocene.

extensive evidence of widespread aridity with
sandstones, devoid of fossils, in close proximity
to evaporate deposits. The aridity was probably
associated with warm conditions that continued
into the Jurassic and Cretaceous. Coral limestones
extended well into present-day middle latitudes
while rich and varied flora have been found in
Greenland. This is the time of widespread
saurians and, as already noted, these cold-
blooded animals thrived in warm climates. Their
widespread extent is indicated by fossil evidence
found as far north as Canada. The high average
global temperatures are also indicated by the
predominance of chemical weathering; bauxites,
an end product of such weathering, are widely
dispersed in the lower Cretaceous.

Oxygen isotopes have been used to derive the
prevailing temperature conditions and show that
they were appreciably higher than today. Estimated sea temperature for Jurassic rocks in
Scotland give the prevailing temperatures as between 17°C (62.6°F) and 23°C (73.4°F), a marked

contrast to the 7°C (44.6°F) to 13°C (55.4°F) of
today; similarly, analysis of fossils in the Upper
Cretaceous of England shows that prevailing
temperatures were between 16°C (60.8°F) and
23°C (73.4°C) compared with the 5°C (41.0°F)
to 15°C (59.0°F) of today.

THE CENOZOIC

Climates of the Cenozoic have been reconstructed in more detail than those of earlier eras.
The beginning of the Cenozoic dates back some
63 million years, a much shorter period of time
than the Cambrian, just one period of the
Palaeozoic Era. As such, the evidence of past
climates is much more widespread and fossil
flora and fauna can be equated much more
readily to present-day organisms. The Cenozoic
consists of the Tertiary and the Quaternary
Periods, with the Quaternary comprising the
Pleistocene and the Holocene epochs. These
later times extend back only a million or so

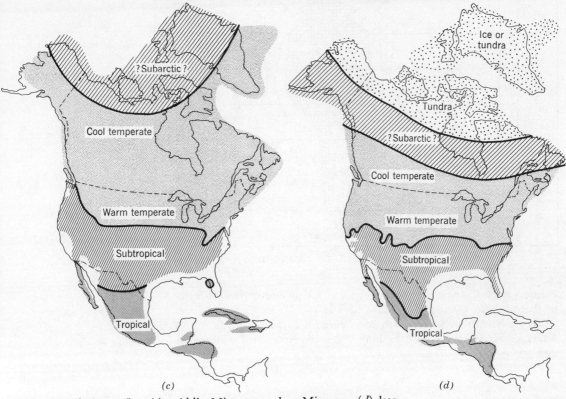

Figure 13-5 (*Continued*). (*c*) middle Micoene to late Miocene; (*d*) late Pliocene. (From Dorf, E., *American Scientist*, 48, 1960. Reprinted by permission, *American Scientist*, Journal of the Society of Sigma Xi.)

years and include the evidence recorded by man's tenure on earth. It is thus possible to visualize these past climates in much more detail and identify minor changes that are not apparent in older rocks.

The Tertiary

The epochs that form the Tertiary period are

Pliocene	"more" recent
Miocene	"less" recent
Oligocene	"few" recent
Eocene	dawn of the recent
Palaeocene	"old" (ancient) recent

To illustrate the changes that occurred in climates over the Tertiary, it is most expedient to describe the climatic evolution of a single continent, in this case, North America. Dorf (1960) has produced a series of maps that provide an excellent survey of reconstructed vegetation-climatic belts; the following account draws upon his interpretation.

The climate of the Palaeocene is not well known, although most indications point to its being a relatively cool epoch. Thereafter, from the Eocene to the early Oligocene a general warming occurred (Figure 13-5a), with evidence of tropical forests found as far north as Tennessee and Missouri. The general warmth of northern realms during these times is further indicated by fossil leaves found in the present-day bleak Disko Island located close to the Greenland ice sheet. The fossil flora indicates that a distinctly temperate climate existed there during the Early Oligocene.

By the end of the Oligocene a cooling trend set in (Figure 13-5b) and with slight modifications continued throughout the Miocene (Figure 13-5c). By late Pliocene times (Figure 13-5d), the climate approximated that of the present.

Such a sequence of events in North America is also shown in the succession of vegetation types found in the northern part of the Great Basin

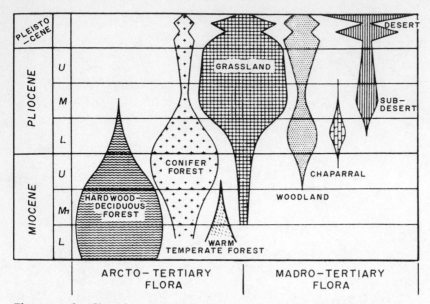

Figure 13–6. Changing vegetation types in the present-day desert region of the northern Great Basin during Middle and Upper Cenozoic times. (After Axlerod, from Schwarzbach, M., *Descriptive Palaeoclimatology*, A.E.M. Nairn, ed., copyright © 1961 by John Wiley and Sons. Used with permission.)

of the western United States. As Figure 13–6 indicates, the progressive dominance of different plant communities points to changes such as those indicated by Dorf. A similar sequence of events is also found in western Europe. Fossil flora of the Eocene London Clay indicate tropical climatic conditions while later deposits indicate a progressive cooling toward the Pliocene. At the end of the Pliocene, the stage was set for the glacial period of the Pleistocene.

The Pleistocene

"The latest epoch of geologic history has witnessed changes in the physical aspect of the Earth and in the distribution of animals and plants on the Earth's surface such as are not recorded in any earlier span of time of comparable length." This opening paragraph of R. F. Flint's well-known *Glacial and Pleistocene Geology* characterizes changes that have occurred since the end of the Tertiary. It is an intriguing time of earth history and one that is highly significant in terms of both the existence of man on earth and the shaping of the face of the earth on which he lives. It is not surprising that it is an epoch about which volumes have been written.

The Pleistocene itself is sometimes referred to as the "Ice Ages" as well it might be, for it was characterized by ice covering in areas far beyond the present-day limits of ice. The advance of ice was not a simple case of a single ice advance and retreat. It has been established in both North America and in Europe that at least four great advances occurred, each separated from the other by a warmer or interglacial period. Table 13-2 shows the divisions of the Quarternary in Europe and North America. The glacial periods in Europe (Gunz, Mindel, Riss, and Würm) are named from Bavarian rivers and represent phases of Alpine glaciation. The interglacial are hyphenated terms, for example, Riss-Würm represents the interglacial between the Riss and Würm advances. In northern Europe other place names are used. The North American glacials are named for states while the interglacials are given place names as indicated.

In Europe there were three major ice centers (Figure 13–7) located in the Alps, Britain, and Scandinavia. The ice of the two later centers often converged. At its maximum, the Scandinavian ice might have been as much as 3000 m thick and the relatively recent retreat has allowed

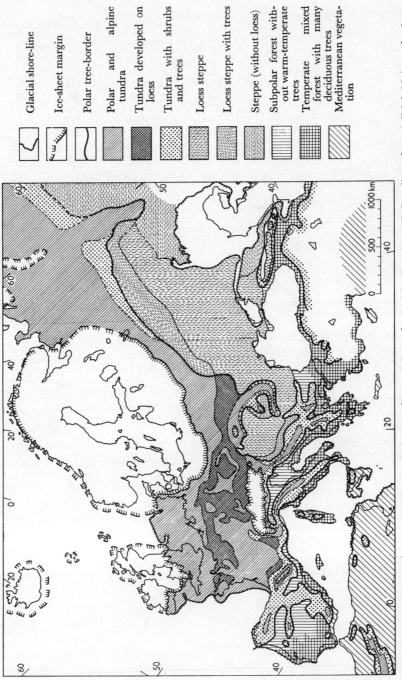

Glacial shore-line

Ice-sheet margin

Polar tree-border

Polar and alpine tundra

Tundra developed on loess

Tundra with shrubs and trees

Loess steppe

Loess steppe with trees

Steppe (without loess)

Subpolar forest without warm-temperate trees

Temperate mixed forest with many deciduous trees

Mediterranean vegetation

Figure 13–7. Europe during the Würm glaciation. (From Schwarzbach, M., *Descriptive Palaeoclimatology*, A.E.M. Nairn, ed., Copyright © 1961 by John Wiley and Sons. Used with permission.)

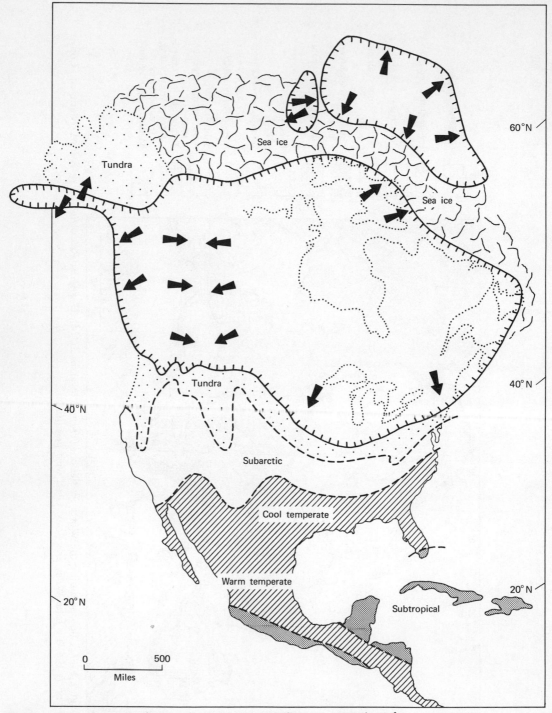

Figure 13–8. Generalized climatic zones representing a composite of Pleistocene glacial conditions in North America. (After various sources.)

Table 13-2

DIVISIONS OF PLEISTOCENE IN EUROPE AND NORTH AMERICA

	Alps	North Germany and Holland	North America
Last glacial	Würm	Weichsel	Wisconsin
Last inter-glacial	Riss-Würm	Eem	Sangamon
Glacial	Riss	Saale	Illinoian
Interglacial	Mindel-Riss	Holstein	Yarmouth
Glacial	Mindel	Elster	Kansan
Interglacial	Gunz-Mindel	Cromer	Aftonian
Glacial	Gunz	Weybourne	Nebraskan
Interglacial	Donau-Gunz	Tegelen	
Glacial	Donau	Pretegelen	

detailed studies to be carried out. The multistage ice advance was first observed in Europe, and much of the groundwork for the interpretation was laid by Penck and Brückner between 1901 and 1909. Further, it was found that each glacial epoch itself contained warmer and cooler periods. Detailed subdivisions of these stadials (cold glacials) and interstadials (warm glacials) have been developed.

In North America, the main ice centers were located over the low-lying parts of the Canadian Shield (Figure 13-8). At their maximum extent, these linked with ice originating in the Cordilleran system to produce an unbroken coverage from Newfoundland to the Aleutians. The terminal moraines of the four major ice advances vary, and as shown in Figure 13-9, they can be traced over wide areas. Less is known about the continental glaciation of Asia. It is believed that much of western Siberia was covered by ice, and this may have been an extension of the Scandinavian ice sheet. Most of the mountain ranges appear to have carried large glaciers.

While a great deal is known about the course and nature of the ice ages, there are still many unsolved problems. Indicative of this is the varied interpretation of the times and prevailing conditions of the Pleistocene as evaluated by oceanic cores. Figure 13-10 shows the results obtained by analysis of the cores through

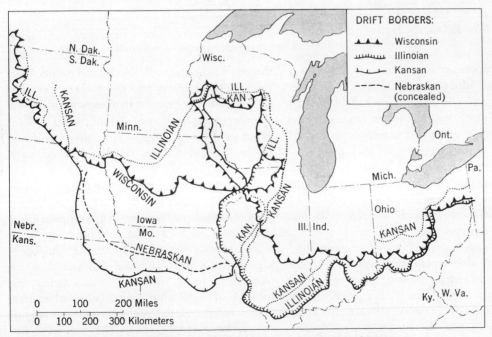

Figure 13-9. Extent of Pleistocene ice advances in central United States as indicated by drift borders, In each glacial stage, the ice reached a different line of maximum advance. (From Strahler, A.N., *Physical Geography*, copyright © 1969 by John Wiley and Sons. Reproduced by permission.)

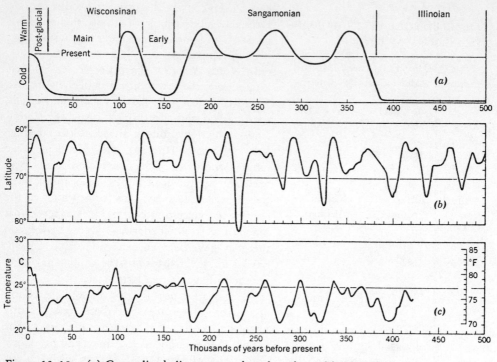

Figure 13–10. (a) Generalized climate curve based on data of foraminifera in deep-sea cores. (After Ericson and Wollin 1968.) (b) Curve of incoming solar radiation, scaled in equivalent latitude of present value at 65°N. (After Van Woerkom, 1953.) (c) Palaeotemperature curve based on oxygen-isotope ratios. (After Emiliani, 1966.) (From Strahler, A.N., *The Earth Sciences*, 2nd. ed., © 1971 by A.N. Strahler. Reprinted with permission of Harper and Row Publishers, Inc.)

isotopic temperature determination and by examination of globigerina content. They differ appreciably and the findings prompted Strahler (1971) to write "that the interpretations of two highly qualified groups of investigators should disagree to such a large extent may prove disconcerting to those who ask scientists for a set of unimpeachable interpretations of the world around them. Despite the advances of basic science, controversies such as this will not diminish in frequency."

As ice advanced, the climates of areas south of the ice sheets also experienced a change. Figure 13–11 shows an estimate of what happened to the climatic zones during the Würm glaciation by comparison to those of today. Ice extended as far south as 55°N and the tundra to 45°N. The climatic zones between these cold extensions and the equatorial climates were, of necessity, compressed. An exception to this, according to

this reconstruction, is in the tropical realms where the equatorial forest and the savanna extended further from the equator than they presently do. Another interpretation of the migrating zones is given in Figure 13–12, which shows the swing of the belts in relation to glacial advance and retreat.

It is clearly evident that during the Pleistocene, the general circulation of the earth must have been modified considerably. The compression of the climatic zones and the enormous area of ice would modify the positions of the subtropical high-pressure cells and the subtropical jet. A number of maps showing, or estimating, the location of the modified wind belts and pressure systems of the earth have been produced. Figure 13–13, for example, shows the suggested prevailing winds that resulted in widespread aridity in Africa during parts of the Pleistocene. In reviewing the nature of climatic zones during

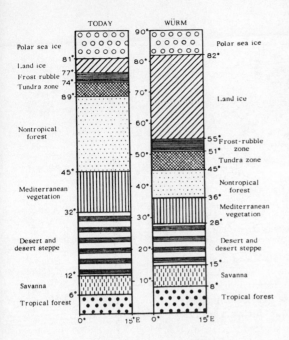

Figure 13–11.   Displacement of the climatic belts of the northern hemisphere during the Würm glaciation. (After Schwarzbach, 1961.)

the Würm glacial age, Suzuki (1971) presents his own interpretation of major frontal zones (Figure 13–14). Of note here is his interpretation of the position of the northern hemisphere Polar Front in winter; it is shown to extend almost to 20°N.

The Holocene

The boundary between the Pleistocene and Holocene is not a precise one. Ice retreated and warmer climates permitted tree growth in northern areas around 12,000 B.C., but this retreat was temporary. An abrupt freezing occurred once again and between 10,000 and 8500 B.C., ice again advanced. Final withdrawal of ice from Scandinavia is estimated from varves at about 8500 B.C. (Figure 13–15). Thereafter a general warming trend took place culminating in the "climatic optimum" that occurred between 5000 and 3000 B.C. It is thus convenient to begin a study of historical climates at about 4000 B.C., a date that approximately coincides with early Egyptian data concerning man's civilization.

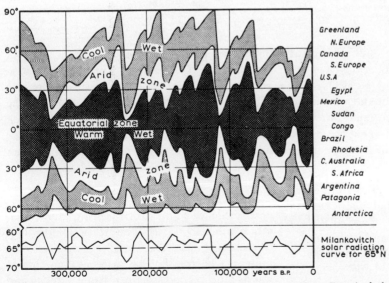

Figure 13–12.   Suggested variability of the earth's major climatic belts over the last few hundred thousand years. The Milankovitch chronology for 65°N is used on the assumption that, because it is the mean latitude for both maximum continentality and maximum temperate belt glaciation, it is significant for world climates .(From Fairbridge, R. W., *Problems in Palaeoclimatology*, A.E.M. Nairn, ed., 1964 copyright © by John Wiley and Sons. Used with permission.)

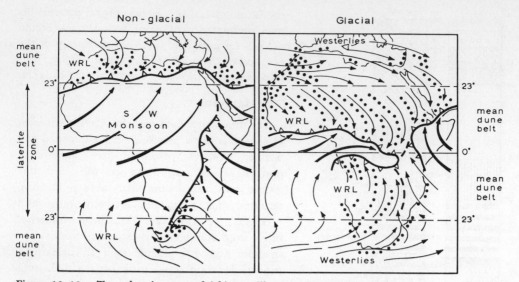

Figure 13–13.   Two sketch maps of Africa to illustrate the hypothical distribution of subtropical fronts and mean July airstreams during a nonglacial (or interglacial) and a glacial stage. Key: Double-shafted arrows indicate moist monsoon and tropical airstreams; single arrows indicate Westerlies and Trade Winds; rows of dots show dune sands; WRL refers to the "wheel round latitude" that marks the Westerlies/Trade Wind transition and thus the mean dune belt. Note that this occurred at about 30°N and S in nonglacial (or interglacial) times but migrated down to as much as 5° in glacial phases, a shift of some 2000 to 3000 km. (From Fairbridge, R.W., *Problems in Palaeoclimatology*, A.E.M. Nairn, ed., Copyright © 1964 by John Wiley and Sons. Used with permission.)

A number of authors have outlined climatic changes that have occurred during the Holocene (Fairbridge, 1968, Lamb, 1969, UNESCO, 1963); while their general interpretations are similar, the dates that they give and the groupings they use are not the same. Drawing upon the varied accounts, it seems that the following times might be differentiated:

4000 to 3000 B.C.	The Climatic Optimum
3000 to 750 B.C.	The Subboreal Phase
750 B.C. to 800 A.D.	The "Classical" (Roman-Greek) and post-Classical Times
800 to 1200 A.D.	The "Little" Climatic Optimum
1200 to 1550 A.D.	The Medieval Climates
1550 to 1880 A.D.	The "Little Ice Age"
1800 to Present	The Secular or Instrumental Period

The terms used to describe these times are obviously not systematic. For example, the Subboreal Phase refers to a pollen zone of Scandinavia; the Medieval Climates refer to a period so designated by historians. They are used merely to facilitate description of the breakdown that is given. Dates have also been rounded and are given in terms of B.C. and A.D. rather than B.P. (Before Present against a datum of 1950) that is often used by geologists.

4000 to 3000 B.C.

All evidence points to this as being a time when mean atmospheric temperature of middle latitudes was about 2.5°C (4.5°F) above that of the present. The tree line on European mountains was as much as 500 m higher than it is today and the tundra climate had practically disappeared from northern Siberia (Schwarzbach, 1963). It is often described as the "Climatic Optimum," a term originating in Scandinavia when the warmth would be favorable for highly varied flora and fauna. It is not a good term from the standpoint of other world areas, for while Scandinavia might

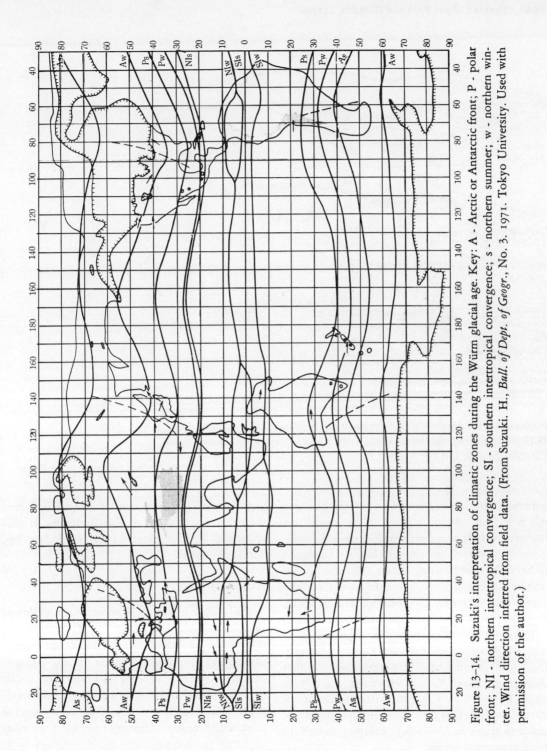

Figure 13–14. Suzuki's interpretation of climatic zones during the Würm glacial age. Key: A - Arctic or Antarctic front; P - polar front; NI - northern intertropical convergence; SI - southern intertropical convergence; s - northern summer; w - northern winter. Wind direction inferred from field data. (From Suzuki. H., *Bull. of Dept. of Geogr.*, No. 3. 1971. Tokyo University. Used with permission of the author.)

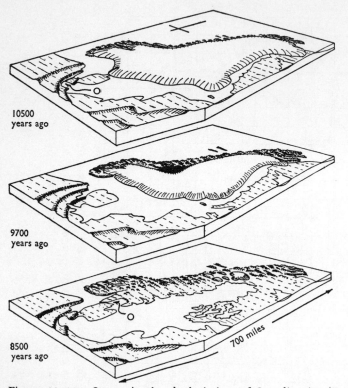

Figure 13–15. Stages in the deglaciation of Scandinavia. (From Dury, G.H., *The Face of the Earth*, 1959, G.H. Dury. Used with permission of Penguin Books, Ltd.)

experience optimum conditions, places elsewhere did not. The semiarid regions of the world, for instance, might have been much drier and conditions deteriorated in terms of living conditions rather than optimizing them. Indeed, other writers have suggested alternate names for the times, for example, Deevey and Flint (1951), suggest that Hypsithermal is more appropriate.

During this optimum, which corresponds to the early Neolithic times of northwestern Europe, the sea level must have been appreciably higher than it is today. Various estimates place it as rising as much as 3 m above the present level. Arctic ice is believed to have disappeared while glaciers, those in Alaska for example, underwent their maximum retreat.

In Egypt this was the time of the Early Kingdom cultures and evidence suggests that the River Nile had about three times its present volume. Fairbridge (1963) reports that a rock cut by the river, dated at 4000 B.C. by radio-carbon dating, shows a flood level 12–20 m above that of

the present. The heavy rainfall in the Nile watershed that contributed to this high flow is substantiated by the great extent of Lake Chad during this time.

3000 to 750 B.C.

Analysis of pollen in Scandinavia has allowed various pollen phases to be established. During this period, pollen of the climatic optimum—comprising the "Main Atlantic" phase with oak and elm—altered to that indicative of cooler climates. This has been termed the "Subboreal" phase, which itself is divided into early (3000 to 2000 B.C.), middle (2000 to 1500 B.C.), and late (1500 to 750 B.C.).

The Early Subboreal was marked by a cooling trend, subsequent to which temperatures have never, for any appreciable period, returned to those of the climatic optimum. The cooling caused marked drops in sea level and the emergence of many islands and coastal areas; the

Bahama Emergence, for example, occurred at this time.

The cooling trend was interrupted between 2000 and 1500 B.C. with milder conditions indicative of the Middle Subboreal. But once again, starting at about 1500 B.C. and lasting until 750 B.C. (450 B.C. in some places) a cooling trend occurred. Colder conditions once again led to renewed ice growth and a sea level drop of between 2 and 3 m below present-day levels.

Much material concerning this cooling trend is available from Egyptian records. In the reign of Amenemhat III, barrages and canals, necessitated by the drop in flood levels, were constructed. About 1250 B.C., Ramses II constructed the Sweetwater Canal, the first Suez Canal to link the Mediterranean and Red Seas. The decreasing flood levels are also recorded in Central Africa where the level of Lake Victoria was progressively lowered.

In western Europe, the climate was cool and moist. Glacial advance occurred and extensive bog formation took place. Many of the glaciers in the Rocky Mountains, south of 50°N, were initiated at this time.

750 B.C to 800 A.D.

The cold climates associated with the Late Subboreal may have persisted in places up to 450 B.C., but, for the most part, the period up to 150 B.C. is indicative of a mild climatic phase. In northern Europe this has been termed the "Subatlantic" phase, indicating a vegetation return to beech forests indicative of a warmer climate.[1] The warm conditions of the Atlantic phase were never realized.

During the Roman Era (here given as 150 to 350 B.C.), a cooling took place and glaciers were enlarged considerably. The resulting drop in sea level is borne out by many cultural features. Port facilities, those that were not subjected to tectonic activity, were constructed and conditioned to a sea level lower than that of today. Evidence from the Mediterranean area is also found in Egypt where the Sweetwater Canal, as a response to lowered sea level, became silted.

[1] Alternatively, the growth of beech forests may not indicate a milder climate. It may be that during the Subatlantic, which certainly began as raw and cold, the beech continued a slow northward migration and that it merely happened to reach Europe during this time.

Roman reports of the climate of Britain around 50 B.C. tended to class it as a chilly, damp place. Whether this can be accounted for by the existing conditions or whether it is the way that the climate was perceived by those accustomed to a Mediterranean climate is a moot point. Julius Caesar in 55 B.C. suggested that the climate was too cold for agriculture; yet in 98 A.D. the historian Tacitus reported that with the exception of the olive and vine, the climate allowed most "ordinary" crops to be grown.

The Mediterranean climate itself differed from that of today. Weather records kept by Ptolemy in the second century A.D. (Table 13-3) show that precipitation occurred throughout the year, in contrast to the winter maximum of today. Whether this moist climate was widespread and facilitated development of wheat and olive agriculture over large areas of North Africa is a point on controversy.

By 350 A.D. the climate had become milder in northern realms while in tropical regions it appears to have become excessively wet. Tropical rains in Africa caused high-level Nile floods and temples built earlier (1250 B.C.) were inundated. At this time, too, Central America experienced heavy precipitation and tropical Yucatan was very wet.

At the close of the eighth century evidence points to a marked cooling. In 800 to 801 A.D., the Black Sea was frozen and in 829 A.D., ice formed on the Nile. The raininess of Yucatan decreased and conditions became favorable for development of the Mayan culture.

800 to 1200 A.D.

This early Medieval period has been called the "Little Climatic Optimum," because it represents the warmest climate that occurred in the northern hemisphere for several thousand years. The Domesday Book records, for example, show that northwestern Europe was warmer and drier than it is now; they indicate widespread growth of vines in areas where climates are now not suitable. This was the time of the Viking settlements in Greenland and Iceland, which clearly indicates the northward recession of drift ice. Tree ring analysis of Alaskan species indicate that temperatures were as much as 2°C higher than today and that the eleventh century was probably quite dry. In the Rockies, the snow may have

Table 13-3

FREQUENCY OF METEOROLOGICAL PHENOMENA AT ALEXANDRIA, EGYPT[a]

| | Frequency (Days) | | | | | | | | | | | |
	Jan.	Feb.	Mar.	Apr.	May	Jun.	July	Aug.	Sept.	Oct.	Nov.	Dec.
Rain												
First century	4	3	0	5	3	1	2	0	4	3	3	2
First century—"fine rain"	1	0	1	3	4	5	0	0	2	0	2	2
Present	11	6	5	1	1	0	0	0	0	1	7	10
Thunder												
First century	1	0	1	1	2	2	1	1	1	0	0	0
Present	0.7	0.5	0.3	0.2	0.3	0.1	0	0	0	0.4	0.7	1.5
Great heat												
First century	0	0	0	0	0	3	8	6	1	0	0	0
Present	0	0	0	2	6	12	0	1	3	2	0	0

[a] After Brooks (1949).

been 370 m above the present snowline. In the southwestern United States, geomorphic evidence points to a much drier climate with marked dessication.

1200 to 1550 A.D.

The "Little Climatic Optimum" was followed by a cooling trend and, in middle-latitude areas of the northern hemisphere, by increased storm activity. Winters were extremely cold with widespread river freezes. Arctic temperatures declined and the Norse settlements of Greenland were abandoned. Pollen analysis in Scandinavia indicates a shift from a mild maritime climate to a cold, dry continental climate.

A great drought is thought to have occurred in the American southwest between 1276 and 1299 (Antevs, 1955), although the geomorphic evidence on which this is based has been questioned by other writers (Fairbridge, 1968). There is not much doubt that the area did become drier later in this period, because both tree ring analysis and shrinkage of developed agricultural areas point to such an event. Early Indian cultures of the Plains regions shifted from agriculture to hunting, a possible response to progressive dryness.

A mild phase was resumed between 1450 and 1550 and extensive flooding occurred in western Europe. The wide range of this warming trend is indicated by glacial retreat in North America;

some glacial moraines were abandoned around 1530 A.D.

1550 to 1850 A.D.

This has been termed a "Little Ice Age"[2] which, as the name implies, experienced cool temperatures with the mean annual of the northern hemisphere about 1°C (1.8°F) below twentieth century levels. Arctic ice expanded, glaciers advanced—many to reach their maximum extent about 1750—and many dry areas suffered desiccation.

The Caspian Sea, often used as a guide to prevailing conditions over much of historic times, reached its highest level (2.5 m above present, 9 to 10 m above the 800 to 1200 A.D. level). The level of the lake is influenced by many factors, but of extreme importance are the varying evaporation rates. Low levels usually indicate

[2] The term "Little Ice Age" has a number of interpretations. First proposed by Matthes (1939) it embraced the concept of climatic deterioration during the late Holocene, a period of about 4000 years, extending to the eighteenth century. Its purpose was to distinguish Holocene oscillations from those of the last glacial—the Würm or Wisconsin. On the other hand, Brooks (1949) identified the "Little Ice Age" as extending from the sixteenth to late nineteenth centuries. It is most often used in this respect, although Schove (1955) recognizes four little ice ages of quite different times.

high prevailing temperatures; during cool periods, the level rises.

The harshness of this cold period is well documented. Bad weather, with resulting crop failure, led to famine in Iceland when between 1753 and 1759 a quarter of the population perished. In New England the year 1816 was called "the year without a summer." Snow fell in June and frost occurred in July.

Unlike the following period to be described, actual climatic data indicating conditions that existed during each of the above periods are practically nonexistent. Despite this, a number of writers have put together the evidence (e.g., Table 13–4) for much of the world and it has become possible for others to reconstruct the prevailing atmospheric conditions that existed. Of note in this respect is the work of Lamb (1969). Figure 13–16 shows how he has reconstructed the circulation conditions that existed during several past epochs. While such representations are somewhat tentative, they provide what might be the ultimate goal of climatic reconstruction, namely the understanding and representation of the pressure and wind systems that once existed. If such maps could be constructed on a global basis, then interpretation of climatic changes over historic time would be a much simpler one.

1850 to Present

This comprises the instrumental period. With the introduction of instruments it would seem that interpretation of climatic changes over time would be a simple task. While it is true that statisticians now have numbers to work with, interpretation of recorded values is not an easy task. Although some climatic elements have been measured at a few stations for a long time, the acquired data must be carefully analysed. Often the data are discontinuous while changes in instruments or station site need be taken into account. Some stations have also become enveloped in urban developments and microclimates might have been significantly altered. Prior to automation, the human element might also have given rise to errors in readings, an instance of which has been suggested quite recently (see p. 245). Before any data can be used to evaluate climatic change, their homogeneity must be tested. The factors leading to long-term change can be, at times, apparent rather than real. Figure 13–17 shows some of the factors that might contribute toward changes in temperature. They are not necessarily a response to a worldwide heating or cooling trend.

Careful studies have, however, been completed by many workers. From all of these the most striking feature to emerge, and generally the

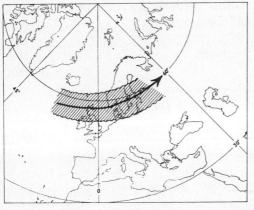

(a)

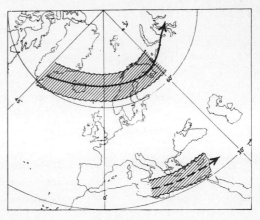

(b)

Figure 13–16. (*a*) Estimated zone of most frequent depressions (extratropical cyclones) in July 1550 to 1600. This circulation coincides with a period of marked glacial advance. (*b*) Estimated zones of most frequent depressions (extratropical cyclones) in July 1000 to 1100. This circulation coincides with the warmest climatic period in the last 2000 years. (From Lamb, H.H., *Problems in Palaeoclimatology*, A.E.M. Nairn, ed., copyright 1964 by John Wiley and Sons. Used with permission.)

Table 13-4

A BRIEF CHRONOLOGY OF THE CLIMATE OF THE LAST 10,000 YEARS [a]

Dates	Region	Climate	Sources
9000–6000 B.C.	Southern Arizona	Warm and arid	Martin (1963)
7800–6800 B.C.	Europe	Cool and moist, becoming cool and dry by 7000 B.C. Ice sheets left Sweden in 6840 B.C.	Brooks (1951) Antevs (1955)
6800–5600 B.C.	North America, Europe	Cool and dry, with possible extinction of mammals, particularly in Arizona and New Mexico. Cochrane readvance in Alaska and southeast Canada	Brooks (1951) Flint and Deevey (1951) Martin (1958) Sears (1958) Terasmae (1961)
5600–2500 B.C.	Both hemispheres	Warm and moist, becoming warm and dry by 3000 B.C. (Climatic Optimum). Intermittent drought in the western United States after 5500 B.C. Start of glacial retreat in the McMurdo Sound region of Antarctica about 4000 B.C. Maximum glacial retreat in Alaska near 3500 B.C.	Brooks (1951) Antevs (1955) Pèivé (1960) Karlstrom (1961) Gentilli (1961)
2500–500 B.C.	Northern hemisphere	Generally warm and dry with periods of heavy rain (in Europe near 1300 B.C.; in the Near East about 1100 B.C. and between 850 and 800 B.C.; and in the western United States after 660 B.C.) and intense droughts (in Europe after 2200 B.C., between 1200 and 1000 B.C., and between 700 and 500 B.C.; in China from 842 to 771 B.C.; and in the western United States near 510 B.C.). Glaciation in Alaska (between 2380 and 1340 B.C. and near 700 B.C.)	Brooks (1951) Flint and Deevey (1951) Flint (1957) Butzer (1958)
500 B.C.–A.D. 0	Europe	Cool and moist. Glacial maximum in Scandinavia and Ireland between 500 and 200 B.C.	Brooks (1951) Flint (1957)
330	United States	Drought in the Southwest	Antevs (1955)
600	Alaska	Glacial advance	Karlstrom (1961)
590–645	Near East, England	Severe drought in the Near East, followed by cold winters. Drought in England	Butzer (1958)
673	Near East	Black Sea frozen	Butzer (1958)
800	Mexico	Start of moist period	Sears (1958)
800–801	Near East	Black Sea frozen	Butzer (1958)
829	Africa	Ice on the Nile	Butzer (1958)
900–1200	Iceland	Glacial recession (Viking period)	Schwarzbach (1963)
1000–1011	Africa	Ice on the Nile	Butzer (1958)
1000–1100	Utah	Snowline 300 m higher than today	Wright (1963)
1200	Alaska	Glacial advance	Karlstrom (1961)
1180–1215	United States	Wet in the West	Schove (1961)
1220–1290	United States	Drought in the West	Schove (1961)

(Continued)

Table 13-4 (*continued*)

Dates	Region	Climate	Sources
1276–1299	United States	"Great Drought" in the Southwest	Antevs (1955)
1300–1330	United States	Wet in the West	Schove (1961)
1500–1900	Europe, United States	Generally cool and dry. Periodic glacial advances in Europe (1541 to 1680, 1741 to 1770, and 1801 to 1890) and North America (1700 to 1750). Drought in the southwestern United States from 1573 to 1593	Brooks (1951) Schove (1961) Schwarzbach (1963)
1880–1940	Both hemispheres	Increase of winter temperatures by 1.5°C. Drop of 5.2 m in the level of the Great Salt Lake. Alpine glaciation reduced by 25% and Arctic ice by 40%. Rapid glacial recession in the Patagonian Andes (1910–1920) and the Canadian Rockies (1931–1938)	Flint (1957) Heusser (1961) Mitchell (1961)
1920–1958	United States	25% decrease in mean annual precipitation in the Southwest	Sellers (1960)
1942–1960	Both hemispheres	Worldwide temperature decrease and halt of glacial recession	Mitchell (1961) Heusser (1961)

[a] From Sellers (1965).

Table 13-5

OBSERVATIONS CONCERNING TEMPERATURE TRENDS USING INSTRUMENTAL DATA GIVEN IN PAPERS PUBLISHED BETWEEN 1945 AND 1961[a]

Author	Area	Findings
Labrijn (1945)	Netherlands	General rise in mean winter temperature from 1790 on. Slow increase in summer mean between 1800 and 1900 followed by a decrease until 1920, thereafter another upward trend
Eythorsson (1949)	Iceland	Steady rise of annual temperature (about 1.1°C) between the decades 1916 to 1925 and 1926 to 1935
Keranen (1952)	Finland	Finland continuously milder from seventeenth century on. Peak of rise in 1930s thereafter a leveling of temperatures
Blüthgen (1952)	Lapland	Mean annual temperature for the period 1935 to 1950 higher by 1°C than that for 1901 to 1930. Means for December and November up by 3°C
Hesselberg with others (1956, 1958)	Norway	Marked rises in temperature, especially for Spitzbergen. Rise slowed by 1922 to reach a peak in the 1930s. A reversal of the trend between 1941 and 1950
Glasspoole (1955)	England and Wales	An increase of over 1°C in spring temperatures of period 1943 to 1952 compared with that for 1923 to 1932. Almost 1°C rise in summer temperatures 1942 to 1951 compared to 1918 to 1927
Rubinshtein (1956)	U.S.S.R.	Winter warming trend over much of U.S.S.R. (1950 to 1955). Peak warming differed in east and west
Langley (1953)	Canada	Warming in most areas between 1880 and 1900, thereafter variations from district to district with many minor fluctuations varying in time and space

(*Continued*)

369

Table 13-5 (*continued*)

Author	Area	Findings
Thomas (1957)	Canada	An increase, although not a steady one, in Ontario from last century. Results more marked in southern than in northern Ontario
Thomas (1955)	Canada	Coastal areas of Nova Scotia and Newfoundland experienced a warming trend after 1890. Most increase in winter with peak temperatures reached in early 1930s
Conover (1951)	Northeast United States	A 3.5°F rise in winter temperatures in the 100-yr series at Bluehole, Mass.
Kincer (1946)	United States	Drawing upon many stations showed a general temperature rise to 1892. Subsequent to a leveling especially between 1903 and 1920, a continued rise to 1940. The temperature of the warmer times as much as 2°C higher than those of cooler
Landsberg (1960)	United States	Using two 25-yr periods (1906 to 1930, 1931 to 1955) found a significant temperature rise in many locations with the most marked being in the Great Lakes region
Yamamoto(1950, 1951, 1957)	Japan	Secular trends quite variable in different parts of the country
Fujiwara and and Ishiguro (1957)	Japan	Rise in the average spring and summer temperatures in Hokkaido after 1910. Temperatures of the period 1940 to 1950 greater than in any of the previous 30 years
Pramanik (1954)	India	Found no general increase or decrease in maximum or minimum temperatures
De Boer and Euwe (1949)	Indonesia	Jakarta temperatures show a fluctuating but continuous increase. Authors suggest this is not due to urbanization
Schmidt et al. (1951)	Indonesia	Jakarta temperatures show overall increase
Wexler (1961)	Antarctica	Near edge of ice shelf a 0.5°C warming trend between 1912 and 1957. No significant trend found in McMurdo Sound
Maksimou (1954)	Arctic	Noted marked decrease in the ice cover of the North Atlantic
Lamb and Johnson (1959)	Arctic	Decrease in ice between 1920 and 1940 followed by an increase in the 1950s
Bjerknes (1959)	Oceans	Using ship information shows a rise in sea surface temperatures from 1800s in the Gulf Stream from Cape Hatteras to the Newfoundland Banks
Callendar (1960)	World	Temperature increases in many areas although, since 1948, a cooling trend marked in sub-Arctic. Between 1957 and 1959 cool European summers caused a recession of glaciers. A similar effect in New Zealand. In Japan, 1948 to 1957 warmer than any decade before 1880
Landsberg and Mitchell (1961)	World	Made note of the role of urbanization in temperature changes although found that it had little effect on warming trends in earlier decades. Warming trend reversed in 1940 although continued to rise in some places. Regions of most rapid pre-1940 rise tended to be those with most rapid cooling in post-1940 period

[a] In part after Veryard (1963), where references to cited papers are given.

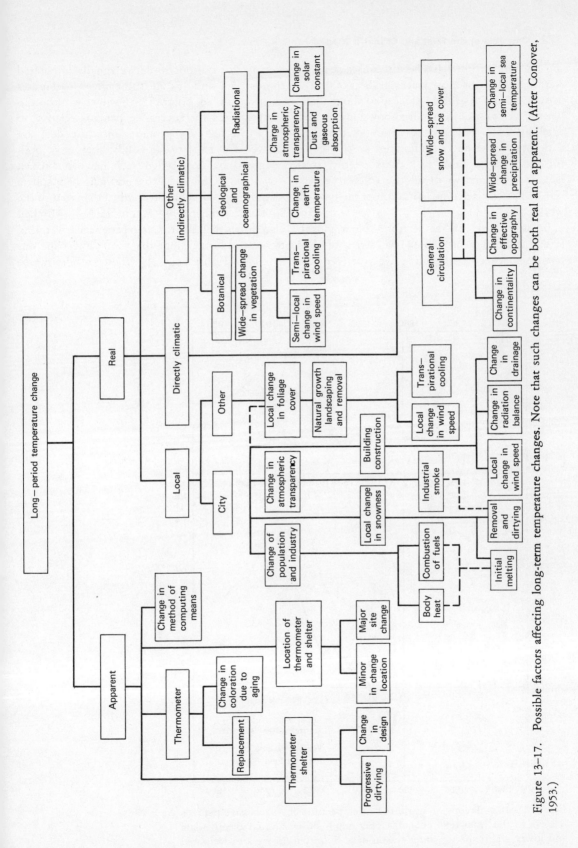

Figure 13–17. Possible factors affecting long-term temperature changes. Note that such changes can be both real and apparent. (After Conover, 1953.)

most widely accepted, has been the discovery of a warming trend for most places in the world that started around 1850 and lasted up to about 1940. Thereafter, most of the studies show a leveling or a falling off of temperatures. Some of the contributory studies and their findings in relation to this are given in Table 13-5. This lengthy table is not representative of all the studies carried out, but does give some indication of the type of work completed.

In assessment of temperatures on a global level, Mitchell (1963) used the series of differences between consecutive five year (pentad) averages of stations to derive zonally integrated trends. His findings are summarized in Figures 13-18 to 13-20. Figure 13-18 shows world mean temperatures for successive pentads relative to the 1880 to 1884 pentad. Since the data are derived for 10° latitude bands, they have been weighted by area. When such data are broken down, it is seen that the northern hemisphere warming was much more marked than that of the southern hemisphere (Figure 13-19). Mitchell also shows that the tropics experienced a warming trend, and furthermore, that the 1940 decline was evident (Figure 13-20). These results agree

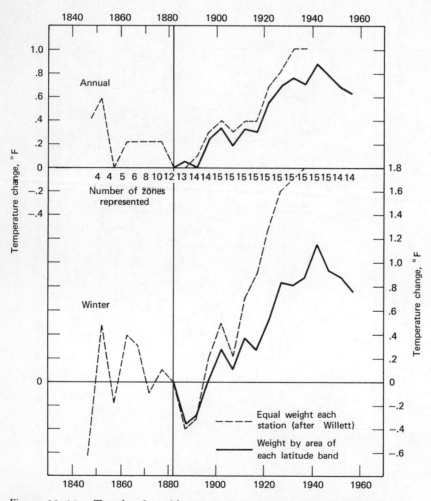

Figure 13-18. Trends of world mean temperature for successive pentads relative to the 1880-1884 pentad. Upper graph shows annual average data and lower graph shows winter season data. Solid curves are area-weighted averages of 10° latitude bands. Dashed curves are unweighted averages for all selected stations. (From J. Murray Mitchell, Jr., *Arid Zone Research*, 20, 1963, UNESCO.)

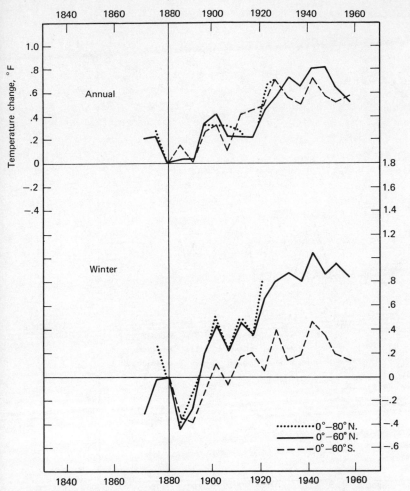

Figure 13-19. Trends of mean temperatures in the northern and southern hemispheres, by pentads, within indicated latitude limits. Upper graph shows annual average data and lower graph winter season data. All curves are area weighted. (From J. Murray Mitchell, Jr., *Arid Zone Research*, 20, 1963, UNESCO.)

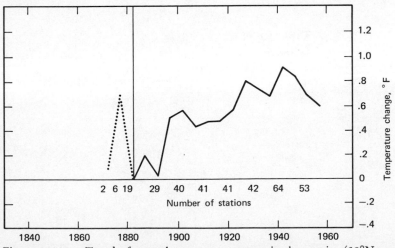

Figure 13-20. Trend of annual mean temperature in the tropics (30°N to 30°S) by pentades. Number of stations used is indicated. (From J. Murray Mitchell, Jr., *Arid Zone Research*, 20, 1963, UNESCO.)

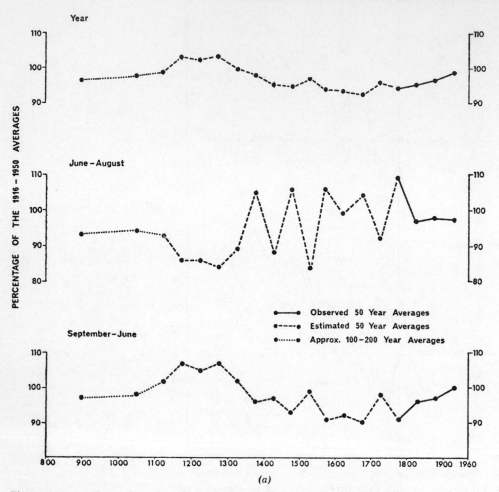

Figure 13–21. Examples of studies of precipitation changes. (*a*) Precipitation trends in England and Wales since 800 A.D. Data shown as 50-year averages expressed as percentages of the mean for the period 1916-1950. (From Barry, R.G., *Water, Earth and Man*, R.J. Chorley, ed., Methuen. Used with permission.)

Table 13-6

THIRTY-YEAR CHANGE OF ANNUAL MEAN TEMPERATURE (°F), 1890–1920 to 1920–1950[a]

Zone	After Callendar		After Willett and Mitchell	
	Inclusive Latitudes	Δ	Inclusive Latitudes	Δ
World	60°N.–50°S.	+0.41	60°N.–50°S.	+0.37
North temperate	60°.–25°N.	+0.70	60°–30°N.	+0.64
			60°–20°N.	+0.57
Tropical	25°N.–25°S.	+0.31	30°N.–30°S.	+0.35
			20°N.–20°S.	+0.39
South temperate	25°–50°S.	+0.25	20°–50°S.	+0.10
			30°–50°S.	+0.08

[a] From Mitchell (1963). Callendar's data are 1891–1920 to 1921–1950; Willett and Mitchell's data are 1890–1919 to 1920–1949.

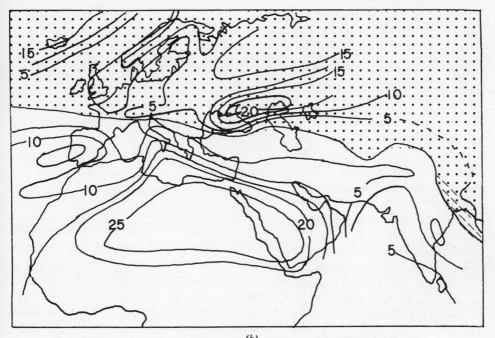

Figure 13-21 (*Continued*). (*b*) Annual precipitation anomalies 1881-1910 to 1911-1940, expressed as percentage deviations from the mean for 1881-1910. Stipple area designates areas of positive anomalies; unshaded area indicates negative anomalies. (After Butzer, 1957.)

fairly well with other derived values, such as those calculated by Callendar (1958) and shown in Table 13-6.

It is much more difficult, in view of short-term variations and year-to-year variability, to obtain such a clear view of precipitation trends. Use of plotted cumulative deviations from the mean and power-spectrum analysis have, however, shown some trends. Kimble (1950), for example, suggests that there was a decline in precipitation in the first half of the twentieth century in North America. Other writers (Dingle, 1955, Landsberg, 1960) disagree and find no significant trend. Examples of some precipitation investigations are illustrated in Figure 13-21. The first graph shows precipitation trends in England and Wales on an annual and seasonal basis, each

expressed as a percentage of the mean for the period 1916 to 1950. Obviously, the older the extrapolation, the more tenuous are the results. The map given in Figure 13-21 shows Butzer's estimation of precipitation changes in Europe and North Africa. The negative anomalies in the Sahara and in Arabia as well as much of the Mediterranean area indicate that this century is appreciably drier than the last. On the other hand, the more northern areas, with positive anomalies, experienced an increase in precipitation up to the end of the record used.

The significance of climatic changes, particularly those in temperature during the twentieth century has caused much discussion over the role of man in inducing such changes. This problem is explored in Chapter 14.

CHAPTER 14

Theories of
Climatic Change

From the wealth of evidence available, it is clear that over time great climatic changes have occurred over the face of the earth. Given this fact, it is natural that the possible cause of such changes has evoked much study. The answer to the intriguing question of what causes climates to change has not been, and perhaps never will be completely ascertained. It is not, however, for lack of ideas; in the 100 years or so since climatic change was first recognized, it has been estimated that for every year that has passed, a new theory has been postulated. Not all have proved satisfactory, because some fail to account for two basic ingredients of a viable theory. First, the theory must be able to explain the onset of ice ages through geologic time (occurrence of long periods of warmth with cooler interruptions); second, the theory should contain some explanation of how warming and cooling periods can occur within an identified glacial epoch. Unfortunately, difficulty in substantiating (or rejecting) those theories that do fulfill these criteria leads to further problems, for, as Menzel (1953) notes, "One of the major difficulties with the problem of ice ages is that we have too many theories, almost any one of which sounds plausible on the basis of qualitative reasoning. Quantitative discussion of the problem proves to be difficult because there are so many unknown factors and uncertain elements."

The quest for theories explaining ice ages is no longer purely academic. At the present time man is actively modifying both the atmosphere and the surface of the earth; there is quite a weight of evidence to indicate that such changes might induce climatic change. To comprehend this change in the total framework of the global climate, the understanding and explanation of "natural" climatic change is required. Bryson (1968) shows this clearly in a short paper titled *All Other Factors Being Constant . . .* which, although essentially concerned with man-induced changes, is really applicable to the whole spectrum of climatic change. He notes that neglecting the very small flow of heat from the earth's interior, the earth's heat budget may be expressed by

$$ScA = KeT^4(4c)$$

or in words "the intensity of sunlight arriving at the earth (S), times the cross-sectional area of the earth (c) times the fraction of that radiation which is absorbed (A) is equal to a constant (K), times the 'effective emissivity' of the earth (e), times the average temperature of the earth raised to the fourth power (T^4), this outward heat flowing from the entire surface of the earth $(4c)$." Excluding constants and earth dimensions, the equation tells the nature of the variables that can be altered to lead to a modified energy budget and ultimately to a changed climate. Either the solar output might vary or the nature in which energy is budgeted in the earth-atmosphere system might change. Changes can be thus related to extraterrestrial causes or changes in the earth-atmosphere system.

EXTRATERRESTRIAL CAUSES

Orbital Variations

In Chapter 1, variations in the earth's motion around the sun were used to explain diurnal and seasonal differences in the amount of solar energy arriving at the surface. The relative positions of the earth's axis and the distance from the sun are not, however, constant values, as they vary over time. In 1842, the French astronomer Adhémar suggested that variations in the earth's orbit around the sun might induce climatic changes at the surface. James Croll, in 1860, extended this idea to include other aspects of orbital variation. Essentially, the combined effects include:

a. The obliquity of the ecliptic; the variation of the angle of the plane through the equator compared with the plane of the earth's orbit around the sun. It has been calculated that, on a cycle of a period of about 40,000 years, the angle varies by about 1.5° about a mean of 23.1°. When the angle of obliquity is large, marked hemispheric contrasts will occur between summer and winter (Figure 14-1*a*).

b. The earth's eccentricity in its orbit around the sun. The earth moves around the sun in an elliptical orbit so that at present there is a time when the earth is most distant (aphelion) and closest to the sun (perihelion). The eccentricity of the orbit is given by

$$e = \frac{le.}{a}$$

as shown in Figure 14-1*b*.

c. Precession of the equinoxes; because of the earth's various motions and its position relative to other planets, the day on which the earth reaches its closest point to the sun varies. At present, perihelion occurs during the northern hemisphere winter, but the date varies throughout the entire year over a period of 21,000 years. Thus, in 10,500 years time, the date of perihelion will have passed from December 21 to June 21 (Figure 14-1*c*).

The dates at which these events occur can be worked out going back thousands of years. It is possible to establish times at which the variations are in or out of phase, so that they might produce very high, or very low, radiation values within the hemispheres. The Yugoslavian astronomer Milankovitch spent many years establishing relationships between the variations and climates of the past. Figure 14-2 shows some of the curves he derived. In developing his theory of the astronomical cause of climatic change, Milankovitch assumed that glaciation would occur during periods when high latitudes receive a minimum of solar radiation. As deduced from the preceding rotational-orbital variations, these minima will occur when (1) the tilt of the earth's axis is smallest and (2) when the summer solstice occurs at or about perihelion. That is, glaciation is favored by mild seasons.

While many workers accept the basic idea of the astronomic control of climate, others have voiced strong objections. Common criticisms include:

a. The computed astronomical dates of when ice ages should occur does not correspond closely to those that have actually been dated.

b. Climatic cycles of less than 21,000 year periodicity cannot be adequately explained. At the same time, derived curves extended back into time do not always closely correspond to the times at which ancient ice ages actually occurred. This would seem to suggest that there must be other contributing effects and the astronomical theory alone cannot account for ice ages.

c. Variations in the earth's orbit should give rise to glacial periods in just one hemisphere. Evidence clearly indicates that glaciation is a worldwide phenomenon.

In recent years the astronomical theory has been revived and reexamined in the light of new evidence and new dating techniques. At the same time, a number of modifications have been introduced to overcome the preceding objections. Panofsky (1956), for example, suggests that alternate hemispheric glaciations would only be about 10,000 years out of phase and, in terms of geologic time, would not be significant. More recently, work by Broecker (1966) and Broecker and Van Donk (1970) has resulted in modification of the Milankovitch curves to derive new correlations.

Broecker notes that Milankovitch chose to weight the effects of the axial tilt and precessional effect ". . . in accord with the insolation changes they induce at latitudes at which continents are covered by large ice sheets. As the tilt effect is prominent at high latitudes, any redistribution of heat between these latitude zones would lead to a smaller tilt effect than that assumed by Milankovitch. The precessional effect, on the other hand is more uniform throughout the hemisphere." Accordingly, Broecker gave different weightings to the effects (Figure 14-3). When, for example, the weighting factor of tilt versus precession is given a value of 3.7, a correlation can be made between the peaks and glacial-interglacial sequences. The comparison between the predicted event and those that actually occurred, dated using O^{18}/O^{16} and Pa^{231}/Th^{230} methods, are shown in Table 14-1. The correlation is not perfect, and Broecker, with Van Donk, revised the findings using newly acquired data. As shown in Figure 14-4, they relate computed insolation differences to both sea level changes and temperatures derived from oxygen isotope analysis.

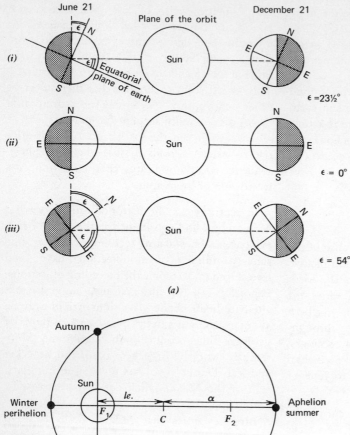

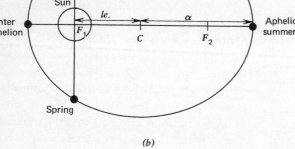

(b)

Figure 14–1. Zeuner's representations of variations in earth-sun relationships. (a) The influence of obliquity of the ecliptic. Diagrams show cross section of the orbit, with earth in midwinter and midsummer positions, when (i) the obliquity is at present value $\epsilon = 23\frac{1}{2}°$, and (ii) at imaginary obliquity, $\epsilon = 0°$. This would result in a lack of seasons but marked climatic zonation. (iii) At imaginary obliquity, $\epsilon = 54°$. This would result in very marked seasonal differences with climatic zonation reduced to a minimum. (b) Eccentricity of orbit. The sun is one focus (F_1) of the elliptical orbit. The distance from the center (C) to aphelion or perihelion is given by half axis major, shown here as a. The distance C to F_1 is termed the linear eccentricity ($le.$). This is used to determine eccentricity (e), which is given by the linear eccentricity divided by half axis major ($e = le./a$).

The Sun as a Variable Star

In most theories of climatic change the output of energy from the sun is considered constant or nearly constant. This is to be expected because if the sun's surface emits energy faster than the core can produce it, then it will contract. Con-versely, if energy is given out too slowly, expansion will occur. A steady-state situation is maintained and the output, and the amount intercepted by the earth, should not alter appreciably. The utilization of the term solar *constant* is further indicative of the climatological

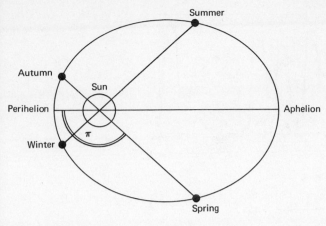

(c)

Figure 14–1 (*continued*). (*c*) The influence of precession of the equinoxes. The four cardinal points move along the orbit. Movement is given by the angle π. (After Zeuner, 1964.)

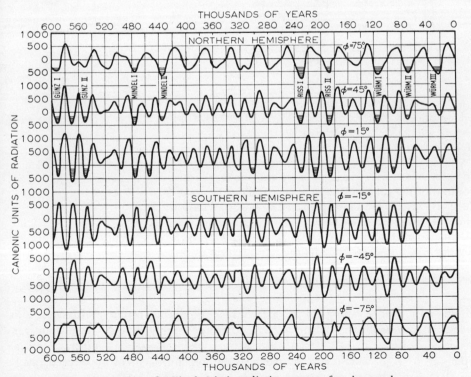

Figure 14–2. Examples of Milankovitch radiation curves for the northern and southern hemispheres at various latitudes. Curves are shown in terms of canonic units obtained by substituting a value of 1 for the solar constant and 100,000 for the sidereal year. (From Schwarzbach, M., *Climates of the Past*, © 1963 by Litton Publishing Co., reprinted with permission of Van Nostrand Reinhold Co.)

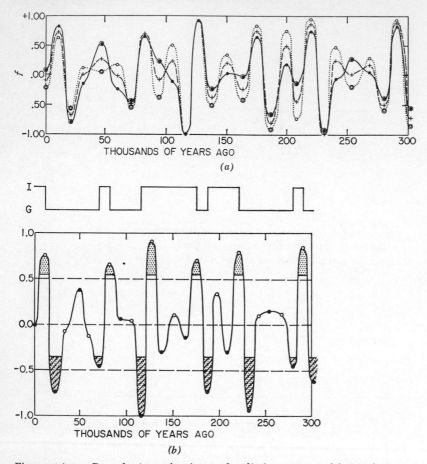

Figure 14-3. Broecker's evaluations of radiation curves. (*a*) Insolation curves constructed for various weighting factors (x) of tilt effect versus precession effect; $x = 5$ (solid curve), $x = 2.5$ (dashed curve) and $x = 1.2$ (dotted curve). (*b*) Northern hemisphere summer insolation based on a weighting factor $x = 3.7$. Dotted areas represent portions of warm peaks capable of triggering the system from glacial to interglacial mode. Hatched areas are those portions of the cold peak capable of returning the system to its glacial mode. The Topset graph shows predicted modes as a function of time derived from the triggers; I represents interglacial conditions, G represents glacial conditions. (From Broecker, W.S., *Science*, 151, 1966. Copyright 1966 by the American Association for the Advancement of Science.)

interpretation of the solar energy falling on a plane at the outer limits of the earth's atmosphere.

When it is considered, however, that a fluctuation of solar output of 8 to 9% of what it now is could account for all climatic changes that have occurred on earth, it is perhaps inevitable that variations in solar output are considered as a possible cause of climatic change. One of the most interesting attempts to use this approach is given in a model suggested by Opik (1958). He

asserts that the output of energy from the sun could alter appreciably if, apart from hydrogen and helium, the sun contains other materials, such as carbon, nitrogen, neon, oxygen, magnesium, silicon, and iron; these are collectively termed "metals" in the model.

The process visualized by Opik is shown in Figure 14-5. Initially a "normal" situation exists (the prevailing situation responsible for warm climates on earth). As time passes, hydro-

Table 14-1

COMPARISON OF (i) DATA PREDICTED FROM BROECKER'S MODEL DESCRIBED WITH (ii) EMILIANI'S ESTIMATES BASED UPON O^{18}/O^{16} TEMPERATURES FOR FOSSIL ORGANISMS IN DEEP-SEA CORES AND ON ABSOLUTE AGE DATA OBTAINED WITH C^{14} AND P^{231}/TH^{230} TECHNIQUES[a]

| | Prediction[b] | | | | Observation | | | |
Period	Beginning (Years Ago)	End (Years Ago)	Length (L) (yr)	Period	Beginning (Years Ago)	End (Years Ago)	Length (L) (yr)	L_{pred}/L_{obs}
W-1	11,200	—	—	1	11,000[c]	—	—	(1.0)
C-1	71,200	11,200	60,000	2-4	65,000[d]	11,000	54,000	1.1
W-2	82,400	71,200	11,200	5	100,000[d]	65,000	35,000	0.3
C-2	116,100	82,400	33,800	6	125,000[d]	100,000	25,000	1.4
W-3	175,700	116,100	59,600	7-9	175,000[e]	125,000	50,000	1.2
C-3	186,900	175,700	11,200	10	195,000[e]	175,000	20,000	0.5

[a] From Broecker, 1966.
[b] $x = 3.7$; boundary limits, -0.35 and $+0.55$ heat units. (See Figure 14-3*b*.)
[c] Value established by C^{14} dating.
[d] Value established by Pa^{231}/Th^{230} dating.
[e] Value established by extrapolation.

gen from the mantle diffuses to the core, leaving behind metals that diffuse more slowly. These accumulate to form a barrier to radiation from the core. In keeping with maintenance of a steady-state condition, the sun contracts.

The metal barrier itself becomes hot and convection currents develop, resulting in the formation of a very large core. This would increase the hydrogen content available for fuel, and energy output would increase. So much heat is produced that it cannot be adequately transported to the surface; the result is that the sun expands. In expanding, energy is expended so that the net effect is to reduce the heat and light output from the sun. The reduced solar radiation would trigger a cooling stage in earth.

Expansion lowers the temperature of the core and the amount of energy it produces. The core shrinks and ultimately the sun returns to its equilibrium conditions. The cooling stage on earth is over and again long, uninterrupted periods of warmth prevail.

Although highly sophisticated, the problem with the Opik model is that it can neither be proved or disproved at the present time. Obviously, the key to the effect is the amount of

elements other than hydrogen and helium in the sun. Estimates of the amounts present are extremely variable.

An alternative approach to the role of the sun as a variable star is through analysis of sunspots. The outer surface of the sun consists of three layers. The outer layer, which defines the visible solar disc seen in white light, is the photosphere. Above this are the chromosphere and the corona, both of which can be observed at solar eclipses or through telescopes that artificially eclipse the sun. Although the photosphere appears as a bright disc to the unaided eye, magnification often identifies circular, dark indentations in its surface. These spots are complicated hydromagnetic phenomena that display strong magnetic fields and exhibit rapid gas motion in and out of the spot. It is a cool region with an average temperature of 4600°K, a marked contrast to the surrounding photosphere at 6000°K. It is the periodic fluctuation of sunspots that has attracted the attention of those looking for a cause of climatic change on earth.

Because of the long history of record and ease of observation, sunspots are used as an indicator of solar activity. A number of formulas to express

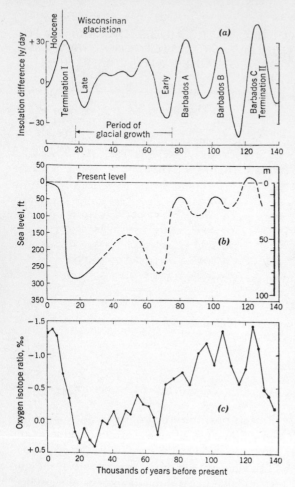

Figure 14–4. Correlation of summer insolation, sea level, and oxygen-isotope ratios for the past 150,000 years. (*a*) Insolation differences at latitude 45°N. (*b*) Inferred sea level changes. (*c*) Oxygen-isotope ratios, modified from core data derived by Emiliani. (From Broecker, W.S. and J. Van Donk, *Reviews in Geophysics*, 8, 1970. Used with permission.)

this activity are available, the most common of which is the Wolf number, given as

$$R = K(10\,g - f)$$

This compares the number of sunspots (*f*) to the number of groups in which they occur (*g*). A calibration constant (*K*) is used depending upon the particular observatory and the time of observation. Current data are given as a daily sunspot number and then are averaged as required, for year or month.

Study of sunspot activity going back many years shows that a number of well-marked cycles can be identified. The best known of these is the 11-year solar cycle and the related 22-year Hale cycle. Other recognized cycles include the 80-year Gleissberg cycle and the 205-year cycle, both representative of cycle strength.

Knowing the differences in solar activity as indicated by sunspot numbers and cycles, a number of workers have attempted to relate climatic cycles to variable solar activity. Most notable is the work of Willett (1953, 1961, 1967). Cited as the physical cause of the relationship is the variable solar emissions of short-wave ultraviolet radiation and charged particles that are related to sunspot activity. The ultraviolet radiation effects are felt in high tropical atmosphere, possibly leading to high-altitude warming, which results in a modification of the zonal circulation. The corpuscular activity, on the other hand, leads to a direct heating of the polar stratosphere. These effects, which are both associated with sunspot activity, produce differing results. Sellers (1965) aptly describes the situation:

"An increase in ultraviolet radiation . . . leads to a strengthening of the zonal circulation pattern and, hence, to a pattern with definite glacial characteristics. On the other hand, the direct heating of the polar stratosphere by excessive solar corpuscular radiation results in 'highly chaotic periods of climatic stress, with markedly contrasting extremes of storminess, rainfall, and temperature in middle latitudes, accompanied by anticyclonic blocking of the circulation pattern and widespread occurrence of hot, dry summers' (Willett, 1953). These periods correspond most closely to Simpson's warm, wet interglacials. The temperate, dry interglacials are generally periods when both ultraviolet and corpuscular radiations are low. Since sunspot activity is related to both ultraviolet and corpuscular radiation, its connection with climatic change is somewhat confused."

Despite such valid criticism, Willett's work has met with much acceptance, probably because it can be verified over short-term periods. Increases in ultraviolet emissions result in a shift of circulation patterns while corpuscular radiation is sometimes followed by a breakdown of zonal circulation. Nonetheless, it remains a major problem in reconciling the two effects in terms of climatic change.

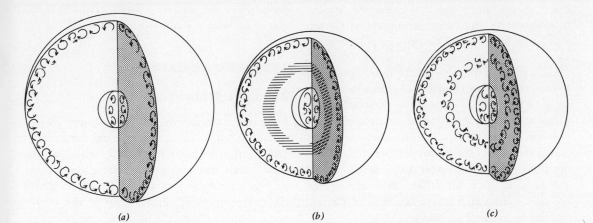

Figure 14–5.   Opik's concept of the theoretical cycle of the sun. (*a*) Cycle starts with normal sun that results in a warm period on earth. Nuclear reactions in the core are promoted by convection currents. (*b*) Hydrogen from the mantle diffuses into the core, leaving behind more slowly diffusing "metallic" elements. These act as a barrier to energy radiated from the core and the sun contracts. (*c*) As the barrier itself becomes hot, it develops its own convection currents. (*d*) The initial barrier spreads to form a large core. Increased hydrogen content results in increased output of energy. Accordingly, the sun expands to use excess energy. Less radiation will strike the earth and an ice age may result. (*e*) the sun then returns to its normal, long-term state. (From Opik, E.J., *Scientific American*, June, 1958. Copyright © 1958 by Scientific American Inc. All rights reserved.)

Evidence is still being accumulated and published relating sunspots to climatic changes. Bray (1970), for example, has noted that the most marked aberration of the past 2000 years, the temperature decline between 1550 and 1900, correlates with an interval of low solar activity. Evidence is also presented that such decreased activity recurs at cyclic periods of around 2400 to 2600 years.

The explanation of climatic change through a decrease in the amount of energy output of the sun has not gone unchallenged. In 1934, Sir George Simpson suggested that an increase in solar radiation instead of a decrease would trigger a cooling period. He maintained that lower energy input would decrease the amount of evaporation and moisture content of the air, eventually leading to a decrease in precipitation. Despite lower temperatures resulting from decreased solar output, ice would simply not form because of lack of snowfall. On the other hand, he maintained that an increase in solar output would increase evaporation, and movement of air poleward would lead to excessive snow and high cloudiness. Even the higher temperatures resulting from the increased solar output would not be sufficient to melt all fallen snow during the summer; season after season accumulation would see the growth of icecaps.

The growth of the ice would terminate when temperatures were so high that most of the precipitation fell as rain rather than snow. An interglacial period would then commence. Based upon such a sequence, Simpson visualized two such cycles in the Pleistocene glaciation. Figure 14-6 is based upon that given by Bell (1953),

who provides an excellent summary of the problems of the glaciation with a cool sun versus glaciation with a warm sun controversy.

TERRESTRIAL CAUSES

Atmospheric Modification

As noted previously, the atmosphere plays a significant role in modifying the amount of insolation reaching the ground, while, at the same time, the greenhouse effect retards the passage of outgoing radiation. It follows that if the composition or constituency of the atmosphere is altered, then the heat and energy budget will be modified. This premise supplies the basis for several theories of climatic change. Recently, studies of the role of the atmosphere in modification of radiation have assumed even more importance, because man's pollution of the atmosphere could accelerate climatic changes that result from such variations.

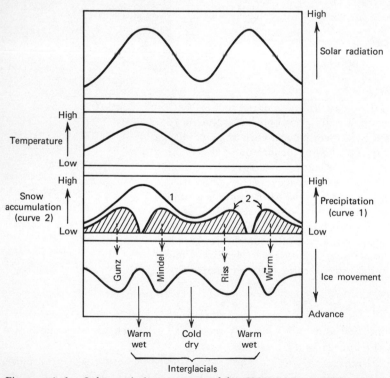

Figure 14-6. Solar variations suggested by Simpson to explain Pleistocene variations. He suggested that four glacial advances would result from two maxima of solar radiation. (After Bell, 1959.)

Particulate Matter

In August 1883, the volcanic island of Krakatoa in the East Indies exploded with such violence that two-thirds of the island "disappeared." It was estimated that some 13 cu miles of dust and ash were injected into the atmosphere. The debris rose more than 20 miles into the air, where it was caught up in the upper air circulation. For some time after the explosion, the haze in the upper air caused the moon to appear in a blue hue, while sunsets and sunrises over wide areas of the earth were brightly colored. At Montpellier Observatory in France, a marked reduction in the intensity of solar radiation was recorded, and for the next three years the value was about 10% below normal (Figure 14–7). A similar reduction in solar radiation was recorded in observatories remote from the large volcanic eruption that occurred at Katmai, Alaska in 1912.

These events prompted much interest into whether, in fact, volcanic dust that resulted in times of marked volcanic activity could have been responsible for a decrease in temperatures in the past. A leading figure in research into this possibility was the American physicist and meteorologist Humphreys, whose book *Physics of the Atmosphere* contains an absorbing account of his attempts to correlate cooling periods and volcanic activity. Unfortunately, the relationship could not be unequivocally substantiated.

The role of volcanoes as a contributing factor to climatic change was revived by Wexler (1956). He drew upon the work of Schaefer to expand the basic theory, for Schaefer had shown that volcanic dust is an effective nucleating agent at below freezing temperatures and that cloud formation resulting from volcanic dust would further modify climate. Another development applicable to the relationship was the study of deep sea cores that show that, in some instances, glacial deposits do occur immediately above volcanic deposits. There are instances, however, when volcanic debris is absent from lower levels of glacial deposits. Wexler explained this by stating that a relatively small amount of dust can modify incoming radiation but that it would be insufficient to account for a well-defined marine deposit.

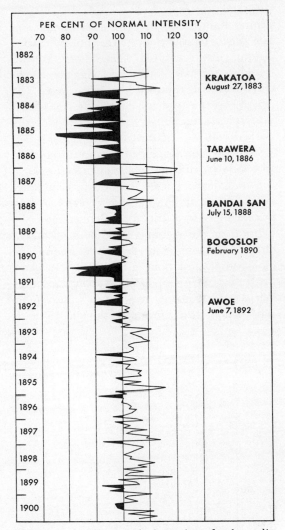

Figure 14–7. Changes in intensity of solar radiation at Montpelier, France, between 1882 and 1900, suggest that five volcanic explosions affected the climate of the earth. The intensity of radiation was measured on clear days at noon. Location of named volcanoes: Krakatoa is between Sumatra and Java; Tarawera in New Zealand; Bandai San in Japan; Bogoslof in the Aleutian Islands; Awoe in the Malay Archipelago. (From Wexler, H., *Scientific American*, April, 1952. Copyright © 1952 by Scientific American Inc. All rights reserved.)

A recent contribution to the effect of volcanoes (Cronin, 1971) stresses the importance of the height to which volcanic material is vented into the atmosphere and the significance of the material that is ejected. The author notes that after the gigantic explosion of Bezymianny, Central Kamchatka, in 1956, no volcanic eruption penetrated the troposphere until 1963 when Agung, Bali, erupted. Thereafter, Trident, Alaska, and Surtsey, off Iceland, also gave rise to eruptions that ruptured the tropopause. The period between 1956 and 1963 was a time of low stratospheric turbidity. After the eruptions, a marked increase occurred (Figure 14-8).

Cronin notes that different types of volcanic eruption have different effects upon stratospheric contamination. Surtsey, for example, was responsible for the intromission of water; Agung provided stratospheric ash. Furthermore, the location of the volcanic eruption is of significance in its long-term effect upon atmospheric modification. Eruptions in the Arctic belt, where the height of the tropopause is about two-thirds that at the equator, may be of much more significance than previously thought. This is further indicated by the fact that observations of volcanoes capable of contributing toward stratospheric turbidity within, for example, the Aleutians, simply may not be noticed.

The possibility of volcanic material inducing changes in climate leads directly to the potential role of other particles in the atmosphere. Sources and amounts of particulate matter, both natural and man-made, are shown in Table 14-2.

There are many interpretations of the effects of man's pollution of the atmosphere; one that has generated much interest recently is the possibility of climatic change resulting from high-flying supersonic transport planes (SSTs). These aircraft will fly at an altitude of about 12 miles and, through fuel combustion, will release water vapor, nitrogen oxides, carbon dioxide, and particulate material high in the stratosphere. The possible effects of development of a fleet of such aircraft might have a marked impact upon

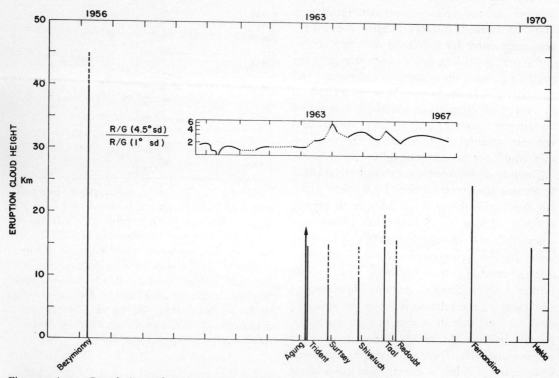

Figure 14-8. Correlation of atmospheric turbidity and recent volcanic eruptions that vented to the stratosphere. Arrow indicates maximum height to be an altitude higher than reported. The inset shows red (R) to green (G) twilight sky radiance measurements at 1° and 4.5° solar depressions, a measure of atmospheric turbidity. (From Cronin, J.F., *Science*, 172, 1971. Copyright 1971 by the American Association for the Advancement of Science.)

Table 14-2

ESTIMATES OF PARTICLES SMALLER THAN 20-μ RADIUS EMITTED INTO OR FORMED IN THE ATMOSPHERE (10^6 METRIC TONS/YEAR)[a]

NATURAL

Soil and rock debris[b]	100– 500
Forest fires and slash-burning debris[b]	3– 150
Sea salt	(300)
Volcanic debris	25– 150
Particles formed from gaseous emissions	
Sulfate from H_2S	130– 200
Ammonium salts from NH_3	80– 270
Nitrate from NO_x	60– 430
Hydrocarbons from plant exudations	75– 200
Subtotal	773–2200

MAN-MADE

Particles (direct emissions)	10– 90
Particles formed from gaseous emissions	
Sulfate from SO_2	130– 200
Nitrate NO_x	30– 35
Hydrocarbons	15– 90
Subtotal	185– 415
Total	958–2615

[a] From SMIC (1971).
[b] Includes unknown amounts of indirect man-made contributions.

climate, the SCEP (1970) report suggests the following:

1. The added SST water vapor in the stratosphere may introduce the following three effects in order of likely importance;

(a) Stratospheric clouds, already observed in the polar night, may increase in frequency, thickness, and extent. The effect of SST water vapor will be heightened by the increasing trends in CO_2 and natural stratospheric water vapor (observed only over Washington, D.C.). (b) Direct radiation effects will (according to Manabe and Wetherald's quasistatic radiation calculations) result in warming of air at ground level in regions of peak moisture concentration by less than a tenth of a degree C on a worldwide basis and cooling in the stratosphere by a few degrees C. (The actual global effect would be smaller than the expected changes due to CO_2 increases.) (c) The reduction of ozone due to water vapor interaction (in a static photochemical model) has been estimated to lie well within the present day-to-day geographical variability of total ozone.

2. The direct role of quantities of CO, CO_2, NO, NO_2, and hydrocarbons in altering the heat budget is small. It is also unlikely that their involvement in ozone photochemistry is as significant as water vapor.

3. The SO_2, NO_x, and hydrocarbons can undergo complex reactions that produce particles. Increased particulate loadings may raise the temperature and play an important role in the stratospheric heat balance.

As described in the discussion of urban climates (Chapter 8), the role of particulates introduced into the atmosphere through combustion of fossil fuels is not completely resolved. It is generally thought, however, that high-altitude aerosols induce surface cooling while low-altitude aerosols, those mostly associated with human activities, have a warming effect (Mitchell, 1972).

Carbon Dioxide

The familiar greenhouse effect is a partial consequence of carbon dioxide in the atmosphere. This gas, like water vapor and ozone, is transparent to visible radiation but opaque to outgoing long-wave radiation at given wavelengths. As long ago as 1861, John Tyndall suggested that variations in the CO_2 content of the atmosphere could result in climatic change. The theory has been extensively reexamined by Plass (1956). As already outlined, the carbon dioxide cycle is a complex one, with exchanges occurring between the three great reservoirs, the oceans, the lithosphere, and the biosphere. The atmospheric content, which contains about 2.3×10^{12} tons of carbon dioxide, depends upon the amounts supplied and withdrawn from these reservoirs.

The amount of CO_2 in the atmosphere can vary naturally. Changes can result because of differing rates of weathering and volcanic activity or because the amount "tied up" in vegetation or other organisms does vary over time. But this variation is a relatively temporary interruption of an equilibrium state between atmosphere and oceans. It is suggested that a decrease of the amount in the atmosphere would initially result in a drop in temperature and possible growth of glaciers. Glacial growth would cause a drop in sea level, and the sea would contain an excess of carbon dioxide in carbonate form. Carbon dioxide would be released to the atmosphere and the equilibrium state would be reestablished; a temperature rise would result, ice would melt, and oceans would fill. The cycle would then repeat itself.

Plass suggests that as long as the total amount of carbon dioxide in the earth-atmosphere system remains constant, the cycle just outlined will continue. He writes "The oscillation is reinforced by the accompanying change in earth's humidity. A colder atmosphere holds less water vapor, and so further reduces the atmospheric absorption of infrared radiation emitted by the earth's surface. At the same time, however, the earth's cloud cover thickens and precipitation increases despite the reduction in water vapor. . . . The top of the cloud is cooled by radiation of heat into space; when there is less carbon dioxide in the atmosphere, cloud tops

lose more heat energy and thus become colder. With a steeper temperature gradient there is increased convection in the cloud. The result is larger clouds and more precipitation." It follows, too, that reflection from the increased cloud cover results in lower earth temperatures.

The theory seems highly plausible, but it is weakest in quantitative terms. Kaplan (1960) has reworked Plass's calculations and, even neglecting the shielding effect of water vapor, finds that a decrease of one-half of the carbon dioxide content of the atmosphere would result in a temperature decrease of only 1.8°C. Thus, to account for changes from glacial to interglacial periods, concentration must vary by one order of magnitude. There is no evidence that this occurs naturally.

More recently, however, the amount of carbon dioxide added to the air through man's activities has increased enormously, and again the specter of climatic change is evoked. The global rise in air temperature between 1870 and 1940 was attributed to addition of carbon dioxide through man's activities. Since the 1940s, however, tem-

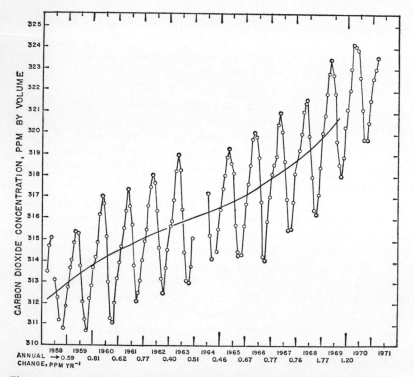

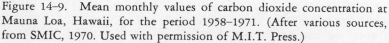

Figure 14–9. Mean monthly values of carbon dioxide concentration at Mauna Loa, Hawaii, for the period 1958–1971. (After various sources, from SMIC, 1970. Used with permission of M.I.T. Press.)

perature has decreased despite the fact that carbon dioxide is still being added to the atmosphere (Figure 14-9). The potential effects of carbon dioxide have been vigorously investigated by a number of researchers and numerical models have resulted. Significant work in this respect has been published by Möller (1963), Manabe and Wetherald (1967), and Rasool and Schneider (1971). Most of these conclude that carbon dioxide will result in a rise in global temperature, the actual amount varying with each model. In terms of what might actually result from changes in carbon dioxide additions, it is necessary to predict both the future concentration and what amount will remain in the atmosphere. As an example of variation and trend, Figure 14-9

shows the increases of carbon dioxide measured at Hawaii. There is a seasonal variation, with lower values in summer, while the long-term values show an upward trend despite the fact that the year-to-year rate of increase is variable. The prediction of what might happen by the year 2000 is shown in Figure 14-10. It is estimated that the concentration may rise to 375 ppm. Given this, the Manabe and Wetherald study previously cited suggests that this will result in a warming of the surface layer by 0.5°C.

Surface Changes

Changes at the surface of the earth can modify the thermal balance and ultimately the general atmospheric circulation. There are obviously

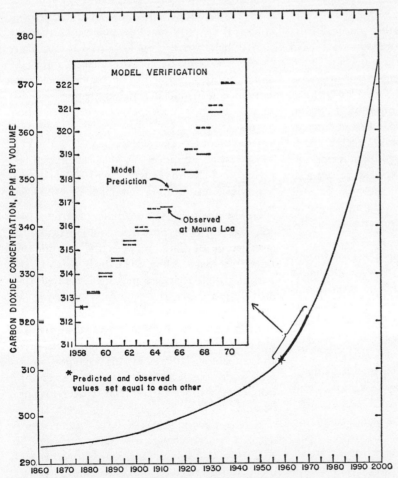

Figure 14-10. Model calculations of atmospheric carbon dioxide from combution of fossil fuels. Observed data from Hawaii (inset) is shown as part of model prediction extending to the year 2000. (From SMIC, 1970. Used with permission of M.I.T. Press.)

many ways in which the surface of the earth, both in terms of the land and sea surfaces, can and have, changed over time. At one time or another most of the major changes have been incorporated into theories of climatic change.

Mountain-building epochs have frequently been cited as the physical mechanism for climatic change. One has only to think of the formation of permanent snow on Mount Kilimanjaro on the equator to visualize the potential role. Geologists have long noted the relationship between great orogenic periods and glacial epochs. For example, both the Carbo-Permian and the Quaternary ice ages were preceded by extensive mountain building. Unfortunately, in relating the two there is a tremendous time lag of millions of years between mountain building and onset of glaciation. The response to this criticism by supporters of the mountain-building theory is that glacier formation is most favored in mountains that have undergone erosion, so that they are best suited for glacial accumulation. Such an explanation would partially account for the time lag between mountain formation and the onset of glaciation.

Other criticisms are not so easily countered. For example, the Caledonian orogenesis that occurred in Europe 370–450 million years ago did not give rise to glaciation. The theory also cannot satisfactorily explain the periodic advances and retreats of ice during a glacial epoch. Nonetheless, most researchers in the field of climatic change agree that mountain building, or at least the presence of high mountains, certainly contributes toward optimum conditions for ice formation. Some have noted, too, that the vulcanicity associated with mountain building may also contribute toward the effect; here, however, the problem of the time lag again produces problems of correlation.

A somewhat different approach to the role of mountains in the causation of ice ages has been suggested by Zeuner (1959). His eustatic-temperature control theory depends upon the fact that, in northern realms, a decrease in temperature of about 1°C for every 100 m ascent occurs. Thus, as part of the theory that assumes that the antarctic ice sheet formed prior to the Pleistocene, a drop in sea level could result in a worldwide cooling trend. The water tied up in the

ice sheet would produce a lowering of sea level, a eustatic change, which would have the effect of increasing altitude, or height above sea level, over large areas of the globe. The effect would be most marked in polar realms, because a drop in sea level that resulted in a lowering of temperature of 1°C at 45°N, might be represented by a drop of 10°C at 80°N.

The role of the ocean is also considered in several other theories of climatic change. One often-cited effect is the modification of oceanic circulation that occurred during Tertiary times. As Figure 14–11 shows, originally there was an east-west axis of oceanic flow because of the extent of the Tethys Sea. During the Tertiary, barriers were formed and the whole circulation pattern was modified.

Another often invoked cause of climatic change is continental drift. As Schwartzbach notes, "If the position of the pole varies, then the climate belts would migrate." As a result, many of the problems of climatic change might be more readily explained. The nature and problems of continental drift have already been discussed, but the concept is reintroduced here because it plays a significant role in a number of ice-age theories. One theory that has attracted considerable attention is that suggested by Ewing and Donn (1956, 1958). Their initial hypothesis rested upon a number of principles:

1. The melting of an Arctic pack-ice cover (such as exists at present) would increase the interchange of water between the Atlantic and Arctic oceans, cooling the North Atlantic and warming the Arctic and making it ice-free, thus providing an increased source of moisture for the polar atmosphere.

2. Two factors would then favor the growth of glaciers; (i) increased precipitation over arctic and subarctic lands and (ii) changes in atmospheric circulation, the latter also resulting from the warmer Arctic and cooler Atlantic oceans.

3. The lowering of sea level would greatly decrease the exchange of water between the Atlantic and Arctic oceans, which together with the cooling effect of the surrounding glaciers, would reduce Arctic surface temperatures until an abrupt freezing occurred. The fairly sudden reversal of conditions favorable to glacial de-

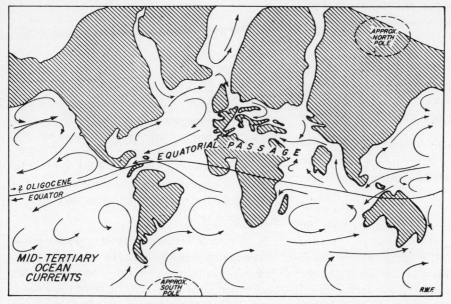

(a)

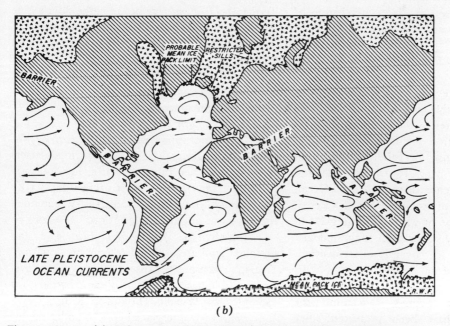

(b)

Figure 14–11. (a) Palaeogeographic map of world oceanic circulation during the middle Tertiary. Note the absence of the present north-south barriers and the world-encircling "equatorial passage." (b) Palaeogeographic map of world oceanic circulation during the later Pleistocene. Note the appearence of modifications caused by (a) the Isthmus of Panama, (b) the narrow Strait of Gibraltar, (c) blocking of the Tethys Sea across northern India, (d) narrowing of the East Indian passages, and (e) the closing of the Torres Strait Islands. The restricted zonal circulation, with more north-south water transfer, is in marked contrast to the upper map. (From Fairbridge, R.W., *Problems in Palaeoclimatology*, A.E.M. Nairn, ed., 1964, John Wiley and Sons. Used with permission.)

velopment would terminate the growth of glaciers abruptly.

4. As continental glaciers waned, the sea level would rise, causing an increased transport of surface waters northward until the Arctic ice sheet melted once again, completing the cycle.

They thus visualize that the temperature changes of Arctic and Atlantic waters are the cause, rather than the effect, of glacial development (Figure 14-12).

For the initiation of this sequence, it is necessary to consider polar wandering. About this the authors write, "The poles are presently located in positions of extreme thermal insolation, in marked contrast to the conditions that would prevail if both were in the open ocean. If the North Pole were located in the North Pacific and the South Pole at the antipodes of this, in the South Atlantic Ocean, the free interchange of water with polar regions would preclude the formation of polar ice caps. . . . It is proposed here that the migration of the poles from an open-ocean environment to the thermally isolated arctic and antarctic regions resulted in the change from the warm equable climate (of the Tertiary) to the glacial climates of the Pleistocene."

Since the postulation of this theory a number of arguments against its validity have been presented. In a 1967 paper, for example, Ku and Broecker offer evidence that the Arctic Ocean has not been free of ice for 150,000 years, a theory that does not fit with the Ewing-Donn hypothesis; similarly, it has been suggested that the distribution of ice cover is not consistent with the concept of an ice-free ocean (SMIC, 1971).

Numerous other theories concerning the changing surface have been presented; they range from the effects of changing surface albedoes (e.g., Hoikes, 1968) to the instability of the antarctic ice sheet as a trigger effect for glacial-interglacial cycles (Wilson, 1964). Other ideas, including possible changes resulting from variations in the ozone layer, are outlined in a good review by Beckinsale (1965). Recently, man-made surface changes have been studied to see how modification through large-scale irrigation schemes, clearance of forests, and so forth, may influence the global climate (SMIC, 1971, SCEP, 1970).

Toward a Compromise of Theory

The preceding account merely presents the better-known theories of climatic change. From these, it is clear that while no single theory is perfect, there are aspects contained within it that probably contribute toward climatic change. Perhaps climatic change is best expressed as a convergence of ideas rather than any single, dominant theory. In presenting what he terms an eclectic theory of ice ages, Fairbridge (1961) brings together many of the ideas that may be represented as follows:

1. In middle Tertiary the most important event in *world oceanography* was the closing of the Tethys Sea, and, soon after the formation of the barrier that closed the Atlantic from the Pacific. Cool ocean currents now move north and south replacing the east-west warm currents. In late Tertiary *polar wandering* brought the South Pole into the mountainous continental mass of Antarctica and the North Pole to the Arctic Ocean.

2. With the build up of ice in Antarctica a worldwide lowering of sea level occurred and the planetary atmosphere cooled by 0.5 to 1.0°C. This is the *eustatic control* of Zeuner.

3. Because of *astronomical solar radiation cycles*, additional cooling occurred permitting glaciers to merge into ice caps. Similarly, low-intensity phases in *longer-period sunspot* oscillations set up atmospheric circulation changes that extended the effects of mountain glaciation.

4. Initially the *Arctic basin was open* and marginal land areas would receive precipitation. Freezing occurred because of drop of sea level and possible *tectonic uplift* of oceanic barriers, for example, the Wyville Thompson Ridge.

5. With each interglacial epoch, under the influence of *Milankovitch radiation oscillations* a rise of temperature occurred in middle latitudes, while *longer-period sunspot cycles* would also modify the general zonal circulation. Periodic readvances of ice would possibly reflect the secondary role of 550-year cycles that led to short-term interruptions of the melting.

6. The rise in sea level occurred with the melting, but *retardation*, both in relation to air temperature with respect to effective radiation, and melting with respect to air temperature, assumed importance. The retardation may have been in

Continental ice
Floating ice
N. Atlantic Drift
Davis–Iceland–Faroes Ridge

(a)

(b)

(c)

(d)

Figure 14–12.   Schematic representation of the Ewing-Donn glacial theory. Maps show the north polar region at (*a*) the beginning of a glacial phase; the Arctic Ocean is ice free; (*b*) the glacial phase when the Arctic Ocean remains ice free; (*c*) the maximum extent of glaciation: with the Arctic Ocean ice covered, this would also represent the beginning of an interglacial phase; and (*d*) the interglacial phase, the present condition, when the Arctic Ocean is ice covered. (After Schwarzbach, 1963.)

the order of several thousand years. Once set in motion, the melting was rapid. Ice of Greenland and Antarctica did not melt during the radiation oscillations of the last half million years because of their basin shapes and the high albedo of high-altitude ice.

The preceding simplified outline indicates that Fairbridge, like many other workers in the field of climatic change, favors solar controls superimposed upon topographic situations to explain ice ages. The outline represents and includes many of the theories described, and perhaps the holistic viewpoint represents the most realistic approach.

The very necessity of a compromise theory points to shortcomings in our total understanding of climate and climatic change. Perhaps it is appropriate to conclude with some remarks by F. Kenneth Hare (1971) on future climates and future environments. He writes, "Explaining past climates and predicting future climates are complementary processes. As Murray Mitchell pointed out . . . (Mitchell, 1968), we are miles away from explaining the past, whose record is so blurred that there remains 'many degrees of freedom for the construction of altogether new hypotheses.' He laid stress where I shall . . . on the view that explaining climatic variation must rest on *an adequate theory of existing climate*. Until we can argue backwards and forwards in time from such a secure foundation we shall continue to be the blind led by the blind."

APPENDICES

APPENDIX 1

Selected Climate
Classification Schemes

MILLER'S CLASSIFICATION

A climate classification widely used in the United Kingdom is that devised by A. A. Miller. It is an interesting system from a number of standpoints. First, it is a simple system devised purely to facilitate general description of world climates; second, it is not entirely empiric, for empirically derived boundaries are essentially defined in terms of only one climatic variable. The system is based upon the division of the globe into temperature zones and their subdivision by precipitation seasonality. The temperature divisions are derived from vegetation zonation and five types are recognized. These are designated hot, warm temperate, cool temperate, cold, and arctic.

The hot zones must have the essential quality of never experiencing temperatures too low for active plant growth. Miller states, "There is much to be said for delineating climates which are continuously hot by means of the coldest month isotherm. . . ." In accordance with this, he selects the 70°F isotherm for the coldest month as the poleward limit of his hot zones. This limit corresponds closely to Koppen's 18°C isotherm delineating the region in which megathermal plant types are found. Outside of the hot zone, where there is sufficient moisture, three types of forests are found: broadleaf evergreen, broadleaf deciduous, and coniferous forests. Miller suggests that broadleaf evergreens do not grow where temperatures fall below 43°F; he uses this to delimit the warm temperate climates. The poleward limit of the zone is thus equated to the 43° annual isotherm. Where temperatures fall below 43°F, either broadleaf deciduous or coniferous forests are found. With the former there is a "resting" period imposed by the low temperature. Miller's study of the limits of the broadleaf deciduous forests and their transition to the coniferous forests shows an approximate correlation to the 43°F isotherm

for six months. This is assumed to be the boundary between the cool temperate and the cold climates. To differentiate the cold from the arctic zone, Miller uses the boundary that corresponds to the 43°F isotherm for three months.

The zonal base of Miller's system is thus based upon vegetation limits and is highly empiric. Further empiricism occurs in the definition of his arid regions. After a discussion of the rather complex formulas devised to define aridity, Miller concludes that a simple formula might prove equally valid. He uses the one first suggested by de Martonne, which defines an arid region as one in which the total annual precipitation is less than one-fifth the average annual temperature.

$$P \leq T/5$$

Clearly, such an estimate, without regard to seasonality of precipitation, leaves much to be desired. But it is perhaps more meaningful than that used in some adaptations of Miller's system, where the boundary is equated to the 10-in. annual isohyet.

Beyond the determination of aridity, no further quantitative boundaries are given. Miller completes the system by superposing seasonal rainfall distribution onto his temperature zones. Three types are differentiated:

1. Areas with rain at all seasons.
2. Areas of marked seasonal drought.
3. Areas of drought ($P \leq T/5$).

Because, with the exception of his dry areas, no actual definitions are provided for the precipitation distribution, the criteria that Miller uses must be assessed from the classification itself (Table A1-1) and a map showing the distribution of the climates differentiated (Figure A1-1). An inspection of these two indicates that precipitation *total* plays no part in the divisions recognized; they are based upon seasonality and,

Table A1-1

MILLER'S CLIMATIC CLASSIFICATION

A. HOT CLIMATES (Mean annual temperature exceeding 70°F)

1. Equatorial: double maxima rain
1m. Equatorial (monsoon variety)
2. Tropical, marine: no marked dry season
2m. Tropical, marine (monsoon variety)
3. Tropical, continental: summer rain
3m. Tropical, continental (monsoon variety)

B. WARM TEMPERATE CLIMATES (No cold season, i.e., no month below 43°F)

1. Western Margin (Mediterranean): winter rain
2. Eastern Margin: uniform rain
2m. Eastern Margin (monsoon variety): marked summer maximum rain

C. COOL TEMPERATE CLIMATES (Cold season of one to five months below 43°F)

1. Marine: uniform rain or winter maximum
2. Continental: summer maximum of rain
2m. Continental (monsoon variety): strong summer maximum rain

D. COLD CLIMATES (Long cold season of six months or more below 43°F)

1. Marine: uniform rain or winter maximum
2. Continental: summer maximum of rain
2m. Continental (monsoon variety): strong summer maximum rain

E. ARCTIC CLIMATES (No warm season. Three months or less above 43°F)

F. DESERT CLIMATES (Less than 10 in. rain/yr or $R = T/5$)

1. Hot Deserts: no cold season, no month less than 43°F
2. Midlatitude Deserts: One or more month below 43°F

G. MOUNTAIN CLIMATES

as suggested by the groupings presented, ultimately upon cause. For example, the hot zone includes subdivisions that depend upon seasonal pressure distribution (the monsoon variety, A3m), migration of the ITC (double maxima rainfall, A1), and location in relation to prevailing air masses (marine versus continental, A3). Other precipitation controls, essentially based upon cause, can be identified in each of the major zones.

It is evident that Miller's system is a teaching device that permits an analysis of climatic distribution on a world scale. As such, it can and has been—(Monkhouse, 1961, Bucknell, 1966)—put to good use in introductory discussions concerning world climatic patterns. Obviously, so general a system must have shortcomings. For example, the division allows no place for semiarid climates; the F (hot desert) climates pass directly into the neighboring humid realms. Such an omission shows up in the United States, where much of the continent is in one climatic region, the C2. Concentration on the cause of precipitation, with no concern for total, means that, for example, stations within the A3m variety vary from very wet, (e.g., Darjeeling, 122 in.p.a.) to quite dry (e.g., Mogadiscio, 16 in.p.a.). Such problems might be expected in so simple a world classification and do not necessarily detract from the basic world pattern of climates that Miller is attempting to define.

PÉGUY'S CLASSIFICATION

The classification devised by the French climatologist Charles Péguy is, like Miller's, an attempt to organize the climatic environment so that world climates can be discussed and relative locations assessed. The system is quantitative "updating" of de Martonnes classification with, of course, many modifications and innovations.

The system is based upon a delineated *climogram* (Figure A1-2) and is illustrated by a series of *climatic diagrams* (Figure A1-3). The climogram delineates the climatic character of each month on the basis of temperature and precipitation. Each of the world's climates can be characterized by a particular combination of monthly climate types; this is reflected in a climatic diagram that represents a 12-month succession of climatic conditions. The use of the climogram is significant because, as Péguy points out, by plotting temperature and precipitation data simultaneously, the system reflects a combination of both climatic elements.

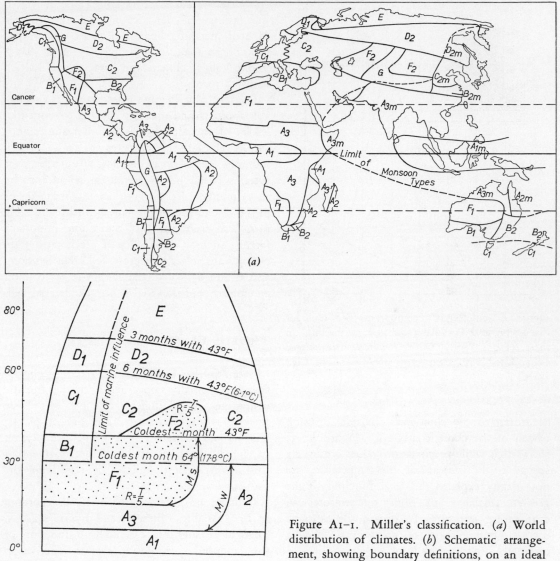

Figure A1-1. Miller's classification. (*a*) World distribution of climates. (*b*) Schematic arrangement, showing boundary definitions, on an ideal continent. (After Bucknell, 1966.)

The climogram is divided into five sections:

G   frigid months (mois *g*laciaux)

F   cold and humid months (mois *f*roids et humides)

O   temperate months (mois temperes –*o*ptimum)

A   arid months (mois *a*rides)

T   tropical months (mois *t*ropicaux)

According to Péguy, the climogram boundaries used to establish these groups were derived empirically. It is unfortunate that, with the exception of the arid (*A*) group, there is no way of determining how the boundaries were so derived. This lack of information is further compounded by the fact that the inset of Figure A1-2 shows the tropical area divided into two, very humid tropical and tropical months; no reason for the location of the boundary is given. The lack of criteria makes it difficult to assess the meaningfulness of the divisions shown on the climogram.

The representation of monthly data on the climatic diagrams permits stations to be classified

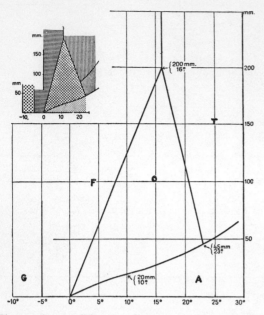

Figure A1-2. Péguy's climogram. See text for discussion and Fig. A1-3 for key to shading on inset. (From Péguy, C. P., *Précis de Climatologie*, 1961. Used with permission of Masson et Cie., Paris.)

according to the climogram divisions. Examination of the climatic diagrams (Figure A1-3) shows that while precipitation values are evenly spaced on the vertical axis, the temperatures are not. Péguy explains this as a function of the greater efficiency of high temperatures in evaporating moisture. To obtain the ratio he used a modification of an aridity formula originally suggested by Birot.

By relating monthly data plotted on the climatic diagrams to the zones delineated on the climogram, Péguy constructs a classification of world climates. He differentiates 26 types, each characterized by a letter of the alphabet. (There is no *i*, but there are two *y*'s, *y* and *y*'.) Each of the 26 types is distinguished by the mixture of monthly climates it describes so that, for example, if six months of the combined temperature and precipitation data fall within the *O* sector of the climogram, and six months in the *A* sector, then the station would occur in the Temperate Climate of the classification and would be represented by the letter "*r*." The

complete system is shown in Table A1-2; in the fashion of deMartonne, Péguy provides each of the differentiated climates with a type location so that, for example, Tasmania has a "Loire Valley" climate and Patagoina a "Baltic" climate.

Conceptually, Péguy's classification is a creditable approach and equating climates to sequential letters of the alphabet is an interesting innovation. The fact that climates are analyzed on a monthly basis is a distinct advantage over systems based purely upon annual means. Its shortcomings derive from some discrepancies in the system as it is presented in map and table form and from the fact that some important factors, such as seasonality of precipitation, are not always clearly represented. Furthermore, and as already noted, the system is based upon an empiric approach that is not readily analyzed.

TROLL'S CLASSIFICATION

Carl Troll has introduced a system of climate that is essentially concerned with the climatic definitions of biological systems. His basic premise is the fact that "The life of plants, animals and humans . . . is subject to rhythm of climatic phenomena," and the treatment of this aspect is best attained by classifying the *seasonal climates of the earth*.

Climates are differentiated by Troll according to two variables: thermal and hygric seasons. Thermal seasons are essentially a function of the variation of insolation that is effective on both a diurnal and seasonal basis. In order to assess the effects of both factors, Troll uses the thermoisopleth diagram as an integral part of his system. The diagram (examples are shown in Figures A1-4 and A1-5) consists of isotherms drawn so that both diurnal and seasonal temperature variations are shown simultaneously. The horizontal axis represents months of the year and the vertical axis represents hours of the day. By plotting hourly and monthly observations of temperature on the chart, isotherms can be constructed. The resulting isopleth distribution is analagous to a contour map, where the valleys represent low temperatures

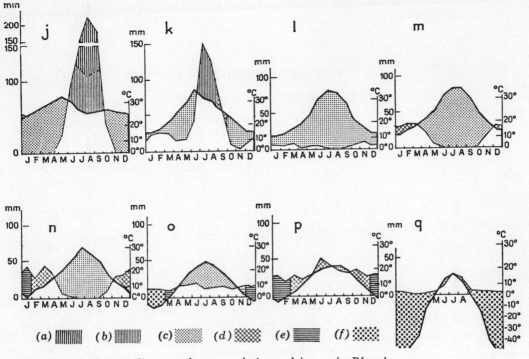

Figure A1-3. Climatic diagrams for types designated j to q in Péguy's classification. Key to shading: (*a*) very humid tropical months; (*b*) tropical months; (*c*) arid months; (*d*) temperate (optimum) months; (*e*) cold months; (*f*) frigid months. Stations depicted: j - Kayes; k - Lahore; l - Tougourt; m - Baghdad; n - Teheran; o - Astrakhan; p - Odessa; and q - Verkhoiansk, (From Péguy, C.P., *Précis de Climatologie*, 1961. Used with permission of Masson et Cie., Paris.)

and the peaks and ridges represent areas of warmth.

The diurnal and seasonal thermal pattern can be interpreted in terms of temperature gradient in vertical and horizontal directions respectively. The diagram for Irkutsk (Figure A1-4) shows a very steep gradient from January to July, indicating marked seasonal change. The thermo-isopleth diagram for Belem, (Figure A1-5), shows that diurnal range is greater than annual range, being typically representative of low-latitude climates.

The thermoisopleth diagram is extremely useful for considering temperature variation over time. Its utility is enchanced by the fact that interpretation of the diagram may be completed in terms of isopleth pattern rather than absolute temperatures. Some authors, for example Tre-

wartha (1968), use the diagram to facilitate their world description of climatic types. The distinctive pattern of the isopleths is of considerable aid in analyzing Troll's thermal regions. These are indicated in Table A1-3 which shows the complete system.

In the treatment of his hygric seasons, Troll considers the tropical and extratropical realms separately. Under the relatively stable high-temperature conditions of the tropics, he assumes that the dominant seasonal variation depends upon the hygric seasons rather than the thermal seasons, which are frequently absent. In effect, he equates the wet-dry seasons of the tropics to the thermally generated seasons of extratropical zones, in that they both provide the seasonal pattern to which plants and animals respond. For the tropical realm, the hygric seasons are

Table A1–2
CLIMATES OF THE WORLD[a]

Number of Months					
G	F	O	A	T	
TROPICAL CLIMATES—WITH SOME DRY MONTHS					
o	o	o	0–3	9–12	*a.* Indonesian
o	o	o	4–5	7–8	*b.* Guinean
o	o	0–1	5–6	6	*c.* Sudanian
CLIMATES OF THE TROPICAL—TEMPERATE TRANSITION—EXCLUDING ARID CLIMATES					
o	o	4–5	o	7–8	*d.* Formosan
0–2	6–8	o		3–6	*e.* Chinese
3–4	x	o		4–6	*f.* Japanese
4–6	x	o		2–3	*g.* Manchurian
DRY CLIMATES WITH EITHER TEMPERATE OR TROPICAL MONTHS					
o	o	2–5	x	2–5	*h.* Argentinian
o	o	o	7–10	2–5	*j.* Senegalian (Kayes)[b]
o	o	0–1	9–10	2–3	*k.* Punjabian (Lahore)
o	o	0–3	9–12	0–1	*l.* Saharan (Touggourt)
o	o	4–5	7–8	o	*m.* Syrian (Bagdad)
TEMPERATE CLIMATES WITH BOTH ARID AND COLD MONTHS, TRANSITION EXCLUDING TROPICAL MONTHS					
o	1–2	x	4–6	o	*n.* Iranian (Téhéran)
	1–4	x	4–8	o	*o.* Aralian (Astrakhan)
o	2–5	x	2–5	o	*p.* Ukranian (Odessa)
1–6	x	x	1–3	o	*q.* "Yakoute" (Verkhoiansk)
TEMPERATE CLIMATES WITH DRY OR COLD MONTHS					
o	o	6	6	o	*r.* "Hellène"
o	o	7–8	4–5	o	*s.* Castillian
o	o	9–10	2–3	o	*t.* "Provençal"
o	0–1	11–12	0–1	o	*u.* "Ligérien"
2–3	9–10	o	o	o	*v.* Danubian
4–6	6–8	o	o	o	*w.* Baltic
COLD CLIMATES					
4	x	6–8	o	o	*x.* Manitobian
1–5	x	4–5	o	o	*y.* Finnish
o	x	4–5	o	o	*y.'* Icelandic
1–5	x	0–3	o	o	*z.* Greenland

[a] After Péguy (1961).

[b] Stations representing climates j to q are shown in Figure A1-3.

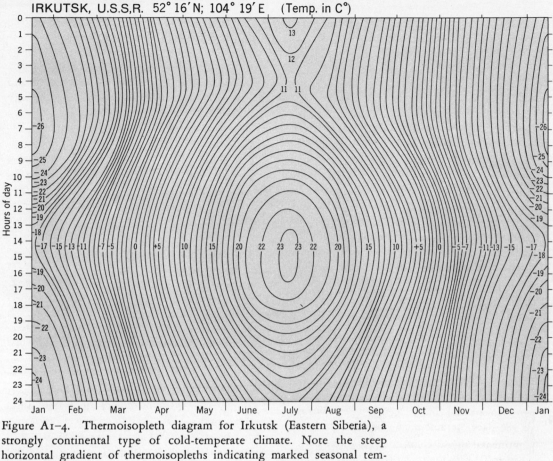

Figure A1–4. Thermoisopleth diagram for Irkutsk (Eastern Siberia), a strongly continental type of cold-temperate climate. Note the steep horizontal gradient of thermoisopleths indicating marked seasonal temperature change. (From Trewartha, G.T., *Introduction to Climate*, 4th ed., 1968. Used with permission of McGraw-Hill Book Company.)

determined by the number of humid months in relation to dry months.

Humid Months		Arid Months
12–9½	Belt of tropical rain forest and transitional wood	0 – 2½
9½–7	Humid savanna belt	2½– 5
7 –4½	Dry savanna belt	5 – 7½
4½–2	Thorn savanna belt	7½–10
2 –1	Semidesert belt	10 –11
1 –0	Desert belt	11 –12

Outside of the tropical zone the hygric seasons are of secondary importance; the thermal seasons provide the major determining criteria. Obviously, Troll does recognize the importance of

seasonal variations of precipitation and uses humid versus arid months in his consideration of, for example, the warm temperate subtropical zones. For the most part, however, hygric seasons are treated qualitatively in marked contrast to the specific temperature definitions of the extratropical zones.

The complete system (Table A1–3) provides graphic evidence of the seasonal approach. In the criteria defining each climate type and subtype, Troll gives evidence of the resulting vegetation association. The classification is essentially a vegetation scheme, because no reference is made to animal or human life patterns. Clearly there will be other biologic responses to the seasons that Troll describes so adequately, but they are so complex that they cannot be considered in a framework such

The climatic levels of mountains should be interpreted as altitudinal variations of the climatic zone concerned.

I. Polar and Subpolar Zones

1. High-polar ice-cap climates; polar ice-deserts.

2. Polar climates with little solar heat (warmest month below +6°C): polar frost-debris belt.

3. Subarctic tundra climates with cool summers (warmest month 6° to 10°C) and great winter cold (coldest month below −8°C): tundra.

4. Highly oceanic subpolar climates with moderately cold winters, poor in snow (coldest month −8° to +2°C), and cool summers (warmest month +5°C to +12°C; annual fluctuation <13°C, often <10°C): subpolar tussock grassland and moors.

II. Cold-Temperate Boreal Zone

1. Oceanic boreal climates (annual fluctuation 13 to 19°C) with moderately cold winter, with, however, relatively prolific snow (coldest month +2°C to −3°C; winter precipitation maximum), moderately warm summers (warmest month +10° to +15°C), and a vegetation period of 120 to 180 days; oceanic humid coniferous woods.

2. Continental boreal climates (annual fluctuation 20 to 40°C) with long, very cold winters, prolific in snow, but short, relatively warm summers (warmest month +10° to +20°C) and a vegetation period of 100 to 150 days: continental coniferous woods.

3. Highly continental boreal climates (annual fluctuation >40°C) with permanently frozen soils, very long, extremely cold, and dry winters (coldest month below −25°C), short but sufficient warming up in summertime (warmest month +10°C to +20°C) and deep thawing soils: highly continental dry coniferous woods.

III. Cool-Temperate Zones

Woodland Climates

1. Highly oceanic climates (annual fluctuation <10°C) with very mild winters (coldest month +2° to +10°C) high winter precipitation maximum, and cool to moderately warm summers

(warmest month below +15°C): evergreen broad-leaved and mixed woods.

2. Oceanic climates (annual fluctuation <16°C) with mild winters (coldest month above +2°C) autumn and winter maxima of precipitation and moderately warm summers (warmest month below 20°C): oceanic deciduous broad-leaved and mixed woods.

3. Suboceanic climates (annual fluctuation 16° to 25°C) with mild to moderately cold winters (coldest month +2° to −3°C), autumn to summer maxima of precipitation, moderately warm to warm and long summers and a period of vegetation of more than 200 days: suboceanic deciduous broad-leaved and mixed woods.

4. Subcontinental climates (annual fluctuation 20° to 30°) with cold winters (coldest month −3° to −13°C) and distinct winter break in vegetative process, with moderately warm summers (warmest month generally below +20°C), summer maximum of precipitation and vegetation period of 160 to 210 days: subcontinental deciduous broad-leaved and mixed woods.

5. Continental climates with cold, slightly dry winters (annual fluctuation 30° to 40°; coldest month −10° to −20°C) and moderately warm and moderately humid summers (warmest month 15° to 20°C) and a vegetation period of 150 to 180 days: continental deciduous broad-leaved and mixed wood as well as wooded steppe.

6. Highly continental climates with cold and dry winters (annual fluctuation generally >40°C; coldest month −10° to −30°C) and short, warm, and humid summers (warmest month above 20°C): highly continental deciduous broad-leaved and mixed wood as well as wooded steppe.

7. Humid and warm summer climates (annual fluctuation 25° to 35°C) with moderately cold, but dry winters (coldest month 0° to −8°C; warmest month 20° to 26°C): deciduous broad-leaved and mixed wood and wooded steppe favored by warmth, but withstanding cold and aridity in winter.

7a. Dry and warm summer climates with a mild to moderately cold, but slightly humid winter half year (coldest month +2° to −6°C; warmest month 20° to 26°C): dry wood and

(Continued)

wooded steppe which favors warmth, but withstands mild-temperature to hard winters.

8. Permanently humid, warm summer climates (annual fluctuation 20° to 30°C) with mild to moderately cold winters (coldest month +2° to −6°C; warmest month 20° to 26°C): humid deciduous broad-leaved and mixed wood which favors warmth.

Steppe Climates

9. Humid steppe climates with cold winters and 6 or more humid months, vegetation period in spring and early summer (coldest month below 0°C): high grass-steppe with perennial herbs.

9a. Humid steppe climates with mild winters (coldest month above 0°C).

10. Steppe climates with cold winters, arid summers, and less than 6 months of humidity (coldest month below 0°C): steppe with short grass, dwarf shrubs, and thorns.

10a. Dry steppe climates with cold winters and arid summers (coldest month 0° to +6°C); steppe with short grass, dwarf shrubs, and thorns.

11. Humid-summer steppe climates with cold and dry winters (coldest month below 0°C): Central and East-Asian grass and dwarf shrub steppe.

12. Semidesert and desert climates with cold winters (coldest month below 0°C): semidesert and desert with cold winters.

12a. Semidesert and desert climates with mild winters (coldest month 0° to +6°C): semideserts and desert with mild winters.

IV. Warm-Temperate Subtropical Zones

[All plains and hill country climates with mild winters (coldest month +2° to +13°C, from +6° to +13°C in the southern hemisphere).]

1. Dry-summer Mediterranean climates with humid winters (mostly more than 5 humid months): subtropical hard-leaved and coniferous wood.

2. Dry-summer steppe climates with humid winters (mostly less than 5 humid months): subtropical grass and shrub-steppe.

3. Steppe climates with short summer humidity and dry winters (less than 5 humid months): subtropical thorn and succulents-steppe.

4. Dry-winter climates with long summer humidity (generally 6 to 9 humid months): subtropical

steppe with short grass, hard-leaved monsoon wood, and wooded-steppe.

5. Semidesert and desert climates without hard winters, but frequent transient or night frosts (generally less than 2 humid months): subtropical semideserts and deserts.

6. Permanently humid grassland-climates of the southern hemisphere (10 to 12 humid months): subtropical high-grassland.

7. Permanently humid climates with hot summers and a maximum of precipitation in summer: subtropical humid forests (laurel and coniferous forests).

V. Tropical Zone

1. Tropical rainy climates with or without short interruptions of the rainy season (12 to 9½ humid months): evergreen tropical rain forest and half deciduous transition wood.

2. Tropical humid-summer climates with 9½ to 7 humid and 2½ to 5 arid months: rain-green humid forest and humid grass-savannah.

2a. Tropical winter-humid climates with 9½ to 7 humid and 2½ to 5 arid months: half deciduous transition wood.

3. Wet and dry tropical climates with 7 to 4½ humid and 5 to 7½ arid months: rain-green dry wood and dry savannah.

4. Tropical dry climates with 4½ to 2 humid and 7½ to 10 arid months: tropical thorn-succulent wood and savannah.

4a. Tropical dry climates with humid months in winter.

5. Tropical semidesert and desert climates with less than 2 humid and more than 10 arid months: tropical semideserts and deserts.

IV–V. Littoral Climates with Seasonal Mists

IV/V a/b. Seasonally atmospherically humid coastal climates in regions of tropical—subtropical desert climates and alternately humid climates caused by coastal mist.

(*a*) In summer,

(*b*) In winter: types of coastal and mountainous coastal vegetation abundant in epiphytes, mist-green to evergreen, more humid than in the corresponding regional climate.

[a] From Troll (1963).

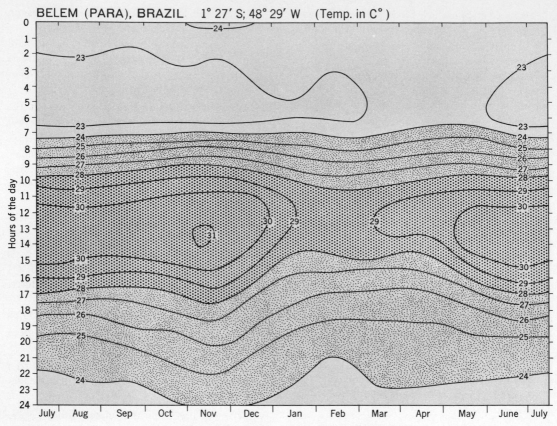

Figure A1-5. Thermoisopleth diagram for Belem, an equatorial station in the Amazon Basin. Note that the vertical thermoisopleth gradient is steeper than the horizontal gradient, showing that the diurnal temperature variations are greater than the seasonal variations. (From Trewartha, G. T., *Introduction to Climate*, 4th ed., 1968. Used with permission of McGraw-Hill Book Company.)

as this. So while Troll initially set out to consider the seasonal effect of climate upon many factors, the ultimate solution is one that points to a relationship between climate and one other factor, in this case, vegetation.

U.S. ARMY WORLD DESERT CLASSIFICATION

In the problem of establishing the climatic regions of the world, the differentiation of the desert regions has provided the substance for much disagreement. The problem relates to the question of what criteria should be used to define a desert and how best to express the concept of precipitation effectiveness. Many definitions of aridity have been formulated. A UNESCO report (1960) lists 19 different formulas, proposed between 1900 and 1952, that purport to define aridity. The classification outlined here defines a desert in relation to the significance of problems that would be experienced by military personnel and equipment within such a climatic region. It provides an example of a classification that has a definitive object at its base; furthermore, it provides an example of the systems that are concerned with the world distribution of a single climatic type.

The basic concept underlying the classification is that the frequency of climatological events is highly significant in estimating existing conditions. As such, the system uses an index of aridity and a thermal grouping that is a function of the frequency of precipitation and the frequency of significant temperatures respectfully.

In terms meaningful to military operations, the influence of precipitation is differentiated according to whether it is sufficient (1) to cause wetting, (2) to accumulate on and in the ground, and (3) to maintain a condition of dampness. If sufficient precipitation falls on a given day so that wetting occurs, then that day might be considered as a rainy day. But the threshold value sufficient to cause wetting, which is dependent upon numerous factors, varies from 0.01 to 0.1 in. for any given storm. Experiments completed by the U. S. Corps of Engineers suggests, however, that if precipitation is less than 0.1 in. in a single rain period, then it is insufficient to wet the surface enough so that trafficability is affected. In this system, the value of 0.1 in. is therefore used to define a *rainy day*. To facilitate the calculation of the number of rainy days from monthly temperature and precipitation data, a nomograph can be used (Figure A1–6).

Having determined the value required to define a rainy day, it is necessary to determine the definition of a wet or dry month. This is approached through the use of the frequency of rainy days according to the following definitions:

	Number of Rainy Days/Month
Wet month	10 or more
Moderately wet month	7 yo 9
Moderately dry month	4 to 6
Dry month	0 to 3

Using this method of differentiation for wet or dry months, the following annual regimes are distinguished:

Extremely arid	10 to 12 months with no more than 1 rainy day per month.
Arid	10 to 12 months with no more than 3 rainy days per month.
Semiarid	6 to 9 months with no more than 3 rainy days per month.
Demiarid	4 to 5 months with no more than 3 rainy days

per month and 10 to 12 months with no more than 6 rainy days per month.

The thermal regimes of the classification depend upon the effects of temperature on military personnel, supplies, and equipment. Clearly, extremes are much more significant than mean values in this respect, particularly where very high and very low temperatures are concerned. The criteria used to establish the thermal regions are shown in Table A1–4.

It is evident that the system outlined has practical use for military purposes; it might even be supposed that it has utility in characterizing other aspects of the desert environment. But as a classification, it has marked shortcomings. The system is actually two unit classifications put onto one map. The map of Australia (Figure A1–7) shows that the thermal regions are not correlated in any way to the arid zones. The system merely consists of a precipitation distribution map overprinted by an isotherm map; the holistic view of the desert realm is completely lacking. Further criticism might be leveled at the numerical values used in definitional boundaries: there appears to be no precise rationale behind their selection. Nevertheless, the method does make an important contribution to the problem of the differentiation of deserts. There is little doubt that mean values are almost meaningless in characterizing the desert climate and the use, in this system, of frequency data is certainly an important factor in its utility.

KÖPPEN SYSTEM—MODIFICATIONS

The Köppen system is discussed in Chapter 6. Figure A1–8 provides a summary of the system giving the definition of letters used and graphs for the determination of dry and monsoon climates.

The reasons given for modifying the Köppen system are quite variable. In some instances the modifications were concerned with "line-fitting," that is the selection of a more appropriate boundary to fit the climatic features of the region under study. Russell (1926), for example, preferred the temperature of 32°F (rather than 26.6°F) in differentiating the C and D climates

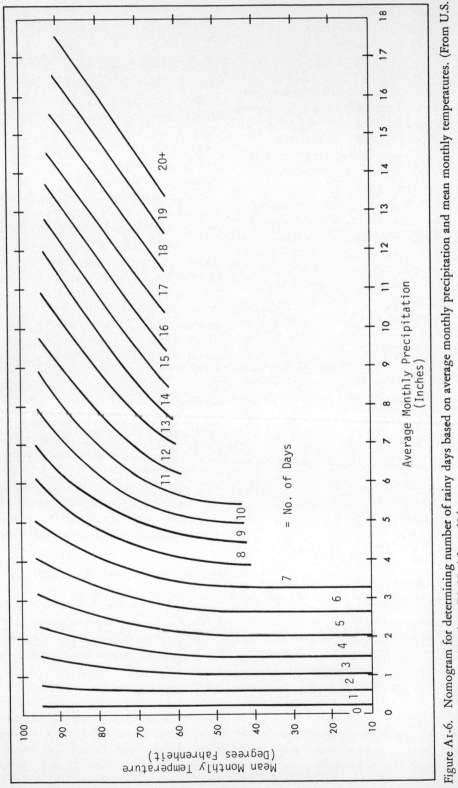
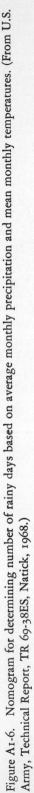

Figure A1-6. Nomogram for determining number of rainy days based on average monthly precipitation and mean monthly temperatures. (From U.S. Army, Technical Report, TR 69-38ES, Natick, 1968.)

Table A1-4

THERMAL SUBDIVISIONS (THESE FREQUENTLY INCLUDE PARTS OF TWO OR MORE ADJACENT ARIDITY REGIONS; IDENTIFYING SYMBOL APPLIES TO ENTIRE AREA DELINEATED BY THE HEAVY DASHED LINE)[a]

EH	Extremely Hot	At least 2 months with mean daily maximum temperature 105°F or higher (this implies temperatures exceeding 115°F up to one-half the days and exceeding 95°F on most days).
VH	Very Hot	At least 2 months with mean daily maximum temperature 95–104°F (temperatures exceeding 95°F at least one-half of the days and exceeding 115°F occasionally).
H	Hot	At least 2 months with mean daily maximum temperature 85–94°F (temperatures exceeding 95°F up to one-half the days).
C	Cold	At least 2 months with mean daily minimum temperature 25–44°F (this implies that temperature will drop to and below freezing up to two-thirds of the nights).
VC	Very Cold	At least 2 months with mean daily minimum temperature 0–24°F (temperature dropping below freezing from two-thirds to most of the nights and below zero as many as half of the nights).
EC	Extremely Cold	At least 2 months with mean daily minimum temperatue below 0°F (temperature dropping to −25°F or colder from occasionally to more than half of the nights).
H/C		Denotes thermal subdivision which is Hot part of the year and Cold another part. Many combinations of EH, VH and H with C, VC, and EC occur somewhere in the world's arid and semiarid climates.
M	Moderate Temperature	At least 10 months neither Hot nor Cold; no more than 1 month with mean daily maximum temperature 85°F or above and/or no more than 1 month with mean daily minimum temperature 44°F or below.

[a] From U.S. Army (1968). See Figure A1-7.

of North America. He suggested, too, that the mean temperature of 32°F for the coldest month was more meaningful in determining the location of *Bh* and *Bk* desert climates. Ackerman (1941) utilized such modifications in his map showing North American climate types.

In other cases the modifications were not so much concerned with changing boundary definitions as with overcoming inconsistencies in the system itself. As previously noted, the boundaries of the *B* climates became a focal point of criticism and new criteria were suggested.

More recently, James Shear has proposed a number of modifications to help clarify the system. In one (Shear, 1964) he suggests the recognition of a Polar Marine Climate as part of the *ET* of Köppen. Some areas within the *ET* realm experience marked maritime climates in that although the temperature does not rise above 50°F in any one month the temperature seldom goes below 32°F for extensive periods. Such cool and highly equable climates need to be differentiated from the *ET* climates, and Shear suggests that they be given the designation *EM*.

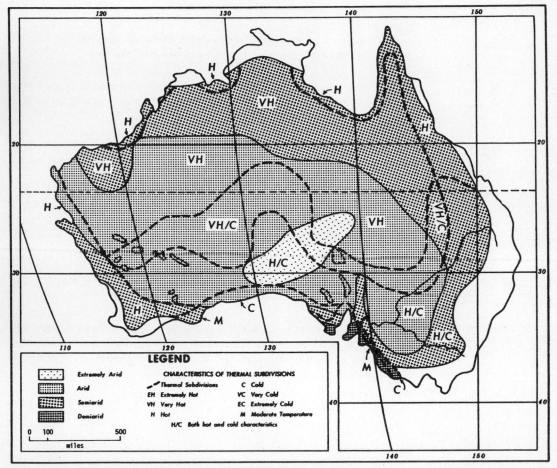

Figure A1–7. Distribution of climatic types in Australia according to the United States Army (1968) classification. (From U.S. Army, Technical Report, TR 69-38ES, Natick, 1968.)

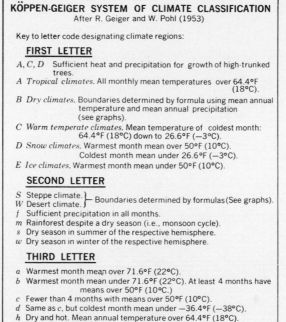

KÖPPEN-GEIGER SYSTEM OF CLIMATE CLASSIFICATION
After R. Geiger and W. Pohl (1953)

Key to letter code designating climate regions:

FIRST LETTER

A, C, D Sufficient heat and precipitation for growth of high-trunked trees.

A Tropical climates. All monthly mean temperatures over 64.4°F (18°C).

B Dry climates. Boundaries determined by formula using mean annual temperature and mean annual precipitation (see graphs).

C Warm temperate climates. Mean temperature of coldest month: 64.4°F (18°C) down to 26.6°F (−3°C).

D Snow climates. Warmest month mean over 50°F (10°C). Coldest month mean under 26.6°F (−3°C).

E Ice climates. Warmest month mean under 50°F (10°C).

SECOND LETTER

S Steppe climate. ⎫ Boundaries determined by formulas (See graphs).
W Desert climate. ⎬
f Sufficient precipitation in all months.
m Rainforest despite a dry season (i.e., monsoon cycle).
s Dry season in summer of the respective hemisphere.
w Dry season in winter of the respective hemisphere.

THIRD LETTER

a Warmest month mean over 71.6°F (22°C). At least 4 months have means over 50°F (10°C).
b Warmest month mean under 71.6°F (22°C). At least 4 months have means over 50°F (10°C.)
c Fewer than 4 months with means over 50°F (10°C).
d Same as *c*, but coldest month mean under −36.4°F (−38°C).
h Dry and hot. Mean annual temperature over 64.4°F (18°C).
k Dry and cold. Mean annual temperature under 64.4°F (18°C).

(a)

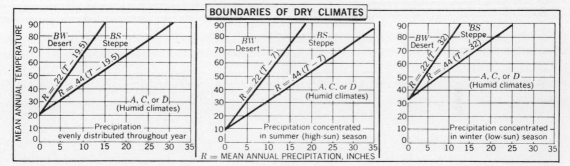

(b)

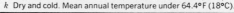

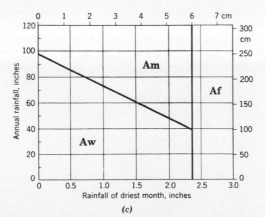

(c)

Figure A1–8. Summary of boundary definitions in the Köppen classification. (*a*) Key to letter code. (*b*) Graphs for determination of the dry climates. (*c*) Graph for the determination of Am, Aw, and Af climates.

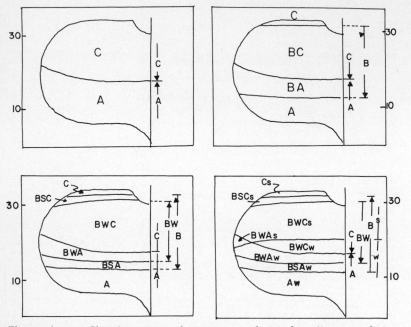

Figure A1–9. Shear's suggested treatment of the dry climates of the Köppen system. (After Shear, 1966.)

Table A1–5

SUMMARY OF DEPARTURES OF TREWARTHA'S (1954) CLASSIFICATION FROM THE
STANDARD KÖPPEN SYSTEM[a]

Köppen	Trewartha	Reason
(a) *Bh/Bk* boundary mean annual isotherm of 64.4°F (18°C)	32° isotherm for the coldest month	In, for example, the United States the January 32°F (0°C) isotherm is more significant
(b) *C/D* boundary coldest month isotherm of 26.6°F (−3°C)	Same as above	As above (see also *Geog. Rev.*, Vol. 31, pp. 108–111 for elaboration)
(c) Letter order of *C* climates three types *Cs*, *Cw*, *Cf*, based upon seasonal distribution of rainfall	(Rainfall versus temperature) three types one based initially on rainfall distribution, other two on temperature—*Cs*, *Ca* (*f* and *w*), *Cb* (*f* and *w*)	Seasonal rainfall only critical in *Cs* climates. Others more varied because of temperature
(d) Letter order of the *D* climates As above—with *D* for *C*	(rainfall versus temperature) As above—with *D* for *C*	As above
(e) Mountain-highland climates differentiated by climate type using boundary criteria as for lowland climates	Not differentiated	Not enough data and problems of great complexity of types within mountain areas

[a] After Trewartha (1954).

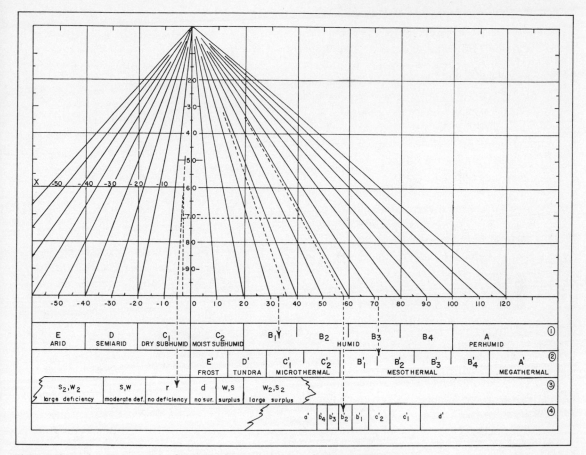

Figure A1-10. Nomogram for the determination of Thornthwaite's symbolization for climatic classification. See text for use and application of graph. (From Basile, R. M. and S. W. Corbin, *Annals A.A.G.*, 59,1969. Reproduced by permission from the *Annals* of the Association of American Geographers.)

A second proposal by Shear (1966) concerns the nomenclature of the B climates. Using Venn diagrams as his base of inquiry, he demonstrates how dry climates might be recognized and depicted in relation to their neighboring humid areas (Figure A1-9). Thus, the *BShs* climate of the northern Sahara might become a *BSCsa* and the *BShw* of the southern Sahara might become a *BSAw* climate.

The modifications of the Köppen system by Trewartha are treated in Chapter 6; here it merely serves to present in table form the varying boundaries utilized (Table A1-5).

THORNTHWAITE SYSTEM— UTILIZATION

To facilitate classification using the Thornthwaite system a number of writers have suggested methodologies. One example is that given by Basile and Corbin (1968) and the following outlines one of the two methods they propose. The nomograph they present is given in Figure A1-10. In the following analysis parts of the nomogram have been redrawn to facilitate a step by step approach. These appear following step 4.

Step 1. *Determination of the Moisture Index.*

i. On the vertical axis locate the annual *PE*. On the horizontal axis locate the annual surplus (use positive values).
Note the intersection of the two points. From the intersection follow the diagonal line to the horizontal axis. *This is Ih.*

ii. Locate *PE* on the vertical axis.
Locate annual deficiency on the horizontal axis using negative values.
Note intersection of two.
From intersection follow the diagonal line to the horizontal axis. *This is the Ia.* (Actually it is 0.6 *Ia* for nomograph accounts for weighting.)

iii. From derived *Ih* and *Ia* values determine *Im.* ($Im = Ih - Ia$).
Find this value on the horizontal axis and drop a perpendicular to the first row. This locates the Moisture Region.

Step 2. *Determination of Temperature Efficiency Index (TE).*

On horizontal axis locate annual *PE*.
Drop perpendicular from this point to second row.
This is the *Thermal Efficiency* Class.

Step 3. *Determination of seasonal regime of moisture.*

i. If *Im* (derived in Step 1) is positive use *Ia* values.
If *Im* (derived in Step 1) is negative use *Ih* values.

ii. Locate annual *PE* on vertical axis.
Locate *Ia* (or *Ih*) on horizontal.
Note intersection. From intersection go diagonally to horizontal axis.
Drop a perpendicular to the third row.
This is the *seasonal moisture regime.*

Step 4. *Determination of summer concentration of TE.*

Locate annual *PE* on vertical axis.
On horizontal axis locate sum of *PE* for the three summer months. Note the intersection.
From the intersection go diagonally to horizontal axis.
Drop perpendicular to fourth row. This is the *summer concentration of TE.*

As Basile and Corbin note, the method seems long and clumsy at first. After a short time, however, it becomes quite straightforward.

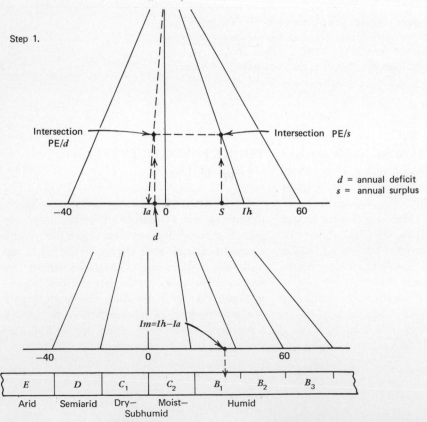

Step 1.

Intersection PE/*d* Intersection PE/*s*

d = annual deficit
s = annual surplus

−40   *Ia*   0   *S*   *Ih*   60

d

$Im = Ih - Ia$

−40   0   60

| *E* | *D* | C_1 | C_2 | B_1 | B_2 | B_3 |

Arid   Semiarid   Dry—Subhumid   Moist—   Humid

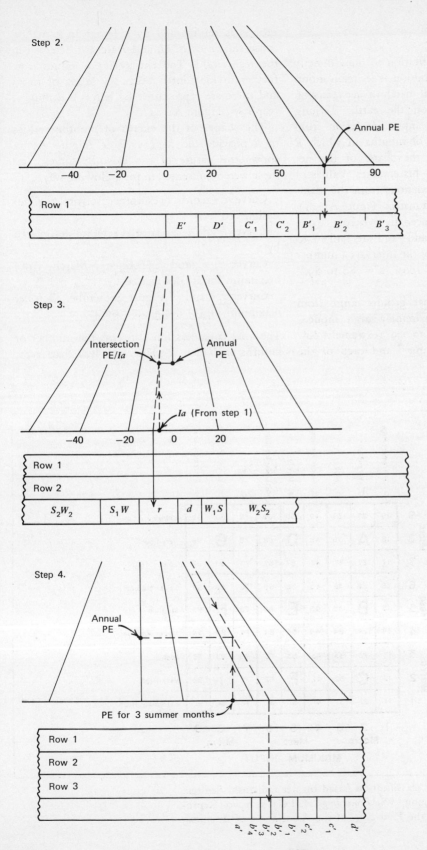

Step 2.

−40   −20   0   20   50   90

Annual PE

Row 1

| | E' | D' | C'_1 | C'_2 | B'_1 | B'_2 | B'_3 |

Step 3.

Intersection PE/Ia

Annual PE

−40   −20   0   20

Ia (From step 1)

Row 1

Row 2

| S_2W_2 | S_1W | r | d | W_1S | W_2S_2 |

Step 4.

Annual PE

PE for 3 summer months

Row 1

Row 2

Row 3

a' b'_4 b'_3 b'_2 b'_1 c'_2 c'_1 d'

TERJUNG'S SYSTEM

This is the first classification to draw directly upon solar energy distribution as its main input. By graphing the seasonal march of net radiation (R) for 1123 points on the earth, Terjung analyzes the maximum input, deviation from maximum, the number of months in which R is less than zero, and the shape of the net radiation curve to group his climates. Values of maximum input and deviations from the maximum are used to derive a matrix (Figure A1–11). The values are first reduced to two bases (one for water and one for land) and are ranked to 10% intervals. For each of the intervals a number is used with 1 = 90 to 100%, 2 = 80 to 89%, and so on.

The matrix shows that groups range from those designated 11 (extremely high input—extremely high range) to 99 (extremely low input—extremely low range) and each of the groups is represented by a letter. In actuality, those shown as F, H, and I are not detected in the real world. The energy input values were further divided into subenergy levels of input and these are represented by letters a through i (top row, Figure A1–11).

The shapes of the curves of monthly values of R provide the next variable. Figure A1–12 shows the nature of the existing curves, and these were represented in the following way:

Curve e (zenith or summer solstice regime) designated by 1.

Curves a and b (equatorial regime) designated by 2.

Curves c, h, and f (February, March, April maximum input) designated by 3.

Curves d, i, and g (August, September, October maximum input) designated by 4.

The final consideration concerns the number of months in which R values are less than zero.

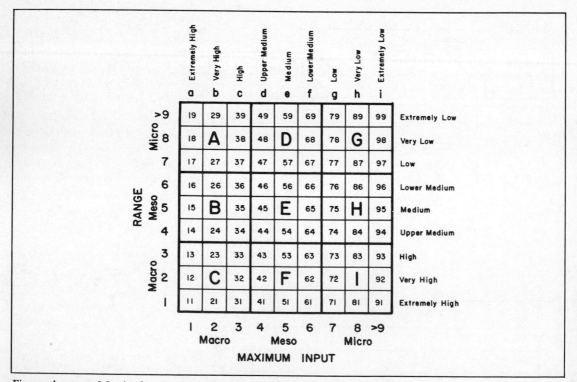

Figure A1–11. Matrix for classification based on net radiation. See text for discussion. (From Terjung, W., *Proceedings A.A.G.*, 2, 1970. Reproduced by permission from the *Proceedings* of the Association of American Geographers.)

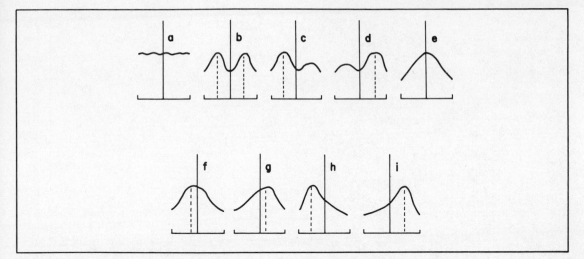

Figure A1-12.   Curves used in identification of climates based on net radiation. See text for discussion. (From Terjung, W., *Proceedings A.A.G.*, 2, 1970. Reproduced by permission from *Proceedings* of the Association of American Geographers.)

Terjung dealt with this by the following grouping:

R values less than zero for less than 20% of the time $= -1$

R values less than zero for from 21 to 40% of the time $= -2$

R values less than zero for from 41 to 60% of the time $= -3$

R values less than zero for less than 60% of the time $= -4$

Thus in a grouping of world climates a place in equatorial Africa might be designated $Ad2$, meaning that it occurs in an area of macroinput of R with microrange; the subenergy input described as falling in the upper medium level. The numeral 2 shows that it experiences an equatorial regime in relation to energy input. By comparison, a place located in Greenland might be designated $Gi4$-4, which could be analyzed in a similar way. The distribution of the climates devised by Terjung is shown in Figure A1-13.

AIR MASS SYSTEMS

Strahler's system (1951) is described in the text of Chapter 6. Table A1-6 and Figure A1-14 outline the classification in more detail.

Another approach to a genetic system is through the use of a proposed air mass frequency model (Oliver, 1970). If it is assumed that each of the recognized air mass types (mE, mT, mP, cT, cP) may be equated to a set within the universe of the earth's surface, then the climates of the earth may be described in terms of the union and intersection of those sets. Analysis is facilitated through the use of Venn diagrams. For example, Figure A1-15 shows two air masses represented as sets. Within the dynamic setting of the atmosphere, the air masses differ in their extent at different times of the year. This is represented in the diagram; in moving from the core the dominance of each air mass is shown to decrease in regular intervals. Clearly, climates within the core of each set experience a single air mass as dominant all the year. These comprise the *dominant regimes*. Areas between the sets are influenced seasonally by the adjacent air masses. Thus, Point A in Figure A1-15 experiences mT air masses for slightly less than six months, and cT air masses for slightly more than six months. Its climate is markedly seasonal and it forms part of a *seasonal regime*. Using representation on the thermohyet diagram, it is possible to show how the shapes of plotted stations representative of the transition appear. The upper part of Figure A1-16 shows the theoretical shapes. As is seen

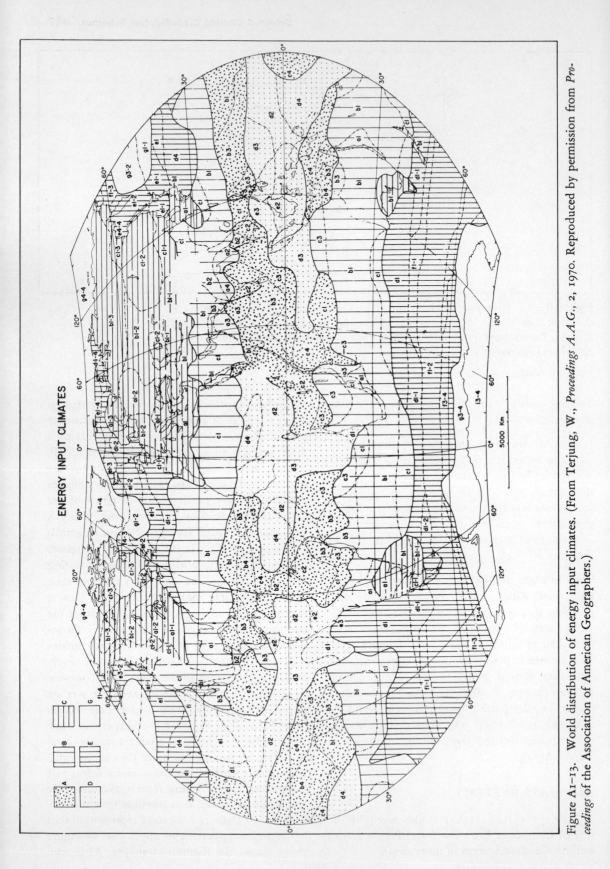

Figure A1-13. World distribution of energy input climates. (From Terjung, W., *Proceedings A.A.G.*, 2, 1970. Reproduced by permission from *Proceedings of the Association of American Geographers*.)

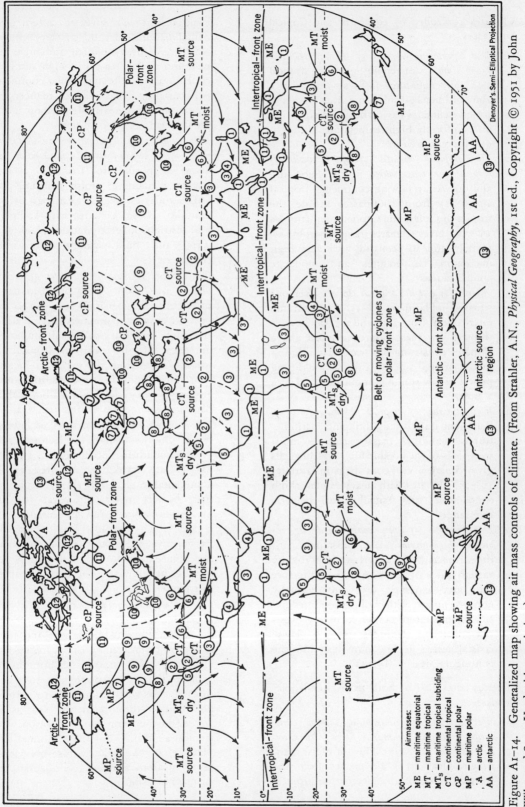

Figure A1–14. Generalized map showing air mass controls of climate. (From Strahler, A.N., *Physical Geography*, 1st ed., Copyright © 1951 by John Wiley and Sons. Used with permission.)

A. Climates controlled by equatorial and tropical air masses.

1. *Wet equatorial climates of the intertropical-front zone.* Converging maritime-tropical (MT) and equatorial (ME) air masses in a zone of intense insolation astride the equator give climates having heavy precipitation and uniformly high temperatures throughout the entire year.

2. *Dry tropical climates of the continental-tropical air mass source regions.* Subsiding and outwardly flowing continental-tropical air (CTs) from landmasses astride the tropics of Cancer and Capricorn in latitudes 15°–35° gives climates ranging from extremely arid to semiarid with very high maximum temperatures and moderate annual ranges.

3. *Alternately wet-dry tropical climates.* Regions situated in latitudes 5°–15°, between the intertropical-front zone and the continental-tropical air mass source regions, have climates consisting of a wet season during period of high sun and a dry season during period of low sun. A combination of Types 1 and 2.

4. *Wet climates of tropical windward coasts.* Persistent trade winds, bringing maritime-tropical (MT) air masses from the moist west sides of the oceanic high-pressure cells give to narrow east-coast belts in latitudes 10°–25° climates with heavy rainfall, high temperatures, and a small annual temperature range.

5. *Dry climates of west coasts bordering the oceanic high-pressure cells.* Subsiding maritime tropical (MTs) air masses from the dry eastern sides of the oceanic high-pressure cells give extremely dry, often virtually rainless climates with cool temperatures and small temperature ranges to narrow west-coast belts in latitudes 15°–30°. These coastal deserts grade continuously into the continental deserts, Type 2.

B. Climates controlled by both tropical and polar air masses.

6. *Wet climates of subtropical eastern continental margins dominated largely by maritime-tropical air masses.* Flow of moist maritime-tropical (MT) air masses from the western sides of the oceanic subtropical high-pressure cells gives to eastern continental margins in latitudes 20°–35° climates which have heavy rainfall and high temperatures during season of high sun but a cool winter with frequent polar-continental (CP) air mass invasions and reduced precipitation. Middle-latitude cyclonic storms are frequent.

7. *Wet climates of windward, middle-latitude west coasts exposed to maritime-polar air-masses.* Exposure to frequent cyclonic storms containing moist maritime-polar (MP) air masses gives to west-coast belts in latitudes 40°–60° climates with abundant, well-distributed precipitation, cool summers, and mild winters with much cloudiness.

8. *Wet-winter, dry-summer climates of middle-latitude west coasts.* A seasonal alternation of Types 5 and 7 occurs in latitudes 30°–40° on west coasts. In summer extreme aridity and high temperatures prevail under the influence of dry maritime-tropical air masses (MTs) from the expanded oceanic subtropical highs; in winter maritime-polar air masses and cyclonic storms bring abundant rain and moderately low temperatures.

9. *Dry middle-latitude climates of complex origin.* On the eastern, or leeward, sides of north-south mountain belts in latitudes 35°–50° are dry climates visited by dry continental-tropical (CT) air masses in summer, by dry polar-continental (CP) air masses in winter, and by maritime air masses which have lost their moisture in passing over the mountain barriers. Ranging from extremely dry to semiarid, these climates have a very great annual temperature range.

10. *Humid middle-latitude continental climates in the battleground of the polar and tropical air masses.* Repeated invasions and interactions of tropical and polar air masses having contrasting characteristics bring highly variable climates to central and eastern continental areas in latitudes 35°–50°. Ample precipitation throughout the year is increased in the warm summer season when tropical air masses dominate. Annual temperature range is great, with severe winters resulting from frequent continental-polar air mass invasions.

(continued)

Table A1–6ᵃ (*continued*)

C. Climates controlled by polar and arctic air masses.

11. *Cold climates of the continental-polar air mass source regions.* In the source regions of the continental-polar air masses, in latitudes 55°–65° N., are climates characterized by long, severely cold winters and short warm summers, making an enormous annual temperature range. Low moisture content of cold continental air masses results in small total precipitation, but evaporation is also low.

12. *Cold climates of the arctic-front zone along the northern continental fringes.* Control by arctic and polar air masses interacting in

the arctic-front zone throughout the year gives to coastal margins north of 90° severely cold humid climates with no warm season which can be termed summer. Moderating influence of arctic ocean waters prevents extreme winter severity found in the continental interiors.

13. *Cold climates of the arctic and antarctic air mass source regions.* Dominated largely by arctic or antarctic air masses, the great ice expanses of the Greenland and antarctic ice caps and the North Polar Sea have climates whose temperatures average far below all other climates and show no above-freezing monthly averages.

ᵃ From Strahler (1951).

in the lower part of the figure such transitions occur in the real world; the example illustrates stations representative of a transect from the desert of Australia to the equatorial climate of New Guinea.

Where more than two sets intersect (Figure A1–17), there is a region that is influenced by

three (or more) air masses. The resulting climate is a compound of the characteristics of these three air masses, and the resulting climate forms part of the *compound regime*.

By representing climatic regimes in this way it is possible to show their interaction on a world basis. Figure A1–18 shows the sets as

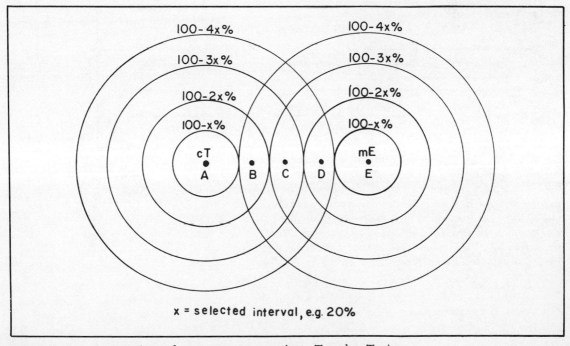

Figure A1–15. Intersection of two sets, representing cT and mT air mass dominance, of given arbitrary boundaries. (After Oliver, 1970.)

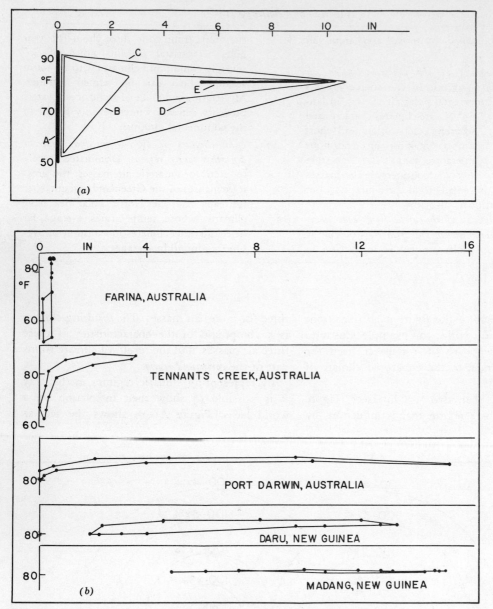

Figure A1-16. (a) Theoretical climograph representation of the climatic variations between two dominant climatic regimes. As indicated in Fig. A1-15, Station A is dominated by cT air masses and Station E by mT air masses. Stations B, C, and D are influenced by seasonally dominant air masses. (b) Such a variation in climograph representation is found in the real world. A traverse from mE dominant station (Madang) to a station dominated by cT air masses (Farina) shows similar shapes to those in the theoretical representation. (After Oliver, 1970.)

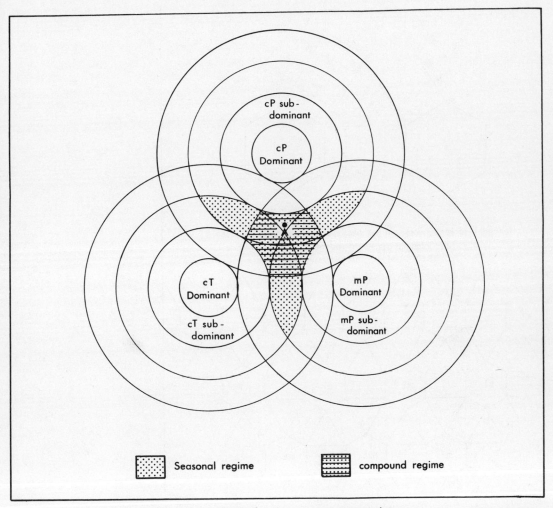

Figure A1-17. Relationship between three dominant sets representing three different air masses. Compound regimes occur where air mass interaction does not allow recognition of either dominant or seasonally dominant air masses. (After Oliver, 1970.)

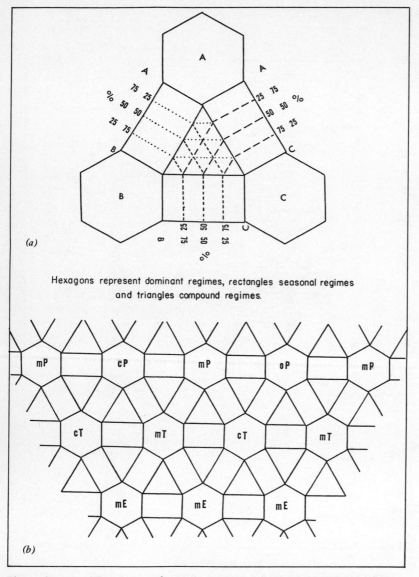

Hexagons represent dominant regimes, rectangles seasonal regimes
and triangles compound regimes.

Figure A1–18.   The air mass frequency model. (a) Graph for determination
of dominant, seasonal, or compound climatic regimes based on annual
percentage frequency of air masses. (b) Schematic representation of possible
relationships between air mass regimes in an area representing the northern
hemisphere. (After Oliver, 1970.)

hexagons in lattice form. The upper part of the diagram shows how each of the three regimes is determined; the lower part shows the lattice superposed on to a cylindrical projection of the northern hemisphere. Using the lattice, it is possible to evaluate all of the possible combinations that occur. These, with tentative descriptive names, are shown in Table A1–7.

In order to locate a climatic station within the air mass frequency model, it is necessary to establish air mass dominance on a monthly basis. Recognition of air masses on a daily scale is basic synoptic meteorology, but the rigorous methodology—using upper air data—cannot be followed in a world system because long-term data is not always available. In order to establish monthly air mass dominance, monthly date for stations of known air mass dominance were plotted on the T.H. diagram and the resulting distribution was analyzed using isopleths. Such analysis enables the construction of air mass realms on the T.H. diagram. The realms can be used to establish the monthly air mass dominance for any set of climatic data, where temperature and precipitation serve as inputs (Figure A1–19).

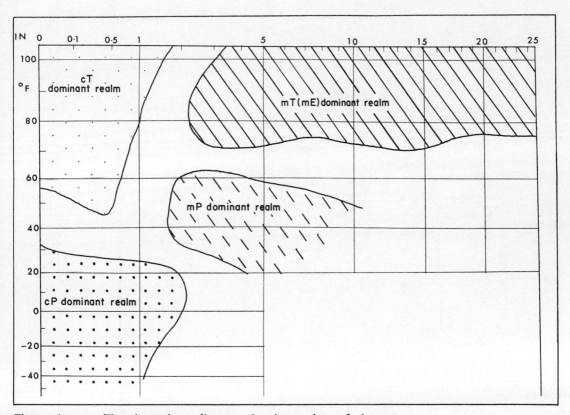

Figure A1–19. The thermohyet diagram showing realms of air mass dominance. It is assumed that plotted monthly station data falling within the realms designated are influenced by the air mass named. (After Oliver, 1970.)

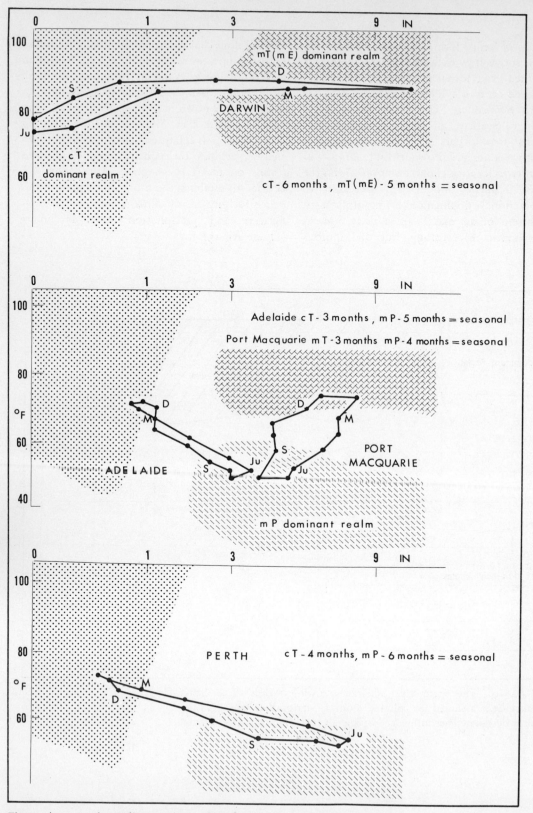

Figure A1–20. Australian stations plotted on the thermohyet diagram (air mass realms shaded) are classified according to the number of months that fall into each air mass realm. (After Oliver, 1970.)

Table A1-7

CLIMATES DETERMINED FROM THE AIR
MASS FREQUENCY MODEL

Air Mass	Name
Dominant Regimes	
mP dominant	Polar maritime equable
cP dominant	Polar continental
cT dominant	Tropical desert
mT_s dominant	Littoral desert
mE dominant	Equatorial
Subdominant Regimes	
mP subdominant	Middle latitude equable
cP subdominant	Cold continental
cT subdominant	Mediterranean or tropical wet-dry steppe
mT_s subdominant	Littoral mediterranean steppe
mE subdominant = mT realm	Tropical littoral
Seasonal Regimes	
cP-mP alternating	Cool continental
mP-mT_s alternating	Littoral mediterranean
mP-cT alternating	Continental mediterranean
mP-mT alternating	East coast maritime
cP-cT alternating	Continental desert
cP-mT alternating	Humid continental
cT-mT/mE alternating	Tropical wet-dry

Compound Regimes

Designated according to relationship to adjacent
seasonal regime by addition of prefix *modified*.

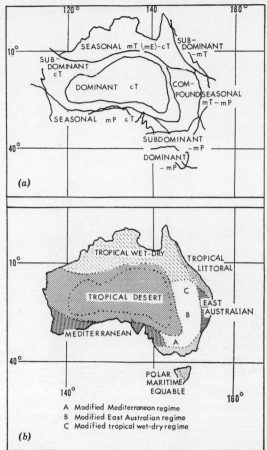

Figure A1-21. Classification of the climate of
Australia according to the air mass frequency model.
(*a*) Air masses and their effects. (*b*) The resulting
classification. (After Oliver, 1970.)

Figure A1-20 shows four stations plotted on
the diagram. Station data that fell into each of
the air mass realms were tabulated and relative
air mass dominance established.

Using such an approach it is possible to
analyze air mass dominance in a number of ways.
For example, Figure A1-21 shows the climates
of Australia according to the air mass frequency
model where the climatic regions are established
according to established criteria (Oliver, 1970).
Figure A1-22 illustrates how monthly air mass
dominance can be used to show the relative
influence of air masses at different times of
the year.

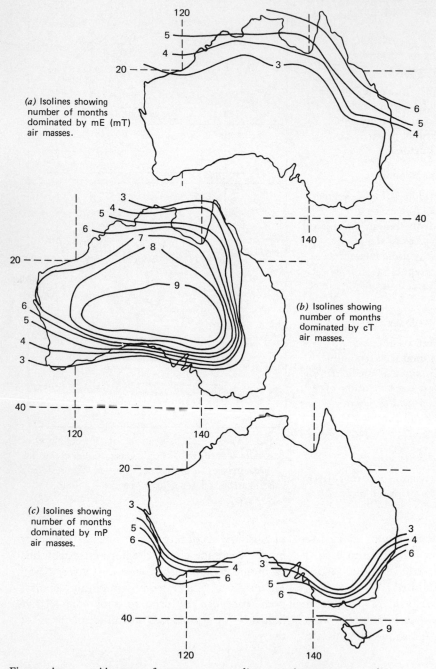

Figure A1–22. Air mass frequency, according to air mass types distinguished using the air mass frequency model, in Australia. (*a*) Isolines showing the number of months dominated by mE (mT) air masses. (*b*) Isolines showing the number of months dominated by cT air masses. (*c*) Isolines showing the number of months dominated by mP air masses.

APPENDIX 2

Climate-Architecture Analysis[1]

THE MAHONEY TABLES

The Mahoney tables provide a guide to design in relation to climate using readily available climatic data. By following a step by step procedure the designer is led from the climatic information to specifications for optimal conditions of layout, orientation, shape and structure needed at the sketch design stage.

The analysis requires the use of five tables. To retain the orderly nature of the original work, the tables in this section of the appendix are numbered 1 through 5 for each of the cases presented.

Table 1. Air Temperature

NOTE. All recordings should be to the nearest 0.5°C.

(a) Record in Table 1 the monthly mean maxima and minima of temperature;

(b) Enter to the right of the air temperature figures the highest of the monthly mean maxima and the lowest of the monthly mean minima;

(c) Find the "annual mean temperature" (AMT) by adding the highest of the monthly mean maxima to the lowest of the monthly mean minima, and dividing by two. Enter the result in the box marked AMT at the right of Table 1;

[1] Reproduced with slight modifications from United Nations, Department of Economic and Social Affairs, Design of Low-Cost Housing and Community Facilities, Vol. 1, *Climate and House Design*, United Nations, New York, 1971.

(d) Find the "monthly mean range" (MMR) of temperatures by deducting the monthly mean minima from the maxima and entering the results for each month in the bottom line of Table 1;

(e) Find the "annual mean range" (AMR) of temperatures by deducting the lowest of the monthly mean minima from the highest of the monthly mean maxima and entering the result in the box marked AMR.

Table 2. Humidity, Rain and Wind

(a) Record in Table 2 the monthly mean maxima and minima of relative humidity (RH) for each month (early morning and early afternoon readings);

(b) Record below these maxima and minima the average relative humidity for each month;

(c) Note below this the "humidity group" (HG) for each month, using the following code:

Average RH	Humidity Group
Below 30 per cent	1
30 to 50 per cent	2
50 to 70 per cent	3
Above 70 per cent	4

(d) Record in Table 2 the monthly rainfall figures in millimetres and add them up to find the annual rainfall;

(e) Record for each month the direction of the prevailing wind and of the secondary wind selected from first and second peaks of frequency figures. (Compass points N, NNE, NE, ENE, E, etc. are sufficient.)

Table 1

AIR TEMPERATURE (°C)

	J	F	M	A	M	J	J	A	S	O	N	D	Highest	AMT
Monthly mean max.														
Monthly mean min.														
Monthly mean range													Lowest	AMR

Table 2

HUMIDITY, RAIN AND WIND

RH (percentage)		J F M A M J J A S O N D	
Monthly mean max. a.m.			
Monthly mean min. p.m.			
	Average		
Humidity group			Total
Rainfall (mm)			
Wind:	Prevailing		
	Secondary		

Table 3. Diagnosis of Climatic Stress

(*a*) Repeat in Table 3 for each month the humidity groups from Table 2;

(*b*) Note the AMT from Table 1;

(*c*) Enter into Table 3 the day and night comfort limits taken from the chart below, using the appropriate humidity group and relevant AMT range; i.e., over 20°C, between 15 and 20°C or under 15°C;

(*d*) Compare the monthly mean maxima with the day comfort limits and compare the monthly mean minima with the night comfort limits and enter the following symbols into the last two lines of Table 3 under the rating of thermal stress (day and night):

Above comfort limits	H (Hot)
Within comfort limits	— (Comfort)
Below comfort limits	C (Cold)

COMFORT LIMITS

Average RH (percentage)	HG	AMT over 20°C		AMT 15–20°C		AMT under 15°C		HG
		Day	Night	Day	Night	Day	Night	
0–30	1	26–34	17–25	23–32	14–23	21–30	12–21	1
30–50	2	25–31	17–24	22–30	14–22	20–27	12–20	2
50–70	3	23–29	17–23	21–28	14–21	19–26	12–19	2
70–100	4	22–27	17–21	20–25	14–20	18–24	12–18	4

Table 3

DIAGNOSIS

	J F M A M J J A S O N D
Humidity group	
Temperature (°C)	
Monthly mean max.	
Day comfort: Max.	
Min.	
Monthly mean min.	
Night comfort: Max.	
Min.	
Thermal stress	
Day	
Night	

Table 4. Indicators

Certain groups of symptoms of climatic stress indicate the remedial action the designer can take. We refer to them as indicators. They tend to be associated with humid or arid conditions. One indicator by itself does not automatically lead to a solution. Recommendations can be framed only after adding the indicators for a whole year and completing Table 4.

Humid Indicators

H 1 indicates that air movement is essential. It applies when high temperature (day thermal stress = H) is combined with high humidity (HG = 4) or when the high temperature (day thermal stress = H) is combined with moderate humidity (HG = 2 or 3) and a small diurnal range (DR less than 10°C);

H 2 indicates that air movement is desirable. It applies when temperatures within the comfort limits are combined with high humidity (HG = 4);

H 3 indicates that precautions against rain penetration are needed. Problems may arise even with low precipitation figures, but will be inevitable when rainfall exceeds 200 mm per month.

Arid Indicators

A 1 indicates the need for thermal storage. It applies when a large diurnal range (10°C or more) coincides with moderate or low humidity (HG = 1, 2 or 3);

A 2 indicates the desirability of outdoor sleeping space. It is needed when the night temperature is high (night thermal stress = H) and the humidity is low (HG = 1 or 2). It may be needed also when nights are comfortable outdoors but hot indoors as a result of heavy thermal storage (i.e., day = H, humidity group = 1 or 2 and when the diurnal range is above 10°C);

A 3 indicates winter or cool-season problems. These occur when the day temperature falls below the comfort limits (day thermal stress = C);

Tick in Table 4 the months when these indicators apply and add the total number of months for each indicator.

Table 4
INDICATORS

	J	F	M	A	M	J	J	A	S	O	N	D	Totals
Humid													
H1 Air movement (essential)													
H2 Air movement (desirable)													
H3 Rain protection													
Arid													
A1 Thermal storage													
A2 Outdoor sleeping													
A3 Cold-season problems													

Recommendations

After completion of Table 4 the designer is ready to lay down specifications. His recommendations depend on the number of months during which one or several of the indicators A and H apply.

Table 5 helps him to formulate recommendations for those features of his building that must be decided during the sketch design stage.

The recommendations are grouped under the following eight subjects:

Layout	Openings
Spacing	Walls
Air movement	Roofs
Outdoor sleeping	Rain protection

Table 5
SKETCH DESIGN RECOMMENDATIONS

Indicator totals from table 4 — Humid: H_1, H_2, H_3; Arid: A_1, A_2, A_3

H_1	H_2	H_3	A_1	A_2	A_3	Recommendations
						Layout
			0–10			1. Buildings orientated on east-west axis to reduce exposure to sun
			11 or 12	5–12		
				0–4		2. Compact courtyard planning
						Spacing
11 or 12						3. Open spacing for breeze penetration
2–10						4. As 3, but protect from cold/hot wind
0 or 1						5. Compact planning
						Air movement
3–12						6. Rooms single banked. Permanent provision for air movement
1 or 2		0–5				
1 or 2		6–12				7. Double-banked rooms with temporary provision for air movement
0	2–12					
0	0 or 1					8. No air movement requirement
						Openings
			0 to 1		0	9. Large openings, 40-80% of N and S walls
			11 or 12		0 or 1	10. Very small openings, 10–20%
		Any other conditions				11. Medium openings, 20–40%
						Walls
			0–2			12. Light walls; short time lag
			3–12			13. Heavy external and internal walls
						Roofs
			0–5			14. Light insulated roofs
			6–12			15. Heavy roofs; over 8 hours' time lag
						Outdoor sleeping
				2–12		16. Space for outdoor sleeping required
						Rain protection
		3–12				17. Protection from heavy rain needed

Instructions for the Completion of Table 5

(a) Transfer the indicator totals from Table 4 to Table 5;

(b) Deal with the eight subjects one by one, i.e., layout, spacing, air movement, etc.;

(c) Examine the indicator columns for each subject to find the appropriate recommendation;

(d) There can be only one recommendation per subject. It is the first you come across while scanning from left to right;

(e) A further alternative exists in a few cases, namely, recommendations 1 or 2, 6 or 7, and 7 or 8. In these cases, the choice is made by proceeding with the scanning of the indicator columns to the right and deciding according to the range of months given in the table.

A worked example of the method is given using Islamabad, Pakistan. Tables 1 through 4 are completed using the necessary data and their application in Table 5 is indicated by the use of dashed lines and arrows.

WORKED EXAMPLES

Location—Islamabad, Pakistan; Longitude—73°07′ E; Latitude—33°36′ N; Altitude—518 metres.

Table 1

Temperature (°C)	J	F	M	A	M	J	J	A	S	O	N	D	Highest / Lowest	AMT / AMR
Monthly mean max.	18	18.5	25.5	36.5	38	39	38	33.5	33.5	32	26.5	20	Highest 39	AMT 19
Monthly mean min.	1	3.5	10	13.5	18.5	24	24	24	20	13	4.5	−1	Lowest −1	AMR 40
Monthly mean range	17	15	15.5	23	19.5	15	14	9.5	13.5	19	22	21		

Table 2

	J	F	M	A	M	J	J	A	S	O	N	D	Total
Humidity (percentage): Monthly mean max.	82	80	63	51	38	65	65	75	60	58	68	85	
Monthly mean min.	46	50	40	30	17	30	48	53	37	30	30	40	
Average	64	65	52	41	28	34	57	64	49	44	49	63	
Humidity group	3	3	3	2	1	2	3	3	2	2	2	3	
Rainfall (mm)	61	58	76	74	31	51	203	228	89	10	5	31	917
Wind: Prevailing	SW	SW	SW	SW	SW	SE	SE	SE	SW	SW	SW	SW	
Secondary	W	W	NW	NW	NW	SW	SW	SW	NW	NW	W	W	

Table 3

Temperature (°C)	J	F	M	A	M	J	J	A	S	O	N	D
Monthly mean max.	18	18.5	25.5	36.5	38	39	38	33.5	33.5	32	26.5	20
Day comfort: Max.	28	28	28	30	32	30	28	28	30	30	30	28
Min.	21	21	21	22	23	22	21	21	22	22	22	21
Monthly mean min.	1	3.5	10	13.5	18.5	24	24	24	20	13	4.5	−1
Night comfort: Max.	21	21	21	22	23	22	21	21	22	22	22	21
Min.	14	14	14	14	14	14	14	14	14	14	14	14
Thermal stress: Day	C	C	—	H	H	H	H	H	H	H	—	C
Night	C	C	C	C	—	H	H	H	—	C	C	C

Table 4

	J	F	M	A	M	J	J	A	S	O	N	D	Total
H1 Air movement (essential)								√					1
H2 Air movement (desirable)													0
H3 Rain protection							√	√					2
A1 Thermal storage	√	√	√	√	√	√	√	√		√	√	√	11
A2 Outdoor sleeping				√	√	√							3
A3 Cold-season problems	√	√										√	3

433

Table 5
SKETCH DESIGN RECOMMENDATIONS FOR ISLAMABAD, PAKISTAN

Indicator totals from table 4							Recommendations
Humid			Arid				
H1	H2	H3	A1	A2	A3		
1	0	2	11	3	3		
							Layout
			0–10				1. Buildings orientated on east-west axis to reduce exposure to sun
			11 or 12		5–12		
					0–4	✓	2. Compact courtyard planning
							Spacing
11 or 12							3. Open spacing for breeze penetration
2–10							4. As 3, but protect from cold/hot wind
0 or 1						✓	5. Compact planning
							Air movement
3–12							6. Rooms single banked. Permanent provision for air movement
1 or 2			0–5				
			6–12			✓	7. Double-banked rooms with temporary provision for air movement
0	2–12						
	0 or 1						8. No air movement requirement
							Openings
			0 or 1		0		9. Large openings, 40–80% of N and S walls
			11 or 12		0 or 1		10. Very small openings, 10–20%
Any other conditions						✓	11. Medium openings, 20–40%
							Walls
			0–2				12. Light walls; short time lag
			3–12			✓	13. Heavy external and internal walls
							Roofs
			0–5				14. Light insulated roofs
			6–12			✓	15. Heavy roofs; over 8 hours' time lag
							Outdoor sleeping
			2–12			✓	16. Space for outdoor sleeping required
							Rain protection
3–12						✗	17. Protection from heavy rain needed

SUMMARY OF RECOMMENDATIONS FOR THE SKETCH DESIGN STAGE

Layout

1. Buildings should be oriented on an east-west axis with the long elevations facing north and south to reduce exposure to the sun, if thermal storage (A 1) is required for up to ten months or if thermal storage is required for eleven or twelve months including more than four winter months (A 2). The buildings may be turned slightly to catch the prevailing breeze (see recommendation No. 6 and the wind directions for the high humidity months in Table 2) or to allow limited solar heating during the cold season (A 3).

2. Buildings should be planned around small courtyards, if thermal storage (A 1) is required for eleven or twelve months and the cold season (A 3) is less than five months.

Spacing

3. Buildings should be well spaced to allow for breeze penetration, if air movement (H 1) is

essential for eleven or twelve months. As a rough guide, the space between long parallel rows of buildings should be five times the height of the buildings or more.

4. If air movement (H 1) is needed between two and ten months of the year, spacing for breeze penetration is still needed, but buildings and planting should also be planned to give protection against dusty hot or cold winds (see Table 3 for conditions and Table 2 for wind directions).

5. Compact planning is essential if air movement (H 1) is needed for not more than two months.

Air Movement

6. Rooms should be single banked with windows in the north and south wall if air movement (H 1) is essential for more than two months. Single banking is desirable if air movement is needed for one or two months and thermal storage (A 1) for zero to five months.

7. Rooms may be double banked if air movement (H 3) is needed for not more than one or two months. If there are months when air movement is not essential but desirable (H 2), the plan should allow temporary cross-ventilation (e.g., it could be double banked with large interconnecting doors). If the prevailing wind is unreliable or if site limitations restrict planning for air movement, ceiling fans should be considered. This must be done at the sketch design stage, because it implies minimum room heights of not less than 2.75 metres.

8. Rooms should be double banked if air movement (H 1) is never required to *achieve* comfort or is required to *maintain* comfort (H 2) for one month or less.

Openings in Walls

9. Openings should be large (between 40–80 per cent of the north and south walls) if thermal storage (A 1) is required for less than two months and there is no cold season (A 3). Large openings need not be fully glazed, but should be protected from sun, sky glare and rain, preferably by horizontal overhangs.

10. Small openings (less than 25 per cent) should be used if thermal storage (A 1) is needed for eleven or twelve months and the cold season (A 3) is less than two months.

11. In all other conditions, medium-sized openings should be used (from 25 per cent to 40 per cent of the area of the north and south walls). Openings in the east walls are desirable only if there is a long cold season (A 3). Openings in west walls are desirable in cold and temperate climates, but must be avoided in the tropics.

Walls

12. External walls should be light with a small heat capacity if thermal storage (A 1) is needed for less than three months. Internal walls should be heavy if the annual range is high (over 20°C).

13. External *and* internal walls should be heavy with high heat capacity if thermal storage (A 1) is needed for three to twelve months.

Roofs

14. A light but well-insulated roof should be used if thermal storage (A 1) is needed for less than six months.

15. A heavy roof should be used if thermal storage (A 1) is needed for six to twelve months.

NOTE. Glazed skylights or roof lights should never be used in the tropics.

Outdoor Sleeping

16. Space for outdoor sleeping should be provided if indicator A 2 applies for more than one month of the year. Sleeping spaces on roofs or balconies or in patios should be exposed to the coldest part of the night sky (the zenith) to permit heat loss by out-going radiation.

Rain Protection

17. Special protective measures are needed if rain is frequent and heavy (H 3), e.g., deep verandahs, wide overhangs and covered passages.

Recapitulation

The climatic analysis of the sketch design stage ends with the completion of Table 5 and the recommendations which follow from it.

After completing Tables 1 to 5 and noting his recommendations, the designer should pause for a few moments and reflect on the process he has followed. It is important that he should be conscious of the logic of this process and not consider it as either arbitrary or mysterious.

What he has done is this: he has recorded the dominant features of his climate one by one noting for each the period during which it is operative. Certain combinations of climatic features indicate that forms of layout, construction, fabric or surface treatment are appropriate.

Different seasons bring different combinations of climatic features. In many cases, the designer can find answers that are appropriate for more than one season. In others, he has to decide according to the season that lasts longest. If he does this he must be aware that there will be periods when the over-all concept of his building will be less than perfect. This does not invalidate the decisions he has arrived at through the use of the Mahoney tables. It means merely that climatic design must not end with the completion of the sketch design stage.

The plan development and element design stages will provide opportunities of emphasizing favourable and of mitigating unfavourable features of the initial concept.

EXAMPLES OF USE OF THE MAHONEY TABLES

To illustrate the use of the Mahoney tables, completed examples are given for three different locations:

Belém, Brazil (warm humid)
Baghdad, Iraq (hot dry)
New Delhi, India (composite)

The Mahoney tables are obviously needed in complex climates such as that of New Delhi, but prove useful also in "one-season climates" like that of Belém. The designer must be prepared to encounter borderline situations, particularly in composite climates. Cases may occur where a major decision, such as the choice of a heavy roof in place of a light insulated roof, may depend on a difference of 0.5°C in the monthly mean maximum. Such cases are fortunately rare. If they occur, a compromise is required. As neither type of roof will be perfect for the whole year, the decision will have to follow local tradition or else other criteria such as availability of materials or cost.

Example A. Belém. Brazil
Climate

Belém lies close to the equator at the mouth of the Amazon River. The climate is characteristic of the warm, humid equatorial zone. High rainfall and high humidity are associated with a low diurnal range and a relatively high and even temperature throughout the year. The only discernible change during the year is the direction of the prevailing wind which comes from the north-east when the sun is to the south and from the south-east when the sun is to the north.

Comfort

The Belém climate is too warm for comfort by day and on most nights. The indicators show that air movement is essential to reduce discomfort throughout the year.

Recommendations

Rooms should be single banked and well spaced to allow breeze to penetrate at low level. Buildings should be designed on an east-west axis to reduce exposure to solar heat. This orientation will also allow the south-east and north-east winds to penetrate buildings. To encourage air movement through rooms, openings should be as large as possible. Protection against rain and insets is essential. The materials used for walls and roof should not store heat as this would increase discomfort at night. The roof should have a light colour or a bright metal surface to reflect solar radiation. A roof cavity and lightweight insulation will reduce the transmission of any solar heat that is absorbed. External circulation spaces should be protected against rain.

Example B. Baghdad, Iraq

Climate

Baghdad is situated on the River Tigris in the plains of Iraq. It has a long hot dry season and a shorter cool season. There is no rain during the hot season and only occasional flash rains in the cool season. The prevailing wind throughout the year comes from the north-west.

Comfort

From May to October it is too hot by day. Thick walls are needed to protect the interiors. The thick walls store heat, which makes the interiors too hot at night. Outdoor sleeping spaces are needed. Other hot-season requirements

EXAMPLE A

Location—Belem, Brazil; Longitude—48°27′ W; Latitude—1°28′ S; Altitude—12.8 metres.

Table 1

Temperature (°C)	J	F	M	A	M	J	J	A	S	O	N	D		
													Highest	AMT
Monthly mean max.	31	30	30	31	32	32	32	32	32	32	32	32	32	27
Monthly mean min.	23	23	23	23	23	23	22	22	22	22	22	22	22	10
													Lowest	AMR
Monthly mean range	8	7	7	8	9	9	10	10	10	10	10	10		

Table 2

	J	F	M	A	M	J	J	A	S	O	N	D	
Humidity Monthly mean max.	97	98	98	98	98	98	98	97	97	97	96	96	
(percentage): Monthly mean min.	87	89	90	89	86	82	82	81	83	83	81	84	
Average	92	93.5	94	95.5	93	90	90	99	90	90	88.5	91	
Humidity group	4	4	4	4	4	4	4	4	4	4	4	4	
Rainfall (mm)	334	422	455	397	278	196	149	117	120	106	94	205	Total 2,855
Wind: Prevailing	NE	NE	NE	NE	NE	SE	SE	SE	NE	NE	NE	NE	
Secondary	SE	NW	SE	NW	NE	NE	NW	NW	SE	SE	SE	SE	

Table 3

Temperature (°C)	J	F	M	A	M	J	J	A	S	O	N	D
Monthly mean max.	31	30	30	31	32	32	32	32	32	32	32	32
Day comfort: Max.	27	27	27	27	27	27	27	27	27	27	27	27
Min.	22	22	22	22	22	22	22	22	22	22	22	22
Monthly mean min.	23	23	23	23	23	23	22	22	22	22	22	22
Night comfort: Max.	21	21	21	21	21	21	21	21	21	21	21	21
Min.	17	17	17	17	17	17	17	17	17	17	17	17
Thermal stress Day	H	H	H	H	H	H	H	H	H	H	H	H
Night	H	H	H	H	H	H	H	H	H	H	H	H

Table 4

	J	F	M	A	M	J	J	A	S	O	N	D	Total
H1 Air movement (essential)	✓	✓	✓	✓	✓	✓	✓	✓	✓	✓	✓	✓	12
H2 Air movement (desirable)													0
H3 Rain protection	✓	✓	✓	✓	✓							✓	6
A1 Thermal storage													0
A2 Outdoor sleeping													0
A3 Cold-season problems													0

437

Table 5
SKETCH DESIGN RECOMMENDATIONS FOR BELÉM, BRAZIL

Indicator Totals from Table 4						Recommendations
Humid			Arid			
H1	H2	H3	A1	A2	A3	
12	0	6	0	0	0	
						Layout
			[0–10]			✓ 1. Buildings orientated on east-west axis to reduce exposure to sun
			11 or 12	5–12		
				0–4		2. Compact courtyard planning
						Spacing
[11 or 12]						✓ 3. Open spacing for breeze penetration
2–10						4. As 3, but protect from cold/hot wind
0 or 1						5. Compact planning
						Air movement
[3–12]						✓ 6. Rooms single banked. Permanent provision for air movement
1 or 2			0–5			
			6–12			7. Double-banked rooms with temporary provision for air movement
0	2–32					8. No air movement requirement
	0 or 1					
						Openings
			[0 or 1]		[0]	✓ 9. Large openings, 40–80% of N and S walls
			11 or 12		0 or 1	10. Very small openings, 10–20%
Any other conditions						11. Medium openings, 20–40%
						Walls
			[0–2]			✓ 12. Light walls; short time lag
			3–12			13. Heavy external and internal walls
						Roofs
			[0–5]			✓ 14. Light insulated roofs
			6–12			15. Heavy roofs; over 8 hours' time lag
						Outdoor sleeping
				2–12		✕ 16. Space for outdoor sleeping required
						Rain protection
		[3–12]				✓ 17. Protection from heavy rain needed

include shading from the sun and protection against dust storms. In the cool season, thermal storage in thick walls and the penetration of the sun through south-facing windows can be used to improve indoor conditions.

Recommendations

Air movement is not needed for human comfort. Closely packed courtyard housing and narrow shaded streets are assets throughout the year. In traditional houses, wind scoops are used to direct air through thick-walled ducts and over an earthenware water jug into ground floor and basement rooms. The air which is thus introduced is cooled by evaporation and humidified. Traditional houses in Baghdad have different rooms for different seasons. In the hot season the roof is used as a sleeping space at night and the shaded courtyard and thick-walled rooms on the ground floor are used during the day. In the cooler season the first floor, where the sun can penetrate, becomes the main living space.

EXAMPLE B

Location—Baghdad, Iraq; Longitude—44°24′ E; Latitude—33° 20′ N; Altitude—34 metres.

Table 1

Temperature (°C)	J	F	M	A	M	J	J	A	S	O	N	D	Highest	AMT
Monthly mean max.	16	18.5	22	29	36	41	43.5	43.5	40	34	24.5	17.5	43.5	23.5
Monthly mean min.	4	5.5	9	14.5	20	23.5	25	24.5	21	16	10.5	5	4	39.5
Monthly mean range	12	13	13	14	16	17.5	18.5	18	19	18	14	12.5	Lowest	AMR

Table 2

		J	F	M	A	M	J	J	A	S	O	N	D	Total
Humidity	Monthly mean max.	87	78	74	68	46	34	32	38	50	67	89		
(percentage):	Monthly mean min.	50	41	35	27	18	13	13	15	21	39	51		
	Average	68.5	59.5	54.5	47.5	32	23.5	22	23.5	26.5	35.5	53	70	
Humidity group		3	3	3	2	2	1	1	1	1	2	3	3	
Rainfall (mm)		24	25	28	15	7	0	0	0	0	3	22	26	150
Wind:	Prevailing	NW	NW	NW	NW	NW	NW	NW	NW	NW	NW	NW	NW	
	Secondary	SE	SE	N	N	N	N&W	N	N	N	N	N	SE	

Table 3

Temperature (°C)		J	F	M	A	M	J	J	A	S	O	N	D
Monthly mean max.		16	18.5	22	29	36	41	43.5	43.5	40	34	24.5	17.5
Day comfort:	Max.	29	29	29	31	31	34	34	34	34	31	29	29
	Min.	23	23	23	25	25	26	26	26	26	25	23	23
Monthly mean min.		4	5.5	9	14.5	20	23.5	25	24.5	21	16	10.5	5
Night comfort:	Max.	23	23	23	24	24	25	25	25	25	24	23	23
	Min.	17	17	17	17	17	17	17	17	17	17	17	17
Thermal stress	Day	C	C	C	—	H	H	H	H	H	H	—	C
	Night	C	C	C	C	—	H	—	—	—	C	C	C

Table 4

	J	F	M	A	M	J	J	A	S	O	N	D	Total
H1 Air movement (essential)													0
H2 Air movement (desirable)													0
H3 Rain protection													0
A1 Thermal storage	✓	✓	✓	✓	✓	✓	✓	✓	✓	✓	✓	✓	12
A2 Outdoor sleeping					✓	✓	✓	✓	✓				5
A3 Cold-season problems	✓	✓	✓									✓	4

Table 5
SKETCH DESIGN RECOMMENDATIONS FOR BAGHDAD, IRAQ

Indicator Totals from Table 4						Recommendations
Humid			Arid			
H1	H2	H3	A1	A2	A3	
0	0	0	12	5	4	
						Layout
			0–10			1. Buildings oriented on east-west axis
			11 or 12		5–12	to reduce exposure to sun
					0–4	✓ 2. Compact courtyard planning
						Spacing
11 or 12						3. Open spacing for breeze penetration
2–10						4. As 3, but protect from cold/hot wind
0 or 1						✓ 5. Compact planning
						Air movement
3–12						6. Rooms single banked. Permanent
1 or 2			0–5			provision for air movement
			6–12			7. Double-banked rooms with temporary
	2–12					provision for air movement
0	0 or 1					✓ 8. No air movement requirement
						Openings
			0 or 1		0	9. Large openings, 40–80% of N and S walls
			11 or 12		0 or 1	10. Very small openings, 10–20%
Any other conditions						✓ 11. Medium openings, 20–40%
						Walls
			0–2			12. Light walls; short time lag
			3–12			✓ 13. Heavy external and internal walls
						Roofs
			0–5			14. Light insulated roofs
			6–12			✓ 15. Heavy roofs; over 8 hours' time lag
						Outdoor sleeping
				2–12		✓ 16. Space for outdoor sleeping required
						Rain protection
		3–12				✗ 17. Protection from heavy rain needed

Example C. New Delhi, India

Climate

New Delhi, situated in the plains of northern India, has three seasons: hot and dry from March or April to June, warm and humid from July to October and cool and dry from November to February. In the two dry seasons the daily range of temperature is large. It decreases with the influx of moist air by the end of June and remains small during the monsoon season.

Comfort

From May to October it is too hot for comfort during the day and the night. During the warm humid season, air movement is needed to encourage evaporation. During the hot dry season, thick walls should be used to maintain cooler indoor temperatures by day. The cool season is pleasant by day, but it can be cold by night.

Recommendations

Buildings must reflect a compromise between the conflicting requirements of a composite

EXAMPLE C

Location—New Delhi, India; Longitude—77°12′ E; Latitude—28°35′ N; Altitude—217 metres.

Table 1

Temperature (°C)	J	F	M	A	M	J	J	A	S	O	N	D		
													Highest	**AMT**
Monthly mean max.	21	24	30.5	36	40.5	39	35.5	34	34	34	29	23	40.5	23.5
													Lowest	**AMR**
Monthly mean min.	6.5	9.5	14.5	20	26	28.5	27	26	24	18.5	11	8	6.5	34
Monthly mean range	14.5	14.5	16	16	14.5	10.5	8.5	8	10	15.5	18	15		

Table 2

		J	F	M	A	M	J	J	A	S	O	N	D	Total
Humidity	Monthly mean max.	41	35	23	19	20	36	59	64	51	32	31	42	
(percentage):	Monthly mean min.	72	67	49	35	35	53	75	80	72	56	51	69	
	Average	57	51	36	27	28	45	67	72	62	44	41	56	
Humidity group		3	3	2	1	1	2	3	4	3	2	2	3	
Rainfall (mm)		23	18	13	8	13	74	181	172	118	10	3	10	643
Wind:	Prevailing	NW	NW	NW	NW	NW	W	W	E	SE	NW	NW	NW	
	Secondary	W	W	W	W	W	W	E	SE	N&W	N	N	W	

Table 3

Temperature (°C)		J	F	M	A	M	J	J	A	S	O	N	D
Monthly mean max.		21	24	30.5	36	40.5	39	35.5	34	34	34	29	23
Day comfort:	Max.	29	29	31	34	31	31	29	27	29	31	31	29
	Min.	23	23	25	26	25	25	23	22	23	25	25	23
Monthly mean min.		6.5	9.5	14.5	20	26	28.5	27	26	24	18.5	11	8
Night comfort:	Max.	23	23	24	25	24	24	23	21	23	24	24	23
	Min.	17	17	17	17	17	17	17	17	17	17	17	17
Thermal stress	Day	C	—	—	H	H	H	H	H	H	H	—	—
	Night	C	C	C	—	H	H	H	H	H	—	C	C

Table 4

	J	F	M	A	M	J	J	A	S	O	N	D	Total
H1 Air movement (essential)							√	√	*				2+
H4 Air movement (desirable)													0
H3 Rain protection							*						0
A1 Thermal storage	√	√	√	√	√	√			*	√	√	√	9+
A2 Outdoor sleeping					√	√			√	√			4
A3 Cold-season problems	√											*	+1

* Indicates a borderline case.

441

Table 5
SKETCH DESIGN RECOMMENDATIONS FOR NEW DELHI, INDIA

Indicator Totals from Table 4						Recommendations
Humid			Arid			
H1	H2	H3	A1	A2	A3	
2+	0	0	9+	4	1+	
						Layout
			0–10			✓ 1. Buildings orientated on east-west axis to reduce exposure to sun
					5–12	
			11 or 12		0–4	2. Compact courtyard planning
						Spacing
11 or 12						3. Open spacing for breeze penetration
2–10						✓ 4. As 3, but protect from cold/hot wind
0 or 1						5. Compact planning
						Air movement
3–12						6. Rooms single banked. Permanent provision for air movement
1 or 2			0–5			✓ 7. Double-banked rooms with temporary provision for air movement
			6–12			
0	2–12					
	0 or 1					8. No air movement requirement
						Openings
			0 or 1		0	9. Large openings, 40–80% of N and S walls
			11 or 12		0 or 1	10. Very small openings, 10–20%
Any other conditions						✓ 11. Medium openings, 20–40%
						Walls
			0–2			12. Light walls; short time lag
			3–12			✓ 13. Heavy external and internal walls
						Roofs
			0–5			14. Light insulated roofs
			6–12			✓ 15. Heavy roofs; over 8 hours' time lag
						Outdoor sleeping
				2–12		✓ 16. Space for outdoor sleeping required
						Rain protection
		3–12				✗ 17. Protection from heavy rain needed

climate. They should be well spaced to benefit from the east and south-east breeze during the humid season. Plans may be compact and rooms double banked, but internal openings must allow for air movement during the three months of the humid season. Heavy walls and roofs should be used to give protection against extremes of temperature. These thick walls will store heat for the night. During the cool season this is an asset, but during the hot dry season this stored heat will make outdoor sleeping essential. Windows should be of medium size to give good internal air movement, but provided with well-made thick shutters to reduce the flow of cold air in the cool season and the flow of heat and dusty air during the hot dry season.

SUN PATH DIAGRAM AND SHADOW ANGLE PROTRACTOR

Solar radiation is welcome when the weather is cold, unwelcome when it is hot. The information in this annex explains how the architect can protect buildings, especially the openings,

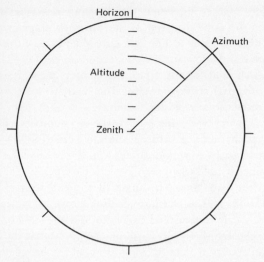

Figure A2–1. A simplified sun path diagram (solar chart).

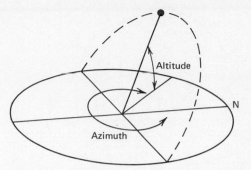

Figure A2–2.   Schematic representation of altitude and azimuth in relation to the sun's path across the sky (dashed line).

against direct solar radiation and how he can encourage sun penetration when it is needed to warm the interior.

The Sun Path Diagram

The sun's position as it moves across the sky can best be shown on a map of the sky, the sun path diagram. This diagram consists of a circle, the periphery of which represents the horizon while the centre represents the zenith directly overhead (Figure A2–1).

Two co-ordinates are needed to locate a position in the sky. They are called azimuth and altitude. In the sun path diagram the azimuth is shown on an angular scale 0–360° around the circle. It is measured clockwise from the north. The altitude of the sun's position is shown by a series of concentric rings, and is measured upward from the horizon (0°) to the zenith (90°) (Figure A2–2).

Sun's Path and Hour Lines

The path of the sun across the sky is shown by a series of lines which start at the eastern edge of the circle (sunrise) and finish on the western edge (sunset). The northernmost line represents the sun's path on June 22 (the summer solstice) and the southernmost line represents it on December 22 (the winter solstice). The lines between these represent the sun's path at intervals throughout the year. Each of these lines represents the sun's

path for two days of the year, one day during the period from January to June, when the sun's path moves further to the north each day, and the second during the period from June to December when the sun's path moves back to the south (Figure A2–3).

The shorter lines that cross the sun's path represent the hours of the day. They show that the sun rises around six o'clock in the morning, crosses the line due north-south at mid-day and sets in the evening around six o'clock. The times given are solar times, which may vary slightly from local time, but the designer can safely ignore this difference. The effect of the difference is negligible. Figure A2–4 shows the sun's path

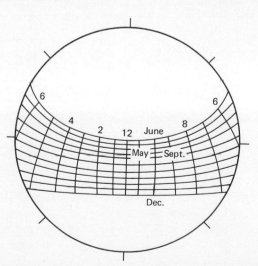

Figure A2–3. Sun path diagram with hour lines added. Eastern edge of circle represents sunrise and western edge of the circle represents sunset.

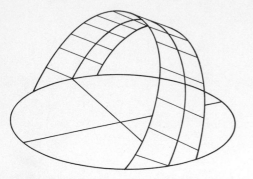

Figure A2–4.   Sun's path across the sky using the perspective given in Fig. A2–2.

across the sky in terms of azimuth and altitude as indicated in Figure A2–2.

In theory, a different sun path diagram should be used for each degree of latitude. In practice, one can manage with fewer diagrams. The solar charts for 4° intervals will provide sufficient accuracy for most applications. (Examples for 0° and 4°N are given in Figure A2–13 and A2–14).

The Shadow Angle Protractor

The shadow angle protractor (Figure A2–5) is used to find the sizes of vertical and horizontal projections (or reveals) which are required to exclude the sun when it is not needed. The shadow angle protractor consists of two series of lines marked on a transparent semi-circle which has the same diameter as the sun path diagram. The first series of lines are curved and show the vertical shadow angles. The second series of lines, which radiate out from the centre, show the horizontal shadow angles. The diameter of the protractor is called the base line.

The curved lines represent a number of hypothetical sun paths. If the sun were to follow these paths it would always appear to have the same altitude when seen in section perpendicular to the base line. The angle of the sun seen in section is the vertical shadow angle (Figure A2–6). It is measured from the horizon (0°) up to the zenith (90°). 'It should be noted that the sun's vertical shadow angle is equal to the solar altitude only when the sun's rays are perpendicular to the base line.

The shadow angle protractor is placed on the sun path diagram and rotated so that the base line and the curved line, which represents the vertical shadow angle, cover the area of sky obscured by a horizontal projection. The extent of the projection is determined by the vertical shadow angle which is measured in a vertical section at right angles to the wall. It is the angle between the horizontal and a line drawn from the edge of the projection to the sill or the lower edge of the opening (Figure A2–7).

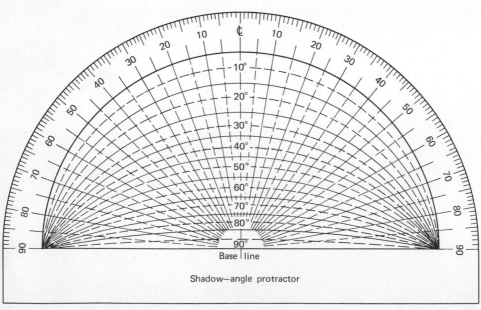

Figure A2–5.   The shadow angle protractor.

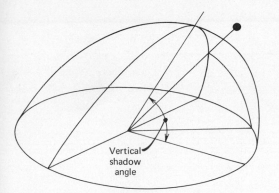

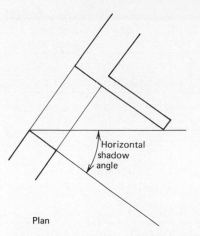

Figure A2–6. Diagrammatic explanation of the vertical shadow angle (VSA).

Figure A2–8. Plan of shade projection related to horizontal shadow angle (HSA).

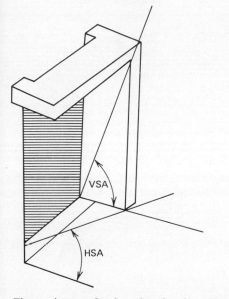

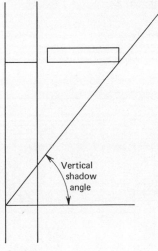

Figure A2–7. Section showing how the extent of shade protection is determined from the vertical shadow angle (VSA).

Figure A2–9. Three-dimensional sketch showing how building projections (both horizontal and vertical) depend on both the VSA (vertical shadow angle) and the HSA (horizontal shadow angle).

It is not always possible to exclude the sun by horizontal projections alone. The second series of lines can be used to find the horizontal shadow angles. The horizontal shadow angle is the angle shown on the plan between a line from the inner edge of the opening to a vertical projection beyond it and a line perpendicular to the base line (Figure A2–8). The relationship between horizontal shadow angle (HSA) and vertical shadow angle (VSA) are shown by example in Figure A2–9.

For actual computational purposes, a plastic shadow angle protractor is available. It is appended to the UNESCO (1971) publication on which this account is based. Solar charts within that publication are appropriately scaled to facilitate use of the protractor.

Shading Periods

When the sun's path for the correct latitude has been found, it is necessary to determine the periods when protection is needed. There are places with some hot and some cold months

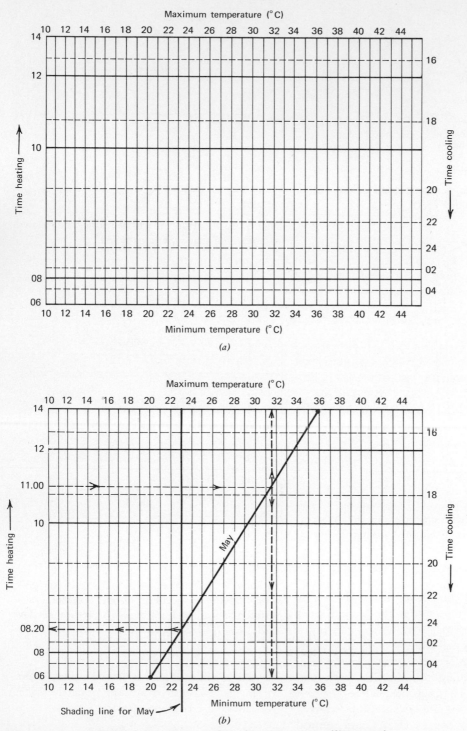

Figure A2-10. (*a*) Diurnal variation and shading time chart. (*b*) Example of a completed chart for Baghdad, Iraq, data for which are given in the Mahoney tables. Refer to text for explanation.

where a simple inspection of a completed Table 3 (of the Mahoney tables) will show the months when sun is needed. These months can be marked on the sun path diagram.

There may be other cases where maximum temperatures during the day are comfortable or hot but the nights are cold and some sun may be welcome in the mornings when the air temperature and the interior of the buildings are still cool. This situation is often found in tropical upland climates.

If the sun is to be excluded when it causes discomfort and allowed to penetrate into the interior when it aids comfort, the temperature for each hour of the day has to be found. As hourly readings are not normally published by meteorological observatories, the daily shading time chart (figure A2–10) can be used to find the times when complete sun protection is needed. Using the Mahoney tables, the steps to find the shading time are explained below:

(*a*) Find the lower limit of the day comfort zone for the month from Table 3. This is the shading temperature. A line should be drawn vertically joining this shade temperature on the top and bottom scales;

(*b*) Find the monthly mean maximum and the monthly mean minimum temperature for the month from Table 1;

(*c*) Mark the maximum on the top scale and the minimum on the bottom scale and join the two points with a diagonal line;

(*d*) Find the point where the diagonal maximum-minimum line crosses the shading time line. From this point draw a horizontal line parallel to the hourly lines to meet the time scale line on the left-hand side. The point of intersection gives the time when complete shading should start.

If the maximum-minimum line does not meet the shading line, it is necessary to differentiate between two possible situations:

(*a*) If the shading line is to the left of the maximum-minimum line, then shading is always needed. The air temperature is never below the comfort zone during this month;

(*b*) If the shading line is to the right of the maximum-minimum line, then complete shading is not essential as the air temperature is always below the comfort zone for this particular month.

The time when shading may end can be found by continuing to the right the line already drawn from the point of intersection. There are many cases where the sun sets before the temperature drops below the comfort zone.

Figure A2–11 gives the shading times for the different months and these times can be marked on the sun path diagram. Those areas of the sky across which the sun moves when complete shading is needed should be hatched.

When designing shading devices it should be remembered that the change in the path of the

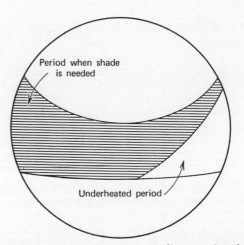

Figure A2–11. The sun path diagram showing the period when shade is required.

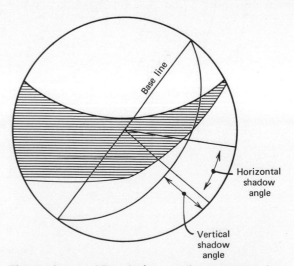

Figure A2-12. The shadow angle protractor is placed over the hatched sun path diagram to determine HSA and VSA.

sun is symmetrical to the solstices. Each sun path line represents the path of the sun across the sky at two different times of the year.

The same cannot be said about the air temperature. In the northern hemisphere the highest air temperature does not always coincide with the summer solstice nor the lowest with the winter solstice. In many regions, there is a time lag of a month or more. It may be too cold in March but too hot in September or, in composite climates, it may be still too hot in November but distinctly cold in February. In such cases, it is always preferable to provide adequate shading for the overheated period rather than to allow sun to penetrate during the cold period. In more expensive houses adjustable louvres are used to provide a potentially better solution; potentially, because its effectiveness depends on good maintenance and correct use of the adjustable louvres.

When air temperatures are low, the designer should not automatically assume that he needs as much sun penetration as possible. Overheating can occur when the sun shines through large

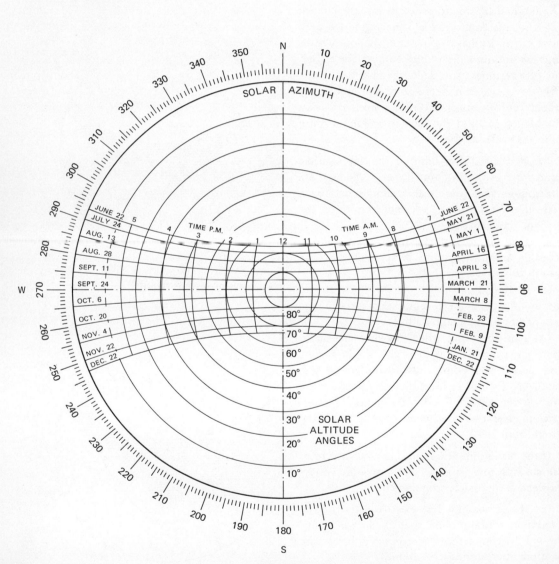

Figure A2-13. Solar chart for the equator. (Based on those prepared by Building Research Station, Watford, England.)

windows even when the external air temperature is well below the comfort zone.

The Design of Shading Devices

The protractor is placed over the hatched sun path diagram to decide the horizontal and vertical shadow angles that will exclude the sun when it is not needed (Figure A2-12). As a rule, the designer can choose from a number of combinations which achieve the same effect, as far as sun protection is concerned. He can then take other considerations into account. In the warm humid tropics it may be desirable to use horizontal projections, such as large overhangs, to protect openings from rain and sky glare. In the hot dry tropics, vertical sun shades are often preferred because they allow a view of the blue sky. There is also the possibility of using a number of small louvres rather than one large one. Egg-crate louvres are often used to give combined vertical and horizontal shading. The possibilities are endless and the method for designing shading devices for windows outlined

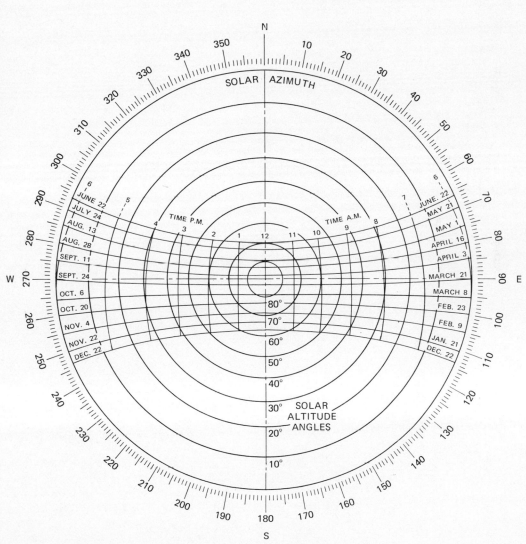

Latitude 4° North

Figure A2-14. Solar chart for 4°N. (Based on those prepared by the Building Research Station, Watford, England.)

here gives the designer a large degree of flexibility.

Three further points should be mentioned in conclusion. First, the shading device affects other window functions such as lighting and air flow. Secondly, the material used should be reflective and as light as possible to avoid the storage and reradiation of heat through the opening. Finally, it is extremely difficult, and often impossible to design effective sun shades for west-facing windows in tropical and subtropical regions.

The solar charts in Figures A2-13 and A2-14 are marked for the northern hemisphere, but can be used also for locations south of the equator by reading the charts upside down, reversing the N-S and E-W signs and the hour lines.

APPENDIX 3

Analytic Methods[1]

The study of climatology is based on analysis and interpretation of meteorological data collected over many years. To analyze such data meaningfully, an awareness of basic statistical methods is a prime requisite, and anyone seriously concerned with extensive use of climatic data should have a basic knowledge of statistical techniques. This appendix outlines some of the available methods. For the most part it provides a statement of the formulas used and the data to which they are applied. Because of the brevity of presentation, some rather sophisticated methods are given only cursory treatment. For further explanation, the reader is referred to fuller accounts provided in basic statistical texts and, for application of the methods to climatic data, to Conrad and Pollak (1950), Brooks and Carruthers (1953), and Gregory (1964).

Studies pertaining to examples given in this annex are cited by author and number in the text. These are keyed to references given at the end of this section.

Climatological analysis uses principles and techniques of meteorological, numerical, and statistical analysis. Previously formulated techniques are applied wherever possible but quite often unique methods must be devised to meet unusual problems and special data characteristics. Adaptations of methods from both meteorological and numerical analysis are applied to both raw and summarized data.

Analytical treatments of individual observations regularly include the following:

(a) Conventional meteorological (isopleth) analysis of historical synoptic daily weather maps.

(b) Interpolation of missing observations.

(c) Extrapolation of incomplete upper air soundings.

(d) Non-routine verification of observations of questionable accuracy.

(e) Interpretation of observations incompletely identified, annotated, or documented.

(f) Establishment of methods for proper combining of similar observations coded differently.

(g) Interpretation of portions of codes that are ambiguous (to a computing machine).

Analytical treatments of summarized observations include:

(a) Isopleth analysis for climatic maps.

(b) Graphical analysis of frequency distributions, scattergrams, etc.

(c) Interpolation to fill gaps in summarized observations. (Monthly averages, etc.)

(d) Nomogram preparation. (Code and climatological units, conversions, verification guides, etc.)

(e) Verification of tabular and graphical summaries.

(f) Ratio and differential analysis for reduction of "single" station data from changing sites to a common or single location.

(g) Selection of stations or areas representative of larger regions.

CLIMATOLOGICAL STATISTICS

Statistical principles and techniques are presently applicable to some of the analytical treatments cited above. As research and development work continues, statistical methods will be adapted more and more to problems in climatology. In fact, greater understanding of the consistencies and variations in the atmosphere is dependent largely upon more accurate observations, more precise documentation of the observations, and further developments in the field of statistics. Most standard statistical methods and their theoretical bases, explained fully in many

---

[1] Reproduced with slight modifications from *Climatology at Work*, U.S. Department of Commerce, Weather Bureau, Washington, D.C., 1960. Reprinted by permission of the U.S. Department of Commerce, National Oceanic and Atmospheric Administration.

textbooks, are merely mentioned in this chapter. Methods of specific climatological use are discussed in somewhat greater detail, including selected examples.

CLIMATOLOGICAL SERIES

The methods of statistical analysis apply to climatological data because, to a large extent, if these data are properly taken, sequences of such data behave like random variables. Since statistical analysis only applies to samples from populations of data the sequences of climatological data must be defined so as to be samples from populations. To accomplish this we define a climatological series as a sample series of data consisting of one climatolgial value each year of the record being considered. Thus the thirty January average temperatures for a thirty year record form a climatological series. The 30 January-1st precipitation amounts form a climatological series. The 90 February, March, and April monthly precipitation amounts *do not* form a climatological series but are samples through different populations and are therefore different climatological series, hence they must be dealt with as three separate series. The series of 3720 hourly temperatures for a five year record during March *does not* form a climatological series because there are 24 × 31 different populations and hence really 744 different climatological series are involved. Under certain circumstances such populations can be mixed together, as were the February, March, and April series above, but the individual climatological series and populations must be first defined so that the exact meaning of the mixture of populations is defined in advance of statistical analysis.

Climatological series variables may be either discrete or continuous. Discrete series variables are usually counted values such as the number of days with precipitation greater than 0.10 inch for each of 30 Junes or the number of times the visibility is less than 1 mile, during each of 30 Julys. Continuous series variables are usually measured values such as temperature and precipitation, e.g., the series of 30 totals of spring precipitation (each the total of March, April, and May).

A climatological series is never more than a sample from a *single population* assumed to behave as if it were infinite in extent and having the climatic properties such that the observed climatological series is a random sample from that infinite population, i.e., a sample drawn in a manner independent of the individual magnitudes of the members of the infinite population.

The Frequency Distribution

The frequency distribution is the basic tool for describing and analyzing the population. This is accomplished by estimating the characteristics of the population frequency distribution from the sample or climatological series as tallied in class intervals which are divisions of the range of the climatological variable. The number of class intervals is best taken to be between 10 and 20. This divides the difference between the largest and smallest value or range of the climatological series into from 10 to 20 equal divisions. The procedure for division into class intervals is best illustrated by the following example for August precipitation amounts (in mm) for Geneva, Switzerland: The 30-year record for 1927–56 given in the following table is used.

Table A3-1

AUGUST PRECIPITATION (MM), GENEVA, SWITZERLAND

Year	p	Year	p	Year	p
1927	250	1937	78	1947	54
28	147	38	79	48	72
29	83	39	85	49	49
30	108	40	18	50	110
31	171	41	105	51	100
1932	62	1942	48	1952	125
33	67	43	41	53	57
34	119	44	44	54	206
35	157	45	133	55	107
36	23	46	158	56	144

To find a class interval for this climatological series we follow our rule: The highest value is 250 mm and the lowest 18 mm. This gives a range of 232 mm. Since 20 mm is a convenient division and gives 13 divisions, this is a suitable class interval. Tallying these by classes we obtain the following table of precipitation p and frequency f:

Table A3–2
FREQUENCY DISTRIBUTION OF AUGUST PRECIPITATION (MM), GENEVA, SWITZERLAND

p	f	p	f
0–19	1	100–119	6
20–39	1	120–139	2
40–59	6	140–159	4
60–79	5	160–179	1
80–99	2	180–199	0
		200–219	1
		220–239	0
		240–259	1

If these frequencies are plotted as blocks proportional to f on the scale of precipitation, the histogram of precipitation for Geneva is obtained, as in Figure A3–1. The f's may be divided by 30, the number of years in the climatological series, to obtain the relative frequencies in each class interval. These sample values are estimates of the probabilities in the population of precipitation amounts in the various class intervals.

Cumulative Distribution

Usually the climatologist is more interested in estimates of probabilities over several class intervals which are more conveniently obtained from the cumulative distribution. Also, the cumulative distribution provides better estimates of the probabilities since the arbitrary division into class intervals, as in Table A3–2, tends to waste some of the information on the population given by the climatological series.

To obtain the cumulative distribution the data are first put in order as in the following table:

Table A3–3
CUMULATIVE DISTRIBUTION OF AUGUST PRECIPITATION (MM), GENEVA, SWITZERLAND

m	p	F	m	p	F	m	p	F
1	18	.032	11	72	.355	21	119	.677
2	23	.065	12	78	.387	22	125	.710
3	41	.097	13	79	.419	23	133	.742
4	44	.129	14	83	.452	24	144	.774
5	48	.161	15	85	.484	25	147	.806
6	49	.194	16	100	.516	26	157	.839
7	54	.226	17	105	.548	27	158	.871
8	57	.258	18	107	.581	28	171	.903
9	62	.290	19	108	.613	29	206	.935
10	67	.323	20	110	.645	30	250	.968

The F's are the cumulative relative frequencies or estimates of the cumulative population probabilities and are obtained by the formula $F = m/(n + 1)$ where m is the mth value in order of magnitude of the climatological series and n is the number of terms in the climatological series, in this case 30. The division by $(n + 1)$ instead of n gives a better estimate of population probabilities, especially at the ends of the distribution. It can be shown that $m/(n + 1)$ gives the best simple estimate of the probabilities.

The F's give the probabilities that precipitation is less than any value shown in the table. For example, the probability that p is less than 62 mm is 0.290 and greater than 62 mm is $1 - F = 0.710$. Note that when probabilities are estimated for a continuous random variable such as precipitation, it is a misunderstanding of sampling principles to use the wording "equaled or exceeded" or "less than or equal to," for the probability of any exact value occurring is zero. The probability that it is between 62 and 125 mm

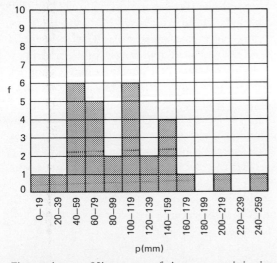

Figure A3–1. Histogram of August precipitation at Geneva, Switzerland.

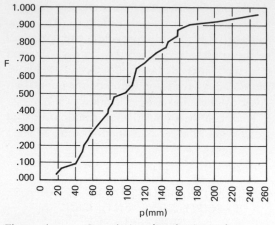

Figure A3-2. Cumulative distribution of August precipitation at Geneva, Switzerland.

is 0.710 − 0.290 = 0.420. Thus the cumulative distribution gives all the information available from histograms, and much in addition, since it uses every value of the climatological series individually to obtain the probability estimates. The sample cumulative distribution may also be put in graphical form by plotting F on the ordinate against p on the abscissa and connecting the points by straight lines, as in Figure A3-2. Climatological series with discrete variables may also be treated in a similar manner.

The average temperatures (°C) for August for Geneva shown in Table A3-4, may be analyzed in a similar fashion as another example. The series has been arranged in order of magnitude in Table A3-4.

Table A3-4
AVERAGE AUGUST TEMPERATURE (°C), GENEVA, SWITZERLAND

m	t	F	m	t	F	m	t	F
1	16.9	.032	11	18.6	.355	21	19.8	.677
2	17.4	.065	12	18.7	.387	22	19.9	.710
3	17.5	.097	13	18.7	.419	23	20.3	.742
4	17.8	.129	14	18.9	.452	24	20.4	.774
5	17.9	.161	15	18.9	.484	25	20.7	.806
6	17.9	.194	16	19.2	.516	26	20.8	.839
7	18.1	.226	17	19.3	.548	27	20.9	.871
8	18.3	.258	18	19.5	.581	28	20.9	.903
9	18.5	.290	19	19.5	.613	29	22.0	.935
10	18.6	.323	20	19.7	.645	30	22.9	.968

Note that since the record length is the same, the F's are the same as in the previous table, and hence have the same interpretation as previously. The estimated probability that the average temperature for August at Geneva is less than 20.3°C is 0.742 and that it is greater than 20.3 is 1 − 0.742 = 0.258. The mean recurrence interval or return period (i.e., the average time between occurrences) for values exceeding any value t is $1/(1 − F)$. Hence for temperatures exceeding 20.3° the mean recurrence interval is $1/0.258$ or about 4 years.

Homogeneity of Data Series

A data series is said to be homogeneous if it is a sample from a *single* population. Hence by definition a climatological series is homogeneous and elementary probability analysis must be applied only to climatological series. The previous temperature and precipitation series were, of course, analyzed on the assumption of homogeneity. If a series is not homogeneous, adjustments must be made so that statistical estimates will be valid estimates of the *population* parameters applying to the last terms in the series or such that they are estimates obtained from a hypothetical homogeneous series including the latest data as elements.

In cases where instrument exposures have changed it is necessary to make a statistical test to insure homogeneity. Many of the older methods of testing for homogeneity were incomplete in the sense that they provided inadequate criteria for accepting or rejecting the hypothesis of homogeneity. The valid test of homogeneity is a statistical test of hypothesis which provides an hypothesis of homogeneity (null hypothesis) and a rule for accepting or rejecting this hypothesis on the basis of probability of occurrence. Thus if the probability of the evidence for homogeneity is small, it is concluded that the series is heterogeneous; if it is large, the decision is for homogeneity. The rule specifies the probability limit (significance limit) beyond which the hypothesis of homogeneity would be rejected and some alternative to homogeneity accepted. In most instances distributions on the null hypothesis and the alternatives to homogeneity are difficult to specify, hence the so-called non-parametric tests ordinarily must be used.

The alternatives to homogeneity in a series of meteorological data are usually slippage of the mean, trend, or some form of oscillation. Since these alternatives, especially the latter, may be difficult to specify exactly, it is best to use a nonparametric test which does not require exact specification of these alternatives or the null distribution. A well-known non-parametric test which is sensitive to all of these alternatives is the run test provided by Swed and Eisenhart (1). This test is made by counting the number of runs, u above and below the median or middle value in a naturally ordered series and testing this by means of a table of the distribution of u. The test is best illustrated by applying it to the Geneva August average temperatures. These are given in their historical order in Table A3-5 and it is seen that the median or middle value is between 18.9 and 19.2. It may be taken as half way between these two values or 19.05. Using this value the entries in Table A3-5 may be marked with a B if they are below this value and with an A if above this value. The runs then are marked as sequences of A's and B's. The total number of runs is seen to be $u = 15$.

Table A3-5
RUNS FOR OBSERVED TEMPERATURE SERIES (°C), GENEVA, SWITZERLAND

1927	17.4 B	1937	19.5 A	1948	18.9 B
1928	20.9 A	1938	18.5 B	1949	20.7 A
		39	18.6 B	50	19.7 A
1929	18.7 B	40	17.9 B	51	19.5 A
30	18.7 B	41	17.8 B	52	20.3 A
31	16.9 B			53	19.8 A
		1942	19.9 A		
1932	20.8 A	43	20.9 A	1954	18.3 B
33	20.4 A	44	22.9 A		
				1955	19.3 A
1934	17.9 B	1945	18.9 B		
35	18.1 B			1956	17.5 B
36	18.5 B	1946	19.2 A		
		47	22.0 A		

It is clear that too many runs would be an indication of oscillation while too few runs would be an indication of a trend or a shift in the median during the sample record. Hence, if the probability of a u being exceeded is small an oscillation would be suspected while if the probability of being less than a sample u is small a trend or shift in median would be suspected. If the probability of being either greater than or less than u is large then neither oscillation nor trend is suspected and the series is said to be homogeneous or from a single population. To make this test a distribution table of u is required. This is given below: Since the median was chosen, the number of values above the median N_A will equal the number of values N_B below the median; hence the table is for $N_A = N_B$.

Table A3-6
DISTRIBUTION TABLE OF NUMBER OF RUNS (u), $N_A = N_B$

	P			P	
N_A	0.10	0.90	N_B	0.10	0.90
10	8	13	19	16	23
11	9	14	20	16	25
12	9	16	25	22	30
13	10	17	30	26	36
14	11	18	35	31	41
15	12	19	40	35	47
16	13	20	45	40	52
17	14	21	50	45	57
18	15	22			

Table A3-6 gives the lower and upper 0.10 significance limits i.e., for probabilities P of 0.10 and 0.90. These 0.10 significance limits are most satisfactory for many meteorological applications, for, because of frequent high variability, it is desirable to increase the significance limit probabilities since this will increase the changes of accepting the alternative hypothesis. Since u is discrete the u values shown in the tables are those corresponding to the probability closest to 0.10 and 0.90. The maximum divergence from exact probability values is +0.03. If a sample u is below the lower limit, heterogeneity is due to trend or mean slippage, if above it is due to oscillation.

It was seen in Table A3-5 that $u = 15$ for $N_A = N_B = 15$. The upper and lower limits from Table A3-6 for $N_A = 15$ are 12 and 19. $u = 15$ is within this range; hence u is not significantly different from u's expected from homogeneous series, and the series is concluded to be homogeneous.

In order to further illustrate application of the runs test the series has been deliberately made heterogeneous by subtracting 1°C from each of the first 12 years of record and subtracting 0.5°C from each of the next eight years. The heterogeneous series is shown in Table A3-7.

Table A3-7
RUNS FOR HETEROGENEOUS TEMPERATURE SERIES (°C), GENEVA, SWITZERLAND

1927	16.4	1944	19.4
		43	20.4
1928	19.9	44	22.4
1929	17.7	1945	18.4
30	17.7		
31	15.9	1946	18.7
		47	22.0
1932	19.8	48	18.9
33	19.4	49	20.7
		50	19.7
1934	16.9	51	19.5
35	17.1	52	20.3
36	17.5	53	19.8
37	18.5		
38	17.6	1954	18.3
39	18.1		
40	17.4	1955	19.3
41	17.3		
		1956	17.5

The number of runs is reduced to 11 by the two shifts of the mean which in effect produce a kind of trend. Referring to Table A3-6 at $N_A = 15$ it is seen that the probability of less than 12 runs is 0.10, and since Table A3-7 has only 11 runs, the heterogeneity was found by the test. Of course, it was already known that the heterogeneity was there because it was introduced deliberately. It will naturally be suspected, and correctly so, from this example that the ability of such tests to find heterogeneities when the exact alternatives to homogeneity are not known will not be very good. This brings out the very important point that the best way to determine heterogeneities is to determine their cause in the history of the record. If the history of a record shows changes which could cause heterogeneities and which can be described according to period and character, more powerful parametric tests such as "Student's" t-test may be employed to determine the significance of the heterogeneities. Such tests, however, may only be employed where the periods and character of the heterogeneities are known *a priori*.

Adjustment of Climatological Means

Heterogeneity in climatological data series is usually due to some disturbing factor such as change in station location or change in exposure. Although in the past attempts have been made to homogenize series having such disturbances, it must be made very clear that it is *not possible* to homogenize a series in the sense that a new series of individual values is derived with the same properties as a sample from the proper hypothetical population. In other terms if the data from a particular station are unavailable for a particular period of record, it is impossible to reproduce the individual items of the series for that period. The reason for this is that any adjustment disturbs the variability of the series and hence changes the scale or dispersion of the frequency distribution. It is possible, however, to adjust certain *statistics* of the series so that these adjusted values are, in effect, like those estimated from samples taken from the proper hypothetical population. The most common application of such adjustments is to the means of data series for the purpose of obtaining normals. It is recommended that such adjustments be made if possible only on the basis of *a priori* known heterogeneities.

It may be shown by theoretical analysis that the classical difference and ratio methods are close to optimum for the adjustment of temperature and precipitation means. Such adjustments are often made to compensate for missing record and to remove heterogeneities. The difference method employs the difference between temperature means of two concurrent homogeneous series as an additive factor on the available series mean. The ratio method employs the ratio of precipitation totals or means of two concurrent homogeneous series as a multiplying factor on the available series total or mean. The adjustments are best illustrated by examples.

The method involves using a supplementary station with a concurrent homogeneous record. This station should be as close as possible to the station to be adjusted as the effectiveness of the adjustment depends on the correlations between the two stations. Usually a station less than 50 miles from the station to be adjusted and in the same climatic regime will serve the purpose. Several supplementary stations may be averaged and used as the supplementary record, but this usually does not increase the correlation greatly. If a supplementary station does not have a complete record, the adjustment may have to proceed by stages using a different supplementary station for each period of record.

(a) The Difference Method

In Table A3–7 deliberate heterogeneity was introduced into the average temperature record by subtracting 1.0°C from each of the first 12 years, 0.5°C from the next 8 years, and leaving the last 10 years unchanged. It is now assumed that during each of the first two periods the station was moved or the exposure of instruments changed, and that it is desired to adjust the 30-year mean to the exposure during the last 10 years. This is a typical adjustment problem. Other arrangements of the heterogeneities in a record are easily taken into account by a simple variation in the adjustment procedure.

To adjust the means of temperature and precipitation of the Geneva record, given the dates of heterogeneous periods, and therefore also the dates of homogeneous periods, it has been found convenient to use Lausanne as the supplementary station. It is not presumed that Lausanne is the best supplementary station. It is only used because it serves the purpose at hand well to illustrate the adjustment of a known heterogeneity. The adjustment formula for temperature is

$$\bar{y} = a + \bar{x} \tag{1}$$

Here $\bar{x}$ is the mean for the homogeneous period at the supplementary station corresponding to the heterogeneous period at the station whose record is being adjusted, and $\bar{y}$ is the adjusted mean. The adjustment constant a is estimated by the equation

$$a = \bar{v} - \bar{u} \tag{2}$$

Here $\bar{u}$ and $\bar{v}$ are the means from concurrent periods of homogeneous record at the supplementary station and station being adjusted, respectively. The process of adjustment for temperature then consists of estimating a, using concurrent homogeneous records at the supplementary station and the station to be adjusted, and substituting this value in turn in equation (1) to obtain the adjusted mean $\bar{y}$. The $\bar{y}$'s for the various parts of the 30-year record are then weighted according to length of period in years and averaged to obtain the adjusted 30 year record.

The means for each period were obtained from Table A3–7 which is artificially heterogeneous. These are shown in Table A3–8.

Table A3–8

MEAN TEMPERATURE ADJUSTMENT (°C), FOR GENEVA, SWITZERLAND

	Lausanne $\bar{x}$	Geneva-Unadjusted Means	Geneva $\bar{y}$
1927–38	17.9	(17.9)	19.3*
1939–46	18.4	(19.0)	19.8*
1947–56	18.2	19.6	19.6
Adjusted Record Mean			19.5*

Substituting the homogeneous values for $\bar{u}$ and $\bar{v}$ in equation (2) gives an estimate of the adjustment factor $a = 19.6 - 18.2 = 1.4$. Inserting this in equation (1) and substituting successively the homogeneous values 17.9 and 18.4 gives the adjusted values $\bar{y} = 17.9 + 1.4 = 19.3*$ and $\bar{y} = 18.4 + 1.4 = 19.8*$. Next multiplying the values $\bar{y}$ by 12, 8, and 10, their respective lengths of record, summing these and dividing by 30 gives the weighted mean 19.5*. This is the estimated adjusted mean of August average temperature for Geneva. Note that this compares favorably to the actual value for the undisturbed record 19.3. The procedure provides the best estimate of the hypothetical mean for the 1927–56 record at Geneva based on the homogeneous period 1947–56.

(b) The Ratio Method

In order to illustrate the application of the ratio method of adjustment which must be used for precipitation, the Geneva precipitation record for 1927–56 was made heterogeneous as follows: The record was subjected to a scale change by multiplying the precipitation for each of the first 12 years by 1.20 and each of the ensuing 8 years by 0.90, the last 10 years being left undisturbed. The resulting heterogeneous series is shown in Table A3-9.

Before proceeding with the adjustment it is easy to test the homogeneity of the series to provide a further illustration of the use of the run test. Of course, this test is really unnecessary, for the heterogeneities are known *a priori*. In this instance the median may be readily found, by ordering the data, to be 97.5 mm. The runs of values above and below the median may be marked as shown in Table A3-5. This is seen to give the number of runs $u = 9$. Since $N_A = N_B = 15$ as with Table A3-5, the upper and lower significance limits 12 and 19 are the same as previously. The value 9 lies outside of this range; hence the series is not homogeneous. As would be expected u has been made too small by slippage of the mean values for the periods 1927–38 and 1939–46.

Table A3-9

HETEROGENEOUS AUGUST PRECIPITATION SERIES (MM), FOR GENEVA, SWITZERLAND

1927	300	1936	28	1947	54
28	176	37	94	48	72
29	100	38	95	49	49
30	130	39	77		
31	205	40	16	1950	110
		41	95	51	100
1932	74	42	43	52	125
33	80	43	37		
		44	40	1953	57
1934	143				
35	188	1945	120	1954	206
		46	142	55	107
				56	144

Since heterogeneities in precipitation series are scale changes in the frequency distribution, it is proper to adjust for heterogeneities by scale adjustment, i.e., by using the ratio of homogeneous totals. This is seen to be equivalent to adjusting by the ratio of homogeneous means.

By this principle, if y is the precipitation for one unit of the year on the station to be adjusted, and x is the corresponding value for the supplementary station, then

$$\sum y = b \sum x \tag{3}$$

where the summations are over a period heterogeneous at the station to be adjusted. Thus the estimated total precipitation on a unit of the year for a period of record is equal to the total for the same unit and period at the supplementary station times the adjustment constant b. The adjustment constant b is estimated by the equation

$$b = \frac{\sum v}{\sum u} \tag{4}$$

where Σv is the sum of precipitation over the homogeneous period at the station to be adjusted and Σu is the sum for the corresponding period at the supplementary station. This, of course, should be the latest period of record for active stations since it is desired to adjust to a population from which observed values at the active station location will be obtained. The process of adjustment consists in estimating b for a homogeneous period by means of equation (4) and applying equation (3) with this statistic to the heterogeneous periods. The results are shown in Table A3-10.

Table A3-10

MEAN AUGUST PRECIPITATION ADJUSTMENT (MM), FOR GENEVA, SWITZERLAND

Lausanne	Σx	Geneva Unadjusted Totals	Geneva Σy
1927–38	1602	(1613)	1295*
1939–46	753	(570)	609*
1947–56	1267	1024	1024
Adjusted Record Mean			97.6

Substituting the values of Σx and Σy from table 12 for the homogeneous period 1947–1956 for Σu and Σv in equation (4) gives $b = 1024/1267 = 0.8082$. Inserting this value for b in equation (3) and successively substituting the homogeneous totals 1602 and 753 gives $\Sigma y = 0.8082 \times 1602 = 1295*$ and $\Sigma y = 0.8082 \times 753 = 609*$ the adjusted values. Finally applying linear weighting and averaging, as in the example for temperature, yields

$$\bar{y} = (1295 + 609 + 1024)/30 = 97.6 \text{ mm.}$$

This is a near optimum estimate of the mean total precipitation for August at Geneva.

ESTIMATION OF STATISTICAL PARAMETERS

A statistical parameter is a fixed value which is a function of all of the population values. Thus the mean for a population would be the average of all the values in that population. Since the entire population of values is never known in climatology, it is only feasible to estimate population parameters from samples or climatological series. Such an estimate of a population parameter is called a statistic. A statistic is a function of the sample or climatological series. Statistical parameters may be dealt with only in theory while in practice statistics or estimates of the parameters must always be used.

Since every function of a random variable is also a random variable, statistics are random variables and are therefore subject to random variation similar to that in a climatological series. Every climatological statistic is therefore a random variable which forms a population for which there is a frequency distribution. The variability of this frequency distribution about the population parameter is called the dispersion of the statistic. There are always a number of functions of the sample, or statistics, which estimate the same population parameter. The best of these estimates will have the smallest dispersion. The estimate with the least dispersion will in general extract the most information from the sample on the value of the population parameter. The dispersion of a statistic decreases with increase in sample size hence statistics for long climatological series have less dispersion than those for short climatological series. Since poor statistics have greater dispersion the use of such statistics in effect discards climatological record, hence is wasteful of usually scarce record length and is to be avoided if possible. An example is the use of the median to estimate the center of a normal (Gaussian) distribution (e.g., a climatological series of temperature which has a distribution close to normal). Both the median and the mean are statistics for the center of a normal distribution. The median, however, has a larger dispersion than the mean and in fact requires about a one-third longer climatological series than the mean to obtain an equally good estimate of the center of the distribution. There are a number of other inefficient statistics used in

climatology. Examples are the use of the mean absolute deviation as an estimate of the standard deviation and certain short cut estimates of the correlation coefficient. Statistics with the smallest dispersions are called efficient. It is naturally advantageous to employ either efficient statistics or those with high efficiencies in climatological analysis. If the distribution form is not known little exact information can be inferred about the efficiency of a statistic.

While it is always desirable to use the most efficient statistic available, it is sometimes also desirable, but not necessarily essential in all problems, to have the statistic be mean unbiased or what is commonly known as unbiased. A statistic is said to be (mean) unbiased if the mean of the statistic for m samples of size n approaches its parameter value as m increases without limit or mn approaches the number of values in the whole population. Efficiency and unbiasedness do not naturally occur together. In statistical analysis it is common practice to choose an efficient statistic and make it unbiased if the latter property is necessary, such as in cases where statistics are to be added or averaged.

There are in general two kinds of statistics: (a) Those which are direct estimates of the parameters of a frequency distribution, and (b) those which are estimates of other population properties. The mean and standard deviation are estimates of the population or distribution parameters of the normal distribution. The mean is also an estimate of the population mean or expected value independent of the distribution form.

Common Statistics of Climatological Variables

The mode is defined as the value of the random variable where the density of probability is a maximum. If the analytical form of the frequency distribution is known efficient estimates of the mode (the peak of the curve) may be obtained by substituting efficient estimates of the distribution parameters and obtaining the maximum of the frequency curve by differentiation. If the analytical form of the frequency distribution is not known, there is no good method of estimating the mode. If the sample is large the center of the class with the highest frequency may be taken as an estimate of the mode. In general the mode is not recommended for use in climatology.

There has been a good deal written about multimodal distributions in climatology. Most of the multimodality observed is caused by mixing small samples from several populations which falsely gives the impression that large samples have been used. In these cases the multimodality is not real but only an effect resulting from improper statistical analysis.

The median of a population is defined as the value of the random variable below which the probability of occurrence is 0.50. If the frequency distribution is known, it may be obtained by integrating up to the value of the random variable where the probability reaches 0.50. If the distribution is not known the median is best obtained by reading the 0.50 value from cumulative distributions plotted from data such as those shown in Tables A3–1,–2. Rough estimates of the median may be obtained by taking the middle value of an ordered series or if there are two middle values they may be averaged to obtain the median. The median is one of a class of quantities called quantiles which are defined as X_F where F is the probability of X being less than X_F. The median is then the 0.50 quantile. Quantiles should be estimated from fitted analytical distributions where possible as those obtained either from the empirical cumulative distributions or from ordered series tend to be more variable.

The mean is the most used climatological parameter. In most cases it is best to obtain it by summing the climatological series and dividing by the number of years of record. It has two properties: First it is an estimate of the well-known expected value or mathematical expectation, i.e., the mean of the population. This is important in applied climatology for the mean of any linear function of the climatological series is also a linear function of the mean of the series. Secondly, the mean is the center of the normal distribution and is therefore the center of the distribution for climatological series having this distribution. The mean, as computed above, is generally optimum for estimating the expected value for precipitation and optimum for both the expected value and the center of the distribution for temperature.

The moments about the mean, or central moments, are also commonly employed in statistical-climatological work. These are defined for the population R by

$$\mu_r = \int_R (x - u)^r f(x) dx. \qquad (1)$$

Here μ_r is the rth moment, u is the mean, $f(x)$ is the probability density function or frequency curve, and R is the population interval or region over which $f(x)$ is defined. The unbiased estimate of the second moment or variance is

$$s^2 = \frac{\sum_{}^{n} (x - \bar{x})^2}{n - 1} \qquad (2)$$

The square root of this value is the standard deviation. The higher moments may be estimated by

$$m_r = \frac{\sum_{}^{n} (x - \bar{x})^r}{n} \qquad (3)$$

The third moment is often used to measure the skewness and the fourth moment the flatness or kurtosis of frequency distributions. For these purposes the statistics

$$g_1 = m_3/s^3 \quad \text{and} \quad g_2 = (m_4/s^4) - 3$$

which are estimates of the parameters γ_1 and γ_2 may be employed. For the normal distribution $\gamma_1 = \gamma_2 = 0$. The statistic $a = \sum_{}^{n} |x - \bar{x}|/(ns)$ is often substituted for g_2 since it has a simpler distribution and tends toward normality faster. Tabular values of a, also a related skewness ratio

$$\sqrt{b_1} = m_3/s^{3/2}$$

are available in Pearson and Hartley's tables (2). Moments higher than the 4th are ordinarily not recommended for climatological work since they are highly variable for the short climatological series usually available.

Again it should be stated that if good estimates of the distribution parameters are available formula (1) should be used directly for estimating the moments. Another statistic occasionally used is the range. This statistic is not recommended except for very crude work since it has a high variability. Related to the range are the extreme values of record. These are even more highly variable than the range and depend greatly on the length of record. The extreme values for each year may, of course, be fitted by appropriate frequency distributions. Statistics of these distributions give a much better appraisal of individual extremes. For example, quantiles from these distributions are independent of the length of record used; hence, they give valid information about unusual values.

The coefficient of variation (or variability) or relative standard deviation also has been used in climatology. It is defined as the ratio of the standard deviation to the mean, $s/\bar{x}$. The statistic in absolute value depends on the interpretation which can be given the standard deviation. If the distribution is not normal the standard deviation has no simple meaning and hence an individual relative standard deviation has little value. However, it is useful for comparison to other relative standard deviations from populations having the same analytical form of distribution. In this case the ordinary estimate may be an inefficient statistic. A better estimate could be obtained using the proper functions of the estimated parameters in equation (1).

GENERAL STATISTICAL METHODS

The basic problems of climatological analysis may be classified into three general types: (1) Problems of specification which occur in the choice of the analytical form of the population. (2) Problems of inference which arise in the estimation of population parameters and in testing hypotheses and establishing confidence intervals on the population parameters. (3) Problems of relationship which occur in relating several climatological variables and in relating climatological variables to non-climatological variables.

The problem of specification is solved by specifying the frequency distribution in the population of the climatological variable. This may be done either empirically or using theoretical reasoning. An empirical specification of the population usually consists simply in assuming the existence of a distribution of probability whose cumulative distribution has the characteristic ogive form. This was the approach followed previously in obtaining the distribution of August precipitation for Geneva. Occasionally, on the basis of examination of numerous samples,

a mathematical form of distribution may be specified for convenience of computation. A theoretical specification of the population distribution is always expressed in mathematical form. This form is derived from a consideration of the bounds of the variable; scale, location, and shape behavior; behavior in convolution; etc. A theoretical specification of the normal distribution may result from an application of the central limit theorem.

The estimation part of the inference problem is solved by providing the most satisfactory statistics for estimating the population parameters. As was seen previously, the most satisfactory statistics or estimators will be those having a small dispersion in their distributions. Usually maximum likelihood estimates will provide the best estimates of the parameters.

Confidence intervals for the parameter estimates should always be provided to give a measure of their accuracy. Tests of hypothesis may also be made to ascertain whether the population meets certain prescribed conditions or whether the parameters differ from other sets of parameters of similar character. Previously, for example, tests were made to examine the homogeneity of temperature and precipitation series. Confidence interval and test of hypothesis problems are similar in that they both involve distributions of the estimates or statistics.

The relationship problem may involve only climatological variables or it may involve climatological and other variables. The first problem arises when functions of climatological variables are needed to replace climatological variables which are not available or to form a new variable which has some special properties. For example, statistics on daily temperatures may be impossible or too expensive to obtain directly and it is required to obtain estimates of these from monthly statistics. The heating degree-day variable is a simple example of a function of temperature which has special useful properties not possessed by temperature. The second type of problem, where climatological variables are related to non-climatological variables, is encountered in every problem in applied climatology. The basic objective in such problems is to develop a relationship which will transform a frequency distribution on the climatological variable to one on the applied variable. A simple example would be a relationship between degree-days and heat consumption in a building which would give the distribution of heat consumption from the distribution of degree-days.

Since many of the inference problems of climatology are closely associated with specification problems, these will be discussed together. The test of hypothesis problem has already been introduced in connection with tests of homogeneity, and space will not allow further treatment. More detail on this subject is readily available in the statistical literature. The relationship problem will be treated separately.

FREQUENCY DISTRIBUTIONS

An example of specification of the population has already been introduced where the empirical distribution was specified for August precipitation at Geneva. The only theory employed there was to assume the existence of a population and a random variable and hence the set of cumulative probabilities. In many instances of climatological analysis the specification of an empirical distribution is all that is necessary or justified. It is only where the theory is strong or where several distributions are to be fitted and comparison or smoothing of their statistics is required that theoretical distributions are fitted. A mathematical fit adds little in other circumstances.

Frequency distributions are of two general types, discrete and continuous. In discrete distributions the probability density is a function of a discrete random variable, i.e., one that varies in steps. The most common discrete climatological variable is frequency, e.g., the number of hail storms, days with rain, etc. In continuous distributions the probability density is a function of a continuous random variable. Temperature, pressure, precipitation, or any element measured on a continuous scale has a continuous random variable. Often for convenience a discrete random variable may be treated as continuous. Also, for special application, continuous random variables may be transformed to discrete random variables. Cloud height, for example, is a continuous variable which may be transformed into a discrete variable consisting of heights below and above an arbitrary height h.

While there has been a good deal of consideration given to fitting frequency distributions to meteorological data, much of this has been empirical in nature. Often also the fitting has been done to improperly defined populations such as mixtures of several climatological series which have led to quite anomalous interpretations. Because of lack of space only the most common distributions can be discussed.

The Normal Distribution

The most important continuous distribution in climatological analysis, and, of course, statistical analysis, is the normal or Gaussian distribution. Its frequency or probability density function is

$$f(x) = \frac{1}{\sqrt{2\pi}\sigma} \exp - \frac{1}{2} \left(\frac{x - \mu}{\sigma} \right)^2$$

where μ is the population mean and σ is the population standard deviation. The parameter μ is best estimated by $\bar{x}$ and σ by s. These are obtained from the sample values x by the relationships

$$\bar{x} = \sum_{}^{n} x/n$$

and

$$s = \sqrt{\sum \frac{(x - \bar{x})^2}{n - 1}}$$

The normal distribution function cannot be expressed in terms of simple functions but must be evaluated by means of function expansions. Many tables of the normal distribution function and related functions have been prepared using the variable $u = (x - \mu)/\sigma$ as argument; u is called a standardized variable. Using this variable the distribution function becomes

$$F(t) = \frac{1}{\sqrt{2\pi}} \int_{-\infty}^{t} \exp - \frac{1}{2} u^2 du$$

which can be converted to any desired normal distribution simply by varying μ and σ. Thus a single normal table with argument t, which is also a table of the distribution with mean zero and standard deviation unity, may be used to obtain the probabilities for any normal distribution. $F(t)$, of course, gives the probability that u is less than t, $1 - F(t)$ the probability that u is greater than t, and $F(t_2) - F(t_1)$ the probability that u is between t_1 and t_2.

The importance of the normal distribution in

climatology stems, to considerable extent, from the central limit theorem. This causes means and sums of a sufficient number of climatological values to be normally distributed. For example, rainfall climatological series for short periods for which the mean rainfall is small would have very skewed distributions. As the period increases several shorter periods are added together and an increase in the mean occurs. Thus the size of the mean is some measure of how many periods have been added together; hence, as the mean value gets larger the sum of the several component periods approaches a normal distribution. It may be shown that under average conditions, periods with a mean rainfall of 500 mm or more will be close to normally distributed, the greatest discrepancy in probability being about 0.01 at the median. Even for 250 mm means under ordinary conditions the largest discrepancy in probability is only about 0.02.

The normal distribution also provides good fits in most instances to climatological variables which are unbounded above or below, such as temperature and pressure. The sample of data fitted must, of course, be a sample from a homogeneous climatological series. It must not be a sample from mixed populations which in the past has led to erroneous conclusions such as frequency distributions having several modes, etc.

The normal distribution has found wide application in determining the probability of a freezing air temperature occurring before or after a given date in fall or spring. The data in Table A3–11 are taken from Shaw, Thom, and Barger (3). Geary and Pearson's (4) tests for skewness and kurtosis indicate no significant deviation from normality at the 10% and 20% levels, respectively. Actually, in studying a large number of Iowa locations the number of significant cases was found to be no more than expected with a selected probability level.

Utilizing the coded dates in Table A3–11 the sample mean and standard deviation(s) are 34 (May 5) and 11.4 respectively. Using normal probability paper with, say, May 5 ± 1.96s plotted at 0.975 and 0.025, respectively, the probability of the last occurrence of 32°F or lower being before or after a given date in northwestern Iowa can be read directly from the abscissa in Figure A3–3.

Table A3–11

DATES OF LAST OCCURRENCE IN SPRING OF 32°F OR LOWER AT ALTA, IOWA

Year	Date	Code	Year	Date	Code
1893	May 2	= 31	1923	May 12	= 41
1894	Apr. 21	= 20	1924	May 24	= 53
1895	May 19	= 48	1925	May 17	= 46
1896	Apr. 21	= 20	1926	May 3	= 32
1897	May 30	= 60	1927	Apr. 24	= 23
1898	Apr. 14	= 13	1928	Apr. 27	= 26
1899	May 13	= 42	1929	May 16	= 45
1900	May 3	= 32	1930	May 16	= 46
1901	May 12	= 41	1931	May 22	= 51
1902	Apr. 23	= 22	1932	Apr. 27	= 26
1903	May 3	= 32	1933	Apr. 27	= 26
1904	Apr. 26	= 25	1934	Apr. 27	= 26
1905	May 26	= 55	1935	May 3	= 32
1906	May 9	= 38	1936	Apr. 22	= 21
1907	May 27	= 56	1937	May 14	= 43
1908	May 3	= 32	1938	May 8	= 37
1909	May 10	= 39	1939	Apr. 21	= 20
1910	May 4	= 33	1940	May 2	= 31
1911	May 3	= 32	1941	Apr. 24	= 23
1912	May 14	= 43	1942	May 4	= 33
1913	Apr. 27	= 26	1943	May 8	= 37
1914	May 12	= 41	1944	May 6	= 35
1915	May 18	= 47	1945	May 10	= 39
1916	May 2	= 31	1946	May 12	= 41
1917	May 4	= 33	1947	May 29	= 58
1918	May 11	= 40	1948	Apr. 13	= 12
1919	Apr. 25	= 24	1949	Apr. 24	= 23
1920	Apr. 28	= 27	1950	May 7	= 36
1921	May 14	= 43	1951	Apr. 23	= 22
1922	Apr. 19	= 18	1952	Apr. 15	= 14

Mean = May 5
s = 11.4

The Gamma Distribution

Since there are a number of zero bounded continuous variables in climatology, it is important to give a distribution which may be used for such variables. The gamma distribution which has a zero lower bound has been found to fit several such variables well. It is defined by its frequency or probability density function

$$g(x) = \frac{1}{\beta^\gamma \Gamma(\gamma)} x^{\gamma-1} \exp - x/\beta$$

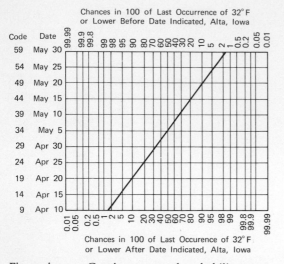

Chances in 100 of Last Occurrence of 32° F
or Lower Before Date Indicated, Alta, Iowa

Figure A3-3. Graph, on normal probability paper, for determining the probability of the last occurrence of 32°F (0°C) before or after a given date in northwestern Iowa.

where β is a scale parameter, γ is a shape parameter, and $\Gamma(\gamma)$ is the ordinary gamma function of γ, i.e., $\Gamma(\gamma) = (\gamma - 1)!$

The moments in this instance give poor estimates of the parameters. Sufficient estimates are, however, available and these are closely approximated by

$$\hat{\gamma} = (1 + \sqrt{1 + 4A/3})/4A$$

and

$$\hat{\beta} = \bar{x}/\hat{\gamma}$$

where A is given by $A = \ln \bar{x} - \left(\sum\limits_{}^{n} \ln \bar{x} \right)/n$ and ln signifies the natural logarithm.

The distribution function, from which probabilities may be obtained, is

$$G(x) = \int_0^x g(t)dt$$

Pearson's "Tables of the Incomplete Γ-Function" (5) give $G(u)$ where $u = x/\sigma$, $\sigma = \beta\sqrt{\gamma}$, $\gamma = p + 1$ and $\beta = 1/\gamma'$; γ' is the same as Pearson's γ which is his scale parameter, whereas we use γ as the shape parameter.

The Gamma distribution has been found to give good fits to precipitation climatological series. In case these contain zeros the mixed distribution function of zeros and continuous precipitation amounts may be employed. This is given by

$$H(x) = q + pG(x)$$

where q is the probability of a zero and $p = 1 - q$. Thus when $x = 0$, $H(0) = q$ as it should be. If m is the number of zeros in a climatological series, q may be estimated by m/n.

Most extensive utilization of the Gamma distribution has been made by Thom and Barger et al. (6 and 7). These list many distribution parameters and probabilities of selected amounts of precipitation, as well as maps and graphs pertaining to each region and nomograms for estimating probabilities from the parameters.

The Extreme Value Distributions

Often in design problems the climatological variable of interest is the annual extreme, either upper or lower. This arises from the fact that if a designed structure can withstand the highest (lowest) value in a year it can also withstand all other values in the year. Hence, a distribution of annual extreme values furnishes the proper climatological prediction. Up to the present the Fisher-Tippett Type I distribution has been of main interest. It has been widely applied by Gumbel (8). Its distribution function is given by

$$F(x) = \exp\left[-e^{\pm(x-\alpha)/\beta}\right]$$

Here the negative of the double sign holds for maximum values and the positive sign applies for minimum values. The Type II distribution, which is an exponential transformation of the Type I distribution, also has been employed by Thom (9) but the satisfactory fitting of this distribution is too complicated to give here.

As with most other skewed distributions the moments give poor estimates of the parameters. Lieblein (10) provides a simple method of fitting the Type I distribution which gives estimates of the quantiles with minimum variance. This is a desirable property for climatological work, for our ultimate objective is always to obtain quantiles or probabilities.

The Lieblein fitting procedure involves carefully maintaining the original time order of the climatological series and dividing into suitable subgroups for the computations. The following table of weights is needed in the computations.

TABLE OF ORDER STATISTICS WEIGHTS

m		$x_{.1}$	$x_{.2}$	$x_{.3}$	$x_{.4}$	$x_{.5}$	$x_{.6}$
2	$a_{.j}$	0.91637	0.08363				
	$b_{.j}$	−0.72135	0.72135				
3	$a_{.j}$	0.65632	0.25571	0.08797			
	$b_{.j}$	−0.63054	0.25582	0.37473			
4	$a_{.j}$	0.51100	0.26394	0.15368	0.07138		
	$b_{.j}$	−0.55862	0.08590	0.22392	0.24880		
5	$a_{.j}$	0.41893	0.24628	0.16761	0.10882	0.05835	
	$b_{.j}$	−0.50313	0.00653	0.13045	0.18166	0.18448	
6	$a_{.j}$	0.35545	0.22549	0.16562	0.12105	0.08352	0.04887
	$b_{.j}$	−0.45928	−0.03599	0.07319	0.12673	0.14953	0.14581

As previously, the sample climatological series is assumed to have n values. Retaining the original time order these n-values are to be divided into subgroups of size m. It will be noted that the table of weights allows m to be chosen from 2 to 6. It is best to choose m as large as possible. Thus, if the sample size is 30, $m = 6$ would be chosen rather than $m = 5$. If n is not divisible by $m = 4$, 5, or 6, an additional weighting will be necessary. First consider that $n = 30$. The sample is maintained in original time order and divided into $k = 5$ subgroups of $m = 6$. The values within the subgroups are then arranged in order according to increasing magnitude. The ith subgroup would then appear as x_{i1}, x_{i2}, x_{i3}, x_{i4}, x_{i5}, x_{i6}. All ordered subgroups are then arranged in a table as follows:

x_{11}	x_{12}	x_{13}	x_{14}	x_{15}	x_{16}	
x_{21}	x_{22}	x_{23}	x_{24}	x_{25}	x_{26}	
x_{31}	x_{32}	x_{33}	x_{34}	x_{35}	x_{36}	
x_{41}	x_{42}	x_{43}	x_{44}	x_{45}	x_{46}	
x_{51}	x_{52}	x_{53}	x_{54}	x_{55}	x_{56}	
$S_{.1}$	$S_{.2}$	$S_{.3}$	$S_{.4}$	$S_{.5}$	$S_{.6}$	
$a_{.1}$	$a_{.2}$	$a_{.3}$	$a_{.4}$	$a_{.5}$	$a_{.6}$	
$a_{.1}S_{.1}$	$a_{.2}S_{.2}$	$a_{.3}S_{.3}$	$a_{.4}S_{.4}$	$a_{.5}S_{.5}$	$a_{.6}S_{.6}$	$\displaystyle\sum^{6} a_{.j}S_{.j}$
$b_{.1}$	$b_{.2}$	$b_{.3}$	$b_{.4}$	$b_{.5}$	$b_{.6}$	
$b_{.1}S_{.1}$	$b_{.2}S_{.2}$	$b_{.3}S_{.3}$	$b_{.4}S_{.4}$	$b_{.5}S_{.5}$	$b_{.6}S_{.6}$	$\displaystyle\sum^{6} b_{.j}S_{.j}$

Each column of x's is first summed to obtain the $S_{.j}$. These are multiplied by $a_{.j}$ and summed to obtain the row sum. Next the $S_{.j}$ are multiplied by the $b_{.j}$ and summed to obtain the second row sum.

In the Type I distribution function the exponent $(x - \alpha)/\beta$ is a standardized variable, i.e., it is a variable located at α and scaled in β. If x_p is a quantile in x (a value of x corresponding to $F = p$), then

$$y_p = (x_p - \alpha)/\beta$$

and

$$x_p = \alpha + \beta y_p$$

Lieblein showed that a minimum variance estimated for a given y_p is given by

$$x_p{}^* = \sum^{m} a_{.j}S_{.j}/k + \left(\sum^{m} b_{.j}S_{.j}/k \right) y_p$$

Thus the minimum variance estimates for α and β are

$$\alpha^* = \sum^{m} a_{.j}S_{.j}/k$$

and

$$\beta^* = \sum^{m} b_{.j}S_{.j}/k$$

For the sample of 30 under consideration they are

$$\alpha^* = \sum^{6} a_{.j}S_{.j}/5$$

and

$$\beta^* = \sum^{6} b_{.j}S_{.j}/5$$

When these values are substituted in the Type I distribution function estimated probabilities are obtained.

In case $m = 5$ or 6 is not an even multiple of the sample size n, a further simple computation is necessary. Suppose that $n = 33$ instead of 30. The last three values of the sample climatological series then form an additional subgroup of $m' = 3$. These values are also arranged in order of increasing magnitude giving x_{61}, x_{62}, and x_{63}. A similar table is formed with the weights for $m' = 3$ as follows:

x_{61}	x_{62}	x_{63}	
$a_{.1}$	$a_{.2}$	$a_{.3}$	
$a_{.1}x_{61}$	$a_{.2}x_{62}$	$a_{.3}x_{63}$	$\sum^{3} a_{.j}x_{6j}$
$b_{.1}$	$b_{.2}$	$b_{.3}$	
$b_{.1}x_{61}$	$b_{.2}x_{62}$	$b_{.3}x_{63}$	$\sum^{3} b_{.j}x_{6j}$

The estimator for this sample is then as before

$$u_p^* = \sum^{3} a_{.j}x_{6j} + \left(\sum^{3} b_{.j}x_{6j}\right)y_p$$

Lieblein has shown that the estimator for v_p the quantile for the variable in the sample $n = 33$ is

$$v_p^* = \frac{km}{n} x^*_p + \frac{m'}{n} u_p^*.$$

This gives for the final estimates

$$\alpha^* = \frac{km}{n} \sum^{6} a_{.j}S_{.j}/5 + \frac{m'}{n} \sum^{3} a_{.j}S_{.j}$$

and

$$\beta^* = \frac{km}{n} \sum^{6} b_{.j}S_{.j}/5 + \frac{m'}{n} \sum^{3} b_{.j}S_{.j}.$$

The fitting of any sample size is a simple variation of the above procedures. For minimum values or lower extremes the magnitude order arrangement in the rows of the computation tables is reversed, i.e., instead of going from low to high values they should go from high to low values. All other parts of the tables remain the same.

The Binomial Distribution

This distribution does not in general fit climatological data well because of correlations which occur when the probabilities of occurrence are high enough to meet one of its requirements for application. It is important, however, because it is related to the Poisson and negative binomial distributions which apply respectively for small probabilities (rare events, often uncorrelated) and for correlated events. Because of this relation it has occasionally been used to give simple rough probability estimates to replace the more crude observed extreme relative frequencies. The most important aspect of the binomial distribution for climatological analysis is that it is the distribution of the estimated probabilities obtained from any distribution function, empirical or theoretical. This makes it possible to obtain confidence limits for estimated probabilities and quantiles.

The binominal probability function is given by

$$f(x) = \binom{m}{x} p^x(1 - p)^{m-x}$$

where p is the probability of an event occurring, $1 - p$ is the probability of the event not occurring, x is the frequency of occurrence, and x can take the values $0, 1, \ldots, m$. The distribution function is given by

$$F(x) = \sum_{t=0}^{x} \binom{m}{t} p^t(1 - p)^{m-t}, \quad t = 0, 1, \ldots, m.$$

This, of course, gives the probability that the frequency is x or less. The probability p is usually estimated by $\Sigma x/n$ where n is the total number of occurrences and non-occurrences of the event. The climatological events which might be considered in this category are widely varied. Examples are hail and no-hail days, rain and no-rain days, days with rainfall less than an amount u and those with rainfall greater than u, observations with visibility less than V and those greater than V, etc. Most of these variables have the limitation that they are correlated and therefore the binomial distribution can be used only for rough biased estimates of probabilities for use where only summarized data are available or quick results are needed.

The important application of the binomial distribution in climatological analysis is in obtaining confidence bands for estimated probabilities. It may be seen that when an estimate $F(h)$ of the probability that $x < h$ is obtained from any distribution function, theoretical or empirical, the probabilities in random sampling are divided into those less than h and those greater than h. These form a binomial distribution. If the sample size is m from which $F(h)$ has been estimated, then the probability that values of $F(h) = c/m$ in random sampling will be below $F(a) = p_L$ is

$$\alpha = \sum_{x=0}^{c} \binom{m}{c} p_L{}^c (1 - p)^{m-c} \quad .$$

Likewise, if the probability that values of $F(h)$ will be above $F(b) = p_U$ is also made α, then

$$\alpha = \sum_{x=c}^{m} \binom{m}{c} p_U{}^c (1 - p)^{m-c} \quad .$$

It is now seen that the probability that $F(a)$ lies between p_L and p_U is $1 - 2\alpha$. Thus the probability relationship

$$P(p_L \leq F(\eta) \leq p_U) = 1 - 2\alpha$$

defines a confidence interval for $F(a)$ with confidence probability $1 - 2\alpha$ where $F(\eta)$ is the true or population value of $F(h)$. "Biometrika Tables for Statisticians" (2) gives convenient graphs for $1 - 2\alpha = 0.95$ and 0.99, 0.95 being the smallest confidence probability recommended for climatological work. A level of $1 - 2\alpha = 0.90$ is better and may be obtained approximately from the National Bureau of Standards "Tables of the Binomial Probability Distribution" (11).

If the inverted function notation $h = F^{-1}(c/m)$ is employed, the confidence interval for a quantile h_F may be expressed as the probability relationship

$$P[F^{-1}(p_L) < \eta_F < F^{-1}(p_U)] = 1 - 2\alpha$$

This is obtained by simply finding the x values corresponding to $F = p_L$ and $F = p_U$ in the probability confidence interval.

It should be noted that both confidence intervals are independent of the functional form of F which in a sense makes them non-parametric. If the functional form of F is known, parametric confidence intervals may be available which will be shorter than those above; however, some authors simply assume that p and the corresponding quantiles are normally distributed. This can only give a good approximation at values near the middle of $F(x)$. For values of $F(x)$ near o or 1, it is better to use the binomial confidence intervals. They are slightly too broad but they reflect the right shape for the distribution of $F(h)$.

The Poisson Distribution

When m becomes large and p approaches zero with the mean $\mu = mp$ constant, the binomial distribution approaches the Poisson distribution. Thus the Poisson distribution fits events with a small probability. Since this also means for climatological series that a small number of events on the average is found in the annual time interval or a portion of it, the correlation between successive events will ordinarily be small. The distribution, therefore, fits annual hail frequency when the mean frequency is not too high, excessive precipitation events, annual tornado and typhoon frequency, etc.

The Poisson probability function is given by

$$f(x) = \mu^x e^{-\mu}/x!$$

The distribution function is then

$$F(x) = \sum_{t=0}^{x} \mu^t e^{-\mu}/t!$$

Here the only parameter is the mean μ which is best estimated by $x = (\Sigma x)/n$. Probabilities may be obtained readily from $F(x)$ with the aid of tables of exponentials and factorials.

The Negative Binomial Distribution

The negative binomial distribution is useful in fitting discrete dichotomous random variables in which the individual events tend to be correlated. Thus, when too many events are packed on the average into an annual time interval, this distribution tends to fit better than the Poisson distribution. For example, annual hail days and annual frequency of typhoons tend to be fitted better by the negative binomial distribution when the mean annual occurrence is high. Continuous data should in general not be fitted with theoretical discontinuous distributions unless a simple transformation to a discrete variable is made e.g., to a dichotomous variable. There are a number of bad examples of such misfitting in the meteorological literature. On the other hand the fitting of continuous distributions to discontinuous data is often useful.

A test of hypothesis is available to test the adequacy of the Poisson distribution. Thus, if

$$\chi_{n-1}^2 = n(\sum x^2 / \sum x) - \sum x,$$

where n is the number of years of record, is not greater than the 0.05 value in a chi square table with $n - 1$ degrees of freedom, the Poisson distribution is adequate. If it exceeds the 0.05 value the negative binomial distribution should be fitted.

The negative binomial probability function is

$$f(x) = \frac{\Gamma(x + k)}{\Gamma(x + 1)\Gamma(k)} \frac{p^x}{(1 + p)^{k+x}}$$

The distribution function is given by

$$F(x) = \sum_{t=0}^{x} f(t).$$

The moment estimates of p and k are

$$p^* = (s^2 - \bar{x})/\bar{x}$$

and

$$k^* = x^2/(s^2 - \bar{x}).$$

where x is the sample arithmetic mean and s^2 is the sample variance.

The moment estimates are not always adequate, i.e., efficient enough. Fisher (12) has given a criterion which suggests the use of a better fitting procedure if the efficiency falls below 90%. Thus if

$$(1 + 1/p^*) (k^* + 2) > 20$$

the method of maximum likelihood should be used. This method of fitting is too complex to consider here. For details on the method see Thom (13).

Correlation and Regression Analysis

The most important use of correlation methods in climatological analysis is in connection with the correlation between climatological series caused by the natural persistence of the meteorological variable within the year. Correlation problems also occur in connection with compound variables i.e., where two or more variables are combined into a single variable and in connection with the propagation of variability in relationships of theoretical or applied problems. Most other applications of correlation are supplementary to regression analysis.

Regression analysis is applied whenever the objective is to estimate a functional relationship for predicting the values of a variable from one or more others. Its main uses are in relating one or more meteorological variables so that one may be substituted for one or more others and in relating applied variables to meteorological variables. There is also some application to the study of systematic variation of climatological variables in time. However, this aspect is largely of special interest and will not be considered here. In any case the regression analysis in this instance is only a variation of that considered here except that the independent variable is time and the regression terms may be harmonic functions or of some other form.

Correlation Analysis

In a strict sense correlation analysis in climatology consists largely of accounting for the effect of correlation between climatological series. For example, if the temperature climatological series for the average temperature series for May 1 and 2 have sample variances s_1^2 and s_2^2 then the series for the average of May 1 and 2 has a variance which is affected by the correlation between the May 1 and May 2 series. Similarly the variance of the average of the May 1, 2, . . . , m series will be affected by the correlations among the m climatological series. Clearly the climatological series could also be for weeks, months, or any other portion of the year.

Just as it is necessary always to work with climatological series so is it necessary to work with the proper correlation coefficients in the present aspect of climatological analysis. The only correlation coefficients useful in the type of analysis considered here are those computed between the two series in any pair of climatological series. If the two series are for the same element, they will be displaced in time within the year; hence it will be possible to have a whole sequence of such correlations. The pairs of climatological series may be separated by different units of time and so there will be a time lag between them. Because of the time sequential nature of these correlation coefficients and to differentiate them from autocorrelation coefficients they will be called sequence correlation coefficients. The sequence correlation between the ith and jth climatological series is defined as

$$\rho(x_i, x_j) = E(x_i - \mu_i)(x_j - \mu_j)/\sigma_i \sigma_j.$$

The numerator is the expected value of the product of the departures of the x_i and x_j from their respective population means and is called the *covariance*. The denominator is the product of the population standard deviations of x_i and x_j. The sample estimate of the sequence correlation coefficient is given by

$$r(x_i, x_j) = \sum_{k=1}^{m} (x_{ik} - \bar{x}_i)(x_{jk} - \bar{x}_j)/n s_i s_j.$$

Here x_{ik} is the kth term (year) in the ith climatological series and x_{jk} is the kth term (year) in the ith climatological series and x_i, s_i, and $\bar{x}_j$, s_j are their respective means and standard deviations.

The sequence correlation coefficient should be carefully differentiated from the autocorrelation coefficient (sometimes called serial correlation coefficient). The sequence correlation coefficient is really a single correlation with a time displacement so that the effect of variation in the mean and standard deviation through the year is removed. The autocorrelation coefficient, on the other hand, includes the variation in the mean and standard deviations. In the methods discussed here it is always wrong to use an autocorrelation coefficient.

For the May 1, 2, . . . , m climatological series considered above there are $m(m - 1)$ possible pairs of series. Since $\rho(x_i, x_j) = \rho(x_j, x_i)$, there are only $m(m - 1)/2$ different sequence correlations. All of these must be considered in obtaining the variance of the sum and average series formed by summing or averaging for each year. If i and j both run over the same sequence of

series, the sample variance of the sum may be expressed by

$$v\left(\sum_{i}^{m} x_i\right) = \sum_{i}^{m} s_i^2 + 2 \sum_{i=1}^{m} \sum_{i>i}^{m} s_i s_j r(x_i, x_j).$$

This is the variance of the linear function $y = \Sigma x_i$. If the x_i have different weights k_i so that the linear function is $y = \Sigma k_i x_i$, the variance becomes

$$v\left(\sum_{i}^{m} k_i x_i\right) = \sum_{i}^{m} k_i^2 s_i^2 + 2 \sum_{i=1}^{m} \sum_{i>i}^{m}$$
$$\times k_i k_j s_i s_j r(x_i, x_j)$$

It may be noted that when the $r(x_i, x_j) = 0$, the relationship reduces to the simple variance formula

$$v\left(\sum_{i}^{m} k_i x_i\right) = \sum_{i}^{m} k_i^2 s_i^2.$$

If $m = 2$ and k_2 has a negative sign the formula gives

$$v(k_1 x_1 - k_2 x_2) = k_1 s_1^2 + k_2 s_2^2$$
$$- 2k_1 k_2 s_1 s_2 r(x_1, x_2)$$

If $k_1 = 1$ and $k_2 = -1$

$$v(x_1 - x_2) = s_1^2 + s_2^2 - 2s_1 s_2 r(x_1, x_2).$$

For $r(x_1, x_2) = 0$, $k_1 = 1$, $k_2 = -1$

$$v(x_1 - x_2) = s_1^2 + s_2^2$$

If $k = 1/m$, so that the linear function is a simple average $\Sigma x_i/m$, the variance becomes

$$v\left(\sum_{i}^{m} x_i/m\right)$$

$$= \frac{1}{m^2}\left[\sum_{i}^{m} s_i^2 + 2 \sum_{i=1}^{m} \sum_{i>i}^{m} s_i s_j r(x_i, x_j)\right].$$

Thus the *average* temperature for June has a variance formed from the daily variances and sequence correlations given by

$$v\left(\sum_{i}^{30} x_i/30\right)$$

$$= \frac{1}{30^2}\left[\sum_{i}^{30} s_i^2 + 2 \sum_{i=1}^{30} \sum_{i>i}^{30} s_i s_j r(x_i, x_j)\right].$$

The variance of the *total* precipitation for June based on the individual daily variances and sequence correlations is

$$v\left(\sum_{}^{30} x_i\right) = \sum_{i=1}^{30} s_i^2 + 2\sum_{i=1}^{30} \sum_{i>i}^{30} s_i s_j r(x_i, x_j).$$

Since monthly total precipitation is not very near to normally distributed, there would be more interest in the variance of the mean or normal for n years

$$\sum_{}^{30} x_i/n. \text{ This is } v\left(\sum_{}^{30} x_i\right)/n^2.$$

All of the formulas also apply where the x_i are the variables of different elements which are observed simultaneously or otherwise. This makes them useful in applied problems where the relationship with the applied variable is linear. For example, the outside air cooling load for an air conditioning system may be closely approximated by the linear relationship

$$q = -k_1 t + k_2 t' + k_3$$

where t is dry bulb temperature, t' is wet bulb temperature and the k's result from purely physical considerations. Since t and t' are nearly normally distributed around ordinary design levels, the variance of q is important. By means of the formulas given above

$$v(q) = k_1^2 v(t) + k_2^2 v(t') - k_1 k_2 r(t,t').$$

The standard deviation of q is therefore $s(q) = \sqrt{v(q)}$ and the mean of q is given by

$$\bar{q} = -k_1 \bar{t} + k_2 \bar{t}' + k_3.$$

Thus the normal distribution function $N[q;\bar{q},s(q)]$ gives the probabilities for climatological predictions based on the distributions of t and t'.

Correlation analysis enters in other ways into climatological analysis, but most of these are closely connected with regression analysis. In fact, wherever relationships are desired between random variables, regression analysis is the proper tool to employ.

Regression Analysis

A regression is a functional relationship between an independent random variable and one or more dependent random variables. For a given set of values of the independent variables the regression gives a mean value of the dependent variable. Regression analysis is used in climatology to estimate the constants in functional relationships where these are not given directly as physical quantities. It is used for the establishment of relationships both between climatological series and between climatological series and applied variables. The latter may often be accomplished without climatological series by employing sets of values of the independent variables which are *simply uncorrelated* within each set and which vary over a range of values equal to the range of values in the climatological series. Thus the relationship between an applied variable and climatological variables can often be established with a short simultaneous record of the two sets of variables.

The first problem in regression analysis is to estimate the constants. This is commonly done by the least squares method applied to the residuals about the regression function obtained when the values of the independent variables have been substituted. The minimization of the residuals of the dependent variable alone requires that the values of the independent variables be fixed or be measured essentially without error. If this condition is not met biases will be introduced in the regression constants. As mentioned above the values of each variable must also be mutually independent. The least squares estimates have certain optimum properties which make the method a desirable one for fitting regressions.

The least square principle is very general and may be applied to almost any type of function. If the regression function is of the form

$$y = R(x_1, \ldots, x_k; \beta_0, \beta_1, \ldots, \beta_k),$$

the sum of the square residuals may be expressed as

$$\sum_{}^{n} e_j^2 = \sum_{}^{n} [y_j - R(x_{1j}, \ldots, x_{kj};$$

$$\beta_0, \beta_1, \ldots, \beta_k)]^2 = \sum_{}^{n} (y_j - R_j)^2$$

where j runs over the sample values from 1 to n. The "least square" is obtained by minimizing the sums of squared residuals through differentiating and setting to zero. This gives the so-called normal equations

$$\frac{\partial}{\partial \beta_0} \sum^n (y_j - R_j)^2 = 0$$

$$\frac{\partial}{\partial \beta_1} \sum^n (y_j - R_j)^2 = 0$$

$$\cdot \qquad \cdot \qquad \cdot$$
$$\cdot \qquad \cdot \qquad \cdot$$
$$\cdot \qquad \cdot \qquad \cdot$$

$$\frac{\partial}{\partial \beta_k} \sum^n (y_j - R_j)^2 = 0.$$

The simultaneous solution of the normal equations gives the least squares estimates of β_0, $\beta_1, \ldots, \beta_k$.

The regression function R can of course take an infinite variety of forms. As usual the linear forms are the most used. Linear regressions for one and two independent variables are considered here. More complicated functions may be analyzed by finding the proper normal equations by the process given above.

The linear regression equation in one independent variable is best written as

$$y = \alpha + \beta(x - \mu).$$

Since measuring x from the mean μ makes the least squares estimate of α independent of that of β. The least squares estimates of α and β are

$$a = \bar{y}$$

and

$$b = \frac{\sum y(x - \bar{x})}{\sum (x - \bar{x})^2}$$

where the summation is over the sample values. The regression equation may then be written as

$$y_c = a + b(x - \bar{x})$$

Frequently it is known by physical means that $\alpha = 0$. In this case the regression equation becomes

$$y_c = bx.$$

There is now only one normal equation which gives the least squares estimate

$$b = \sum xy / \sum x^2.$$

It is often necessary to test the fitted regression for reality and for linearity. This is best done by the analysis of variance which is a technique devised by R. A. Fisher (12) to analyze the mean squares due to several components of the variation. For the linear regression given above it may be observed that there is a total variability of the y's which is divided into a variability accounted for by the regression, and a variability unaccounted for by the regression or residual variability. This may be expressed conveniently by an analysis of variance table:

ANALYSIS OF VARIANCE

	Sum of Squares	Degrees of Freedom	Mean Square
Accounted for by Regression	$\sum(y_c - \bar{y})^2$ $= Q_R$	1	$Q_R/1$
Unaccounted for by Regression (Residual)	$\sum(y - y_c)^2$ $= Q_T - Q_R$	$n - 2$	$(Q_T - Q_R)/$ $n - 2$
Accounted for by Mean(Total)	$\sum(y - \bar{y})^2$ $= Q_T$	$n - 1$	

"Degrees of freedom" is a term used by R. A. Fisher to express the whole number the sum of squares is to be divided by to give the mean square. When the mean has been estimated and therefore fixed only $n - 1$ of the observations may then vary since once the mean is fixed and $n - 1$ of the observations are chosen the nth value is automatically fixed by the fact that the n values must average to the mean. One degree of freedom is therefore taken up by fitting the mean or $n - 1$ degrees of freedom remain for estimating the total mean square which involves the mean. It will not be needed and so is not computed. A further degree of freedom is lost in estimating b; hence there are $n - 2$ degrees of freedom left for estimating the residual mean square. It is seen that the degrees of freedom of the components of variation in an analysis of variance table add to the total degrees of freedom. The sum of squares Q_T and Q_R are obtained from

$$Q_T = \sum^n y^2 - \left(\sum^n y \right)^2 / n$$

and

$$Q_R = \left(\sum_{}^{n} y(x - \bar{x}) \right)^2 \sum_{}^{n} (x - x)^2.$$

The squared correlation coefficient is given by

$$r^2 = Q_R/Q_T.$$

From this it is seen that r^2 gives the proportion of the sum of squares or variability explained by the regression. Thus in using the correlation coefficient as a measure of the goodness of relationship it is best to square it to obtain a realistic estimate of the amount of variability the linear relationship explains. This will, of course, always be less than r.

The analysis of variance tables also provides a test of significance of the linear regression. The statistic F is given by

$$F(1, n - 2) = \frac{Q_R/1}{(Q_T - Q_R)/n - 2}.$$

This is to be compared to an F or variance ratio table with 1 and $n - 2$ degrees of freedom at the 0.10 or 0.05 significance level to determine whether a linear relationship really exists. Or, in other terms, whether the mean square explained by the linear regression is large enough in comparison to the residual mean square to decide that the regression is due to a real effect rather than to random sampling.

After learning about tests of significance or tests of hypotheses there has been some tendency to attribute too much importance to them. Thus it might be concluded that if a regression is significant this is all that is necessary, but this is far from true for there are two kinds of significance, *practical* and *statistical*. If a regression is *not practically significant* it is little use to test its statistical significance. However, if it is practically significant then the test of hypothesis must be made to test for reality. For the linear relation, practical significance is measured by the squared correlation coefficient, i.e., by what proportion of the total variability is explained by the regression. It may be observed that if $r < 0.50$ i.e., $r^2 < 0.25$, the regression is of doubtful practical use.

If the sample values of the independent variable x can be divided into, say, four or more classes or columns with at least two y- values in each class, a second analysis of variance table may be prepared which will lead to a test of linearity. Such a test will tell whether it might be worth while to fit additional terms of higher degree.

With the data arranged into classes or columns with n_j in the jth column, the total variability may be divided into variability between column means arranged according to increasing x and variation within columns or residual. This leads to a second analysis of variance table:

ANALYSIS OF VARIANCE

	Sum of Squares	Degrees of Freedom	Mean Square
Column Means	$\sum\limits_{j=1}^{k} \sum\limits_{i=1}^{n.j} n_{.j}(\bar{y}_{.j} - \bar{y})^2 = Q_M$	$k - 1$	$Q_M/(k - 1)$
Residual	$\sum\limits_{j=1}^{k} \sum\limits_{i=1}^{n.j} (y_{ij} - \bar{y}_{.j})^2 = Q_T - Q_M$	$n - k$	$(Q_T - Q_M)/(n - k)$
Total	$\sum\limits_{j=1}^{k} \sum\limits_{i=1}^{n.j} y_{ij}{}^2 - \left(\sum\limits_{i=1}^{k} \sum\limits_{j=1}^{n.j} y_{ij} \right)^2 \Big/ n = Q_T$	$n - 1$	

An F-test may be made on this table by computing

$$F(k - 1, n - k) = \frac{Q_M/k - 1}{(Q_T - Q_M)/n - k}.$$

If this F is not significant, then there is no relation

between y and x, linear or otherwise. Had there been doubt about both linearity and whether there were a relationship at all, this test could have been made first.

It will be seen from the first analysis of variance table that the fitting of the linear regression leaves $Q_T - Q_R$ of the variability expressed as a sum of squares unexplained by the regression. If a more complicated function is to provide an improved fit, the improvement must come by removing or reducing this residual variability. Hence, this residual sum of squares may become the total for a third analysis of variance table. Since by the least squares principle a maximum amount of variability will be explained by fitting the column means, the residual from this fitting will be the smallest possible. If this residual is subtracted from the residual left by linear regression, the remainder is the amount explained by the column means over what was explained by the linear regression. The analysis of variance is as follows:

ANALYSIS OF VARIANCE

	Sum of Squares	Degrees of Freedom	Mean Square
Column Means about Regression	$Q_M - Q_R$	$k - 2$	$(Q_M - Q_R)/k - 2$
Column Mean Residual	$Q_T - Q_M$	$n - k$	$(Q_T - Q_M)/n - k$
Linear Regression Residual	$Q_T - Q_R$	$n - 2$	

The test for linearity is now made by comparing

$$F(k - 2, n - k) = \frac{(Q_M - Q_R)/(k - 2)}{(Q_T - Q_M)/(n - k)}$$

to the value corresponding to $k - 2$ and $n - k$ degrees of freedom of an F- table. If this is significant the linear regression does not explain all of the variability and it may be desirable to fit higher degree terms.

SELECTED APPLICATIONS OF STATISTICAL METHODS

More and more, statistical methods are being utilized in analyzing weather measurements and frequencies. Only a few additional examples, plus references to more complete descriptions in the literature, will be cited here.

Contingency Tables

Contingency tables and the related χ^2 test for determining if the cell frequencies depart significantly from the expected number of occurrences in each class are illustrated in Tables A3–12 and A3–13. The former shows frequencies of wind in a two-way classification by direction and speed. In the last three lines, Table A3–12 also lists the estimates of the wind distribution parameters as defined. Appropriate totals for each row and column are included.

The simple two-way classification shown in Table A3–13 makes it possible to judge if thunderstorms are associated with concurrent high winds. The parenthetical frequencies are obtained by multiplication of the row and column subtotals and division by the grand total. The occurrence of thunderstorms with high winds is greater than the expected frequency and the occurrence of no high winds with no thunderstorms is greater than expected. The differences between the observed and the expected frequencies are 33 in all cells. The sum of these differences squared and divided by their respective theoretical or expected frequencies is

$$\chi^2 = (32.5)^2 \left(\frac{1}{15} + \frac{1}{57} + \frac{1}{175} + \frac{1}{673} \right) = 96.5$$

We use 32.5 instead of 33 to correct for the biased probability of χ^2 which results from relatively small sample numbers. Actually, our sample is large enough that the bias is extremely small but the reduction in absolute difference between the observed and expected frequencies is included for illustration. There is only one degree of freedom in Table A3–13 and the value of $\chi^2 = 96.5$ is seen to be much greater even than the value expected with a probability of 0.001. There seems to be little reasonable doubt that the relationship between high winds and thunderstorms is significant.

Winds as Vectors

As seen above, wind is described in terms of both direction and magnitude, i.e., wind can be described as a vector. The square of a vector is a scalar; therefore, the vector variance and the vector mean (the mean resultant) are two im-

Table A3–12

WINDS ALOFT SUMMARY BY DIRECTION AND SPEED

U.S. DEPT. OF COMMERCE WEATHER BUREAU

$\left\{ \begin{array}{c}\text{PERCENTAGE}\\ \text{FREQUENCY}\end{array} \right\}$ OF DIRECTIONS BY SPEED GROUPS

WINDS ALOFT SUMMARY

94326 Station	ALICE SPRINGS, AUSTRALIA Station Name	RAWINS Type of Observation	35000 Level	FEET
Years	50 51 52 53 54 55 56 57 58	APR Mo. or Season		

Speed M/S	1-4	5-10	11.15	16.20	21.25	26.30	31.38	39.51	52-77	78-102	≥103	Total All Obs.		Speed (Knots)	
Knots	1-9	10.19	20.29	30.39	40.49	50.59	60.74	75.99	100-149	150-199	≥200	Obs.	%	Sum	Mean
M.P.H. Dir.	1-10	11-22	28-33	34-45	46-56	57-68	69-85	86-114	115-172	173-229	≥230				
N															
NNE															
NE															
ENE															
E	1											1	1.0	6	5.8
ESE															
SE															
SSE															
S	1											1	1.0	4	3.9
SSW			1	1	1	1						4	3.8	169	42.3
SW			1	1	2	5	1					10	9.6	462	46.2
WSW		3	1	2	6	4	5	1	1			23	22.1	1121	48.7

	1	2	3	5	5	9	10	8	2	Total	Percent		ΣV	Mean
W	I	2	3	5	5	9	I0	8	2		45	43.3		2562	56.9
WNW				I	3	4	I	3	3		I2	II.5		670	55.8
NW		I	I	I	I		3				7	6.7		321	45.8
NNW						I					I	I.0		51	50.5
Calm															
Totals	3	7	6	II	I7	24	2I	I2	3		I04	I00.0		5365	51.6
Percent	2.9	6.7	5.8	I0.6	I6.3	23.I	20.2	II.5	2.9		I04	I00.0			
Mean speed (Knots) by Group	5.8	I5.8	25.9	35.5	44.I	52.7	65.2	8I.7	I08.8						

θ	v_r	σ_r		Σxy	Σy^2	$n\sigma_y/\Sigma x$	σ_v/V_r	σ_r	σ_v	σ_e
265	46.7690	30.869		13563.416	325634.647	.438—	.66	23.455	20.069	21.776

n	$\Sigma x/n$	σ_x	Sum of E. Components (Σx)	Σx^2	$n\sigma_x/\Sigma x$	σ_v/σ_x	r	e
104	46.621—	23.157	4848.627—	281285.007	.497—	1.42	-.091	1.169

ψ	$\Sigma y/n$	σ_r	Sum of N. components (Σy)	Σy^2	$n\sigma_y/\Sigma y$	$V_r/\bar v$	ΣV	$\bar v$
162	3.712—	20.412	386.095—	44349.633	5.498—	.91	5365	51.6

σ_x Standard deviation of east components

σ_r Standard deviation of north components

σ_v Standard vector deviation of wind-velocity

r Correlation coefficient of north and east components

$\bar v$ Average wind speed

V Scalar Wind Speed

σ_x Standard deviation of wind components along the major axis of the distribution

σ_n Standard deviation of wind components perpendicular to the major axis of distribution

ψ Angle of rotation of the major axis of the wind distribution counter-clockwise from E-W direction

θ Resultant wind direction

V_r Resultant wind speed

σ_x Standard deviation of wind speeds

e $\sigma_\cdot/\sigma_-$

Table A3-13

CONTINGENCY TABLE OF FREQUENCY OF OCCURRENCE OF THUNDERSTORMS WITH HIGH WINDS AT WILLIAMSPORT, PA., DURING JULY, AUGUST, AND SEPTEMBER (TEN YEAR PERIOD)

	T	NT	Total
HW	48(15)	24(57)	72
NHW	142(175)	706(673)	848
Total	190	730	920

where   HW indicates High Wind
      NHW indicates No High Wind
      T indicates Thunderstorm
      NT indicates No Thunderstorm

portant wind parameters which can be treated fairly easily. To obtain the vector mean of the wind, individual wind vectors are summed and averaged. To obtain the vector variance of the winds, the wind speeds are squared and summed and the square of the vector mean is subtracted from this sum. Although the vector variance is constant for a particular sample, no real knowledge of the wind distribution is known unless two additional facts are determined or assumed. If the variances along any set of orthogonal axes are equal and there is no correlation between the components, the bivariate circular normal distribution applies. A circle centered at the origin of the vector mean with a radius equal to the vector standard deviation will contain 63% of the wind observations. If, however, the component variances are not equal or the components are correlated, the wind distribution is elliptical. This case and other problems encountered in the vector analysis of winds have been described by Crutcher (14) and Brooks and Carruthers (15) in some detail.

REFERENCES CITED

1. Swed, F. S. and C. Eisenhart (1943), Tables for testing randomness of grouping in a sequence of alternatives, *Annals of Mathematical Statistics*, **14**, 66–87.
2. Pearson, E. S. and H. O. Hartley (1956), *Biometrika Tables for Statisticians*, Vol. 1. Cambridge Univ. Press, London.
3. Shaw, R. H., H. C. S. Thom, and G. L. Barger (1954), *The Climate of Iowa, I. The Occurrence of Freezing Temperatures in Spring and Fall*, Special Report No. 8, Iowa State College.
4. Geary, R. C. and E. S. Pearson (1938), *Tests of Normality*, Separate No. 1 from Biometrika, 22, 27, 28, Biometrika Office, University College, London.
5. Pearson, K., et al. (1951), *Tables of the Incomplete Γ-Function*, Cambridge Univ. Press, London.
6. Thom, H. C. S. (1957), *A Statistical Method of Evaluating Augmentation of Precipitation by Cloud Seeding*. Tech. Report No. 1, Vol. II, Final Report of the President's Advisory Comm. on Weather Control, Washington, D.C.
7. Barger, G. L., R. H. Shaw, and R. F. Dale (1959), *Chances of Receiving Selected Amounts of Precipitation in the North Central Region of the United States*, First Report to North Central Region Tech. Comm. on Weather Information for Agriculture, Agric. and Home Econ. Exper. Sta., Iowa State Univ., Ames, Iowa.
8. Gumbel, E. J. (1958), *Statistics of Extremes*, Columbia University Press, New York.
9. Thom, H. D. S. (1954), Frequency of maximum wind speeds, *Proc. Amer. Soc. Civ. Engin.*, 80 Separate No. 539.
10. Lieblein, J. (1954), A new method of analyzing extreme—value data, *U.S. Nat. Adv. Comm. for Aero.*, Tech. Note No. 3053.
11. U.S. National Bureau of Standards (1950), Tables of the binomial probability distribution, *Applied Math*, Series 6.
12. Fisher, R. A. (1938), *Statistical Methods for Research Workers*, Oliver and Boyd, London.
13. Thom, H. C. S. (1957), The frequency of hail occurrence, *Archiv für Meteorologie, Geophysik, u. Bioklimatologie*, Band 8, 185–194.
14. Crutcher, H. L. (1956), Wind aid from wind roses, *Bull. Amer. Met. Soc.* **37,** 391–402.
15. Brooks, C. E. P. and N. Carruthers (1953), *Handbook of Statistical Methods in Meteorology*, London, M. O. 538, Air Ministry Met. Off., Her Majesty's Stationery Office, pp. 1–412.

APPENDIX 4

Climatological Publications[1]

Publishing climatological information has been part of the overall program of the national weather service almost as long as there has been such a service. The number of people seeking climatological information is so great and the uses made of the information so variable that it would be almost impossible to furnish everyone (through publications) with the exact information he wants. Rather than to serve specific applications, climatological publications make basic data available on a broad scale.

Those publications in greatest demand are described below. Additional information on those included here and on other climatological publications may be found in "Selective Guide to Published Climatic Data Sources," Key to Meteorological Records Documentation No. 4.11. It is available through the Superintendent of Documents, U.S. Government Printing Office, Washington, D.C. 20402.

All of the types of data mentioned in the descriptions which follow may not be available for every station included in the particular publication under discussion. Moreover, the data may not be presented exactly as you might want it. The surest way to obtain precise and complete information on what data are available for a particular station of interest and the forms in which data for that station are published is to contact the National Climatic Center (NCC). If the data are not published in the form you need them, in most cases they can prepare the data presentation to your specifications.

Subscription Publications

Subscription publications are issued on a scheduled periodic basis. Orders for subscriptions should be sent to the National Climatic Center, Attention: Publications, Federal Building, Asheville, N. C. 28801. Back issues for recent years are usually available at the NCC.

"Local Climatological Data"

This monthly publication is prepared separately for approximately 300 cities and towns, all with government-operated weather stations. It contains daily information on temperature (including heating degree days), dew point, precipitation (including snowfall), pressure, wind, sunshine, and sky cover. Monthly averages or other appropriate summarizations of these elements are presented. The month's accumulation of rainfall is shown in hourly amounts. If the particular station operates at all hours, eight almost complete meteorological observations per day at 3-hourly intervals are listed. A monthly average for certain times of the day is indicated for some elements.

"Local Climatological Data" is generally uniform in format for all stations. Subscription to this publication includes the annual summary described below, but, as is the case with all climatological publications, single editions are available on request.

"Local Climatological Data with Comparative Data"

This annual publication is prepared for the same localities as the monthly publication described above. It summarizes by month and for the year the data recorded during the past calendar year. A table of normals, means, and extremes, based on long periods of record, is presented for the same elements. In addition, tables of average monthly and annual temperature, precipitation, snowfall, and heating degree days cover a long period of record. Also contained in the annual issue are a brief description of the climate of the locality and a detailed history of the weather station.

[1] Reproduced with slight modifications from *The National Weather Records Center*, U.S. Department of Commerce, Environmental Science Services Administration, Washington, D.C., 1967, and updated using *The National Climatic Center*, U.S. Department of Commerce, National Oceanic and Atmospheric Adminstration, Washington, D.C., 1970. Reprinted by permission of the U.S. Department of Commerce, National Oceanic and Atmospheric Administration.

"Climatological Data"

This monthly publication contains primarily daily measurements and monthly summaries of precipitation and temperature extremes at co-operative stations. Issued separately for each State (or a combination of States), a station listing is included. A locator map is included when space permits. Observations from government-operated stations are used to fill data gaps. Supplemental data on evaporation, wind, relative humidity, sunshine, or soil temperature are included for stations making any of these measurements. Seasonal tables for heating degree days and snowfall are published in the July issue.

The stations for which data are presented in the annual issue of this publication are the same as for the monthly. The annual issue contains monthly and annual averages and departures from long-term means of temperature, precipitation, and evaporation; total wind movement; soil temperature and soil moisture table; a table of temperature extremes and freeze data; a station index; and a locator map.

"Climatological Data, National Summary"

Issued monthly, this publication contains pressure, temperature, precipitation, and wind data for selected U.S. stations. A general summary of weather conditions over the country is presented, and special articles describe hurricanes, unusual weather, and river and flood conditions. Severe storm damage is summarized by States. Monthly averages of rawinsonde data are presented in tabular form. Daily and monthly values of solar radiation are included. Charts graphically portray temperatures, precipitation, snowfall, percentages of sunshine; tracks of cyclones and anticyclones, solar radiation, monthly average upper air winds, and heights of constant pressure surfaces.

The annual issue presents summaries of all the above data for the year, and includes information on excessive rainfalls, hurricane tracks, and tornado paths.

"Hourly Precipitation Data"

This publication is issued monthly and separately for each State (or a combination of States) except Alaska. Hourly and daily precipitation values are presented for stations equipped with automatic recording gages. The annual issue contains monthly and annual totals of precipitation.

"Storm Data"

Issued monthly and including data for each of the 50 States, this publication presents the place, time, character, and estimated damage of all reported severe storms or unusual weather phenomena. Data are divided by State and place within the State. No annual summary is issued in this series.

"Monthly Climatic Data for the World"

This publication contains monthly mean values of surface and upper air measurements from a large number of selected stations throughout the World. The surface elements included are pressure, temperature, relative humidity, and precipitation. The upper air data consists of the height, the temperature, the dew point, the wind direction, and the wind speed at standard constant pressure levels.

"Monthly Climatic Data for the World" is sponsored by the World Meteorological Organization in cooperation with the Environmental Science Services Administration. There is no annual issue of this publication.

"Northern Hemisphere Sea Level Charts and 500-Millibar Charts"

This is a series of daily synoptic weather charts, each volume consisting of charts of the Northern Hemisphere for one month. One sea level chart and one 500-millibar chart are presented for each day. Both are prepared from data observed at 1200 GMT. The series begins with data for January 1899, but some of the back issues are in limited supply or out of print.

"Weekly Weather and Crop Bulletin"

Especially valuable to agriculturists, this publication is prepared by the parent organization, the Environmental Data Service, and distributed from Washington, D.C. about noontime each Tuesday. Crop data are collected by the U.S. Department of Agriculture and State agricultural agencies. In addition to the narrative summaries of the weather over the country and its effects on crops, condensed summaries furnish this information for each State. In season, small grains, pastures, corn, cotton, soybeans, and other crops are discussed separately. Tabular weather information consists of summaries of temperature, heating degree days, precipitation, and snow depth on the ground. Ice thickness in rivers, harbors, and lakes is included during the winter season. During the spring, a summary of ice conditions on the Great Lakes is featured prior to the opening of the shipping season. Special articles on subjects of general interest to agriculture, such as droughts, are added from time to time, as well as charts and tabulations of current importance. Drought analyses are made weekly for any areas having a deficiency of precipitation for an extended period of time, using a formula known as the Palmer Index. These analyses are presented in chart form in terms of drought severity one week after the temperature and precipitation measurements are made.

A monthly chart portraying weather highlights over the country and the Monthly Weather Outlook, prepared by the Extended Forecast Division of the Weather Bureau, are also included in this publication.

Unscheduled Publications

The first two of the non-subscription publications described here are updated at roughly ten-year intervals with no specific deadlines for their release. The other four are updated and expanded as workload and funds permit. All of these publications generally are available through either the Superintendent of Documents or the NCC. Requests for information on their availability should be directed to the National Climatic Center, Federal Building, Asheville, N.C. 28801. Those publications that are out of print, in most cases, can be obtained from the NCC for the cost of duplication.

"Climatic Summary of the United States—Supplement for 1951-1960"

This summary is issued separately for each State (or a combination of States) and contains monthly total precipitation, monthly mean temperature, and monthly total snowfall for the ten-year period, as well as averages for the entire period of record. These data are presented for 100 to 200 locations within the State. Normals are included for those stations having 30 years

or more of continuous record. Miscellaneous tables provide additional information on temperatures, precipitation, and evaporation. An index shows the history of each station location. Early issues of this publication in 1909, 1920, and 1930 were commonly referred to as Bulletin W, although its official title was "Summary of Climatological Data for United States by Sections." A supplement was issued in the mid-fifties for the period 1931–1952.

"World Weather Records"

This publication contains tables of mean temperatures, mean pressures, and precipitation at all stations throughout the World for which complete data for the ten-year period are available. There are earlier issues covering every ten-year period back to about 1900.

"Climates of the States"

This unscheduled publication is issued for each State (or a combination of States) and contains a local climatological data summary for each Weather Bureau station in the State, a freeze data table for 100 to 200 locations, depending on the size of the State, and a narrative summary describing the climatological features of the State.

"Climatological Substation Summaries"

Climatological summaries similar to the annual issues of "Local Climatological Data" described previously, have been issued for various cooperative stations in each State. These stations are at locations where there is no full-time Weather Bureau office. The number of these summaries varies from a few for some States to as many as a hundred in others.

"Climatic Guides"

Climatic guides have been issued for some of the larger cities. They contain a wide variety of tables and charts of particular interest to the locality. A narrative summary of general climatic conditions in the surrounding metropolitan area is also included.

"Climatography of the United States No. 82—Summary of Hourly Observations"

These summaries have been published for 138 Weather Bureau stations where 24 hourly observations are recorded daily. Of the total number of summaries completed, 102 cover the ten-year period 1951–1960. Another 36 are based on the five-year period 1956–1960.

Tables give percentage frequencies of wind directions and speeds, ceilings and visibilities, and weather conditions of various types. Additional tables show hourly and daily occurrences of precipitation amounts by category, and total occurrences of various ranges of relative humidity under categorized temperature and wind conditions. A narrative description of the location and the topography of the station, together with one pertaining to smoke sources, is included. Where available, a smoke source map of the local area is shown.

This series supersedes the series entitled "Climatography of the United States No. 30—Summary of Hourly Observations," a five-year summary published in 1956.

SPECIAL SERVICES

Data Tabulations

While publications may be considered the standard method of presenting climatological data, special hand and machine tabulations often are necessary to bridge the gap between published data and the original observations. An important part of the NCC's mission is to help solve individual problems by furnishing data in the form and quantity needed.

Special tabulations or summaries may be prepared on request. The customer pays the cost of such service. It is important before any work is begun that NCC personnel and each customer agree as to exactly what data are needed and the specifications of the finished product. Among the items which need to be clearly specified are the elements, the stations, the period(s) of record, the manner in which the data are to be presented, etc. For example, the customer may want a machine printout of a bivariate (two elements) or a trivariate (three elements) frequency distribution. An example of the latter for which the NCC has frequent requests is the distribution of temperature by groupings of wind speed and relative humidity values. Such information has a variety of applications. It relates to human comfort, for example, and is used in designing buildings, special-use clothing, air conditioning, etc. Another application of this information

is in planning aviation and space operations. Jet aircraft and rocket engines are especially sensitive to temperature and wind factors.

Although one may thoroughly understand his need to apply climatological information to his problem, if he is unfamiliar with the peculiarities of the basic data, it may be difficult for him to clearly specify what he wants. To illustrate the situation he may encounter, let us examine one element, surface air temperature, for its vagaries. We will start with the assumption that all weather stations record the surface air temperature at regular intervals in degrees Fahrenheit. Now, this question arises: What are the "regular intervals"? The intervals at which temperature measurements are made vary with the type of station involved. We collect data from stations which report surface air temperature one, four, eight, twelve, or any number of times up to and including twenty four times each day. Some Weather Bureau stations obtain a continuous record of temperature.

The next question might well be: What other kinds of temperatures are recorded? Among the many other temperatures recorded are wet-bulb, dew-point, maximum, minimum, soil, sea-surface, etc.

The next logical question would be: Do all stations record all of the temperature measurements mentioned above? The answer to this is "no." In fact, a data survey is the first requirement of a large-scale job. Not only must the availability of the data for the job be determined, but just as important is a determination of whether or not the data are in a readily suitable form for machine processing.

If the data are not in suitable form for machine processing, the NCC still can fulfill a request by hand tabulation. In cost estimating special data tabulations, the decision of whether to use a machine, hand tabulate, or use a combination of the two, hinges upon such factors as (1) the form(s) of available input data; (2) the size of the job as indicated by the number of elements, the number of stations, and the period of record involved; and (3) the statistical operations required. Every effort is made to furnish the specified tabulation or summary at the lowest cost possible.

The NCC has prepared thousands of special data tabulations and summaries for customers.

Several examples of the categories of application of these data and some of the elements involved follow:

AGRICULTURE: Drought; evaporation; freeze data; precipitation; temperatures; wind speeds.

AIR CONDITIONING: Dry-bulb temperatures; wet-bulb temperatures.

AIR POLLUTION: Inversions; lapse rates; mixing depths; wind directions and speeds.

AVIATION: Cloud-visibility distributions; cross-wind components; humidity and runway temperatures; low ceiling-visibility wind roses; temperatures aloft; winds aloft.

HIGHWAYS: Fog; freeze data; snow or ice cover; strong winds.

INSURANCE: Destructive winds; drought; excessive rainfall (including floods); hail; hurricanes; tornadoes.

MEDICINE: Cloud cover; humidities; pressures; solar radiation; temperatures; winds.

OFFSHORE PETROLEUM EXPLORATION: State of sea, wave height and direction; storm frequency; winds.

RADIO AND TELEVISION: Refractivity; maximum winds; temperature extremes in connection with construction of towers.

SPACE EXPLORATION: Densities aloft (includes humidities); temperatures aloft; winds aloft.

TRANSMISSION (petroleum and natural gas): Duration of critical temperatures; maximum temperatures; minimum temperatures.

A copy of each resulting tabulation or summary is retained in NCC files. New requests are reviewed in the light of these previous summaries. When one or more of them is pertinent to the customer's request, copies are provided at cost of reproduction. Most special jobs are unique, however. Previous summaries still have considerable value because often a method used in developing a previous summary can be adapted to the customer's needs. When this is possible, the customer's job usually costs less.

Data tabulations and summaries are furnished to anyone requesting them according to mutually agreed upon specifications. The NCC assists the customer in determining his climatological data requirements but avoids infringement upon the domain of the private meterological consultant in those cases where a private customer needs assistance in interpreting climatological information and applying it to his specific problem. All private customers needing special assistance are referred to the American Meteorological Society for a list of qualified meteorological consultants. In these cases, the consultant is encouraged to define what is needed from the NCC and then interpret the data tabulation for his customer.

From time to time the NCC participates in special data handling projects of particular benefit to groups of users. An example of this is a special collection of precipitation and temperature data via telephone each week. From these data, drought intensity indexes are computed and promptly sent to the Environmental Data Service. A map of the eastern United States appears each week in the "Weekly Weather and Crop Bulletin" one week after the observations and shows the current status of drought over this part of the nation.

Climatological Investigations

Climatological investigations are distinguished from data tabulations and summaries by their scope. These investigations usually are major jobs requiring tabulations and summarizations of climatological data, but they go beyond the results of machine processing and technical review. Interpretations of the results are applied to specific problem areas presented to the National Climatic Center by our parent organization and other Federal agencies for solution. The National Climatic Center usually carries these climatological investigations to their conclusions and often prepares reports on projects for publication. Numerous chart analyses and technical illustrations often are involved.

A few of the climatological investigations we have made for other agencies of the Federal Government are described below:

Environmental Studies for Aviation Operations

While polar flights to European cities from terminals on the West Coast and in Alaska are now considered routine, the inception of air travel over the Antarctic continent is a comparatively recent and still newsworthy accomplishment. An historic flight of two U.S. Navy aircraft was made from Cape Town, South Africa to McMurdo, Antarctica during the Fall of 1964. These planes flew directly over the South Pole. As with previous flights of significant length over unusual routes, the NCC was called upon to assist in the pre-flight planning. Upper air observations and weather maps of the general area were studied. Meteorologists interpolated wind and temperature information over the vast ocean and continental areas where no observations were available. Electronic computers assimilated this information and correlated it with aircraft speeds and altitudes. The end product enabled selection of the best probable two-week period to accomplish the flight.

Studies of the type described above are essential when fuel capacities and aircraft loads are pushed to their maximum limits. The NCC have already completed similar studies for selected routes proposed for use by U.S. supersonic transport aircraft under development. Investigations of environmental conditions in the 50,000 to 85,000-foot stratum for New York-Paris and San Francisco-Stockholm routes were made in this project conducted for the Federal Aviation Agency. Studies revealed that an aircraft flying at 2,000 knots would seldom encounter winds strong enough to deviate the flying time more than 2 or 3 percent at any time, even in Winter, on either route. Temperatures actually are more significant than winds in supersonic aircraft operations because the rate of fuel consumption is directly affected by temperature. It was found that summer stratospheric temperatures over the two selected routes were relatively constant. However, in winter, and especially over the polar area, temperatures fluctuated greatly at high altitudes within short periods of time.

The NCC also have made a number of climatological investigations for the Federal Aviation Agency in connection with proposed instrument landing systems at air terminals. Briefly stated, incidences of low ceilings and visibilities, together with the volume of air traffic, largely determine if a particular airport needs to be equipped with an instrument landing system. Once a decision is made to thus equip the airport, studies of ceilings, visibilities, winds, and other climatological factors determine the particular runway, if one is suitable, for which the installation is made. An important consideration is the most frequent wind direction at times when ceilings and/or visibilities are low. Among many other considerations is the wind speed. The runway oriented closest to that of the prevailing wind direction under low ceiling and visibility conditions may be a short one relative to other runways at the airport. Wind speeds may become a more important consideration than wind directions in such cases.

Environmental Studies for Space Exploration

Due to the significant effects of meteorological factors on space vehicles, the National Weather Records Center has a continuing research program in support of America's space adventure. The Earth's atmosphere plays an important role in the design, launch, and flight of aerospace vehicles. Wind is especially important because the vehicle is designed to withstand certain critical wind speeds and wind shears at the launching pad and in flight. In preparation for a space flight, the vehicle is placed on its launching pad as long as two week prior to lift-off, to prepare it for launching. Winds near the ground, if strong enough, may damage or even topple the vehicle during this risk period of its exposure to the elements. In order to assess the risk for these critical periods, the National Aeronautics and Space Administration has obtained the probabilities of various wind speeds for different hours of each day, different days, months, and seasons. The NCC also furnishes wind probabilities for various exposure periods (four days, ten days, etc.) that the vehicle may be on the launching pad prior to lift-off.

A similar concern to the aerospace engineer involves the winds at different altitudes over the launching site. Of particular concern is the area of maximum wind speeds and wind shears where the vehicle might be deflected from its planned path if winds aloft exceed the design and operational criteria. Here again, probabilities of unfavorable wind conditions are important considerations in the planning for space flights and future vehicle design. NASA's requirements for upper wind studies at first presented a real challenge for the NCC because the data for many levels were incomplete and radio wind soundings often did not reach levels as high as was needed to complete the information of concern to the space program. Although rocket soundings are becoming increasingly common, most upper winds still are determined by tracking a rising balloon with a radio transmitter attached. Wind information often cannot be obtained to desired altitudes for such reasons as the balloon bursting, a malfunctioning radio transmitter, loss of radio signal due to the distance between the transmitter and the ground receiver when very high wind speeds are encountered, and failure of ground tracking equipment. NASA sought complete wind observations up to altitudes of approximately 100,000 feet. Initial input data consisted of six-hourly winds aloft observations from slected stations for a number of past years. Current data are processed on a continuing basis to extend the period of record and thus provide a broader base for statistical applications. In order to fulfill this requirement, it was necessary to develop a method of meteorological analysis in which missing wind data could be estimated with reasonable accuracy. Of particular importance are estimations when high wind speeds cause termination of observations. The method developed by the NCC to serially complete upper wind observations to 100,000 feet has resulted in wind data of sufficient accuracy to adequately meet NASA's requirements.

Other studies made for NASA involve the probable cloud cover for various launching times, the temperature and humidity environment the vehicle will encounter on its ascent, and the probabilities of suitable meteorological conditions being present over "splashdown" areas.

Air Pollution Studies

Air pollution, especially in industrial and urban areas, is a problem of growing national concern. In cooperation with the U.S. Public Health Service, the Weather Bureau Research Station at Cincinnati, Ohio has devised a method of computing and forecasting air pollution accumulation and dispersion. The role of the NCC in this effort is to conduct statistical studies on vertical temperature distributions and wind flow at various major cities. The project requires that the NCC compute "mixing depth" values for each city. The mixing depth is the thickness of the layer of air above the surface in which mixing is restricted by specific conditions of wind speed and/or temperature during a given period. Normally, temperature decreases with height. Mixing of the layer of air near the surface often is restricted by very light wind speeds and a temperature inversion (temperature increasing with height) based a few hundred feet above the ground. Such conditions hold pollutants within the thin layer above the surface and allow continued accumulations until either the wind speed increases sufficiently to carry pollutants away, or until the surface temperature increases until it is warmer than temperatures aloft, thus wiping out the inversion. Once the warmest air is at the surface, then the fundamental principle that warm air rises comes into play. The rising warm air lifts the pollutants and leaves the air near the ground almost free of pollution.

A more serious pollution problem than the above may occur with light wind speeds when the temperature inversion is ground-based. In this situation the concentrated pollutants reach the ground and may seriously affect breathing in extreme cases. Such situations are apt to develop during the night and early morning when radiational cooling of the Earth is greatest. Although ground-based inversions are very persistent at times in some localities, in the typical case, warming of the Earth's surface by the late morning sun destroys this type of inversion by noon.

Also involved in this project are studies to determine the length of time certain wind and temperature conditions may exist in the lower atmospheric layers and how frequently they occur. This information may be used by authorities to control air pollution by placing restrictions on industries releasing various gases and combustion wastes when measurements of air pollution reach a specified limit. Restrictions would be lifted as soon as the meteorological conditions for dispersion had been met and the original pollution no longer remained.

Other special climatological investigations have been concerned with pollution of the air and ground by fallout of radioactive particles from nuclear explosions. Wind models were developed for the Office of Civil Defense for use in designating suitable areas to which large numbers of people could be evacuated in the event of a nuclear attack. Sites for nuclear reactors also are selected largely on the basis of climatology.

REQUESTING CLIMATOLOGICAL SERVICES

The variety of users of climatological information is large. They want information for a specific application, information which either is unavailable through published sources, or contained in sources unknown to them. The main differences are the organizations represented, the types of data wanted, the forms in which the data are to be summarized, and the manner in which the jobs are financed. This section provides a discussion on the ways in which non-federal agencies, private organizations or individuals obtain services of the NCC, and the importance of users clearly specifying what they want.

In addition to serving the general public through climatological publications, the National Climatic Center furnishes special services for private clients under the authority of an act of Congress which permits these services at the expense of the requester. The amount the customer is charged in all cases is intended solely to defray the expenses incurred by the government in satisfying his specific requirements to the best of its ability. Note that advance payment is no longer required. A purchase order or its equivalent, committing funds for payment in advance of the job, with billing to the customer on completion is now in order. The types of financial arrangements follow:

Jobs Estimated to Cost More Than $500.00

The private customer establishes in advance an individual trust fund if the NCC estimates the cost of his job to exceed $500.00. Actual costs of the job are charged, including supporting services. Any funds remaining at the conclusion of the job are refunded to the customer.

In some instances the cost estimates are lower than the amount needed to complete a job. Additional funds are requested from the customer as soon as this becomes evident. The customer generally can expect the total cost of his job to be within plus or minus 15 percent of the cost estimate.

Jobs Estimated to Cost $500.00 or Less

If the NCC estimates the cost for a private customer's job to be $500.00 or less, the estimated cost is established as a firm cost. Payment of the estimated cost, therefore, is accepted as payment in full. Regardless of what the job actually costs the NCC, neither a refund nor a request for additional funds is made to the customer.

Unit costs have been established by the Department of Commerce for reproduction of data in various forms. If your job request is of a type to which unit costs can be applied, the total cost for the service consists of these, plus the cost of personnel services for search and refile.

Open Accounts

Many of the private customers who have continuing need for our services make advance deposits to cover their requirements as they arise. The amount each requester deposits varies with the number and size(s) of the jobs that he anticipates he will require. The deposit may be in any reasonable amount, either less than or more than $500.00, as the customer chooses. Customers who have continuing requirements for special climatological services find open accounts to be advantageous, because such a funding arrangement permits the NCC to start on their jobs at once. Thus the delay that is frequently involved in arranging for proper financing is avoided.

Job Requests and Their Specifications

Three basic categories of job requests are received at the National Climatic Center. Briefly stated, they are:

1. Those which specifically require copies of basic observational data in various forms (magnetic tape, punched cards, microfilm, original records, etc.).

2. Those which can be satisfied by one or several of the previously published works or unpublished tabulations and summaries in our repository of climatological information.

3. Those dealing with a unique problem which can be satisfied only by a non-routine analysis or through scientific exploration.

It is essential that the customer provide a precise statement of his problem with any type of request he is making. In connection with requests of the third type, a mutual understanding about the exact specifications of the job may be especially difficult to attain through telephone conversations and correspondence alone. When the customer has a unique and complex problem which involves the application of non-standard climatological summarizations and analyses, it is most advantageous for him to visit the NCC. In this way, NCC scientists, data processing specialists, and records specialists can explore the problem with the customer as a team.

When the customer finds a visit to the NCC impracticable, he may obtain guidance in stating precisely his climatological needs from the following sources:

Private meteorological consultants

Weather Bureau Regional and State Climatologists

Weather Bureau State User Services Representatives

Personnel in university meteorology departments

Specifications or other information essential to the attainment of pertinent results often are overlooked in requesting climatological services. In making a request, the customer, if at all possible, should include precise information on the following questions:

1. What is the *FINAL USE* to be made of the product(s) you will obtain from us (if this information is not classified)? The answer to this question enables the climatologist to apply his knowledge and experience concerning the pertinence of available data as well as its limitations.

2. What are the *GEOGRAPHICAL AREA(S)* and *PERIOD(S) OF RECORD* of interest?

3. Is information desired on an *ANNUAL* (one composite answer), *SEASONAL* (answers for each portion of the annual cycle), *MONTHLY* (answers for each calendar month), or *OTHER* basis? If seasonal, will the usual Winter (December, January, February), Spring (March, April, May), Summer (June, July, August), and Fall (September, October, November) grouping be suitable, or is some other grouping desired?

4. In what *UNITS* of measurement do you want the values of elements? For example, do you wish wind speeds expressed in miles per hour? meters per second? of knots?

5. Is *DIURNAL VARIATION* (change from one time of day to another) *SIGNIFICANT* for the study? If so, what hours of the day can be grouped? If diurnal variation is unimportant in your study, should we consider all observations available for each day?

6. In what *FORMAT* do you wish the final results? Whenever applicable, the customer should include a rough sketch of the form of presentation he visualizes as a solution to his problem. This simple device usually replaces hundreds of words.

7. How much *TIME* is available to complete your study? The complexity of the study and the workload at the NCC interrlate to determine the time required to make the product available to you. This may vary from a minimum of a few hours to telephone or TWX the data from a selected few observations, to a few weeks to prepare a "Standard Ceiling-Visibility Wind Rose," to perhaps several months to develop a computer program and process large masses of data in a complex fashion. It is NCC policy to estimate the time for completion of each requested task when preparing cost estimates.

References

Ackerman, A. E. (1941), The Köppen classification of climates in North America, *Geogr. Rev.*, **31**, 105–111.

Adolf, E. F. (1947), *Physiology of man in the desert*, Wiley, New York.

Advisory Committee on Weather Control (1957), *Final Report*, I, U.S. Govt. Printing Office, Washington, D.C.

Alissow, B. P. (1954), *Die Klimate der Erde*, Deut. Ubers., Berlin.

Ambroggi, R. P. (1966), Water under the Sahara, *Scientific American*, **214**, 21–29.

Andrewes, C. H. (1958), Climate, weather and season in relation to respiratory infection, in S. W. Tromp, and W. H. Weihe (eds.). *Biometeorology*, Pergamon Press, Symposium Publication Division, New York, 56–62.

Angus, D. E. (1958), Frost prevention by wind machines, UNESCO, *Arid Zone Research*, **XI**.

Antevs, E. (1955), Geochronology of the deglacial and neothermal ages, *J. Geol.*, **61**, 195–230.

Arakawa, H. (1957), Three great famines in Japan, *Weather*, **12**, 211–217.

Arakawa, H. (1959), Hydroelectric power generation and the climate of Japan—a case study of engineering meteorology, *Bull. Am. Met. Soc.*, **40**, 416–422.

Aronin, J. E. (1956), *Climate and architecture*, Reinhold, New York.

Ashton, F. M. (1956), Effects of a series of cycles of alternating low and high soil water contents on the rate of apparent photosynthesis in sugar cane, *Plant Physiology*, **31**, 266–274.

Atkinson, B. W. (1969), *The weather business*, Doubleday Science Series, Garden City, N.Y.

Atkinson, B. W. (1970), *The reality of the urban effect on precipitation—A case study approach.* W.M.O., Tech. Note No. 108, 342–360.

Atkinson, B. W. (1971), The effect of an urban area of the precipitation from a thunderstorm, *J. of Applied Meteorology*, **10**, 47–55.

Auerbach, H. (1961), *Geographic variation in incidence of skin cancer in the U.S.*, Public Health Reports No. 76, 345–348.

Bach, W. et al. (1970), Variations of atmospheric turbidity with height over an urban area, *Proceedings*, Association of American Geographers, **2**, 4–8.

Bach, W. (1972), *Atmospheric Pollution*, McGraw-Hill, New York.

Bagnouls, F. and H. Gaussen (1957), Les climats écologiques et leur classification, *Ann. Geogr.*, **66**, 193–220.

Bailey, H. P. (1958), A simple moisture index based upon a primary law of evaporation, *Geographiska Annaler*, **40**, 196–215.

Bailey, H. P. (1960), Toward a unified concept of the temperate climate, *Geogr. Rev.*, **54**, 516–545.

Bailey, H. P. (1962), Some remarks on Köppen's definitions of climate types and their representations, *Geogr. Rev.*, **52**, 444–447.

Baird, P. D. (1964), *The polar world*, Wiley Interscience, New York.

Baker, J. J. W. and G. E. Allen, (1965), *Matter energy and life*, Addison—Wesley, Reading.

Ball, J. (1910), Climatological Diagrams, *Cairo Scientific J.*, **4**, Cairo.

Banham, R. (1969), *The architecture of the well-tempered environment*, Architectural Press, London.

Barrows, J. S. (1966), Weather modification and the prevention of lightning-caused forest fires, in W. R. D. Sewell, (ed.), *Human dimensions of weather modification*, Univ. of Chicago, Dept. of Geography, Research Paper No. 105, 169–182.

Barry, R. G. and R. J. Chorley, (1970), *Atmosphere, weather and climate*, Holt, Rinehart and Winston, New York.

Basile, R. M., and S. W. Corbin, (1969), A graphical method for determining Thornthwaite climatic classifications, *Annals A.A.G.*, **59**, 561–572.

Battan, L. J. (1969), *Harvesting the clouds—advances in weather modification*, Doubleday, Garden City, N.Y. (Am. Met. Soc. Science Study Series).

Beckinsale, R. P. (1965), Climatic change: A critique of modern theories, in J. B. Whittow and P. D. Woods (eds.), *Essays in Geography for Austin Miller*, Univ. of Reading Press, 1–38.

Beckworth, W. B. (1966), Impacts of weather on the airline industry: The value of fog dispersal systems, in W. R. D. Sewell (ed.), *Human dimensions of weather modification*, Univ. of Chicago, Dept. of Geography, Research Paper No. 105, 195–207.

Beebe, R. G. (1967), The construction industry as related to weather, *Bull. Amer. Met. Soc.*, **48**, 409.

Bell, B. (1953), Solar variations as an explanation of climate change, in H. Shapley (ed.), *Climate change*, Harvard Univ. Press, Cambridge, 123–136.

Belous, R. (1970), Unsolved problems of Alaska's North Slope, *The Environmental Journal, National Parks and Conservation Magazine*, November 1970, 19–21.

Berg, H. (1944), Zum Begriff der Kontinentalität, *Meteor. Zeit.*, **61**, 283–284.

Bergeron, T. (1930), Richlinien einer dynamischen Klimatologie, *Meteor. Zeit.*, **47**, 246–262.

Bergeron, T. (1935), On the physics of cloud and precipitation, *Meteorologie und Geodetik Geophysik International*, Part II, 156–178.

Berry, W. B. N. (1968), *Growth of a prehistoric time scale*, W. H. Freeman, San Francisco.

Bickert, C. Von E. and T. D. Browne (1966), Perception of the effect of weather on manufacturing: A study of five firms, in W. R. D. Sewell (ed.), *Human dimensions of weather modification*, Univ. of Chicago, Dept. of Geography, Research Paper No. 105, 307–322.

Billings, W. D. (1970), *Plants, man, and the ecosystem*, Wadsworth, Belmont, Calif.

Bingham, H. (1922), *Inca lands, exploration in the Highlands of Peru*, Houghton-Mifflin, Boston.

Black, R. F. (1950), Permafrost, in P. D. Track (ed.), *Applied sedimentation*, Wiley, New York, 247–275.

Black, R. F. (1954), Permafrost—a review, *Geol. Soc. Am.*, **65**, 839–856.

Blackwelder, E. (1933), The insolation hypothesis of rock weathering, *Am. Sci.*, **26**, 97–113.

Blaney, H. F. and W. D. Criddle (1950), *Determining water requirements in irrigated areas from climatological data*, Soil Conservation Service, Tech. Publ. No. 96, U.S.D.A.

Bloom, A. L. (1969), *The surface of the earth*, Prentice-Hall, Englewood Cliffs, N.J.

Blumenstock, D. I. and C. W. Thornthwaite (1941), Climate and the world pattern, *1941 Yearbook of Agriculture*, 98–127.

Blüthgen, J. (1964), *Allgemeine Klimageographie*, Walter de Gruyter, Berlin.

Bolin, B. (1970), The carbon cycle, in *The biosphere*, Scientific American book, W. H. Freeman, San Francisco.

Borisov, A. (1968), *Climates of the U.S.S.R.*, Aldine Press, Chicago.

Bornstein, R. D. (1968), Observations of the urban heat island effect in New York City, *J. of App. Met.*, **7**, 575–582.

Bowen, I. S. (1926), The ratio of heat losses by conduction and by evaporation from any water surface, *Physical Review*, **27**, 779–87.

Braidwood, R. J. (1960), The agricultural revolution, in *Man and the ecosphere*, Scientific American Book, W. H. Freeman, San Francisco.

Bray, J. R. (1971), Solar-climate relationships in the Post-Pleistocene, *Science*, **171**, 1242–1243.

Broecker, W. S. (1966), Absolute dating and the astronomical theory of glaciation, *Science*, **151**, 299–304.

Broecker, W. S. and J. van Donk (1970), Insolation changes, ice volumes and the o^{18} record in deep sea cores, *Reviews of Geophysics and Space Physics*, **8**, 169–198.

Brooks, C. E. P. (1949), *Climate through the ages*, McGraw-Hill, New York.

Brooks, C. E. P. (1951), Geological and historical aspects of climatic change, in T. F. Malone (ed.), *Compendium of Meteorology*, Am. Met. Soc., Boston, 1004–1018.

Broome, M. R. (1966), Weather forecasting and the contractor, *Weather*, **21**, 406–410.

Bruce, J. P. and R. H. Clark, (1966), *Introduction to hydrometeorology*, Pergamon Press, Oxford.

Brunnschweiler, D. H. (1957), Die Luftmasses der Nordhemisphäre. Versuch einer genetischen Klimaklassifikation auf aerosomatischer Grundlage, *Geographica Helvetica*, **12**, 164–195.

Bryson, R. A. (1967), Possibilities of major climatic modification and their implications,

Northwest India, *Bull. Am. Met. Soc.*, **48**, 136–142.

Bryson, R. A. (1968), All other factors being constant . . . , *Weatherwise*, **21**, 56–61.

Bryson, R. A. and J. E. Kutzbach (1968), *Air pollution*, Assoc. of Amer. Geogr., Comm. on College Geogr., Resource Paper No. 2, Washington, D.C.

Bubeck, R. C. et al. (1971), Runoff of deicing salt: Effect on Irondequoit Bay, Rochester, New York, *Science*, **172**, 1128–1132.

Bucknell, J. (1966), *Climatology*, Macmillian, London.

Büdel, J. (1944), Die morphologischen Wirkungen des Eiszeitklimas im gletscherfreien Gebeit, *Geol. Rundschau*, **34**, 482–519.

Büdel, J. (1948), Die Klima-morphologischen Zonen Polarländer, *Erdkunde*, **2**, 22–53.

Büdel, J. (1968), Geomorphology-principles, in R. W. Fairbridge, (ed.), *Encyclopedia of Geomorphology*, Reinhold, New York, 416–427.

Budyko, M. I. (1956), Teplovoi balans zemnoi poverkhnosti, English trans. Stepanova, N. A., (1958), *The heat balance of the earth's surface*, Office of Technical Services, U.S. Dept. of Commerce, Washington, D.C.

Buettner, K. J. (1962), Human aspects of bioclimatological classification, in *Biometeorology*, Pergamon Press, London, 91–98.

Bunting, B. T. (1965), *The geography of soil*, Aldine, Chicago.

Bunting, B. T. (1970), Concept, class and terminology in studies of tropical soils, *Prof. Geogr.*, **22**, pp. 55–61.

Burton, A. C. and O. G. Edholm (1955), *Man in a cold environment; physiological and pathological effects of exposure to low temperatures*, Monograph of the Physiol. Soc., No. 2, Arnold, London.

Burton, I. (1965), Flood-damage reductions in Canada, *Geogr. Bull.*, **7**, 161–185.

Butzer, K. W. (1957), The recent climatic fluctuations in lower latitudes and the general circulation of the Pleistocene, *Geographiska Annaler*, **39**, 105–113.

Butzer, K. W. (1964), *Environment and archeology: An introduction to Pleistocene geography*, Aldine, Chicago.

Callendar, G. S. (1961), Temperature fluctuations and trends over the earth, *Quart. J. Royal Met. Soc.*, **87**, 1–12.

Carter, D. B. (1965), *Fresh water resources*, Assoc. of American Geographers, High School Geography Project, (preliminary text), Washington, D.C.

Carter, D. B. and J. R. Mather (1966), *Climatic classification for environmental biology*, C. W. Thornthwaite Associates, Laboratory of Climatology, Publications in Climatology, **19** (4), Elmer, N.J.

Carter, G. F. (1968), *Man and the land*, Holt, Rinehart and Winston, New York.

Chandler, T. J. (1962), London's urban climate, *Geogr. J.*, **127**, 279–302.

Chandler, T. J. (1965), *The climate of London*, Hutchinson, London.

Chandler, T. J. (1967), Night-time temperatures in relation to Leicester's urban form, *Meteorological Mag.*, **96**, 244–250.

Chandler, T. J. (1970), Urban climatology—inventory and prospect, in *Urban Climatology*, W.M.O., Tech. Note No. 108, 1–9.

Chang, J-Hu. (1959), An evaluation of the 1948 Thornthwaite classification, *Annals A.A.G.*, **49**, 24–30.

Chang, J-Hu. (1968), *Climate and agriculture: An ecological survey*, Aldine, Chicago.

Chang, J-Hu. (1968), Progress in agricultural climatology, *Prof. Geogr.*, **20**, 317–320.

Chang, J-Hu. (1970), Potential photosynthesis and crop productivity, *Annals A.A.G.*, **60**, 92–101.

Chang, J-Hu. (1971), A critique of the concept of growing season, *Prof. Geogr.*, **23**, 337–340.

Chang, J-Hu. (1971), *Problems and methods in agricultural climatology*, Oriental Pub. Co., Hawaii.

Changnon, S. A. and G. E. Stout (1967), Crop-hail intensities in central and northwest U.S., *J. Appl. Met.*, **6**, 542–548.

Changnon, S. A. and J. C. Neill (1968), A meso-study of corn-weather response on cash-grain farms, *J. Appl. Met.*, **7**, 94–104.

Changnon, S. A. (1970), Recent studies of urban effects on precipitation in the United States, in *Urban climates*, W.M.O., Tech. Note No. 108, 325–341.

Changnon, S. A. et al. (1971), METROMEX: An investigation of inadvertent weather modification, *Bull. Am. Met. Soc.*, **52**, 958–967.

Chappell, J. E., Jr. (1970), Climatic change reconsidered: Another look at the 'pulse of Asia,' *Geogr. Rev.*, **60**, 347–373.

Chorley, R. J., ed. (1969), *Water, earth and man*, Methuen, London.

Chow, V. T. (1962), Determination of waterway areas for the design of drainage structures in small drainage basins, *Univ. of Illinois, Eng. Experimental Station Bull.* No. 442, Urbana, 78–82.

Chow, V. T., ed. (1964), *Handbook of applied hydrology*, McGraw-Hill, New York.

Claiborne, R. (1970), *Climate, man and history*, Norton, New York.

Clarke, J. F. (1969), Nocturnal urban boundary layer over Cincinnati, *Monthly Weather Review*, **97**, 582–589.

Clarke, J. F. and J. L. McElroy (1970), Experimental studies of the nocturnal urban boundary layer, in *Urban climates*, W.M.O., Tech. Note No. 108, 108–112.

Clawson, M. (1966), The influence of weather on outdoor recreation, in W. R. D. Sewell (ed.), *Human dimensions of weather modification*, Univ. of Chicago, Dept. of Geog., Research Paper No. 105, 183–193.

Cloud, P. and A. Gibor (1970), The oxygen cycle, in *The biosphere*, Scientific American Book, W. H. Freeman, San Francisco.

Cohen, P., O. L. Franke, and N. E. McClymonds (1969), *Hydrologic effects of the 1961–66 drought on Long Island, New York*, Geologic Survey Water Supply Paper, 1879-F.

Colbert, E. H. (1953), The record of climatic changes as revealed by vertebrate paleoecology, in H. Shapley (ed.), *Climatic change: Evidence, causes and effects*, Harvard Univ. Press, Cambridge, 249–271.

Cole, F. W. (1970), *Introduction to meteorology*, Wiley, New York.

Cole, L. C. (1958), The ecosphere, reprinted in *Man and the ecosphere*, A Scientific American Book, 1970, W. H. Freeman, San Francisco.

Collins, G. F. and G. M. Howe (1964), *Weather and extended coverage: A final report on the effects and distribution of storms producing extended coverage losses*, Travelers Research Center Service Corp., Hartford, Conn.

Conover, J. H. (1953), Climatic changes as interpreted from meteorological data, in

H. Shapley (ed.), *Climatic Change*, Harvard Univ. Press, Cambridge, 221–230.

Conrad, V. and L. W. Pollak (1950), *Methods in climatology*, Harvard Univ. Press, Cambridge.

Corbel, J. (1964), L'érosion terrestre, étude quantitative (méthodes-techniques-résultats), *Annales de Géographie*, **73**, 385–412.

Cotton, C. A. (1942), *Climatic accidents in landscape making*, Wiley, New York.

Crawford, T. V. (1964), Computing the heating requirements for frost protection, *J. Appl. Met.*, **3**, 750–760.

Critchfield, H. J. (1966), *General climatology*, Prentice-Hall, Englewood Cliffs, N.J.

Crocker, R. L. (1952), Soil genesis and the pedogenic factors, *Quart. Rev. of Biology*, **27**, 139–168.

Cronin, J. F. (1971), Recent vulcanism and the stratosphere, *Science*, **172**, 847–849.

Cross, K. W. (1967), The association of certain meteorological variables with the incidence of respiratory disease in an industrial city, in S. W. Tromp and W. H. Weihe (eds.), *Biometeorology*, Pergamon Press, Symposium Publications Division, New York, 63–69.

Cunningham, G. and J. Vernon (1968), Some extremes of weather and climate, *J. of Geogr.*, **66**, 530–535.

Dansereau, P. (1957), *Biogeography; an ecological perspective*, Ronald Press, New York.

Dasmann, R. F. (1968), *Environmental conservation*, Wiley, New York.

Daubenmire, R. F. (1959), *Plants and environment: A textbook of plant autecology*, Wiley, New York.

Davies, M. (1960), Grid system operation and the weather, *Weather*, **15**, 18–24.

Davis, W. M. (1899), The geographical cycle, *Geogr. J.*, **14**, 481–504.

Davis, W. M. (1909), The geographical cycle in an arid climate, *Geographical Essays*, Ginn and Co., Boston, 296–322. Reprinted in 1954 by Dover Publications, New York.

Davitaya, F. F. (1969), Atmospheric dust content as a factor affecting glaciation and climatic change, *Annals A.A.G.*, **59**, 552–560.

Delwiche, C. C. (1970), The nitrogen cycle, in *The biosphere*, Scientific American Book, W. H. Freeman, San Francisco, 69–80.

Deevey, D. S. (1952), Radiocarbon dating, *Sci. American*, **186**, 24–28.

Deevey, D. S. and R. F. Flint (1957), Postglacial hypsithermal interval, *Science*, **125**, 182–184.

DeMarrais, G. A. (1961), Vertical temperature difference observed over an urban area, *Bull. Am. Met. Soc.*, **42**, 548–552.

DeMartonne, E. (1925), *Traité de Geographie Physique*, Paris.

DeWiest, R. J. M. (1965), *Geohydrology*, Wiley, New York.

Dingle, A. N. (1955), Patterns of change of precipitation in the United States, *J. Met.*, **12**, 220–225.

Dingle, A. N. (1957), Hay fever pollen counts and some weather effects, *Bull. Am. Met. Soc.*, **38**, 465–469.

Donn, W. L. and M. Ewing, (1966), A theory of ice ages, III, *Science*, **152**, 1706–1712.

Dore, R. P. (1968), Climate and agriculture. The intervening social variables, *UNESCO Symposium on Agroclimatological Methods*, Reading, England, 201–208.

Dorf, E. (1960), Climatic changes of the past and present, *American Scientist*, **48**, 341–364.

Douglass, A. E. (1919, 1928, 1936), Climatic cycles and tree growth, *Carnegie Inst. Wash. Pub.* **289**, Vols. I, II, and III.

Dowling, D. F. (1968), The thermoregulatory significance of the hair coat with special reference to cattle, S. W. Tromp and W. H. Weihe (eds.), *Biometeorology*, **1**, Pergamon, 383–386.

Drew, M. and J. Fry (1956), *Tropical architecture in the humid zone*, Reinhold, N.Y.

Droessler, E.G. (1968), First national conference on weather modification—conference summary, *Bull. Am. Met. Soc.*, **49**, 982–986.

Duckworth, F. S. and J. S. Sandberg (1954), The effect of cities upon horizontal and vertical temperature gradients, *Bull. Am. Met. Soc.*, **35**, 198–207.

Dury, G. H. (1960), *The face of the earth*, Penguin Books, Harmondsworth, England.

Dury, G. H. (1969), *Perspectives on geomorphic processes*, A.A.G. Comm. on College Geography, Resource Paper No. 3, Washington, D.C.

Ehrlich, P. R. and A. H. Ehrlich (1972), *Population, resources, and environment*, W. H. Freeman, San Francisco.

Ellison, W. D. (1948), Erosion by raindrop, *Scientific American*, **179**, 40–45.

Emberger, L. (1932), Sur une formule climatique et ses applications en botanique, *Météorologie*, 423–432.

Emiliani, C. (1955), Pleistocene temperatures, *J. Geol.*, **63**, 538–578.

Emiliani, C. (1958), Ancient temperatures, *Scientific American*, **198**, 54–63.

Emiliani, C. (1966), Isotopic paleotemperatures, *Science*, **154**, 851–857.

Ericson, D. B. (1959), Coiling directions of globigerina pachyderma as a climatic index, *Science*, **130**, 219–220.

Ericson, D. B. (1961), Pleistocene climatic record in some deep-sea sediment cores, *Annals N.Y. Acad. Sciences*, **95**, 537–541.

Ericson, D. B. and G. Wollin (1964), *The deep and the past*, Knopf, New York.

Estall, R. C. and R. O. Buchanan (1961), *Industrial activity and economic geography*, Hutchinson, London.

Evans, L. T. (1963), *Environmental controls of plant growth*, Academic Press, New York.

Evelyn, John. (1661), *Fumifugium: Or, the inconvenience of the aer and smoak of London dissipated.* London, W. Godbed for G. Bedel. (Reprinted by the National Society for Clean Air, 1961.)

Ewing, M. and W. L. Donn (1956), A theory of ice ages, *Science*, **123**, 1061–1066.

Ewing, M. and W. L. Donn (1958), A theory of ice ages II, *Science*, **127**, 1159–1162.

Fairbridge, R. W. (1961), Convergence of evidence on climatic change and ice ages, *Ann. N.Y. Acad. Sci.*, **95**, 542–579.

Fairbridge, R. W. (1963), Mean sea level related to solar radiation during the last 20,000 years, *Changes in Climate*, UNESCO-WMO Symposium, Rome.

Fairbridge, R. W., ed. (1967), *Encyclopedia of atmospheric sciences and astrogeology*, Reinhold, N.Y.

Fairbridge, R. W., ed. (1968), *Encyclopedia of geomorphology*, Reinhold, N.Y.

Fairbridge, R. W. (1971), The Sahara Desert ice cap, *Natural History*, **LXXX**, 66–73.

F. A. O. (1966), *European breeds of cattle*, I and II, Agricultural Studies, **67**, Rome.

Ferris, H. B. (1921), Anthropological studies on the Quichua and Machiganga Indians, *Trans. Conn. Acad. of Arts and Sciences*, April 1921.

Field, W. O., Jr. (1955), Glaciers, *Scientific American*, **193**, 84–92.

Fitch, J. M. (1972), *American building: The environmental forces that shape it*, Houghton Mufflin, Boston.

Fitch, J. M. and D. P. Branch (1960), Primitive architecture and climate, *Scientific American*, **208**, 134–144.

Flawn, P. T. (1970), *Environmental geology*, Harper and Row, New York.

Fletcher, R. J. (1969), A proposed modification of Köppen to incorporate seasonal precipitation, *J. of Geogr.*, **68**, 347–350.

Flint, R. F. (1957), *Glacial and Pleistocene geology*, Wiley, New York.

Flohn, H. (1950), Neue Anschauungen über die allgemeine Zirkulation der Atmosphäre und ihre Klimatische Bedeutung, *Erdkunde*, **4**, 141–162.

Flohn, H., ed. (1969), *General climatology*, Vol. 2 in the series *World survey of climatology*, Elsevier, Amsterdam.

Fregri, K. and J. Iverson (1964), *Textbook of of pollen analysis*, Hafner, New York.

French, M. H. (1966), *European breeds of cattle*, *I* and *II*, FAO Agricultural Studies, No. 67, Rome.

Frenkiel, J. (1965), *Evaporation reduction*, Arid Zone Research, No, XXVII, UNESCO.

Fritts, H. C. (1962), An approach to dendrochronology screening by means of multiple regression techniques, *J. Geophys. Res.*, **67**, 1413–1420.

Fritts, H. C. (1963), Computer programs for tree-ring research, *Tree-Ring Bull.*, **25**, No. 3–4, 2–7.

Fritts, H. C. (1965), Tree ring evidence for climatic changes in western North America, *Monthly Weather Rev.*, **93**, 421–443.

Fritts, H. C. (1968), Tree-ring analysis, in Fairbridge, R. W. (ed.), *Encyclopedia of Atmospheric Sciences and Astrogeology*, Reinhold, New York, 1008–1026.

Fuggle, R. F. and T. R. Oke (1970), Infra-red flux divergence and the urban heat island, in *Urban climates*, W.M.O., Tech. Note No. 108, 70–78.

Fuquay, D. M. and R. G. Boughman (1962), *Project skyfire lightning research*, Final Report to N.S.F., Grant NSFG-10309, Intermountain Forest and Range Expt. Sta., Ogden, Utah.

Galway, J. G. (1966), The Topeka tornado of June 8, 1966. *Weatherwise*, **19**, 144–9.

Garner, H. F. (1968), Tropical weathering and relief, in R. W. Fairbridge (ed.), *Encyclopedia of Geomorphology*, Reinhold, New York, 1161–1172.

Garnett, A. and W. Bach (1965), An estimation of the ratio of artificial heat generation to natural radiation heat in Sheffield, *Monthly Weather Review*, **93**, 383–385.

Gates, D. M. (1962), *Energy exchange in the biosphere*, Harper and Row, New York.

Gates, D. M. (1970), Relationship between plants and atmosphere, in P. Dansereau (ed.), *Challenge for Survival*, Columbia Univ. Press, New York, 145–155.

Gates, D. M. (1971), The flow of energy in the biosphere, *Energy and Power*, A Scientific American Book. W. H. Freeman, San Francisco, 43–54.

Geiger, R. (1958), The modification of microclimate by vegetation in open country and in hilly country, UNESCO, *Arid Zone Research*, **XI**, 255–259.

Geiger, R. (1965), *The climate near the ground*, 4th ed., Harvard Univ. Press, Cambridge, Mass.

Gentilli, J. (1958), *A geography of climate*, Univ. of Western Australia Press, Perth.

Gerard, R. D. and J. L. Worzel (1967), Condensation of atmospheric moisture from tropical maritime air masses as a freshwater resource, *Science*, **157**, 1300–1302.

Gleason, H. A. and A. Cronquist (1964), *The natural geography of plants*, Columbia Univ. Press, New York.

Glinka, K. D. (1927), *The great soil groups of the world and their development* (transl. by C. F. Marbut), Edwards Brothers, Ann Arbor, Mich.

Goldsmith, J. R. and N. M. Perkins (1967). Seasonal variations in mortality, in S. W, Tromp and W. H. Weihe (eds.), *Biometeorology*, Pergamon Press, Symposium Publications Division, New York, 97–114.

Gorczynski, W. (1948), Decimal system of world climates, *Przeglad Meteor. Hydrol.*, **1**, 30–43.

Grande, F. (1962), Nitrogen metabolism and body temperature on man under restriction of food and water, in *Arid Zone Research*, No. XXIV, UNESCO, 103–110.

Green, R. (1961), Palaeoclimatic significance of evaporites, in A. E. M. Nairn (ed.), *Descriptive*

Palaeoclimatology, Interscience Publishers, New York, 61–88.

Greenberg, L. et al. (1967), Asthma and temperature change, in S. W. Tromp and W. H. Weihe (eds.), *Biometeorology*, Part 1, Pergamon Press, Symposium Publications Division, New York, 3–6.

Gregory, S. (1964), *Statistical methods and the geographer*, Longmans, London.

Griffiths, J. F. and M. J. Griffiths (1969), *A bibliography of weather and architecture*, U.S. Dept. of Commerce, ESSA, Technical Memorandum EDSTM. 9, Silver Spring, Md.

Griggs, D. T. (1936), The factor of fatigue in rock exfoliation, *J. of Geology*, **46**, 781–796.

Grillo, J. N. and J. Spar (1971), Rain-snow mesoclimatology of the New York metropolitan area, *J. Applied Met.*, **10**, 56–61.

Guilcher, A. (1965), Questions de morphologie climatique en Mélanésie équatoriale, *Bull. de l'Association de Géographes francais*, 28–40.

Halacy, D. S., Jr. (1968), *The weather changers*, Harper and Row, New York.

Hammond, A. L. (1971), Solar energy: A feasible source of power? *Science*, **172**, 660.

Hance, W. A. (1964), *The geography of modern Africa*, Columbia Univ. Press, New York.

Hance, W. A. (1967), *African economic development*, Rev. ed., published for Council on Foreign Relations by Praeger, New York.

Hare, F. K. (1951), Climatic classification, in L. D. Stamp and S. W. Wooldridge (eds.), *London essays in geography*, Longman's Green and Co., 111–134.

Hare, F. K. (1971), Future climates and future environments, *Bull. Am. Met. Soc.*, **52**, 451–456.

Harris, M. (1965), The myth of the sacred cow, in A. Leeds and A. P. Vayda (eds.), *Man, culture and animals*, A.A.A.S., Publication No. 78, 217–228.

Harman, J. R. and W. M. Elton (1971), The LaPorte, Indiana, precipitation anomaly, *Annals A.A.G.*, **61**, 468–480.

Harte, J. and R. H. Socolow (1971), Energy, in J. Harte and R. H. Socolow (eds.), *Patient earth*, Holt, Rinehart and Winston, New York, 276–294.

Harvey, D. (1968), *Explanation in geography*, St. Martins Press, London.

Haynes, J. B. (1972), North Slope oil: Physical and political problems, *Prof. Geogr.*, **24**, 17–22.

Herbertson, A. J. (1905), The major natural regions: An essay in systematic geography, *Geogr. J.*, **1**, 300–312.

Heller, J. L. (1963), The nomenclature of soils or what's in a name? *Soil Sci. Soc. Am.*, **27**, 216–220.

Hendl, M. (1960), Entwurf einer genetischen Klimaklassifikation auf Zirkulationbasis, *Zeit. Meteor.*, **14**, 46–50.

Hendrick, R. L. and D. G. Friedman (1966), Potential impacts of storm modification on the insurance industry, in W. R. D. Sewell (ed.), *Human dimensions of weather modification*, Univ. of Chicago, Dept. of Geography, Research Paper No. 105, 227–246.

Hershfield, D. M. (1961), *Rainfall frequency atlas of the United States*, U.S. Weather Bureau, Tech. Paper No. 40, Washington, D.C.

Hettner, A. (1930), *Die Klimate der Erde*, Leipzig, B. G. Teubner.

Hewes, L. (1958), Wheat failure in western Nebraska, *Annals A.A.G.*, **48**, 375–397.

Hidore, J. (1969), *Geography of the atmosphere*, W. C. Brown, Dubuque, Iowa.

Hoinkes, H. (1968), Wir leben in einer Eiszeit, *Umschau*, **68**, 810–815.

Hoinkes, H. (1968), Glacier variation and weather, *J. Glaciol.*, **7**, 3–19.

Holdridge, L. (1947), Determination of world plant formations from simple climatic data, *Science*, **105**, 367–368.

Holdren, J. P. (1971), Global thermal pollution, in J. P. Holdren and P. R. Ehrlich (eds.), *Global ecology*, Harcourt, New York, 85–88.

Holmes, A. (1961), *Principles of physical geology*, Nelson, London.

Horton, R. E. (1932), Drainage basin characteristics, *Trans. Am. Geophys. Union*, **13**, 350–361.

Horton, R. E. (1933), The role of infiltration in the hydrologic cycle, *Trans. Amer. Geophys. Union*, **14**, 446–460.

Horton, R. E. (1935), *Surface runoff phenomena*, Edwards Bros., Ann Arbor, Mich.

Hosler, C. R. (1961), Low level inversion frequency in the contiguous United States, *Monthly Weather Rev.*, **89**, 319–339.

Holzman, B. G. (1971), LaPorte precipitation fallacy, letter to *Science*, **171**, 847.

Holzner, L. and G. D. Weaver (1965), Geographic evaluation of climatic and climogenetic geomorphology, *Annals A.A.G.*, **55**, 592–602.

Housing and Home Finance Agency, Division of Housing Research (1954), *Application of climatic data to house design*, Government Printing Office, Washington, D.C.

Howell, W. E. (1965), Twelve years of cloud seeding in the Andes of northern Peru, *J. Appl. Met.*, **4**, 693–700.

Hoyt, W. G. and W. B. Langbein (1955), *Floods*, Princeton Univ. Press, Princeton, N. J.

Hutcheon, R. J. et al. (1967), Observations of the urban heat island in a small city, *Bull. Am. Met. Soc.*, **48**, 7–9.

Hummerstone, R. G. (1972), Cutting a road through Brazil's 'Green Hell,' *New York Times Magazine*, March 5, p. 16.

Huntington, E. (1945), *Mainsprings of civilization*, Wiley, New York.

Imbrie, J. and N. Newell, eds. (1964), *Approaches to Paleoecology*, Wiley, New York.

Inman, R. E. et al. (1971), Soil: A natural sink for carbon monoxide, *Science*, **172**, 1229–1231.

Jacobson, T. and R. M. Adams (1958), Salt and silt in ancient Mesopotamian agriculture, *Science*, **128**, 1251–1258.

James, P. E. (1959), *Latin America*, Odyssey Press, New York.

Janick, J. et al. (1970), *Plant Agriculture*, Readings from Scientific American, W. H. Freeman, San Francisco.

Jenkins, I. (1969), Increases in averages of sunshine in Central London, *Weather*, **24**, 52–58.

Jenny, H. (1941), *Factors of soil formation*, McGraw-Hill, New York.

Johnson, R. J. (1968), Choice in classification: The subjectivity of objective methods, *Annals A.A.G.*, **58**, 575–589.

Johnson, S. R. et al. (1969), Temperature modification and costs of electric power generation, *J. Appl. Met.*, **8**, 919–926.

Johnson, W. M. (1963), The pedon and the polypedon, *Soil Sci. Soc. Am.*, **27**, 212–215.

Jones, C. F. and G. G. Darkenwald (1954), *Economic geography*, Macmillan, New York.

Kahan, A. M. et al. (1969), Progress in precipitation modification, *Bull. Am. Met. Soc.*, **50**, 208–214.

Kaplan, L. D. (1960), The influence of CO_2 variations on the atmospheric heat budget, *Tellus*, **12**, 204–208.

Kawamura, T. (1966), Urban climatology in Japan, *Tokyo Journal of Climatology*, **3**, 9–13.

Kazmann, R. G. (1965), *Modern hydrology*, Harper and Row, N. Y.

Kendrew, W. G. (1922), *The climates of the continents*, Oxford Univ. Press, London.

Kellaway, G. P. (1960), *A background of physical geography*, Macmillan, London.

Kimble, G. H. T. (1950), The changing climate, *Sci. Amer.* **182**, 48–53.

King, L. C. (1953), Canons of landscape evolution, *Bull. Geol. Soc. Am.*, **64**, 725–752.

Kirpich, P. Z. and G. R. Williams (1969), Hydrology, in C. V. Davis and K. E. Sorensen (eds.), *Handbook of Applied Hydraulics*, McGraw-Hill, New York.

Knoch, K. and A. Schulze (1954), *Methoden der Klimaklassifikation*, Petermanns Geogr. Mitt., **249**, Gotha.

Köppen, W. and R. Geiger (1930 and later), *Handbuch der Klimatologie*, 5 vols., Gebrüder Borntraeger, Berlin.

Köppen, W. (1936), Das geographische System der Klimate, Vol. 3 of W. Köppen and R. Geiger, *Handbuch der Klimatologie*, Gebrüder Borntraegar, Berlin.

Kormondy, E. J. (1969), *Concepts of ecology*, Prentice-Hall, Englewood Cliffs, N.J.

Kratzer, P. (1956), Das Stadtklima, *Wissenschaft (Braunschwerg)*, No. 90.

Kraus, E. B. (1958), Meteorological aspects of desert locust control, UNESCO, *Arid Zone Research*, **XI**, 211–216.

Ku, T. L. and W. S. Broecker (1967), Rates of sedimentation in the Arctic Ocean, in M. Sears (ed.), *Progress in oceanography*, **4**, 95–104.

Kubiena, W. (1953), *Bestimmungsbuch und Systematik der Böden Europas*, Ferdinand Enke Verlag, Stuttgart.

Küchler, A. W. (1964), *Potential natural vegetation of the coterminous United States*, Maps and manual, American Geogr. Soc., Special Publ. No. 36.

Kukla, J. (1970), Correlations between loesses and deep-sea sediments, *Geologica Föringen i Stockholm Fördlandlingar*, **92**, 148–180.

Kung, E. C. et al. (1964), Study of a continental surface albedo on the basis of flight measurements, *Monthly Weather Rev.*, **92**, 543–564.

Kupfer, E. (1954), Entwurf einer Klimakarte

auf genetischer Grundlage, *Zeitschrift für Erdkundeunterricht*, **6**, 5–13.

Lachenbruch, A. H. (1968), Permafrost, in R. W. Fairbridge (ed.), *Encyclopedia of Geomorphology*, Reinhold, New York, 833–839.

Lamb, H. H. (1966), *The changing climate*, Methuen, London.

Lamb, H. H. (1969), Climatic fluctuations, in H. Flohn (ed.), *World survey of climatology*, **2**, 173–249.

Landsberg, H. E. and W. C. Jacobs (1951), Applied climatology, in T. F. Malone (ed.), *Compendium of meteorology*, American Met. Soc., Baltimore, 976–993.

Landsberg, H. E. (1956), The climate of towns, in W. L. Thomas (ed.), *Man's role in changing the face of the earth*, Univ. of Chicago Press, 584–606.

Landsberg, H. E. (1961), *Weather as a factor in plant location*, U.S. Dept. of Commerce, Weather Bureau, Washington, D.C.

Landsberg, H. E. (1967), Climatology, in R. W. Fairbridge (ed.), *Encyclopedia of atmospheric sciences and astrogeology*, Reinhold, New York, 217–230.

Landsberg, H. E. (1969), *Weather and health: An introduction to biometeorology*, Doubleday Anchor Books, New York.

Landsberg, H. E. (1970), Climates and urban planning, in *Urban climates*, W.M.O., Tech. Note No. 108, 364–371.

Lane, F. W. (1965), *The elements rage*, Chilton Books, Philadelphia.

Langbein, W. G. (1967), Hydroclimate, in R. W. Fairbridge (ed.), *Encyclopedia of atmospheric sciences and astrogeology*, Reinhold, N. Y., 447–451.

Lattimore, O. (1938), The geographical factor in Mongol history, *Geogr. J.*, **91**, 1–20.

Lee, D. H. K. (1951), Thoughts on housing for the humid tropics, *Geogr. Rev.*, **41**, 124–147.

Lee, D. H. K. (1957), *Climate and economic development in the tropics*, Harper, New York.

Lee, D. H. K. (1958), Proprioclimates of man and domestic animals, in *Climatology, Reviews of Research*, UNESCO, Arid Zone Research, **10**, 102–125.

Lee, D. H. K. (1963), Human factors in desert development, in C. Hodge (ed.), *Aridity and man*, A.A.A.S., Washington, D.C.

Leighton, P. A. (1966), Geographical aspects of air pollution, *Geog. Rev.*, **56**, 151–174.

Lemons, H. (1942), Hail in American agriculture, *Econ. Geog.*, **18**, 363–378.

Leopold, L. B. and J. P. Miller (1956), *Ephemeral streams-hydraulic factors and their relation to drainage net*, U.S. Geol. Survey Paper, 282-A.

Leopold, L. B., M. G. Wolman, and J. P. Miller (1964), *Fluvial processes in geomorphology*, W. H. Freeman, San Francisco.

Leopold, L. B. (1968), *Hydrology for urban land planning*, U.S. Geol. Survey Circular, 554.

Lindley, D. V. and J. C. P. Miller (1953), *Cambridge elementary statistical tables*, Cambridge Univ. Press, London.

Longwell, C. R., R. F. Flint, and J. E. Sanders (1969), *Physical Geology*, Wiley, New York.

Loomis, R. S. and W. A. Williams (1963), Maximum crop productivity: An estimate, *Crop Science*, **3**, 67–72.

Loomis, W. F. (1967), Skin-pigment regulation of vitamin-D biosynthesis in man, *Science*, **157**, 501–506.

Lowry, W. P. and H. E. Reiquam (1968), An index for analysis of the buildup of air pollution potential, Paper presented Ann. Meeting, Air Pollution Control Assoc., Pittsburgh, Penn. (Cited and described in Munn, 1970.)

Lowry, W. P. (1969), *Weather and life: An introduction to biometeorology*, Academic Press, New York.

Ludwig, F. L. (1970), Urban temperature fields, in *Urban climates*, W.M.O., Tech. Note No. 108, 80–107.

Lydolph, P. E. (1957), A comparative analysis of the dry western littorals, *Annals A.A.G.*, **47**, 213–230.

Lydolph, P. E. (1959), Federov's complex method in climatology, *Annals A.A.G.*, **49**, 120–144.

Macfarlane, W. V. (1958), Experimental approaches to the functions of tropical livestock, *Arid Zone Research*, **XI**, 227–234.

McCarroll, J. (1967), Measurements of morbidity and mortality related to air pollution, *J. Air Pollution Control Ass.*, **17**, 51–69.

McCormick, R. A. and K. R. Kurfis (1966), Vertical diffusion of aerosols over a city, *Quart. Jour. Met. Soc.*, **92**, 392–399.

McCormick, R. A. and J. H. Ludwig (1967), Climate modification by atmospheric aerosols, *Science*, **156**, 1358-1359.

McDowell, R. E. and J. R. Weldy (1967), Water exchange of cattle under heat stress, in S. W. Tromp and W. H. Weihe (eds.), *Biometeorology*, Pergamon Press, Symposium Publications Division, New York, 414-424.

McElhinny, M. W. and G. R. Luck (1970), Palaeomagnetism and Gondwanaland, *Science*, **168**, 830-832.

McGee, W. J. (1897), Sheetflood erosion, *Geol. Soc. Am. Bull.*, **8**, 87-112.

McHarg, I. (1969), *Design with nature*, Natural History Press, New York.

Machatschek, F. (1969), *Geomorphology*, (transl. by B. J. Davis), Oliver and Boyd, Edinburgh.

Machta, L. and E. Hughes (1970), Atmospheric oxygen in 1967 to 1970, *Science*, **168**, 1582-1584.

Mahadevan, P. (1968), The relations between climatic factors and animal production, in *Agroclimatological Methods*, UNESCO, 115-121.

Makkink, G. F. and H. D. J. Van Heemst (1956), The actual evapotranspiration as a function of the potential evapotranspiration and soil moisture tension, *Neth. J. of Agric. Science*, **4**, 67-72.

Makkink, G. F. (1957), Testing the Penman formula by means of lysimeters, *J. Inst. Water Eng.*, **11**, 277-288.

Malmstrom, V. H. (1969), A new approach to the classification of climate, *J. of Geogr.*, **68**, 351-357.

Matalas, N. C. (1962). Statistical properties of tree-ring data, as cited in UNESCO *Arid Zone Research*, XX, 255-263.

Matthes, F. E. (1939), Report of Committee on Glaciers, April 1939, *Trans. Am. Geophys. Union*, **20**, 518-523.

Manners, I. (1969), The development of irrigation agriculture in the Hashemite Kingdom of Jordon, with particular reference to the Jordon Valley, Unpubl. Ph.D. thesis submitted to the Board of the Faculty of Anthropology and Geography, Oxford University.

Mahringer, W. (1967), A contribution on climates inside court-yards in the city of Vienna, in S. W. Tromp and W. H. Weihe (eds.), *Biometeorology*, Pergamon Press, Symposium Publications Division, New York, 608-611.

Manabe, S. and R. T. Wetherald (1967), Thermal equilibrium of the atmosphere with a given distribution of relative humidity, *J. of Atmosph. Science*, **24**, 241-259.

Mansfield, W. W. (1958), Reduction of evaporation of stored water, *Arid Zone Research*, **XI**, UNESCO, 61-64.

Marbut, C. F. (1925), The rise, decline and revival of Malthusianism in relation to the geography and character of soils, *Annals. A.A.G.*, **15**, 1-29.

Marbut, C. F. (1935), Soils of the United States, Part III of *Atlas of American Agriculture*, U.S. Govt. Printing Office, Washington, D.C.

Mather, J. R. and G. A. Yoshioka (1968), The role of climate in the distribution of vegetation, *Annals A.A.G.*, **58**, 29-41.

Maunder, W. J. (1962), A human classification of climate, *Weather*, **12**, 3-12.

Maunder, W. J. (1970), *The value of weather*, Methuen, London.

Mausel, P. W. (1971), Letter to the editor, *Prof. Geogr.*, **23**, 73-74.

Mayr, E. (1956), Geographical character gradients and climatic adaptation, *Evolution*, **10**, 105-108.

Menzal, D. H. (1953), On the causes of the Ice Ages, in H. Shapley (ed.), *Climatic Change*, Harvard Univ. Press, 117-122.

Meteorological Office (1967), *Tables of temperature, relative humidity and precipitation for the world*, Parts I through VI (M.O. 617a-f), H.M.S.O., London.

Micklin, P. P. (1969), Soviet plans to reverse the flow of rivers: The Kama-Vychegda-Pechora project, *Canadian Geogr.*, **13**, 199-215.

Milankovitch, M. (1930), Mathematische Klimalehre und Astronomische Theorie der Klimaschwankungen, in *Handbuch Klimatologie*, IA, Berlin.

Miller, A. and J. C. Thompson (1970), *Elements of meteorology*, Merrill, Columbus.

Miller, A. A. (1953), Air mass climatology, *Geography*, **38**, 55-67.

Miller, A. A. (1965), *Climatology*, Methuen, London.

Miller, D. H. (1968), *The energy and mass budget at the surface of the earth*, Association of Am. Geogr. Commission on College Geography, Publication No. 7, Washington, D.C.

Mitchell, J. M., Jr. (1961), Changes of mean temperature since 1870, *Annals N.Y. Acad. Science*, **95**, 235–250.

Mitchell, J. M., Jr. (1961), The temperature of cities, *Weatherwise*, **14**, 224–229.

Mitchell, J. M., Jr. (1963), On the world-wide pattern of secular temperature change, *Arid Zone Research*, **XX**, UNESCO, 161–180.

Mitchell, J. M., Jr. (1971), The effect of atmospheric aerosols on climate with special reference to temperature near the earth's surface, *J. of Appl. Meteor.*, **10**, 703–714.

Mitchell, J. M., Jr. (1972), Air pollution and global climate, in R. W. Fairbridge (ed.), *Encyclopedia of geochemistry and environmental Sciences*, Reinhold, N. Y., 11–14.

Möller, F. (1963), On the influence of changes in the CO_2 concentration in air on the radiation balance of the earth's surface and on the climate, *J. Geophys. Res.*, **68**, 3877–3886.

Monge, C. (1948), *Acclimatization in the Andes*, Johns Hopkins Univ. Press, Baltimore.

Monkhouse, F. J. (1961), *Principles of Physical Geography*, London Univ. Press, London.

Monteith, J. L. (1966), Local differences in the attenuation of solar radiation over Britain, *Quart. J. Royal Met. Soc.*, **92**, 254–262.

Moore, G. W., ed. (1960), Origin of limestone caves—A symposium with discussion, *Natl. Speleol. Soc. Bull.* No. 22, Part 1.

More, R. J. (1967), Hydrological models and geography, in R. J. Chorley and P. Haggett (eds.), *Models in geography*, Methuen, London, 145–185.

Morris, E. A. (1966), Institutional adjustment to an emerging technology: Legal aspects of weather modification, in W. D. R. Sewell (ed.), *Human dimensions of weather modification*, Univ. of Chicago, Dept. of Geography, Research Paper No. 105, 279–288.

Muller, R. A. (1966), A critical review of energy and water balance analyses of various land cover types, C. W. Thornthwaite Associates, *Publications in Climatology*, No. 20, 199–214.

Munn, R. E. (1966), *Descriptive micrometeorology*, Academic Press, New York.

Munn, R. E. (1970), Airflow in urban areas, in *Urban climatology*, W.M.O., Tech. Note No. 108, 15–39.

Munn, R. E. (1970), *Biometeorological Methods*, Academic Press, New York.

Murphey, R. (1951), The decline of North Africa since the Roman occupation: Climatic or human? *Annals A.A.G.*, **41**, 116–132.

Murphy, R. E. (1968), *Landforms of the world*, Annals Map Supplement No. 9, *Annals A.A.G.*, **58**.

Musgrave, G. W. (1947), A quantitative evaluation of water erosion: A first approximation, *J. Soil Water Conserv.*, **2**, 133–138.

Myrup, L. O. (1969), A numerical model of the urban heat island, *J. Appl. Met.*, **8**, 854–862.

Nakamura, K. (1967), City temperatures in Nairobi, *Japanese Progress in Climatology*, Laboratory of Climatology, Tokyo Univ., 61–65.

Nairn, A. E. M., ed. (1961), *Descriptive paleaoclimatology*, Interscience Publishers, New York.

Nairn, A. E. M., ed. (1964), *Problems in palaeoclimatology*, Interscience Publishers, New York.

National Academy of Sciences—National Research Council (1964), *Scientific problems of weather modification*, N.A.S.-N.R.C., Publication No. 1236, Washington, D.C.

National Academy of Sciences—National Research Council (1966), *Weather and climate modification: Problems and prospects*, N.A.S.-N.R.C., Publication No. 1350, Vols. I and II, Washington, D.C.

National Center for Atmospheric Research (1962), *Proceedings of the conference on the climate of the eleventh and sixteenth centuries*, N.C.A.R., Technical Note No. 63-1, Boulder, Colorado.

National Science Foundation (1966), *Weather and climate modification*, N.S.F., Report No. 66-3, Washington, D.C.

Neiburger, M. (1970), Diffusion models of urban air pollution, in *Urban climatology*, W.M.O., Tech. Note No. 108, 248–262.

Nelson, H. J. (1959), The spread of an artificial landscape over Southern California, *Annals A.A.G.*, **49**, 80–99.

Newman, M. T. (1953), The application of ecological rules to racial anthropology of the aboriginal world, *Amer. Anthrop.*, **55**, 311–327.

Newman, M. T. (1955), Adaptation of man to cold environments, *Evolution*, **9**, 101–105.

Newell, N. D. (1959), The nature of the fossil record, *Amer. Phil. Soc.*, **103**, 264–285.

Neuberger, H. and J. Cahir (1969), *Principles of climatology*, Holt, Rinehart and Winston, New York.

Niering, W. A. (1970), the ecology of wetlands in urban areas, in P. Dansereau (ed.), *Challenge for survival*, Columbia Univ. Press, New York, 199–208.

Nuttonson, M. Y. (1947), Agroclimatology and crop ecology of Palestine and Trans-Jordan and climatic analogues in the United States, *Geogr. Rev.*, **37**, 436–456.

Odum, E. P. (1962), Relationships between structure and function in the ecosystem, *Japanese Journal of Ecology*, **12**, 108–118.

Odum, E. P. (1968), World circuits and systems stress, in H. E. Young (ed.), *Symposium on primary productivity and mineral cycling in natural ecosystems*, University of Maine Press, Orono, 81–138.

Odum, E. P. (1971), *Fundamentals of ecology*, W. B. Saunders, Philadelphia.

Oke, T. R. (1968), Some results of a pilot study of the urban climate of Montreal, *Climatology Bulletin*, McGill Univ. **3**, 36–41.

Oke, T. R. and F. G. Hannell (1970), The form of the urban heat island in Hamilton, Canada, in *Urban climates*, W.M.O., Tech. Note No. 108, 113–126.

Okita, T. (1960), Estimation of direction of air flow from observations of rime ice, *J. Met. Soc. of Japan*, **38**, 207–209.

Okita, T. (1965), Some chemical and meteorological measurements of air pollution in Asahikawa, *Air and Water Pollution*, **9**, 323–332.

Olgyay, V. (1963), *Design with climate*, Princeton Univ. Press, Princeton.

Oliver, J. E. (1968), The thermohyet diagram as a teaching aid in climatology, *J. Geogr.*, **67**, 554–563.

Oliver, J. E. (1970), A genetic approach to climate classification, *Annals A.A.G.*, **60**, 615–637.

Ollier, C. (1969), *Weathering*, Elsevier, New York.

Omar, M. H. (1968), Potential evapotranspiration in a warm arid climate, in *Agroclimatological methods*, UNESCO, 347–353.

Oort, A. H. (1970), The energy cycle of the earth, in *The biosphere*, Scientific American Book, W. H. Freeman, San Francisco, 13–24.

Opik, E. J. (1958), Climate and the changing sun, *Scientific American*, **198**, 85–92.

Pack, D. H. (1964), Meteorology and air pollution, *Science*, **146**, 1119–1128.

Page, J. K. (1963), The effect of town planning and architectural design in construction on the microclimatic environment of man, in S. W. Tromp, *Medical biometeorology*, Elsevier, New York.

Paige, S. (1912), Rock-cut surfaces of the desert ranges, *J. Geol.*, **20**, 442–450.

Palmer, W. C. (1965), *Meteorological drought*, U.S. Weather Bureau Research Paper No. 45, U.S. Dept. of Commerce, Washington, D.C.

Panofsky, H. A. (1956), Theories of climatic change, *Weatherwise*, **9**, 183–187.

Panofsky, H. A. (1969), Air pollution meteorology, *Am. Sci.*, **57**, 269–285.

Papadakis, J. (1966), *Climates of the world and their agricultural potentialities*, Publ. by author, Buenos Aires.

Parry, M. (1966), Air pollution patterns in the Reading area, in S. W. Tromp and W. H. Weihe (eds.), *Biometeorology*, Symposium Publications Division, Pergamon Press, New York, pp. 657–667.

Pardé, M. (1955), *Fleuves et riviéres*, Librarie Arman Collin, Collection Armand Collin No. 155, Section de Geographie, Paris.

Pasquill, F. (1962), *Atmospheric diffusion*, Van Nostrand, Princeton, N.J.

Patton, C. P., C. S. Alexander, and F. L. Kramer (1970), *Physical geography*, Wadsworth, Belmont, Calif.

Péguy, C. P. (1961), *Précis de climatologie*, Masson et Cie, Paris.

Peltier, L. (1950), The geographic cycle in periglacial regions as it is related to climatic geomorphology, *Annals A.A.G.*, **40**, 214–236.

Pelton, W. L. (1967), The effect of a windbreak on wind travel, evaporation and wheat yield, *Can. J. Plant Sci.*, **47**, 209–214.

Penman, H. L. (1963), *Vegetation and hydrology*, Commonwealth Bureau of Soils, Technical Communication No. 53, Commonwealth Agricultural Bureaux, Farnham Royal, England.

Peterson, J. T. (1969), *The climate of cities: A survey of recent literature*, U.S. Dept. of Health, Education and Welfare, National Air Pollution Control Administration Publication No. AP-59, Washington, D.C.

Petty, M. T. (1963), Weather and consumer sales, *Bull. Amer. Met. Soc.*, **44**, 68–71.

Plass, G. N. (1956), Carbon dioxide and the climate, *Amer. Sci.*, **4**, 302–316.

Plass, G. N. (1959), Carbon dioxide and climate, *Scientific American*, July 1959 (available as Reprint No. 823).

Polunin, N. (1960), *Introduction to plant geography and some related sciences*, McGraw-Hill, New York.

Pooler, F. (1963), Air flow over a city in terrain of moderate relief, *J. Appl. Met.*, **2**, 446–452.

Pounds, N. J. G. (1961), *An introduction to economic geography*, John Murray, London.

Powers, W. E. (1966), *Physical geography*, Appleton, Century, Crofts, New York.

Rahn, P. H. (1967), Sheetfloods, streamfloods and the formation of pediments, *Annals A.A.G.*, **57**, 593–604.

Rainey, R. C. (1951), Weather and movement of locust swarms: A new hypothesis, *Nature*, **168**, 1057–1060.

Rainey, R. C. (1963), *Meteorology and the migration of desert locusts*, W.M.O., Tech. Note No. 54, Geneva.

Rapoport, A. (1970), *House form and culture*, Prentice-Hall, Englewood Cliffs, N.J.

Rasool, S. I. and S. H. Schneider (1971), Atmospheric carbon dioxide and aerosols: Effects of large increases on global climate, *Science*, **173**, 138–141.

Raunkaier, C. (1934), *The life forms of plants and statistical plant geography*, Clarendon Press, Oxford.

Reifsnyder, W. E. and H. W. Lull (1965), *Radiant energy in relation to forests*, U.S. Dept. of Agric., Tech. Bull. No. 1344, Washington, D.C.

Riehl, H. (1965), *Introduction to the atmosphere*, McGraw-Hill, New York.

Revelle, R. (1963), Water, *Sci. Amer.*, September 1963, also in *Man and the ecosphere*, Scientific American Book (1971), W. H. Freeman, San Francisco.

Roach, W. T. (1961), Some aircraft observations of fluxes of solar radiation in the atmosphere, *Quart. J. Royal Men. Soc.*, **87**, 346–354.

Rodda, J. C. (1965), A drought study in southwest England, *Water and Water Engineering*, **69**, 316–321.

Roueché, B. (1950), *Eleven blue men*, Little, Brown, New York. (Appeared originally in the *New Yorker*.)

Rumney, G. R. (1968), *Climatology and the world's climates*, Macmillan, New York.

Russell, E. W. (1968), Climate and soils, in *Agroclimatological Methods*, UNESCO, 193–199.

Russell, R. J. (1926), Climates of California, Univ. of California Publications in Geography, **2**, 73–84.

Russo, J. A. et al. (1965), *The operational and economic impact of weather on the construction industry*, The Travelers Research Center, Inc., Hartford.

Ryd, H. (1970), *Building climatology*, W.M.O., Tech. Note No. 109.

Sargent, F. and S. W. Tromp (1964), *A survey of human biometeorology*, W.M.O., Tech. Note No. 65.

Sapper, K. (1935), *Geomorphologie der feuchten Tropen*, Geographische Schriften, 7, B. G. Teubner, Leipzig.

SCEP (1970), *Man's impact of the global environment: Assessment and recommendations for action*, Report of the Study of Critical Environmental Problems (SCEP), MIT Press, Cambridge.

Schaefer, V. J. (1946), The production of ice crystals in a cloud of supercooled water droplets, *Science*, **104**, 457–459.

Schaefer, V. J. (1956), Artificially induced precipitation and its potentialities, in W. L. Thomas (ed.), *Man's role in changing the face of the earth*, Univ. of Chicago Press, Chicago, 607–818.

Schaefer, V. J. (1969), The inadvertent modification of the atmosphere by air pollution, *Bull. Amer. Met. Soc.*, **50**, 199–206.

Schimper, A. F. W. (1903), *Plant geography upon a physiological basis* (transl. by W. R. Fisher), Clarendon Press, Oxford.

Schimper, A. F. W. and F. C. von Faber (1935), *Pflanzengeographie auf physiologischer Grundlage*, 2 vols, Fischer, Jena.

Scholander, P. F. (1955), Evolution of climatic adaptation in homeotherms, *Evolution*, **9**, 15–26.

Scholander, P. F. (1956), Climatic rules, *Evolution*, **10**, 339–340.

Schove, D. J. (1954), Summer temperatures and tree rings in North Scandinavia, A.D. 1461–1950, *Geogr. Ann.*, **36**, 40–80.

Schove, D. J. (1955), The sunspot cycle, 649 B.C. to A.D. 2000, *J. Geophys. Res.*, **60**, 127–146.

Schreider, E. (1964), Ecologic rules, body-heat regulation and human evolution, *Evolution*, **18**, 1–9.

Schwarzbach, M. (1963), *Climates of the past*, Van Nostrand, New York.

Science Services (1970), *Science News Yearbook*, Scribner's, New York.

Seawall, F. (1971), Waterways and their utilization, in G. H. Smith (ed.), *Conservation of natural resources*, Wiley, New York, 289–314.

Sekiguti, T. and H. Tamiya (1970), Precipitation climatology of Japanese city area, in *Urban climatology*, W.M.O., Tech. Note No. 108, 363.

Sekiguti, T. (1970), Historical dates of Japanese cherry festivals since the 8th century and her climatic changes, *Japanese Progress in Climatology*, No. 3, Laboratory of Climatology, Tokyo University, 38–45.

Sellers, W. D. (1965), *Physical climatology*, Univ. of Chicago Press, Chicago.

Sen, A. R., A. K. Biswas, and D. K. Sanyal (1966), The influence of climatic factors on the yield of tea in the Assam Valley, *J. Appl. Met.*, **5**, 789–800.

Sewell, W. R. D., ed. (1966), *Human dimensions of weather modification*, Univ. of Chicago, Dept. of Geography, Research Paper No. 105.

Shantz, H. L. and R. Zon (1924), *Natural vegetation*, Atlas of American Agriculture, Agriculture Section, U.S. Dept. of Agriculture, Washington D.C.

Shear, J. A. (1964), The polar marine climate, *Annals A.A.G.*, **54**, 310–317.

Shear, J. A. (1966), A set-theoristic view of Köppen dry climates, *Annals A.A.G.*, **56**, 508–515.

Shul'gin, A. M. (1965), *The temperature regime of soils*, Israel Program for Scientific Translations Ltd. Israel.

Silverberg, R. (1970), *The challenge of climate*, Meredith Press, New York.

Simonson, R. (1962), Soil classification in the United States, *Science*, **137**, 1027–1034.

Simpson, Sir G. (1934), World climate during the Quaternary period, *Quart. Jour. Royal Met. Soc.*, **60**, 425–478.

Slager, U. T. (1962), *Space medicine*, Prentice-Hall, Englewood Cliffs, N.J.

SMIC (1971), *Inadvertent climate modification*, Report of the study of man's impact on climate (SMIC), MIT Press, Cambridge.

Soil Survey Staff (1960), *Soil classification: A comprehensive system—7th approximation*, U.S. Dept. of Agriculture, Soil Conservation Service, Washington, D.C.

Spar, J. and P. Ronberg (1968), Note on an apparent trend in annual precipitation in New York City, *Mon. Weather Rev.*, **96**, 169–171.

Spate, O. H. K. (1952), Toynbee and Huntington: A study in determinism, *Geogr. J.*, **118**, 406–428.

Specht, R. L. (1958), Micro-environment (soil) of a natural plant community, *Arid Zone Research*, **XI**, UNESCO, 152–155.

Stamp, L. D., ed. (1963), *A glossary of geographical terms*, Longmans, Green, London.

Stern, A. C., ed. (1962), *Air pollution*, Academic Press, New York.

Stern, A. C. (1967), The changing pattern of air pollution in the United States, *Amer. Indust. Hyg. Assn. J.*, **28**, 161–165.

Stoddart, D. R. (1969), Climatic geomorphology, in *Progress in geography*, International Reviews of Current Research, Arnold, London, 161–222.

Stoddart, D. R. (1969), Climatic geomorphology, in R. J. Chorley (ed.), *Water, earth and man*, Methuen, London, 473–485.

Stokes, W. L. (1955), Another look at the Ice Age, *Science*, **122**, 815–821.

Stokes, W. L. and S. Judson, (1968), *Introduction to geology*, Prentice-Hall, Englewood Cliffs, N.J.

Stone, R. G. (1943), On the practical evaluation and interpretation of the cooling power in bioclimatology, *Bull. Amer. Met Soc.*, **24**, 295–327.

Strahler, A. N. (1951), *Physical geography* (1st ed.), Wiley, New York.

Strahler, A. N. (1952), Dynamic basis of geomorphology, *Geol. Soc. Amer. Bull.*, **63**, 923–938.

Strahler, A. N. (1964), Quantitative geomorphology of drainage basins and channel networks, in V. T. Chow (ed.), *Handbook of Applied Hydrology*, McGraw-Hill, New York, 4–39 to 4–76.

Strahler, A. N. (1968), *Physical geography* (3rd. ed.), Wiley, New York.

Strahler, A. N. (1971), *The earth sciences* (2nd. ed.), Harper and Row, New York.

Suzuki, H. (1971), Climatic zones of the Würm

glacial age, *Bull. of Dept. of Geogr.*, No. 3, Tokyo University.

Swinnerton, J. W. et al. (1970), The ocean: A natural source of carbon monoxide, *Science*, **167**, 984–986.

Takeuchi, H., S. Uyeda, and H. Kanamori, (1967), *Debate about the earth*, Fremman, Cooper, San Francisco.

Tamiya, H. (1969), Night temperatures in a new town, western suburbs of Tokyo, *Japanese Progress in Climatology*, Laboratory of Climatology, Tokyo University, 49–54.

Tanner, W. F. (1961), An alternate approach to morphogenetic climates, *Southeastern Geol.*, **2**, 251–257.

Tatham, G. (1957), Environmentalism and possibilism, in G. Taylor (ed.), *Geography in the 20th century*, Methuen, London, 128–162.

Taylor, G. (1946), *Our evolving civilization*, University of Toronto Press.

Terjung, W. H. (1966), Physiologic climates of the conterminous United States: A bioclimatic classification based on man, *Annals A.A.G.*, **56**, 141–179.

Terjung, W. H. (1970), The energy balance climatology of a city-man system, *Annals A.A.G.*, **60**, 466–492.

Terjung, W. H. (1970), Toward a climatic classification based upon net radiation, *Proceedings*, Assoc. Amer. Geogr., **3**, 140–144.

Terjung, W. H. and S. S-F. Louie (1971), Potential solar radiation climates of man, *Annals. A.A.G.*, **61**, 481–500.

Thom, E. C. (1959), The discomfort index, *Weatherwise*, **12**, 57–60.

Thomas, M. D. and G. R. Hill (1949), Photosynthesis under field conditions, in J. Franck and W. E. Loomis (eds.), *Photosynthesis in plants*, Iowa State College Press, Ames.

Thompson, L. M. (1957), *Soils and soil fertility*, McGraw-Hill, New York.

Thornthwaite, C. W. (1931), The climates of North America according to a new classification, *Geogr. Rev.*, **21**, 633–655.

Thornthwaite, C. W. and B. Holzman (1942), *Measurements of evaporation from land and water surfaces*, U.S. Dept. of Agriculture, Technical Bull. No. 817.

Thornthwaite, C. W. (1943), Problems in the classification of climates, *Geogr. Rev.*, **33**, 233–255.

Thornthwaite, C. W. (1948), An approach toward a rational classification of climate, *Geogr. Rev.*, **38**, 55–94.

Thornthwaite, C. W. (1956), Modification of rural microclimates, in W. L. Thomas (ed.), *Man's role in changing the face of the earth*, Univ. of Chicago Press, 567–583.

Thornthwaite, C. W. and J. R. Mather (1957), *Instructions and tables for computing potential evapotranspiration and the water balance*, Publications in Climatology, **10**(3), Centerton, N.J.

Thornthwaite Associates (1963), *Average climatic water balance data of the continents*, Publications in Climatology, Vols. XVI and XVII (1964), Centerton, N. J.

Tosi, J. A. (1964), Climatic control of terrestrial ecosystems: A report on the Holdridge Model, *Econ. Geogr.*, **40**, 173–181.

Tosi, J. A. and R. F. Voertman (1964), Some environmental factors in the economic development of the tropics, *Economic Geogr.*, **40**, 189–205.

Trewartha, G. T., A. H. Robinson, and E. H. Hammond (1967), *Elements of geography* (5th ed.), McGraw-Hill, New York.

Trewartha, G. T. (1954), *An introduction to climate* (3rd ed.), McGraw-Hill, New York.

Trewartha, G. T. (1968), *An introduction to climate* (4th ed.), McGraw-Hill, New York.

Trewartha, G. T. (1962), *The earth's problem climates*, Univ. of Wisconsin Press, Madison.

Tricart, J. and A. Cailleux (1965), *Introduction à la géomorphologie climatique*, S.E.D.E.S., Paris.

Troll, C. (1954), Seasonal climates of the earth, in H. Landsberg, *World maps of climatology*, Berlin.

Troll, C. (1958), Climatic seasons and climatic classification, *Orient. Geogr.*, **2**, 141–165.

Tromp, S. W. (1964), Influence of weather and climate on human diseases, in *A survey of human biometeorology*, W.M.O., Tech. Note No. 65.

Tromp, S. W. (1967), Biometeorology, in R. W. Fairbridge (ed.), *Encyclopedia of atmospheric sciences and astrogeology*, Reinhold, New York.

Ullman, E. (1954), Amenities as a factor in regional growth, *Geogr. Rev.*, **44**, 119–132.

UNESCO, (1958) *Climatology and microclimatology*, Arid Zone Research, **XI**, Proceedings of Canberra Symposium.

UNESCO (1963), *Changes of climate*, Arid Zone Research, **XX,** Proceedings of Rome Symposium.

UNESCO (1968), *Agroclimatological methods*, Proceedings of the Reading Symposium, Paris.

United Nations, Department of Economic and Social Affairs (1971), *Design of low-cost housing and community facilities*, **1,** Climate and house design, New York.

U.S. Army (1968), *Classification of world desert areas*, Technical Report, T.R. 69-38ES, U.S. Army Natick Labs., Natick, Mass.

U.S. Dept. of Agriculture (1938), Soils and men, *Department of Agriculture Yearbook, 1938*, U.S. Govt. Printing Office, Washington, D.C.

U.S. Dept. of Agriculture, Forest Service (1967), *Harvesting the national forest water crop*, U.S. Govt. Printing Office, Washington, D.C.

U.S. Dept. of Commerce (1964), *Replenishing underground water supplies on the farm*, Leaflet No. 452, Washington, D.C.

U.S. Dept. of Commerce, Weather Bureau (1955) *Tropical cyclones of the North Atlantic Ocean*, Technical Paper No. 55, Washington, D.C.

U.S. Dept. of Commerce, Weather Bureau (1960), *Climatology at work*, U.S. Govt. Printing Office, Washington, D.C.

U.S. Dept. of Commerce, Weather Bureau (1965), *Meteorological drought*, Research Paper No. 45, Washington, D.C.

U.S. Dept. of Commerce, ESSA (1966), *Weather and the construction industry*, Washington, D.C.

U.S. Geologic Survey (1966), *The changing pattern of ground-water development on Long Island, New York*, Geologic Survey Circular **524,** Washington, D.C.

U.S. Soil Conservation Service (1953), *Farm planners' handbook for Upper Mississippi watershed*, S.C.S., Milwaukee.

Van Houten, F. B. (1961), Climatic significance of red beds, in A. E. M. Nairn (ed.), *Descriptive Palaeoclimatology*, Interscience Publishers, New York, pp. 89-139.

Van Riper, J. (1971), *Man's physical world*, (2nd ed.), McGraw-Hill, New York.

Van Royen, W. (1954), *Agricultural resources of the world*, Atlas of world resources, **1,** Prentice-Hall, Englewood Cliffs, N.J.

Van Royen, W. (1970), Letter to editor, *Prof. Geogr.*, **22,** 196.

Veihmeyer, F. J. (1964), Evapotranspiration, in V. T. Chow (ed.), *Handbook of applied hydrology*, McGraw-Hill, New York, Ch. 11.

Veryard, R. G. (1963), A review of studies on climatic fluctuations during the period of the meteorological record, *Arid Zone Research*, **XX,** 3-17.

Visher, S. S. (1944), *Climate of Indiana*, Indiana Univ. Press, Bloomington.

Visher, S. S. (1954), *Climatic atlas of the United States*, Harvard Univ. Press, Cambridge.

Von Hagen, V. W. (1957), *The realm of the Incas*, Mentor Books, New York.

Vonnegut, B. (1947), The nucleation of ice formation by silver iodide, *Jour. of Appl. Physics*, **18,** 593-595.

Wagner, R. H. (1971), *Environment and man*, Norton, New York.

Walker, D. S. (1964), *The Mediterranean lands*, Wiley, New York.

Ward, R. C. (1967), *Principles of hydrology*, McGraw-Hill, New York.

Warburton, F. E. (1967), The purposes of classification, *Syst. Zoology*, **16,** 241-242.

Watson, D. J. (1963), Weather and plant yield, in L. T. Evans (ed.), *Environmental control of plant growth*, Academic Press, New York, 337-349.

Weber, A. (1929), *Theory of the location of industries*, (transl. by C. J. Friedrich), Univ. of Chicago Press, Chicago.

Weisse, L. L. and R. Kresge (1962), Indications of the uniformity of shore and offshore precipitation for Southern Lake Michigan, *J. of Appl. Met.*, **1,** 271-274.

Went, F. W. (1957), Climate and agriculture, *Scientific American*, **196,** 82-94.

Wexler, H. (1951), Spread of the Krakatoa volcanic dust cloud as related to the high-level circulation, *Bull. Amer. Meteor. Soc.*, **32,** 48-51.

Wexler, H. (1956), Variations in insolation, general circulation and climate, *Tellus*, **8,** 480-494.

Wilcock, A. A. (1968), Köppen after fifty years, *Annals A.A.G.*, **58,** 12-28.

Willett, H. C. (1949), Long period fluctuations of the general circulation of the atmosphere, *J. of Meteor.*, **6,** 34-50.

Willett, H. C. (1961), The pattern of solar climatic relationships, *Annals N.Y. Acad. Sci.*, **95,** 89-106.

Willett, H. C. (1965), Solar-climatic relationships in the light of standardized climatic data, *J. Atmosph. Sci.*, **22**, 120–136.

Wilson, A. (1966), The impact of climate on industrial growth: Tucson, Arizona: A case study, in W. R. D. Sewell (ed.), *Human dimensions of weather modification*, Univ. of Chicago, Dept. of Geography, Research Paper No. 105, 249–260.

Wilson, A. T. (1964), Origin of ice ages: an ice shelf theory for Pleistocene glaciations, *Nature*, **201**, 477–479.

Wilson, L. (1968), Morphogenetic classification, in R. W. Fairbridge (ed.), *Encyclopedia of geomorphology*, Reinhold, New York, 717–729.

Wilson, L. (1969), Les relations entre les processus géomorphologique et le climat moderne comme méthode de paléoclimatologie, *Revue de Géographie Physique et de Géologie Dynamique*, **XI**, 303–314.

Wilson, L. (1971), Suspended sediment yield in United States rivers as a function of climate, Ph.D. dissertation, Department of Geology, Columbia University, New York.

Winslow, C. E. A. and L. P. Herrington (1949), *Temperature and human life*, Princeton Univ. Press, Princeton, N.J.

Wischmeier, W. H. and D. D. Smith (1965), *Predicting rainfall-erosion losses from cropland east of the Rocky Mountains*, U.S. Dept. of Agric., Agricultural Handbook N.282.

Wisler, C. O. and E. F. Brater (1959), *Hydrology*, Wiley, New York.

Wissmann, H. V. (1948), Pflanzenklimatische Grendzender warmem Tropen, *Erdkunde*, **2**, 81–92.

Wittfogel, K. A. (1956), The hydraulic civilizations, in W. L. Thomas (ed.), *Man's role in changing the face of the earth*, Univ. of Chicago Press, Chicago, 152–164.

W. M. O. (1966), *Climatic change*, World Meteorological Org., Tech. Note No. 79, Geneva.

Wolman, M. G. and J. P. Miller (1960), Magnitude and frequency of forces in geomorphic processes, *J. Geol.*, **68**, 54–74.

Woodcock, A. H., R. L. Pratt, and J. R. Breckenridge (1952), *Heat exchange in hot environments*, U.S. Army, Office of Quartermaster General, E. P. Branch, Report No. 183.

Woods, K. B., (1960), Frost action and permafrost, in *Highway engineering handbook*, McGraw-Hill, New York, 13.3–13.35.

Woodwell, G. M. (1970), The energy cycle in the biosphere, *Sci. Amer.*, **223**, 64–74.

Wooldridge, S. W. and W. G. East (1967), *The spirit and purpose of geography*, Capricorn Books, New York.

Woollum, C. A. (1964), Notes from a study of the microclimatology of the Washington, D.C. area for the winter and spring seasons, *Weatherwise*, **17**, 262–266.

Woollum, C. A. and N. L. Canfield (1968), *Washington metropolitan area precipitation and temperature patterns*, ESSA, Technical Memo, WBTM-ER-28.

Wright, H. E. and D. G. Frey, eds. (1965), *The Quaternary of the United States*, Princeton University Press, Princeton.

Yoshino, M. M. (1967), Wind-shaped trees as indicators of micro and local climatic wind situation, in S. W. Tromp and W. H. Weihe (eds.), *Biometeorology*, Pergamon Press, 997–1005.

Zeuner, F. E. (1959), *The Pleistocene Period*, Hutchinson, London.

Zobler, L. and G. W. Carey (1969), *Benefits from integrated water management—the case of the New York metropolitan region*. Report submitted to Office of Water Resources Research, Department of Interior, Washington, D.C.

Index